AF574025

NUMERICAL HEAT TRANSFER

Series in Computational Methods in Mechanics and Thermal Sciences

W. J. Minkowycz and E. M. Sparrow, *Editors*

Baker, Finite Element Computational Fluid Mechanics
Patankar, Numerical Heat Transfer and Fluid Flow
Shih, Numerical Heat Transfer

PROCEEDINGS
Shih, Editor, Numerical Properties and Methodologies in Heat Transfer: Proceedings of the Second National Symposium

NUMERICAL HEAT TRANSFER

Tien-Mo Shih

Department of Mechanical Engineering
The University of Maryland, College Park

HEMISPHERE PUBLISHING CORPORATION

Washington New York London

DISTRIBUTION OUTSIDE NORTH AMERICA

SPRINGER-VERLAG

Berlin Heidelberg New York Tokyo

NUMERICAL HEAT TRANSFER

2 3 4 5 6 7 8 9 0 EBEB 8 9 8 7 6 5

This book was set in Press Roman by Hemisphere Publishing Corporation. The editors were Mary Dorfman, Anne Shipman, and Bettie L. Donley; the production supervisor was Miriam Gonzalez; and the typesetters were Shirley J. McNett and Wayne Hutchins.
Edwards Brothers, Inc. was printer and binder.

Library of Congress Cataloging in Publication Data

Shih, T. M.
Numerical heat transfer.

(Series in computational methods in mechanics and thermal sciences)
Bibliography: p.
Includes index.
1. Heat–Transmission. 2. Numerical analysis.
I. Title. II. Series.
QC320.S5146 1984 621.402'2 83-18469
ISBN 0-89116-257-7 Hemisphere Publishing Corporation
ISSN 0272-4804

DISTRIBUTION OUTSIDE NORTH AMERICA:
ISBN 3-540-13051-9 Springer-Verlag Berlin

To my wife,
Joyce L. Shih

CONTENTS

APPENDIXES

PREFACE

This book is directed to readers who belong to one of the following groups: heat transfer engineers who work in the field of numerical analysis, applied mathematicians who work on heat transfer problems, university instructors who teach numerical heat transfer, and university seniors and graduate students who study numerical heat transfer.

In writing it, I was motivated by the following four considerations. First, the area of heat transfer, which is closely related to energy utilization and conservation, has received considerable attention since the worldwide energy crisis arose. In solving heat transfer problems theoretically, the general trend has been toward heavy use of computers because fewer and fewer closed-form solutions can be obtained. Second, as the principal organizer of the National Symposia on Numerical Heat Transfer, which were initiated in September 1979 in College Park, Maryland, I have received widespread enthusiastic response. Although the proceedings of the conferences are published, the papers collected in these proceedings generally present research and development results, and not fundamental concepts and knowledge. For researchers who are less familiar with numerical heat transfer and wish to make progress, the conference proceedings are probably not a good source of information. Third, although courses on numerical heat transfer are taught in most major universities in the United States, there are few textbooks that describe in detail the usage of various numerical methods, their numerical properties, and their applications to conduction, convection, and radiation. Finally, it seems to me that there is a communication gap between heat transfer engineers and applied mathematicians. Some engineers are satisfied with their computations as long as they obtain some results that look physically reasonable; they feel that it is unnecessary to use mathematical theorems to further investigate the numerical properties of the solution obtained. Likewise, some applied mathematicians devote most of their efforts to proving or developing theorems, and show little interest in what engineers really need. As a result, the engineers' solutions may not converge and the mathematicians' theorems may be hardly used for practical applications. A book that can narrow this gap should be useful.

The complete task of solving a problem in numerical heat transfer consists of the following steps:

Step 1: Define the physical problem.
Step 2: Formulate the governing (integro-) differential equations subject to appropriate initial and boundary conditions.
Step 3: Discretize these (integro-) differential equations into a set of algebraic equations.
Step 4: Analyze the numerical properties of the discretization schemes, such as stability, consistency, convergence, and error bounds.
Step 5: If the scheme chosen is found to be satisfactory, use a matrix solver or an iteration method to solve the algebraic equations. If not, go back to step 3 and use another scheme.
Step 6: Interpret and discuss the computed results.

The scope of the present book includes steps 3, 4, and 5. Derivation of the governing equations and physical interpretation of the numerical results are omitted, since they can be found in most heat transfer textbooks.

More specifically, this book consists of four parts. Part 1 provides preliminary information about the use of various numerical methods (Chapters 1-3) and their numerical properties (Chap. 4). Readers who wish to learn the procedures of these numerical techniques should read Part 1. In Part 2, which contains the core material of this book, some simple physical problems of conduction, convection, and radiation are solved by the techniques described in Part 1. Laminar forced convection is stressed because it is a fundamental heat transfer phenomenon (Chaps. 6-9). Chapters 5 and 10 describe the application of numerical methods to conduction and radiation problems, respectively. Part 3 is an extension of Part 2. Some important physical phenomena related to heat transfer, such as free and mixed convection (Chap. 11), turbulent flows (Chap. 12), and combustion (Chap. 13), are briefly described. Without having to search for other reference books, readers will thus be able to follow the solution procedures demonstrated to solve a few simple physical problems. In Part 4, numerical analyses of the error bounds are described (Chap. 14) and finite-difference and finite-element schemes are compared (Chap. 15). Material in this part can serve as a bridge between the engineering and mathematics communities.

Three features of this book deserve special attention. First, the mesh systems constructed throughout the text are generally so coarse that a desk calculator is sufficient to complete a computational problem. The time and possible errors involved in computer programming can thus be avoided. Once readers are aware of the difference (e.g., in accuracy and stability) between coarse systems and fine systems, it will be relatively simple for them to increase the number of grid points when they wish to. Second, most of the numerical methods that may be used in the area of numerical heat transfer are introduced. After learning the procedures involved in these numerical techniques, the readers may be able to choose the one that best suits a particular problem. Third, in each chapter numerous examples and problems are presented to enhance the readers' understanding of the text.

I have been privileged to obtain permission from Prof. D. Brian Spalding, one of the outstanding researchers and pioneers in the area of heat transfer, to present some of his ideas in this preface. Professor Spalding asserts that, whether they operate under the finite-element or finite-difference banner, most practitioners concern themselves with a finite number of field variables associated with a finite number of subdomains, into which the entire space-time domain is meshed. All finite-difference and finite-element methods can thus be considered as finite-domain methods. The characteristics of various finite-domain methods, according to Professor Spalding, differ with respect to (1) shapes of the subdomains; (2) basis (or trial, shape) functions; (3) number of nodal unknowns involved in the algebraic equation; (4) coefficients in the algebraic equation (even the numbers of nodal unknowns are the same); (5) mathematical formulation leading to the algebraic equation; and (6) schemes solving both the linear and nonlinear algebraic equations. I support these ideas and welcome further comments from the readers.

Finally, I wish to express my deep gratitude to the individuals who have contributed to the realization of this book. Professors Yue-Nan Chen and An-Lu Ren, visiting scholars from Zhejiang University of People's Republic of China, thoroughly reviewed the entire manuscript, computed many numerical problems, and checked the algebra. Professor Houde Han from Peijing University of People's Republic of China enhanced my understanding of functional analysis and made valuable suggestions and corrections. Dr. John deRis of Factory Mutual Research, Prof. R. Bruce Kellogg of the University of Maryland, and Dr. Hwa-Ping Lee of NASA-Goddard Space Flight Center reviewed part of the manuscript and made several constructive comments. Thanks are also due to three individuals and several graduate students in my department: Mr. Roy W. Knight read the entire manuscript and performed some computational work; occasional discussions with Prof. Milton E. Palmer were enlightening; Prof. Patrick F. Cunniff, my department chairman, kindly arranged a comfortable teaching load; and students in my numerical heat transfer class made suggestions. The mechanical engineering faculty at the University of California-Berkeley, especially Prof. Patrick J. Pagni, who supervised my Ph.D. dissertation, and Prof. Ralph A. Seban, whose heat convection course I attended, converted a naive student into a conscientious researcher. Support from the Office of Naval Research and the National Science Foundation for organizing the National Symposia on Numerical Heat Transfer is also acknowledged. Mrs. Lucille Ang kindly undertook the awesome task of typing the final draft of this book and sacrificed her weekends and evenings to its completion.

Finally, but most deeply, I thank my wife, Joyce, without whose constant encouragement, support, assistance, and tolerance this manuscript might only have been published in my dreams.

Tien-Mo Shih

PART ONE

PRELIMINARIES

CHAPTER
ONE
NUMERICAL METHODS USED IN HEAT TRANSFER (I)

In the area of heat transfer, several numerical methods have been developed to deal with a great many complicated physical problems. The conventional names of these methods are listed in the table of contents. Although they may differ in mathematical formulation, numerical properties, popularity, merits, and shortcomings, these methods have one common feature: they are designed to seek approximate solutions to governing equations for which closed-form results are unobtainable or nonexistent.

Even as practitioners in the area of numerical heat transfer, we may not be familiar with all the available numerical methods. If we immediately apply an unfamiliar method to a physically complicated problem, we will be busy simply trying to check the correctness of the numerical procedures and will be unable to explore the complexities of the physical phenomena. Furthermore, the numerical results obtained will be of doubtful accuracy, since errors could be made due to our poor understanding of either numerical methods or physical mechanisms. Therefore, as an introduction to numerical heat transfer, this chapter and the next two are devoted to repetitive applications of several numerical methods to a simple one-dimensional heat transfer problem whose analytical solution can be easily obtained. We may gain confidence in using these methods if the numerical results obtained are in good agreement with the analytical solution. After this step, we extend our preliminary analyses to more advanced physical problems as well as numerical properties.

Throughout this chapter, the one-dimensional physical domain is coarsely subdivided into four intervals, with only three nodal unknowns to be sought. With such a coarse-meshed system, a desk calculator is sufficient to obtain the numerical solution; a large computer is not needed. Readers thus do not have to spend much time on computer programming, which can be a task separate from the numerical analysis. The successful application of a numerical method to coarse networks, however, does not guarantee success with finer ones because the numerical properties of a scheme are strongly dependent on mesh size. We will simply give this warning here and postpone the analysis of numerical properties to later chapters.

Consider a flow with uniform velocity moving in the negative $\bar{x}$ direction in a circular tube of length L and radius R, as shown in Fig. 1-1. Let us assume that the radial variation in temperature is negligible so that the energy transport is one-dimensional. Convective heat transfer takes place between the fluid and the wall with a heat transfer coefficient h_c. The temperatures of both the outlet ($\bar{x} = 0$) and the wall of the tube are maintained at T_0, while the inlet ($\bar{x} = L$) is kept at a different temperature T_L.† Taking the energy balance over an infinitesimal control volume $d\bar{x}$, we can derive

$$\pi R^2 k \frac{d^2 T}{d\bar{x}^2} + \pi R^2 \rho c_p u \frac{dT}{d\bar{x}} - 2\pi R h_c (T - T_0) = 0$$

subject to $T(0) = T_0$ and $T(L) = T_L$. For convenience, this governing equation and the corresponding boundary conditions are nondimensionalized to become

$$-\frac{d^2\phi}{dx^2} - \frac{d\phi}{dx} + 2\phi = 0 \qquad \phi(0) = 0 \qquad \text{and } \phi(4) = 1, \tag{1-1}$$

where $\phi = (T - T_0)/(T_L - T_0)$ and $x = \bar{x}/R$. The parameters uR/α, $h_c R/k$, and L/R are chosen to be 1, 1, and 4, respectively. Throughout this and the next chapters, Eq. (1-1) will be our model equation to be solved by several numerical methods. Its exact solution is found to be

$$\phi(x) = \frac{e^x - e^{-2x}}{e^4 - e^{-8}} \tag{1-2}$$

†The outflow boundary conditions at $\bar{x} = 0$ may not be physically realistic, but is mathematically acceptable.

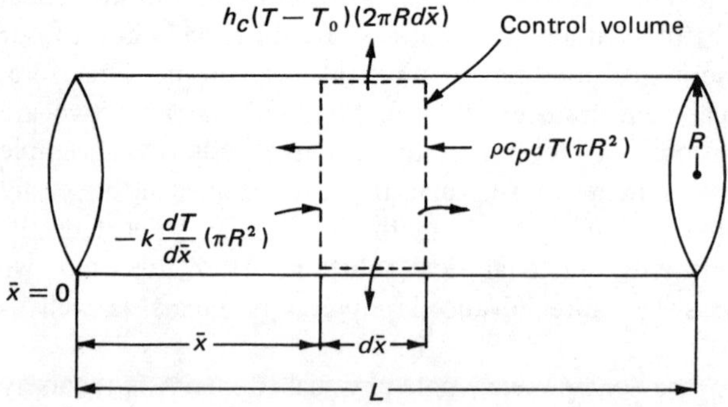

Figure 1-1 Conservation of energy over an infinitesimal control volume disk.

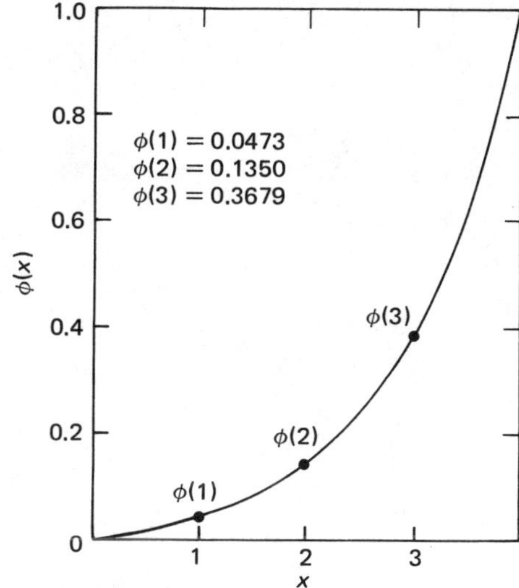

Figure 1-2 Exact solution to Eq. (1-1).

which is plotted in Fig. 1-2. We compute Eq. (1-2) at $x = 1, 2, 3$, to yield

$$\phi(1) = 0.0473 \qquad \phi(2) = 0.1350 \qquad \text{and } \phi(3) = 0.3679.$$

These values can be compared with the numerical result obtained later.

The present chapter will cover the following topics.

In Section 1-1, the finite-difference method will be introduced. Schemes of second-order accuracy and higher-order accuracy are described. One of the two schemes of higher-order accuracy is related to the cubic spline formulation and its discretized equations constitute a tridiagonal 3×3 block system.

In Section 1-2, the Galerkin and variational methods with finite-element discretization are presented. In addition to the use of these methods, it is beneficial to learn the logic behind their mathematical formulations. Because of its popularity and wide applicability, the Galerkin finite-element method is described first. One of the important features of this method is the proper choice of basis functions. The variational finite-element method is then introduced. This method was developed on the basis of variational calculus and therefore should not be categorized as one of the weighted-residual methods. Furthermore, when a first-derivative term is present in the differential equation, the so-called adjoint variational principle must be used, since otherwise the variational principle of the differential equation does not exist.

1-1 FINITE-DIFFERENCE METHOD

The finite-difference method has been one of the most widely used numerical methods for decades. Its popularity may be due to the fact that the mathematical concept of its discretization is relatively simple. Here, discretization is defined as an approximation procedure in which a continuous domain is replaced with a network or mesh of discrete points and the field unknowns are sought only at those discrete

Table 1-1 Finite-difference approximations of first and second derivatives

Derivative	Difference approximation	Order of accuracy
$\frac{d\phi}{dx}$	$\frac{\phi_{j+1}-\phi_j}{h}$ (forward)	$O(h)$
	$\frac{\phi_j-\phi_{j-1}}{h}$ (backward)	$O(h)$
	$\frac{\phi_{j+1}-\phi_{j-1}}{2h}$ (central)	$O(h^2)$
	$\frac{\phi_{j-2}-8\phi_{j-1}+8\phi_{j+1}-\phi_{j+2}}{12h}$ (central)	$O(h^4)$
$\frac{d^2\phi}{dx^2}$	$\frac{\phi_{j-1}-2\phi_j+\phi_{j+1}}{h^2}$	$O(h^2)$
	$\frac{-\phi_{j-2}+16\phi_{j-1}-30\phi_j+16\phi_{j+1}-\phi_{j+2}}{12h^2}$	$O(h^4)$

points rather than everywhere in the domain. The partial derivatives can be approximated by the finite differences in many ways. These approximations are listed in Table 1-1. They were derived by using a Taylor's series expansion. Depending on the truncated errors, all finite-difference schemes fall into two major categories: those of second-order accuracy and those of higher-order accuracy. Being simpler and more popular, the schemes in the former category will be described first.

1-1*a* Schemes of Second-Order Accuracy

Let $\delta^2\phi_j$ and $\delta\phi_j$ denote approximations of $d^2\phi/dx^2$ and $d\phi/dx$, respectively. If

$$\frac{d^2\phi}{dx^2} - \delta^2\phi_j = O(h^2) \quad \text{and} \quad \frac{d\phi}{dx} - \delta\phi_j = O(h^2), \tag{1-3}$$

then we say that the approximations $\delta^2\phi_j$ and $\delta\phi_j$ are of second-order accuracy. From Table 1-1, the second derivative can be written as

$$\frac{d^2\phi}{dx^2} = \frac{1}{h^2}(\phi_{j-1} - 2\phi_j + \phi_{j+1}) + \frac{h^2}{12}\frac{d^4\phi}{dx^4} + O(h^4). \tag{1-4}$$

Upon deletion of the fourth derivative in Eq. (1-4) and adoption of the central difference of $d\phi/dx$, we discretize Eq. (1-1) into

$$-(2-h)\phi_{j-1} + (4+4h^2)\phi_j - (2+h)\phi_{j+1} = 0, \tag{1-5}$$

which is the familiar three-point relation. Experienced readers may immediately pay attention to the magnitude of the mesh size h, wondering if there is any restriction on h in order for Eq. (1-5) to be meaningful. This is indeed an important subject and calls for intensive analysis of numerical stability. For continuity of presentation, however, let us postpone its discussion to later chapters.

For $h = 1$ and $j = 1, 2, 3$, we obtain a set of linear algebraic equations with a tridiagonal coefficient matrix as

$$4\phi_1 - 1.5\phi_2 = 0, \tag{1-6a}$$

$$-0.5\phi_1 + 4\phi_2 - 1.5\phi_3 = 0, \tag{1-6b}$$

$$-0.5\phi_2 + 4\phi_3 = 1.5, \tag{1-6c}$$

which can be easily solved to yield

$$\phi_1 = 0.0582 \qquad \phi_2 = 0.1552 \qquad \text{and } \phi_3 = 0.3944.$$

Since the exact solution Eq. (1-2) is available, we are able to obtain the computational errors at the nodal points such as $e_j = |\phi_j - \phi(x_j)|$. Intuitively, it may be tempting to state that the discretization scheme is considered to be accurate if e_j is small. For most heat transfer problems, however, the exact solution cannot be found (in fact, this is why we bother to study numerical analysis). Furthermore, in some cases we are also interested in the nodal values of the variable gradients m_j or the derivatives of the computational errors, $de_j/dx = |m_j - (d\phi/dx)_{x=x_j}|$. Therefore, examination of the accuracy of a numerical scheme is by no means as easy a task as it may appear intuitively. We will discuss this numerical property in Chapters 4 and 14.

Before turning our attention to other numerical schemes, we remark that it is preferable to write positive signs in front of the diagonal coefficients in Eqs. (1-6a)-(1-6c). The purpose of doing so is to make the coefficient matrix appear positive-definite. A matrix

$$[\mathrm{A}] = \begin{bmatrix} a_{11} & a_{12} & \cdots & a_{1n} \\ a_{21} & a_{22} & \cdots & a_{2n} \\ \cdots & \cdots & \cdots & \cdots \\ a_{n1} & a_{n2} & \cdots & a_{nn} \end{bmatrix}$$

is said to be positive-definite if (1) it is symmetric† and (2) all its principal minors are positive, i.e.,

$$a_{ij} = a_{ji}, \quad a_{11} > 0, \quad \begin{vmatrix} a_{11} & a_{12} \\ a_{21} & a_{22} \end{vmatrix} > 0, \quad \begin{vmatrix} a_{11} & a_{12} & a_{13} \\ a_{21} & a_{22} & a_{23} \\ a_{31} & a_{32} & a_{33} \end{vmatrix} > 0, \cdots,$$

†Sometimes this restriction is relaxed.

and
$$\begin{vmatrix} a_{11} & a_{12} & \cdots a_{1n} \\ a_{21} & a_{22} & \cdots a_{2n} \\ \cdots & \cdots & \cdots \\ a_{n1} & a_{n2} & \cdots a_{nn} \end{vmatrix} > 0.$$

A positive-definite matrix has desirable characteristics that lead to convergence for some iterative schemes [1, 2]. Although the coefficient matrix in Eqs. (1-6*a*)-(1-6*c*) is asymmetric, its principal minors are all positive. We will try to make a habit of writing a matrix with positive diagonal elements so that the matrix is nearly positive-definite.

1-1*b* Schemes of Higher-Order Accuracy

In the previous section, the $\mathbf{O}(h^2)$ term in Eq. (1-4) was truncated. Such a second-order accurate scheme sometimes may not be adequate if high accuracy is desired. One way to increase the accuracy is to retain the $\mathbf{O}(h^2)$ term and rewrite Eq. (1-4) as

$$\left(\frac{d^2\phi}{dx^2}\right)_{x_j} = \frac{1}{h^2}(\phi_{j-1} - 2\phi_j + \phi_{j+1}) - \frac{h^2}{12}\frac{M_{j-1} - 2M_j + M_{j+1}}{h^2} + \mathbf{O}(h^4), \tag{1-7a}$$

where $M_k = (d^2\phi/dx^2)x_k$, $k = j-1,\ j,\ j+1$. With the addition of this term, the accuracy of the new scheme is now raised to fourth order. For consistency, the expression of the derivative $d\phi/dx$ must also be modified into

$$\left(\frac{d\phi}{dx}\right)_{x_j} = \frac{1}{2h}(\phi_{j+1} - \phi_{j-1}) - \frac{h^2}{6}\frac{m_{j-1} - 2m_j + m_{j+1}}{h^2} + \mathbf{O}(h^4), \tag{1-7b}$$

where $m_k = (d\phi/dx)x_k$. At this point, we have two alternatives in proceeding with the discretization of Eq. (1-1). We may directly substitute Eqs. (1-7*a*) and (1-7*b*) into Eq. (1-1) to obtain a discretized expression containing ϕ_k, m_k, and M_k, $k = j-1, j, j+1$. Two additional equations relating ϕ_k, m_k, and M_k to one another are then needed to close the algebraic system.† This method leads to the same result as the cubic spline formulation, which will be described later in Section 7-1. Alternatively, we may wish to avoid introducing additional unknowns m_k and M_k and simply express them in terms of ϕ_k. In this case, Eqs. (1-7*a*) and (1-7*b*) are rearranged into

$$\left(\frac{d^2\phi}{dx^2}\right)_{x_j} = \frac{1}{12h^2}(-\phi_{j-2} + 16\phi_{j-1} - 30\phi_j + 16\phi_{j+1} - \phi_{j+2}) + \mathbf{O}(h^4), \tag{1-8a}$$

†An algebraic system is said to be closed if the number of equations is equal to the number of unknowns.

$$\left(\frac{d\phi}{dx}\right)_{x_j} = \frac{1}{12h}(-\phi_{j+2} + 8\phi_{j+1} - 8\phi_{j-1} + \phi_{j-2}) + O(h^4). \qquad (1\text{-}8b)$$

Equations (1-8*a*) and (1-8*b*) are the five-point relations. If the field variable ϕ is uniformly distributed in the vicinity of nodal point *j*, it is expected that both $(d^2\phi/dx^2)_{x_j}$ and $(d\phi/dx)_{x_j}$ will vanish. Letting $\phi_{j-2} = \phi_{j-1} = \phi_j = \phi_{j+1} = \phi_{j+2}$ and examining the right-hand side of Eqs. (1-8*a*) and (1-8*b*) confirms our expectation. This exercise is a quick way to partially check the algebra of our discretization procedure; it generally works even for more complicated discretized equations. For applications of Eqs. (1-8*a*) and (1-8*b*), see [3, 4].

Although the accuracy of Eqs. (1-8*a*) and (1-8*b*) is of fourth order, there are five nodal unknowns and consequently the computation becomes more cumbersome than that of the three-point relation. Furthermore, Eqs. (1-8*a*) and (1-8*b*) are not applicable at the two grid points adjacent to the boundaries, namely $j = 1, J-1$, since the five-point relation also involves grid points $j = -1, J+1$, which are located outside the domain. Therefore, we are restricted to using the three-point relations at $j = 1, J-1$ and reducing the accuracy to second order there. Substituting Eqs. (1-8*a*) and (1-8*b*) into Eq. (1-1) with $h = 1$ and $j = 2$, we obtain, after algebra,

$$-\frac{2}{3}\phi_1 + \frac{9}{2}\phi_2 - 2\phi_3 = \frac{-1}{6}. \qquad (1\text{-}9)$$

The other two discretized equations at $j = 1, 3$ remain the same as Eqs. (1-6*a*) and (1-6*c*), respectively. Simultaneously solving these three equations yields

$$\phi_1 = 0.0505, \quad \phi_2 = 0.1348, \quad \text{and } \phi_3 = 0.3918.$$

These nodal values as well as those computed from the exact solution and the second-order scheme are listed in Table 1-2 for comparison. It can be seen from the table that the values obtained by using Eq. (1-9) are more accurate than the second-order result. For large mesh systems, a set of algebraic equations having a pentadiagonally banded coefficient matrix will result. While the accuracy is raised, the computation time also increases. Since the emphasis of every computation job is different, we are not in a position to judge which scheme is "better." Nevertheless, it is still possible to make general comparisons of the merits and shortcomings of several numerical schemes. Such comparisons, which are meant to provide only

Table 1-2 Comparison of nodal solutions computed by various schemes

Scheme	ϕ_0	ϕ_1	ϕ_2	ϕ_3	ϕ_4
Exact	0	0.0473	0.1350	0.3679	1
Finite difference (second order)	0	0.0582	0.1552	0.3944	1
Finite difference (fourth order)	0	0.0505	0.1348	0.3918	1
Control volume	0	0.048	0.1344	0.3667	1
Finite element	0	0.044	0.1269	0.3563	1

suggestions and not strict guidelines, can be found scattered throughout this volume, particularly in Chapters 4 and 15.

As mentioned above, it is also possible to solve for m_j and M_j simultaneously with ϕ_j. Two additional equations relating ϕ_k, m_k, and M_k are thus called for. Rearranging Eqs. (1-7b) and (1-7b), we write

$$m_{j-1} + 4m_j + m_{j+1} = \frac{3}{h}(\phi_{j+1} - \phi_{j-1}) \tag{1-10a}$$

and

$$\frac{1}{12}M_{j-1} + \frac{5}{6}M_j + \frac{1}{12}M_{j+1} = \frac{1}{h^2}(\phi_{j-1} - 2\phi_j + \phi_{j+1}). \tag{1-10b}$$

It is noteworthy that Eqs. (1-10a) and (1-10b) representing the three-point relations can also be derived independently from the cubic spline formulation, to be described in Section 7-1. Equations (1-10a) and (1-10b), along with the governing equation in the discretized form, constitute a set of $3 \times J$ linear algebraic equations. This system is closed provided the boundary conditions ϕ_0, m_0, M_0, ϕ_J, m_J, and M_J are completely prescribed. Unfortunately, such a prescription is usually unavailable.† For the second-order ordinary differential equation, which requires only two boundary conditions of either the boundary-value or the initial-value type, four boundary values are missing. Evidently, in order to close the system we must seek a two-point relation that may be used at $j=0$ or $j=J$ to provide information for missing boundary conditions. To proceed, we consider two Taylor's series expansions about the boundary grid points 0 and J, respectively:

$$\phi_1 = \phi_0 + hm_0 + \frac{h^2}{2}M_0 + \frac{h^3}{6}\frac{dM_0}{dt} + O(h^4) \tag{1-11a}$$

and

$$\phi_{J-1} = \phi_J - hm_J + \frac{h^2}{2}M_J - \frac{h^3}{6}\frac{dM_J}{dt} + O(h^4), \tag{1-11b}$$

where $t = x/4$. Generally, for sufficiently small h (say, 0.05), it may be acceptable to assume that

$$\frac{dM_0}{dt} = \frac{1}{h}(M_1 - M_0) + O(h)$$

and

$$\frac{dM_J}{dt} = \frac{1}{h}(M_J - M_{J-1}) + O(h).$$

This assumption implies that M_k is linear and thereby ϕ_k is cubic in x; it reveals the similarity between the present formulation and the cubic spline formulation. For large h (say, 0.5), however, this assumption may be inaccurate, and can be replaced by

$$\frac{dM_0}{dt} = -4M_0 + 32m_0 \tag{1-12a}$$

†If all the boundary conditions are prescribed, the differential equations, in fact, are overdetermined.

and $$\frac{dM_J}{dt} = -4M_J + 32m_J, \tag{1-12b}$$

which are derived by differentiating Eq. (1-1) with respect to t (not x) once. Substituting Eqs. (1-12*a*) and (1-12*b*) into Eqs. (1-11*a*) and (1-11*b*) yields

$$\phi_1 = \phi_0 + \left(h + \frac{16}{3}h^3\right) m_0 + \left(\frac{h^2}{2} - \frac{2}{3}h^3\right) M_0 + O(h^4) \tag{1-13a}$$

and $$\phi_{J-1} = \phi_J - \left(h + \frac{16}{3}h^3\right) m_J + \left(\frac{h^2}{2} + \frac{2}{3}h^3\right) M_J + O(h^4). \tag{1-13b}$$

We note that Eqs. (1-13*a*) and (1-13*b*) remain fourth-order accurate, in contrast to the second-order accuracy of the previous higher-order scheme near the boundaries, and that Eqs. (1-13*a*) and (1-13*b*) contain only two grid points and therefore are classified as two-point relations. At boundary points $j = 0$ and $j = J$, where $J = L/h$, they should be used in order to exclude nonexistent grid points $j = -1, J+1$.

Having derived Eqs. (1-10) and (1-13), we are in a position to tackle the numerical problem described by Eq. (1-1). We replace the function itself and the first and second derivatives with ϕ_j, m_j, and M_j, respectively, to generate $J+1$ equations. Equations (1-10*a*) and (1-10*b*) applied at the interior grid points $j = 1, 2, \ldots, J-1$ and Eqs. (1-13*a*) and (1-13*b*) applied at boundary points $j = 0$ and $j = J$ constitute the remaining $2J$ equations. This solution procedure is illustrated in the following example.

Example 1-1 Use the higher-order scheme described by Eqs. (1-10) and (1-13) to solve Eq. (1-1). Take $h = 1/4$ and $J = 4$.

Solution Let the three interior grid points be designated as 1, 2, and 3. The two boundary points are therefore referred to as 0 and 4, respectively. The governing equation (1-1) can be rewritten as

$$M_j + 4m_j - 32\phi_j = 0, \quad j = 0, 1, 2, 3, 4 \tag{a}$$

in which the independent variable x in Eq. (1-1) has been changed to $t = x/4$.

Equations (1-10*a*) and (1-10*b*) are applicable to interior grid points 1, 2, and 3. For example, for $j = 1$ we have

$$m_0 + 4m_1 + m_2 = 12\phi_2 \tag{b}$$

and $$M_0 + 10M_1 + M_2 = 192(-2\phi_1 + \phi_2). \tag{c}$$

Since there are in total $3 \times 5 - 2$ unknowns, two additional two-point relations are needed to close the system. Applying Eqs. (1-13*a*) and (1-13*b*) to boundary points 0 and 4, we obtain

$$16m_0 + M_0 - 48\phi_1 = 0, \tag{d}$$

and $$24\phi_3 + 8m_4 - M_4 = 24. \tag{e}$$

Now Eqs. (*a*)-(*e*) can be rewritten in matrix form as

$$[\mathbf{A}]\ \{\mathbf{U}\} = \{\mathbf{B}\}, \qquad (f)$$

where

$$[\mathbf{A}] = \begin{bmatrix} 4 & 1 & 0 & 0 & 0 & 0 & 0 & 0 & 0 & 0 & 0 & 0 & 0 \\ 0 & 0 & -32 & 4 & 1 & 0 & 0 & 0 & 0 & 0 & 0 & 0 & 0 \\ 0 & 0 & 0 & 0 & 0 & -32 & 4 & 1 & 0 & 0 & 0 & 0 & 0 \\ 0 & 0 & 0 & 0 & 0 & 0 & 0 & 0 & -32 & 4 & 1 & 0 & 0 \\ 0 & 0 & 0 & 0 & 0 & 0 & 0 & 0 & 0 & 0 & 0 & 4 & 1 \\ 1 & 0 & 0 & 4 & 0 & -12 & 1 & 0 & 0 & 0 & 0 & 0 & 0 \\ 0 & 1 & 384 & 0 & 10 & -192 & 0 & 1 & 0 & 0 & 0 & 0 & 0 \\ 0 & 0 & 12 & 1 & 0 & 0 & 4 & 0 & -12 & 1 & 0 & 0 & 0 \\ 0 & 0 & -192 & 0 & 1 & 384 & 0 & 10 & -192 & 0 & 1 & 0 & 0 \\ 0 & 0 & 0 & 0 & 0 & 12 & 1 & 0 & 0 & 4 & 0 & 1 & 0 \\ 0 & 0 & 0 & 0 & 0 & -192 & 0 & 1 & 384 & 0 & 10 & 0 & 1 \\ 16 & 1 & -48 & 0 & 0 & 0 & 0 & 0 & 0 & 0 & 0 & 0 & 0 \\ 0 & 0 & 0 & 0 & 0 & 0 & 0 & 0 & 24 & 0 & 0 & 8 & -1 \end{bmatrix}$$

is a 13 × 13 coefficient matrix,

$$\{\mathbf{U}\}^T = \{m_0\ M_0\ \phi_1\ m_1\ M_1\ \phi_2\ m_2\ M_2\ \phi_3\ m_3\ M_3\ m_4\ M_4\}$$

is a vector with 13 components, and

$$\{\mathbf{B}\}^T = \{0\ \ 0\ \ 0\ \ 0\ \ 32\ \ 0\ \ 0\ \ 0\ \ 0\ \ 12\ \ 192\ \ 0\ \ 24\}.$$

The superscript T stands for "transpose" of a vector. Equation (*f*) can be solved by matrix decomposition, Gaussian elimination, or some iterative techniques. For continuity here, we will postpone discussion of these techniques to later chapters. At this point it is appropriate to remark that, when several nodal unknowns at the same grid point are sought, it is customary to reduce a large element matrix system to a smaller block matrix system. For example, Eqs. (1-10*a*) and (1-10*b*), along with the governing equation

$$M_j + 4m_j - 32\phi_j = 0 \qquad (g)$$

can be grouped and rewritten as

$$[\mathbf{C}]\ \{\mathbf{V}_{j-1}\} + [\mathbf{D}]\ \{\mathbf{V}_j\} + [\mathbf{E}]\ \{\mathbf{V}_{j+1}\} = \{\mathbf{0}\}, \qquad (h)$$

where

$$[\mathbf{C}] = \begin{bmatrix} \frac{3}{h} & 0 & 0 \\ -\frac{1}{h^2} & 0 & \frac{1}{12} \\ 0 & 0 & 0 \end{bmatrix}, \quad [\mathbf{D}] = \begin{bmatrix} 0 & 4 & 0 \\ \frac{2}{h^2} & 0 & \frac{5}{6} \\ -32 & 4 & 1 \end{bmatrix},$$

$$[\mathbf{E}] = \begin{bmatrix} \frac{3}{h} & 1 & 0 \\ -\frac{1}{h^2} & 0 & \frac{1}{12} \\ 0 & 0 & 0 \end{bmatrix}, \quad \text{and } \{\mathbf{V}_j\} = \begin{Bmatrix} \phi_j \\ m_j \\ M_j \end{Bmatrix}.$$

It can be seen that Eq. (h) constitutes a tridiagonal 3 X 3 block system, which can be solved by a procedure similar to that used to solve a tridiagonal element system. The details of this procedure are presented in Chapter 7. Here we simply list the final results in Table 1-3 and compare them with the exact solution.

So far, the equations that we have discretized have been ordinary differential equations. Finite-difference methods can, of course, be extended to the discretization of partial differential equations as well. After discretization of partial derivatives, subsequent analysis of their numerical properties is needed to examine the adequacy of a particular scheme. Since these are more advanced topics, we will study them in later chapters and now turn our attention to numerical methods that do not require Taylor's series expansions.

Table 1-3 Nodal solutions (function itself and first and second derivatives) of fourth-order finite-difference scheme, $m = d\phi/dt$, $M = dm/dt$

Scheme	$j = 0$ ($t = 0$)	$j = 2$ ($t = \frac{1}{2}$)	$j = 4$ ($t = 1$)
Exact			
ϕ	0	0.1350	1
m	0.2196	0.5444	4.0001
M	−0.8792	2.1440	15.9997
Finite difference			
ϕ	0	0.1343	1
m	0.1889	0.5346	3.9320
M	−0.7557	2.1579	16.2716

1-2 GALERKIN AND VARIATIONAL METHODS WITH FINITE-ELEMENT FORMULATION

The term "finite-element method" has been in the literature for the past three decades. The name was introduced in the early 1950s by researchers in structural mechanics who were working on problems in which solids, such as trusses, rods, and plates, were subject to certain external loads. When analytical solutions were not obtainable for complicated geometries, they divided the physical systems into subdomains and sought approximate solutions in the meshed systems. These subdomains were termed finite elements, the numerical method associated with this subdivision being called the finite-element method. In these formulations, the governing equations under consideration were the *algebraic* equations that follow Hooke's law, and not the *differential* equations that describe transport of field variables such as momentum, energy, and species. Use of the method was limited. Gradually, the concept of finite elements was incorporated into the variational method, the Galerkin method, and others. The resulting method thus became capable of solving differential equations and became popular in most disciplines in the physical sciences. Nowadays, by finite-element method we no longer mean the original algebraic finite-element method. Instead, the name is now commonly considered to represent a combination of finite-element subdivision and one of the variational methods or the method of weighted residuals. Readers who are interested in the history and literature of the finite-element method should see [5-8]. Hereafter in this text finite-element method is a short name that stands for the procedures starting from the differential equations and ending with the algebraic equations involving nodal unknowns. To be more specific, we may call the combination of Galerkin method and *piecewise* polynomial interpolants the Galerkin finite-element method, the combination of variational method and *piecewise* polynomial interpolants the variational finite-element method, and so forth.

The finite-element methods are established on the basis of integral minimization, which is quite unlike the Taylor's series expansion dominantly used in the finite-difference method. In my experience, it is very tempting for engineers to go ahead using a certain method without understanding the logic behind its formulation. Finite-element methods do involve a few logical steps, and although their use has diffused through the engineering community, their mathematical logic may not have been understood by every user. There are several disadvantages of following finite-element procedures rigidly without logical guidance: (1) it is boring, (2) innovative ideas are not likely to arise, and (3) it becomes difficult to understand error bound analyses (Chapter 14). Therefore, in this section it will be beneficial to discuss at length the logic governing the mathematical procedures of the finite-element method before describing its use. Readers who are familiar with this logic may skip pp. 15-19 and continue on p. 20.

Consider the following linear differential equation

$$L\phi = f(x), \quad a \leqslant x \leqslant b \tag{1-14}$$

where L denotes a linear differential operator defined on the interval $[a, b]$. We will ignore the fact that Eq. (1-14) may have an exact solution. Our objective now is to find a solution that approximately satisfies Eq. (1-14). Multiplying Eq. (1-14) by a function $v(x)$ and then integrating both sides with respect to x over $[a, b]$ gives

$$\int_a^b (L\tilde{\phi})v(x)\,dx = \int_a^b f(x)v(x)\,dx, \tag{1-15a}$$

which is commonly written as

$$(L\tilde{\phi}, v) = (f, v), \tag{1-15b}$$

where (A, B) is called the inner product of functions A and B. The notation of the inner product will be used frequently throughout this volume and is explained in more detail in Chapter 14. Now, looking at Eqs. (1-15*a*) and (1-15*b*), we may immediately ask the following qeustions: Why should Eq. (1-14) be integrated at all? Why should it be multiplied by a function $v(x)$ before the integration? What kind of function $v(x)$ should we choose? How shall we proceed to find $\tilde{\phi}(x)$ after Eq. (1-15*a*)?

Let us attempt to answer the first question. It is not necessary to integrate Eq. (1-14) in order to obtain an approximate solution. A typical example without integration is the finite-difference method previously described. Aside from this method, we still have at least one option, in which the function $\tilde{\phi}(x)$ is approximated by a polynomial of degree k. Substituting the polynomial into the differential equation, forcing errors at $k-1$ arbitrary points to vanish, and using the two boundary conditions, we obtain $k+1$ algebraic equations for the $k+1$ undetermined coefficients. This procedure is illustrated in the following example.

Example 1-2 Use the method of power series substitution to solve Eq. (1-1).

Solution We approximate the solution $\phi(x)$ by the following polynomial:

$$\tilde{\phi}(x) = a_0 + a_1 x + a_2 x^2 + a_3 x^3. \tag{a}$$

Substituting Eq. (*a*) into Eq. (1-1) and collecting terms with the same power of x, we obtain

$$2a_2 + a_1 - 2a_0 + 2(3a_3 + a_2 - a_1)x + (3a_3 - 2a_2)x^2 - 2a_3 x^3 = 0. \tag{b}$$

Since Eq. (*b*) should be valid for all $x \epsilon [0, 4]$, we may arbitrarily choose 0 and 1 as two convenient values for x, to obtain, respectively,

$$2a_2 + a_1 - 2a_0 = 0 \tag{c}$$

and

$$7a_3 + 2a_2 - a_1 - 2a_0 = 0. \tag{d}$$

In addition, Eq. (*a*) must satisfy the boundary conditions $\phi(0) = 0$ and $\phi(4) = 1$. Thus, the equations

$$a_0 = 0 \tag{e}$$

and
$$a_0 + 4a_1 + 16a_2 + 64a_3 = 1 \tag{f}$$

are also generated. There are four unknowns a_0, a_1, a_2, and a_3 to satisfy Eqs. (*c*)–(*f*). After arithmetic, the final result of the approximate polynomial can be written as

$$\phi(x) = \frac{7}{100}x - \frac{7}{200}x^2 + \frac{1}{50}x^3. \tag{g}$$

This method is similar to the collocation method, to be described in Section 2-4.

Mathematically, the example above is very straightforward; demanding that the assumed polynomial satisfy the governing equation at only two arbitrary points $x = 0$ and $x = 1$ did not create lengthy arithmetic. It thus seems that this method is quite satisfactory. Unfortunately, this is not true. If higher accuracy is desired, the degree of the assumed polynomial must be raised. Correspondingly, more algebraic equations must be simultaneously solved. It is also likely that there are very few zero elements in the system matrix, which is quite different from the diagonally banded matrices deduced from the finite-difference method or the finite-element method.

But these are only minor reasons why it may be inadequate to solve a differential equation directly. The major reason is as follows. Let us examine Fig. 1-3, in which the solid and dashed curves represent, respectively, the exact solution to a fictitious ordinary differential equation and the approximate solution obtained by the method described in Example 1-2. Since there are five local extrema, a polynomial of quite high degree must have been used and a considerable amount of

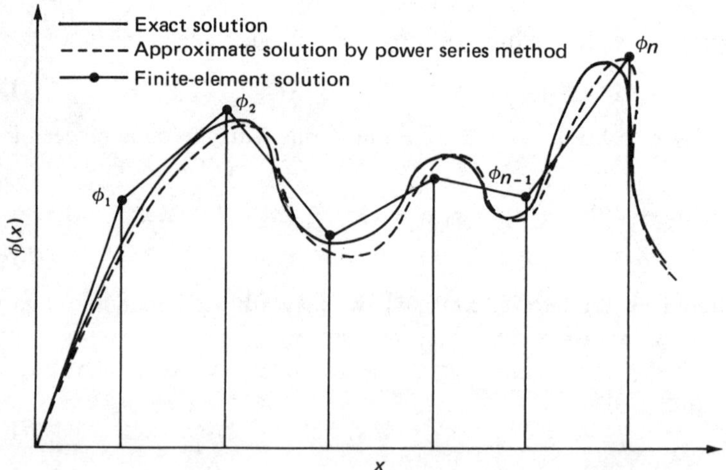

Figure 1-3 A fictitious exact solution and two approximate solutions.

algebra, arithmetic, and computational work must have been performed for the dashed curve to be successfully plotted. In other words, we have paid a high price for the smoothness of the dashed curve, which may not be essential at all. Even if it is, we need to explore some efficient ways other than this elaborate procedure. In Fig. 1-3, piecewise line segments are also shown. Although they lack smoothness, these line segments approximate the exact solution satisfactorily. The procedures that provide such a piecewise linear solution are actually what we need. The finite-element method to be described falls into this category, whereas the method described in Example 1-2 does not.

Alert readers may have noticed that, although the line segment solution shown in Fig. 1-3 is continuous, the first derivative is not and the second derivative does not exist pointwise. This suggests that Eq. (1-14), which contains a second derivative, must be somehow modified. Integration of Eq. (1-14) can eliminate the second derivative and move the approximate solution into a broader space.

The questions of why Eq. (1-14) should be multiplied by a weighting function and what kind of weighting function should be chosen will be answered as we study the following methods.

1-2*a* Galerkin Method

The method initiated by Galerkin in 1915 [9] is probably the most popular of all the finite-element methods. The essence of the Galerkin method is that the weighting (or test) function is chosen to be the same as the basis (or trial, expansion, or shape) function [10–16]. Since we will refer to basis functions throughout this book, we will discuss them at some length here.

Functions $\phi_1(x), \phi_2(x), \ldots, \phi_n(x)$ are said to be basis functions of a linear space X if (1) they are linearly independent of one another and (2) every $\phi \epsilon X$ can be expressed as linear combination of the $\phi_j(x)$. More details concerning basis functions can be found in Chapter 14. Typical examples of basis functions are the power series $1, x, x^2, \ldots, x^n$ and the Fourier series 1, $\sin \pi x$, $\cos \pi x, \ldots, \sin n\pi x$, $\cos n\pi x$ on the interval $[0,1]$. It is known that any bounded function defined on $[0,1]$ can be expressed in terms of one of these two series and that the approximation becomes better as more terms are retained. Two examples involving the definition of the basis functions are presented below.

Example 1-3 Consider the following linear combinations of functions that may or may not be legitimate basis functions on [0,1]:

(*a*)

$$\frac{x}{4} + c_1 x(4-x) + c_2 x(16-x^2) + c_3 x(64-x^3) + \cdots + c_n x(4^n - x^n),$$

(*b*)

$$\frac{x}{4} + d_1 x(4-x) + d_2 x^2(4-x) + d_3 x^3(4-x) + \cdots + d_n x^n(4-x),$$

(*c*)
$$\frac{x}{4}+e_1\frac{x}{4}\left(1-\frac{x}{4}\right)+e_2\frac{x}{4}\left(1-\frac{x^2}{16}\right)+e_3\frac{x}{4}\left(1-\frac{x^3}{64}\right)$$
$$+\cdots+e_n\frac{x}{4}\left(1-\frac{x^n}{4^n}\right).$$

Which linear combination is formed by a legitimate set of basis functions?

Solution The three approximations above all vanish at $x = 0$ and converge to unity at $x = 4$. Also, all the members are linearly independent. However, the last member ϕ_n in (*a*) and (*b*) does not approach a limit as n increases. This trend is depicted in Fig. 1-4*a*, *b*. The peaks of the curves grow in an unbounded way in both cases as n becomes large. It is unlikely that every bounded function on [0,4] can be expressed as a linear combination of these unbounded functions. Now, let us examine Fig. 1-4*c*, in which a family of curves representing members in case (*c*) are plotted. As n increases, these curves asymptotically approach the straight line. Therefore, it can be said that members in case (*c*) are legitimate basis functions on [0,4] and those in cases (*a*) and (*b*) are not.

For one-dimensional problems, it is customary to approximate the function $\phi(x)$ on the interval $[x_{j-1}, x_{j+1}]$ by

$$\tilde{\phi}(x) = N_{j-1}(x)\phi_{j-1} + N_j(x)\phi_j + N_{j+1}(x)\phi_{j+1}, \qquad x\epsilon[x_{j-1}, x_{j+1}],$$

where $N_k(x)$, $k = j-1$, j, $j+1$, are piecewise linear functions, sometimes called pyramid functions (see Fig. 1-5), defined by

$$N_j(x) = \begin{cases} \dfrac{x - x_{j-1}}{h} & x\epsilon[x_{j-1}, x_j], \\ \dfrac{x_{j+1} - x}{h} & x\epsilon[x_j, x_{j+1}], \\ 0 & \text{elsewhere}, \end{cases} \tag{1-16}$$

where h is the uniform mesh size. Naturally, we are interested in seeing whether these popular functions are indeed legitimate basis functions.

Example 1-4 Determine whether the pyramid functions defined in Eq. (1-16) are basis functions on [0,4].

Solution First, let us see if these pyramid functions are linearly independent. The expression $a_1N_1 + a_2N_2 + \cdots + a_nN_n$ cannot be identical to zero for all x unless the coefficients $a_1, a_2, \ldots, a_n$ are all simultaneously equal to zero. On the basis of this mathematical definition, we assert that $N_1, N_2, \ldots, N_n$ are linearly independent. Next, Eq. (1-16) implies that, for $t = (x - x_{j-1})/h\epsilon[0,1]$,

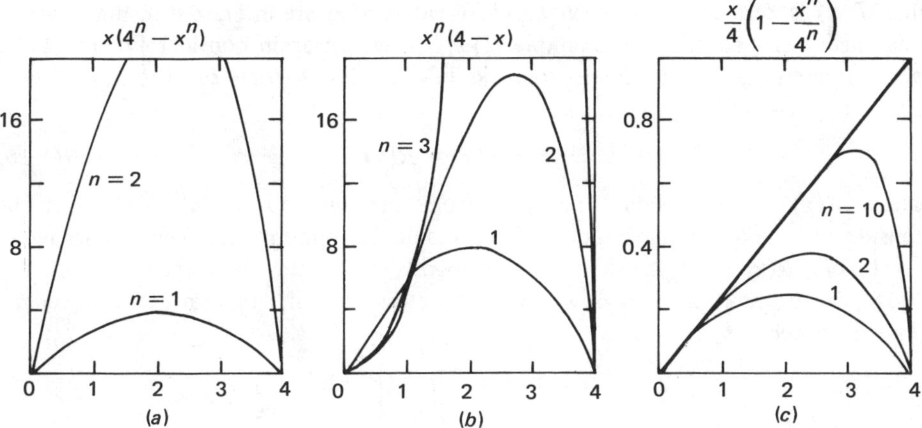

Figure 1-4(*a–c*) Sets of functions that may or may not be legitimate basis functions on [0,4].

$$N_{j-1}(t) = 1 - t \qquad (a)$$

and

$$N_j(t) = t. \qquad (b)$$

It is clear that a linear function on $[x_{j-1}, x_j]$ can always be expressed by a linear combination of $N_{j-1}(t)$ and $N_j(t)$ and that any bounded function on [0,4] can be well approximated by a number of linear functions if the interval $[x_{j-1}, x_j]$ is taken to be sufficiently small. In conclusion, the pyramid functions defined in Eq. (1-16) are basis functions on [0,4].

Having studied the definition of the basis functions, we are now ready to explore an important concept required in the Galerkin formulation. That is, a function $R(x)$ must be identical to zero if it is orthogonal to all the members of a basis in a space. For example, if

$$\int_0^1 R(x)\,dx = \int_0^1 xR(x)\,dx = \int_0^1 x^2R(x)\,dx = \cdots = \int_0^1 x^nR(x)\,dx = \cdots = 0, \qquad (1\text{-}17)$$

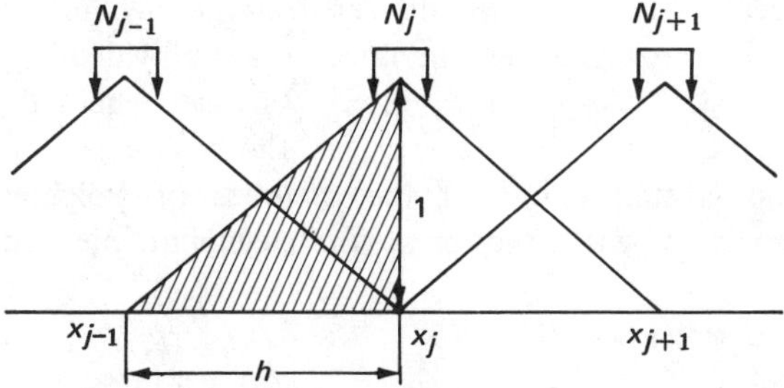

Figure 1-5 Piecewise linear basis functions, also called pyramid functions or roof functions.

then $R(x)$ is zero everywhere on [0,1]. Readers who are interested in the proof of this theory can consult, for example, [17]. Now, we again consider Eq. (1-14). If $\tilde{\phi}(x)$ represents an approximate solution of Eq. (1-14), then substitution of $\tilde{\phi}(x)$ into Eq. (1-14) yields

$$L\tilde{\phi} - f(x) = R(x), \tag{1-18}$$

where $R(x)$ is the residual resulting from the substitution. If $\tilde{\phi}(x)$ is to be considered a good approximation, $R(x)$ should be minimized. Since, according to Eq. (1-17), $R(x)$ must vanish if it is orthogonal to all the members of a basis, Eq. (1-18) can be multiplied by a weighting function $v(x)$ and then integrated over the domain to become

$$(L\tilde{\phi}, v) - (f, v) = (R, v).$$

Then we demand that

$$(R, v) = 0. \tag{1-19}$$

In other words, we seek an approximate solution $\tilde{\phi}(x)$ such that Eq. (1-19) holds, or such that

$$(L\tilde{\phi}, v) - (f, v) = 0. \tag{1-20}$$

In principle, based on Eq. (1-17), we may choose $v(x)$ to be members of any basis in the space. Galerkin brilliantly let $v(x)$ be equal to the basis function that is adopted to express the linear combination. Specifically, he assumed that

$$\tilde{\phi}(x) = \sum_{j=0}^{J} c_j N_j(x), \tag{1-21}$$

where the c_j's are coefficients to be determined. Then he let

$$v_j(x) = N_j(x). \tag{1-22}$$

It turned out that the formulation and the computation based on Eq. (1-22) are very simple in comparison with those based on other legitimate choices. This will be demonstrated in Example 1-6.

So far, we have learned the logic that leads to Eqs. (1-20)–(1-22). Before we continue with the Galerkin method, it is worth mentioning that, for one-dimensional problems, the Galerkin method is said to be a finite-element method only when the basis functions are chosen to be piecewise nonzero† as defined in Eq. (1-16). The reason is that, under such circumstances, the domain is subdivided into a finite number of elements $[0, x_1]$, $[x_1, x_2]$, . . . , $[x_{j-1}, x_j]$. In cases where the basis functions are members of some complete sets such as those discussed in Example 1-3, the Galerkin method is simply one of the methods of weighted residuals and should not be associated with the term finite-element method.

†or, equivalently, chosen to have local support (see Chapter 14).

Hereafter, we define a finite-element method as a scheme that solves governing equations such as Eq. (1-20) with $\tilde{\phi}$ approximated by piecewise functions.

Continuing with the Galerkin finite-element method, we now proceed to solve Eq. (1-1). Corresponding to Eq. (1-20) is the following equation:

$$-\int_0^4 N_j \frac{d^2\tilde{\phi}}{dx^2}\,dx - \int_0^4 N_j \frac{d\tilde{\phi}}{dx}\,dx + 2\int_0^4 N_j\tilde{\phi}\,dx = 0. \tag{1-23}$$

As long as the second derivative $d^2\tilde{\phi}/dx^2$ is present in the first integral of Eq. (1-23), we require that $\tilde{\phi}\epsilon C^2$,† where C^2 denotes the space in which all functions are at least twice differentiable. It is desirable to relax this restriction and move $\tilde{\phi}$ into a larger space C^1 or C^0 $(C^2 \subset C^1 \subset C^0)$. To achieve this, we apply integration by parts to the first integral in Eq. (1-23) and obtain

$$\int_0^4 \frac{dN_j}{dx}\frac{d\tilde{\phi}}{dx}\,dx - N_j\frac{d\tilde{\phi}}{dx}\bigg|_0^4 - \int_0^4 N_j \frac{d\tilde{\phi}}{dx}\,dx + 2\int_0^4 N_j\tilde{\phi}\,dx = 0. \tag{1-24}$$

Frequently, Eq. (1-24) is called the weak (or generalized) form of the original differential equation (1-1). The adjective "weak" is meant to convey the idea that the differentiability of the approximate solution $\tilde{\phi}$ in Eq. (1-24) can be weaker than that of the exact solution ϕ in Eq. (1-1). Examining the definition of $N_j(x)$ in Eq. (1-16) more closely, we find that $N_j(x)$ is equal to zero from 0 to x_{j-1} and from x_{j+1} to 4. Equation (1-24) thus can be simplified to

$$\int_{x_{j-1}}^{x_{j+1}} \frac{dN_j}{dx}\frac{d\tilde{\phi}}{dx}\,dx - \int_{x_{j-1}}^{x_{j+1}} N_j \frac{d\tilde{\phi}}{dx}\,dx + 2\int_{x_{j-1}}^{x_{j+1}} N_j\tilde{\phi}\,dx = 0. \tag{1-25}$$

Even at $j = 1$ and $j = J - 1$, the boundary condition terms in Eq. (1-24) still vanish, since $N_1(0) = N_{J-1}(4) = 0$. Therefore, Eq. (1-25) is applicable to all the grid points. Now, in light of the finite-element formulation, the function $\tilde{\phi}$ is approximated as

$$\tilde{\phi}(x) = N_0(x)\phi_0 + N_1(x)\phi_1 + \cdots + N_{J-1}(x)\phi_{J-1} + N_J(x)\phi_J. \tag{1-26}$$

If there exist the Dirichlet boundary conditions $\phi_0 = 0$ and $\phi_J = 1$, Eq. (1-26) is reduced to

$$\tilde{\phi}(x) = N_1(x)\phi_1 + N_2(x)\phi_2 + \cdots + N_{J-1}(x)\phi_{J-1} + N_J(x). \tag{1-27}$$

Differentiating Eq. (1-27) once and then substituting the result into Eq. (1-25), we are able to derive the form

$$A\phi_{j-1} + B\phi_j + C\phi_{j+1} = 0, \quad j = 1, 2, \ldots, J-1, \tag{1-28}$$

†Actually, we require only that $\tilde{\phi}$ belong to a Sobolev space in which $d^2\tilde{\phi}/dx^2$ is integrable. To avoid complications here, we do not mention the Sobolev space. See Chapter 14 for details.

where

$$A = \int_{x_{j-1}}^{x_j} \frac{dN_{j-1}}{dx} \frac{dN_j}{dx}\, dx - \int_{x_{j-1}}^{x_j} \frac{dN_{j-1}}{dx} N_j\, dx + 2 \int_{x_{j-1}}^{x_j} N_{j-1} N_j\, dx, \tag{1-29a}$$

$$B = \int_{x_{j-1}}^{x_{j+1}} \left(\frac{dN_j}{dx}\right)^2 dx - \int_{x_{j-1}}^{x_{j+1}} \frac{dN_j}{dx} N_j\, dx + 2 \int_{x_{j-1}}^{x_{j+1}} N_j^2\, dx, \tag{1-29b}$$

and

$$C = \int_{x_j}^{x_{j+1}} \frac{dN_j}{dx} \frac{dN_{j+1}}{dx}\, dx - \int_{x_j}^{x_{j+1}} \frac{dN_{j+1}}{dx} N_j\, dx + 2 \int_{x_j}^{x_{j+1}} N_j N_{j+1}\, dx. \tag{1-29c}$$

Since the integrals in Eqs. (1-29*a*)–(1-29*c*) will be repeatedly used throughout this volume, it is convenient to evaluate them once and refer to the tabulated expressions later. This is done in Table 1-4, which shows the integrals, their designated symbols, and their evaluated results, which are obtained in a straightforward manner by using the definition of $N_j(x)$ in Eq. (1-16). We demonstrate the evaluation of an integral in the following example.

Example 1-5 Evaluate the integral

$$\int_{x_{j-1}}^{x_j} \frac{dN_{j-1}}{dx} N_j(x)\, dx.$$

Solution Instead of integrating directly with respect to x, it is convenient to introduce a new variable defined by

$$t = \frac{x - x_{j-1}}{h} \tag{a}$$

Thus,

$$\int_{x_{j-1}}^{x_j} \frac{dN_{j-1}}{dx} N_j(x)\, dx = \int_0^1 \frac{dN_{j-1}}{dt} N_j(t)\, dt, \tag{b}$$

with

$$N_{j-1}(t) = 1 - t \qquad N_j(t) = t. \tag{c, d}$$

Consequently,

$$\int_{x_{j-1}}^{x_j} \frac{dN_{j-1}}{dx} N_j(x)\, dx = -\frac{1}{2}. \tag{e}$$

Table 1-4 Numerical values of integrals involving the piecewise linear basis functions

Expression	Numerical value
$\Lambda_{d(j-1),\,dj} = \int_{x_{j-1}}^{x_j} \frac{dN_{j-1}}{dx}\frac{dN_j}{dx}\,dx, \quad \Lambda_{d(j+1),\,dj}$	$-\frac{1}{h}$
$\Lambda_{dj,\,dj} = \int_{x_{j-1}}^{x_{j+1}} \left(\frac{dN_j}{dx}\right)^2 dx$	$\frac{2}{h}$
$\Lambda_{d(j-1),\,j} = \int_{x_{j-1}}^{x_j} \frac{dN_{j-1}}{dx} N_j\,dx, \quad \Lambda_{d(j+1),\,j}$	$-\frac{1}{2}, \frac{1}{2}$
$\Lambda_{dj,\,j} = \int_{x_{j-1}}^{x_{j+1}} \frac{dN_j}{dx} N_j\,dx$	0
$\Lambda_{j-1,\,j} = \int_{x_{j-1}}^{x_j} N_{j-1}N_j\,dx, \quad \Lambda_{j+1,\,j}$	$\frac{h}{6}$
$\Lambda_{j,\,j} = \int_{x_{j-1}}^{x_{j+1}} N_j^2\,dx$	$\frac{2h}{3}$

We observe that this numerical value is independent of the index j if the mesh size is uniform.

In evaluating the integral in Example 1-5, an important concept regarding the isoparametric element has been used. An element is said to be isoparametric if, after some proper coordinate transformation, the basis (shape, trial, or expansion) function within that element can represent universally those in all the local elements.† Using this definition, we construct Table 1-5, in which the relation between the two typical local elements and the isoparametric element is shown. Other local elements can be related to the isoparametric element in a similar way. The purpose of introducing the isoparametric element is to simplify the evaluation of integrals such as those in Eqs. (1-29*a*)-(1-29*c*). This advantage will become more obvious when higher-order or two-dimensional elements are used.

Another efficient way to evaluate the integrals in Eqs. (1-29*a*)-(1-29*c*) is to utilize Fig. 1-5. Take $\int_{x_{j-1}}^{x_{j+1}} (dN_{j-1}/dx)N_j(x)\,dx$ for example. Since $N_{j-1}(x)$ is linear and its derivative is simply a constant $-1/h$ on $[x_{j-1}, x_j]$, the integral is reduced to

†This is equivalent to saying that the shape functions interpolating $\tilde{\phi}(x)$ are the same as those interpolating the element coordinates, as defined conventionally.

Table 1-5 Correspondence between two local elements and the isoparametric element

Element	Variable domain	Approximate function	Basis function
Line segment between grids x_{j-1} and x_j	$x \epsilon [x_{j-1}, x_j]$	$\tilde{\phi}(x) = N_{j-1}(x)\phi_{j-1} + N_j(x)\phi_j$	$N_{j-1}(x) = \frac{x_j - x}{h}$
Line segment between grids x_j and x_{j+1}	$x \epsilon [x_j, x_{j+1}]$	$\tilde{\phi}(x) = N_j(x)\phi_j + N_{j+1}(x)\phi_{j+1}$	$N_j(x) = \frac{x_{j+1} - x}{h}$
			$N_{j+1}(x) = \frac{x - x_j}{h}$
Isoparametric element between points a and b	$t \epsilon [0,1]$	$\tilde{\phi}(t) = N_a(t)\phi_a + N_b(t)\phi_b$	$N_a(t) = t$ $N_b(t) = 1 - t$

$(-1/h)\int_{x_{j-1}}^{x_{j+1}} N_j(x)\,dx$. But $\int_{x_{j-1}}^{x_{j+1}} N_j(x)\,dx$ is seen in Fig. 1-5 to be equal to the area of the shaded triangle, $h/2$; therefore the given integral is $(-1/h)(h/2) = -\frac{1}{2}$, which agrees with the result obtained with the isoparametric element.

After the coefficients A, B, and C in Eqs. (1-29a)-(1-29c) are evaluated with the aid of Table 1-4, Eq. (1-28) becomes

$$-(6 - 5h)\phi_{j-1} + (12 + 8h^2)\phi_j - (6 + h)\phi_{j+1} = 0 \qquad (1\text{-}30a)$$

or

$$-\phi_{j-1} + 20\phi_j - 7\phi_{j+1} = 0. \qquad (1\text{-}30b)$$

Repeatedly using Eq. (1-30b) at $j = 1$, 2, 3 and simultaneously solving the three algebraic equations yields the following numerical solution: $\phi_1 = 0.0444$, $\phi_2 = 0.1269$, and $\phi_3 = 0.3563$. In comparison with the exact solution, this solution appears to be more accurate than the finite-difference result. The better accuracy, however, is achieved at the expense of lengthy algebra leading to Eq. (1-30). A more thorough comparison of the finite-element method and the finite-difference method is presented in Chapter 15.

This is a good opportunity to repeat the same problem with weighting functions that are different from the basis functions and then examine the efficiency of this numerical scheme. The purpose of such an exercise is to show that it is not absolutely necessary to choose basis functions as the weighting functions.

Example 1-6 Like the pyramid functions, the set $(W_0, W_1, W_2, \ldots, W_J)$ is also complete, where the basis functions are zero everywhere except on the intervals specified as follows:

$$W_0 = 1 - 3t + 2t^2, \quad t = \frac{x}{2h}, \quad x \epsilon [0, x_2],$$

$$W_1 = 4(t - t^2), \quad t = \frac{x}{2h}, \quad x \epsilon [0, x_2],$$

$$W_2 = \begin{cases} W_2^- = 2t^2 - t, & t = \dfrac{x}{2h}, \quad x \epsilon [0, x_2], \\ W_2^+ = 1 - 3t + 2t^2, & t = \dfrac{x - x_2}{2h}, \quad x \epsilon [x_2, x_4], \end{cases}$$

$$W_3 = 4(t - t^2), \quad t = \frac{x - x_2}{2h}, \quad x \epsilon [x_2, x_4],$$

$$W_4 = \begin{cases} W_4^- = 2t^2 - t, & t = \dfrac{x - x_2}{2h}, \quad x \epsilon [x_2, x_4], \\ W_4^+ = 1 - 3t + 2t^2, & t = \dfrac{x - x_4}{2h}, \quad x \epsilon [x_4, x_6], \end{cases}$$

......

$$W_j = 4(t - t^2), \quad t = \frac{x - x_{j-1}}{2h}, \quad x \epsilon [x_{j-1}, x_{j+1}], \quad \text{if } j \text{ is odd,}$$

and

$$W_j = \begin{cases} 2t^2 - t, & t = \dfrac{x - x_{j-2}}{2h}, \quad x \epsilon [x_{j-2}, x_j] \\ 1 - 3t + 2t^2, & t = \dfrac{x - x_j}{2h}, \quad x \epsilon [x_j, x_{j+2}] \end{cases}, \text{ if } j \text{ is even.}$$

The solution $\tilde{\phi}(x)$ remains to be approximated by the pyramid basis functions as shown in Eqs. (1-16) and (1-26). Solve Eq. (1-1) by the procedure leading to Eq. (1-28), except that now the weighting functions are chosen to be W_j.

Solution Multiplying Eq. (1-1) by W_j and integrating the result over the domain, we obtain

$$-\int_0^4 W_j \frac{d^2\tilde{\phi}}{dx^2}\, dx - \int_0^4 W_j \frac{d\tilde{\phi}}{dx}\, dx + 2\int_0^4 W_j \tilde{\phi}\, dx = 0. \quad (a)$$

Using integration by parts for the first integral leads to a weaker form

$$\int_0^4 \frac{dW_j}{dx}\frac{d\tilde{\phi}}{dx}\, dx - W_j \frac{d\tilde{\phi}}{dx}\bigg|_0^4 - \int_0^4 W_j \frac{d\tilde{\phi}}{dx}\, dx + 2\int_0^4 W_j \tilde{\phi}\, dx = 0. \quad (b)$$

Equation (*b*) will be used for $j = 1$, 2, 3 since ϕ_1, ϕ_2, and ϕ_3 are the unknowns. Because W_1 is zero at boundary point $x = 0$ as well as outside the region $[0, x_2]$, Eq. (*b*) can be reduced to

$$\int_0^{x_2} \frac{dW_1}{dx}\frac{d\tilde{\phi}}{dx}\,dx - \int_0^{x_2} W_1 \frac{d\tilde{\phi}}{dx}\,dx + 2\int_0^{x_2} W_1\tilde{\phi}\,dx = 0. \qquad (c)$$

For easier computation, Eq. (*c*) is changed to ($h = 1$)

$$\phi_1 \int_0^{1/2} \left(\frac{dW_1}{dt} - 2W_1\right) dt + (\phi_2 - \phi_1)\int_{1/2}^{1}\left(\frac{dW_1}{dt} - 2W_1\right) dt$$

$$+ 8\phi_1 \int_0^{1/2} W_1 t\,dt + 4\int_{1/2}^{1} W_1\left[(2 - 2t)\phi_1 + (2t - 1)\phi_2\right] dt = 0. \qquad (d)$$

Similarly, for $j = 3$, since W_3 is equal to zero at boundary point $x = 4$ and outside the region $[x_2, 4]$, Eq. (*b*) can be simplified to

$$B_{3,0\to1/2}(\phi_3 - \phi_2) + B_{3,1/2\to1}(1 - \phi_3)$$

$$+ 4\int_0^{1/2} W_3\left[(1 - 2t)\phi_2 + 2t\phi_3\right] dt + 4\int_{1/2}^{1} W_3\left[(2 - 2t)\phi_3 + 2t - 1\right] dt = 0, \qquad (e)$$

where

$$B_{3,0\to1/2} = \int_0^{1/2}\left(\frac{dW_3}{dt} - 2W_3\right) dt \quad \text{and} \quad B_{3,1/2\to1} = \int_{1/2}^{1}\left(\frac{dW_3}{dt} - 2W_3\right) dt. \qquad (f)$$

For $j = 2$, we note that W_2 is nonzero throughout the open domain (0, 4) except at $x = 1$, 3. Unlike Eqs. (*d*) and (*e*), which involve two unknowns and one boundary condition, the following equation, deduced from Eq. (*b*),

$$B_{2,0\to1/2}^{-}\phi_1 + B_{2,1/2\to1}^{-}(\phi_2 - \phi_1) + B_{2,0\to1/2}^{+}(\phi_3 - \phi_2) + B_{2,1/2\to1}^{+}(1 - \phi_3)$$

$$+ 8\phi_1\int_0^{1/2} W_2^{-} t\,dt + 4\int_{1/2}^{1} W_2^{-}\left[(2 - 2t)\phi_1 + (2t - 1)\phi_2\right] dt$$

$$+ 4\int_0^{1/2} W_2^{+}\left[(1 - 2t)\phi_2 + 2t\phi_3\right] dt + 4\int_{1/2}^{1} W_2^{+}\left[(2 - 2t)\phi_3 + 2t - 1\right]$$

$$dt = 0 \qquad (g)$$

involves three unknowns and both boundary conditions. The notations $B_{2,0\to1/2}^{-}$ and $B_{2,1/2\to1}^{+}$ are defined similarly to those in Eq. (*f*) with W_2^{-} and

W_2^+ appearing in the integrand. After evaluation of the integrals, Eqs. (*d*), (*e*), and (*g*) can be rearranged and simplified to

$$22\phi_1 - 7\phi_2 = 0, \qquad (h)$$

$$2\phi_1 - 19\phi_2 + 8\phi_3 = 0, \qquad (i)$$

$$\phi_2 + 22\phi_3 = 7. \qquad (j)$$

Finally, these three equations are solved simultaneously to yield $\phi_1 = 0.0432$, $\phi_2 = 0.1359$, and $\phi_3 = 0.3120$. Compared with the result obtained by the Galerkin method, the solution in this example is less accurate in the Sobolev sense. In addition, the algebraic manipulation is quite cumbersome.†

1-2*b* Mixed-Type Boundary Conditions

Before leaving the subject of the Galerkin method, we will investigate the case where mixed-type boundary conditions exist. A standard example is

$$\phi(0) = 1 \quad \text{and} \left(\frac{d\phi}{dx}\right)_{x=4} = a\phi(4). \qquad (1\text{-}31)$$

Since in such boundary conditions $\phi(4)$ is not explicitly prescribed, ϕ_4 must be sought, in addition to ϕ_1, ϕ_2, and ϕ_3, as a nodal unknown. For $j = 1, 2, 3$, Eq. (1-24) can be reduced to Eq. (1-25) by deleting the boundary-condition term. For $j = 4$, however, neither the basis function N_4 nor the derivative $d\tilde{\phi}/dx$ is zero at $x = 4$. Therefore, the boundary-condition term in Eq. (1-24) remains. With Eq. (1-31), it can be evaluated as

$$N_4 \frac{d\tilde{\phi}}{dx}\bigg|_0^4 = N_4(4)\left(\frac{d\tilde{\phi}}{dx}\right)_{x=4} - N_4(0)\left(\frac{d\tilde{\phi}}{dx}\right)_{x=0} = (1)(a\phi_4) - (0)\left(\frac{d\tilde{\phi}}{dx}\right)_{x=0} = a\phi_4. \qquad (1\text{-}32)$$

The other three terms in Eq. (1-24) are evaluated by the same procedure as mentioned before. Hence there are four algebraic equations and four nodal unknowns ϕ_1, ϕ_2, ϕ_3, and ϕ_4; the algebraic system remains closed.

1-2*c* Variational Principles

Many problems in engineering and the physical sciences can be characterized by variational principles. The variational method was initiated in the 19th century by Rayleigh [18] and in 1908 by Ritz [19] and has been developed since. Although it is similar to the method of weighted residuals in that the assumed solutions in both methods are expressed in terms of a linear combination of basis functions with coefficients to be determined, the variational method is formulated on the basis of

†For the purpose of ensuring numerical stability in the streamwise diffusion equation (see Section 8-3), the weighting function is sometimes intentionally chosen to be different from the basis function. This approach is referred to as the Petrov-Galerkin method.

variational calculus [20, 21]. In the method of weighted residuals, the residuals are minimized, whereas in the variational method, the variational integrals are minimized. In addition, the minimization of the variational integrals in some problems can be precisely interpreted in terms of physics [22, 23]. For example, in the problem of a flexible string under constant tension, the minimization of the time integral of the Lagrangian–i.e., the difference between the kinetic and potential energies–leads to the wave equation of the string. Therefore, it may be desirable to discuss this method at some length.

Most textbooks on variational calculus [20, 21] describe in detail the derivation, starting from minimization of the integral

$$I = \int_a^b F[x, \phi(x), \phi'(x)]\, dx \tag{1-33}$$

with $\phi(a)$ and $\phi(b)$ being prescribed, of the Euler-Lagrange equation

$$\frac{\partial F}{\partial \phi} - \frac{d}{dx}\left(\frac{\partial F}{\partial \phi'}\right) = 0. \tag{1-34}$$

Here we will only present an example to demonstrate this derivation. Consider the following integral (often called the variational principle, the variational integral, or the functional):

$$I = \int_0^4 (\phi'^2 + 2\phi^2)\, dx. \tag{1-35}$$

We seek the solution $\phi(x)$ for which I, a pure number, is minimized†. For the sake of easier discussion, let us assume that $\phi(x)$ is fixed at two endpoints, e.g., $\phi(0) = 0$ and $\phi(4) = 1$. Suppose that $\phi(x)$ is the desired solution. Then we choose a small variation of the function $\phi(x)$, designated as $\delta\phi$, which vanishes at the endpoints $x = 0$ and $x = 4$. For any constant ϵ, the integral

$$I(\epsilon) = \int_0^4 [(\phi' + \epsilon\, \delta\phi')^2 + 2(\phi + \epsilon\, \delta\phi)^2]\, dx \tag{1-36}$$

now becomes a function of ϵ and should have a minimum when $\epsilon = 0$. To minimize $I(\epsilon)$, we must first differentiate it with respect to ϵ, i.e.,

$$\frac{d}{d\epsilon} I(\epsilon) = \int_0^4 \frac{\partial}{\partial \epsilon} [(\phi' + \epsilon\, \delta\phi')^2 + 2(\phi + \epsilon\, \delta\phi)^2]\, dx. \tag{1-37}$$

†Actually, we will work on the extremization of I. The term "extremized" is equivalent to "made stationary," which some authors prefer to use. See Problem 1-9 at the end of the chapter for further details.

Noting that ϕ, $\delta\phi$, ϕ', and $\delta\phi'$ all are functions of x only and are independent of ϵ, we obtain

$$\frac{d}{d\epsilon} I(\epsilon) = \int_0^4 [2(\phi' + \epsilon\,\delta\phi')\delta\phi' + 4(\phi + \epsilon\,\delta\phi)\delta\phi]\,dx. \tag{1-38}$$

Therefore, setting $\epsilon = 0$ gives

$$\left[\frac{dI(\epsilon)}{d\epsilon}\right]_{\epsilon=0} = \int_0^4 (2\phi'\,\delta\phi' + 4\phi\,\delta\phi)\,dx. \tag{1-39}$$

Observant readers may have noticed that Eq. (1-39) appears to be a certain "differentiated" form of Eq. (1-35). Indeed, if we define the differentiation rule for functionals as

$$\delta F(\psi) = \frac{dF}{d\psi}\,\delta\psi, \tag{1-40}$$

where ψ stands for ϕ, ϕ', or other functions, it becomes clear that differentiation of Eq. (1-35) leads to

$$\delta I = \delta \int_0^4 (\phi'^2 + 2\phi^2)\,dx = \int_0^4 (2\phi'\,\delta\phi' + 4\phi\,\delta\phi)\,dx. \tag{1-41}$$

Since $I(\epsilon)$ is to be minimized, it follows that

$$\delta I = \left[\frac{dI(\epsilon)}{d\epsilon}\right]_{\epsilon=0} = 0. \tag{1-42}$$

Stating without rigorous proof that $\delta(d\phi/dx)$ is equal to $d(\delta\phi)/dx$ (also see Problem 1-7), we can integrate Eq. (1-41) by parts to obtain

$$\delta I = \int_0^4 (-\,2\phi'' + 4\phi)\delta\phi\,dx + 2\phi'\,\delta\phi\,\Big|_0^4 = 0. \tag{1-43}$$

Since $\delta\phi$ vanishes at two endpoints and is completely arbitrary, the terms in parentheses must be zero to ensure that the integral be zero. Therefore, for I to be a minimum, it follows that

$$\phi'' - 2\phi = 0. \tag{1-44}$$

Equation (1-44) is the Euler-Lagrange equation corresponding to Eq. (1-35). Since the expression for $f(x, \phi, \phi')$ is available, it is a simple matter to substitute this expression into Eq. (1-34) to derive Eq. (1-44).

Example 1-7 Prove that Eq. (1-44) is the Euler-Lagrange equation corresponding to Eq. (1-35).

Solution From Eq. (1-35), we write

$$F(x, \phi, \phi') = \phi'^2 + 2\phi^2. \tag{a}$$

But

$$\frac{\partial F}{\partial \phi} = 4\phi, \quad \frac{\partial F}{\partial \phi'} = 2\phi'. \tag{b}$$

Hence, Eq. (1-34) is reduced to

$$4\phi - \frac{d}{dx}(2\phi') = 0 \tag{c}$$

or

$$\phi'' - 2\phi = 0.$$

It is important to keep in mind that the solution to Eq. (1-44) is exactly that which minimizes I in Eq. (1-35). To make this more specific, we analytically solve Eq. (1-44) subject to $\phi(0) = 0$ and $\phi(4) = 1$ to obtain

$$\phi(x) = \frac{e^{\sqrt{2}x} - e^{-\sqrt{2}x}}{e^{4\sqrt{2}} - e^{-4\sqrt{2}}}. \tag{1-45}$$

Comparing the value of I obtained by using Eq. (1-45) with those obtained with other arbitrary functions, we will find that the former is always the lowest.

Example 1-8 Evaluate Eq. (1-35) with the $\phi(x)$ specified in Eq. (1-45). Is the value of I thus obtained less than those obtained with (*a*) $\phi(x) = x/4$ and (*b*) $\phi(x) = \sin(\pi x/8)$?

Solution From Eq. (1-45) we obtain

$$\phi' = \frac{ae^{ax} + ae^{-ax}}{A}, \tag{a}$$

where $A = e^{4a} - e^{-4a}$ and $a = \sqrt{2}$. Substitution of Eq. (*a*) and Eq. (1-45) into Eq. (1-35) yields, after algebra,

$$I = \frac{8}{A^2} \int_0^4 \cosh(2\sqrt{2}x)\, dx. \tag{b}$$

Its numerical value is $I = 1.4142$. Next, for case (*a*) $\phi(x) = x/4$, we obtain $I_a = 35/12$. For case (*b*) $\phi(x) = \sin(\pi x/8)$, the numerical value is $I_b = 4.3084$. Therefore, the value of I obtained with Eq. (1-45) is the lowest one in the three cases.

The example described between Eq. (1-35) and Eq. (1-44) is a well-designed one. Starting from a specified variational integral Eq. (1-35), we derived the differential equation (1-44) step by step without obvious difficulty. Furthermore, it is possible to reverse the derivation, starting from Eq. (1-44) and proceeding backward to arrive at Eq. (1-35). In most heat transfer problems, however, the reversing procedure encounters difficulties. What are the difficulties? How do we remove them? Are we able to remove them at all? If we are, how does the numerical method play a role in this problem? These questions are answered in the following paragraphs.

Let us start with the last question. Suppose we wish to minimize, instead of Eq. (1-35), the following variational integral:

$$I = \int_0^4 (3x\phi'^2 + 2\phi^3)\, dx. \tag{1-46}$$

Following the procedure that leads to Eq. (1-44) or utilizing the Euler-Lagrange equation (1-34) directly, we can derive

$$(x\phi')' - \phi^2 = 0. \tag{1-47}$$

We are aware that Eq. (1-47) is a nonlinear second-order ordinary differential equation with variable coefficient and that an analytical solution probably does not exist. A certain numerical method is thus called for to obtain an approximate solution $\tilde{\phi}(x)$ to Eq. (1-47). Since minimizing I in Eq. (1-46) is equivalent to finding $\phi(x)$ in Eq. (1-47), we have two choices: either solve Eq. (1-47) numerically or minimize I in Eq. (1-46) numerically. Although the choice is perhaps a matter of personal preference, we observe that the derivative in Eq. (1-46) is only first order, whereas that in Eq. (1-47) is second order. The existence of a first derivative enables us to search for permissible functions in a larger (or more generalized) space C^0, such as piecewise linear functions that are not twice differentiable at the nodal points. In light of this advantage, we choose to minimize the variational integral provided it exists. Now we will return to Eq. (1-35) to demonstrate how the numerical technique can be used.

As mentioned before, I is a pure number. It becomes a function of ϵ if functions $\phi + \epsilon\, \delta\phi$ are introduced. Based on this argument, if we approximate $\phi(x)$ by

$$\tilde{\phi}(x) = N_0(x)\phi_0 + N_1(x)\phi_1 + \cdots + N_J(x)\phi_J, \tag{1-48}$$

where $N_j(x)$ are the piecewise linear functions, and substitute Eq. (1-48) into Eq. (1-35), then the integral I becomes a function of $\phi_1, \phi_2, \ldots, \phi_{J-1}$, that is,

$$I(\phi_1, \phi_2, \ldots, \phi_{J-1}) = \int_0^4 (\tilde{\phi}'^2 + 2\tilde{\phi}^2)\, dx. \tag{1-49}$$

For $I(\phi_1, \phi_2, \ldots, \phi_{J-1})$ to be minimized, we require that

$$\frac{\partial I}{\partial \phi_1} = \frac{\partial I}{\partial \phi_2} = \cdots = \frac{\partial I}{\partial \phi_{J-1}} = 0. \tag{1-50}$$

Taking the nodal point j for example, we write

$$\frac{\partial I}{\partial \phi_j} = 2\int_0^4 \left(\tilde{\phi}' \frac{\partial \tilde{\phi}'}{\partial \phi_j} + 2\tilde{\phi}\frac{\partial \tilde{\phi}}{\partial \phi_j}\right) dx = 0. \tag{1-51}$$

Since $N_j(x)$ vanishes from 0 to x_{j-1} and from x_{j+1} to 4, Eq. (1-51) can be reduced to

$$\int_{x_{j-1}}^{x_{j+1}} (\tilde{\phi}' N_j' + 2\tilde{\phi} N_j)\, dx = 0. \tag{1-52}$$

Consequently, we are able to derive the form

$$A\phi_{j-1} + B\phi_j + C\phi_{j+1} = 0, \tag{1-53}$$

where

$$A = \Lambda_{d(j-1),dj} + 2\Lambda_{j-1,j}, \tag{1-54a}$$

$$B = \Lambda_{dj,dj} + 2\Lambda_{j,j}, \tag{1-54b}$$

and

$$C = \Lambda_{d(j+1),dj} + 2\Lambda_{j+1,j}. \tag{1-54c}$$

The values of Λ's are listed in Table 1-4. After arithmetic, Eq. (1-53) becomes

$$-(3 - h^2)\phi_{j-1} + (6 + 4h^2)\phi_j - (3 - h^2)\phi_{j+1} = 0. \tag{1-55}$$

Repeated use of Eq. (1-55) at $j = 1, 2, 3$ yields $\phi_1 = 0.0087$, $\phi_2 = 0.0435$, and $\phi_3 = 0.2087$. The exact solution Eq. (1-45) gives $\phi(1) = 0.01352$, $\phi(2) = 0.0589$, and $\phi(3) = 0.2431$. We have learned how the finite-element method can be incorporated to solve Eq. (1-35) numerically. A similar procedure, of course, can also be used to minimize I in Eq. (1-46). This will be left for the reader as an exercise (Problem 1-16). For applications of the variational principle to heat transfer computations, see [24–32].

Let us now turn our attention to the next question: how to remove the possible difficulties encountered when going backward from the Euler-Lagrange equation to the minimization of the variational integral. One difficulty is associated with the presence of the first derivative $d\phi/dx$, which, in physics and engineering, often represent friction or dissipation. When the energy of a system is dissipated by friction, the governing differential equation does not have a corresponding variational principle. For example, the sum of the kinetic energy $m(dy/dt)^2$ and the potential energy $ky^2/2$ of a moving block of mass m attached to a spring of spring constant k will gradually decrease if a damping force $c\,dy/dt$ is present in the system. The equation of motion is analogous to the heat transfer equation (1-1).

Example 1-9 Does Eq. (1-1) possess a variational principle?

Solution We write the inner product as

$$(L\tilde{\phi}, \delta\tilde{\phi}) = \int_0^4 (\tilde{\phi}'' + \tilde{\phi}' - 2\tilde{\phi})\, \delta\tilde{\phi}\, dx. \tag{a}$$

If the reversing procedure from Eq. (1-44) backward to Eq. (1-41) is possible, we should be able to derive, from Eq. (*a*), a form such as

$$\delta I = \delta \int_0^4 G(x, \tilde{\phi}', \tilde{\phi})\, dx. \tag{b}$$

In other words, in order for the variational principle to exist, we should be able to move the variation operator δ out of the integral sign. For the second derivative and the function itself, no difficulty arises since

$$\int_0^4 \tilde{\phi}''\, \delta\tilde{\phi}\, dx = -\frac{\delta}{2}\int_0^4 \tilde{\phi}'^2\, dx \quad \text{and} \quad \int_0^4 \tilde{\phi}\, \delta\tilde{\phi}\, dx = \frac{\delta}{2}\int_0^4 \tilde{\phi}^2\, dx. \tag{c}$$

We consider all the possibilities for a similar operation for the first derivative. Nothing can be achieved but

$$\int_0^4 \tilde{\phi}'\, \delta\tilde{\phi}\, dx = -\int_0^4 \tilde{\phi}\, \delta\tilde{\phi}'\, dx + \tilde{\phi}\, \delta\tilde{\phi}\Big|_0^4, \tag{d}$$

in which the variation operator δ remains embedded inside the integral. Nonrigorously, we have established that differential equations containing first derivatives do not have variational principles. This difficulty motivated the development of the so-called adjoint variational principle, to be described in the next section.

1-2*d* Adjoint Variational Principle

An operator L^* is said to be the adjoint operator of L if

$$(L\tilde{\phi}, \tilde{\psi}) = (\tilde{\phi}, L^*\tilde{\psi}) + \text{boundary-condition terms}, \tag{1-56}$$

where functions ϕ and ψ are differentiable as many times as the highest order of the operators L and L^*. A differential equation $L^*\psi$ is said to be the adjoint equation of $L\phi$, and a function of ψ is said to be the adjoint function of ϕ if Eq. (1-56) holds. In particular, L, $L\phi$, and ϕ are defined as the self-adjoint (also called symmetric) operator, equation, and function, respectively, if

$$L = L^*. \tag{1-57}$$

The following example will help to familiarize us with this definition.

Example 1-10 Consider the following two differential equations:

$$(a)\ L_1\phi = \frac{-d}{dx}\left(x\frac{d\phi}{dx}\right) + \phi = 0 \quad \text{and } (b)\ L_2\phi = -x\frac{d^2\phi}{dx^2} + \phi = 0, \quad x\epsilon[a, b].$$

Find the adjoint equations for both cases. Which equation is self-adjoint?

Solution An inner product of the equation in case (*a*) can be written as

$$(L_1\tilde{\phi}, \tilde{\psi}) = -\int_a^b \frac{d}{dx}\left(x\frac{d\tilde{\phi}}{dx}\right)\tilde{\psi}\, dx + \int_a^b \tilde{\phi}\tilde{\psi}\, dx = 0. \tag{a}$$

After integrating the first term by parts twice, Eq. (*a*) becomes

$$(L_1\tilde{\phi}, \tilde{\psi}) = -\int_a^b \tilde{\phi}\frac{d}{dx}\left(x\frac{d\tilde{\psi}}{dx}\right) dx + \int_a^b \tilde{\phi}\tilde{\psi}\, dx - x\frac{d\tilde{\phi}}{dx}\tilde{\psi}\bigg|_a^b + x\tilde{\phi}\frac{d\tilde{\psi}}{dx}\bigg|_a^b \tag{b}$$

or

$$(L_1\tilde{\phi}, \tilde{\psi}) = (\tilde{\phi}, L_1\tilde{\psi}) + \text{boundary-condition terms.} \tag{c}$$

The problem in case (*a*) is thus self-adjoint. Next let us examine the equation given in case (*b*). Its inner product can be written as

$$(L_2\tilde{\phi}, \tilde{\psi}) = -\int_a^b x\frac{d^2\tilde{\phi}}{dx^2}\tilde{\psi}\, dx + \int_a^b \tilde{\phi}\tilde{\psi}\, dx = 0. \tag{d}$$

Performing integration by parts once for the first term, we obtain

$$(L_2\tilde{\phi}, \tilde{\psi}) = \int_a^b \frac{d\tilde{\phi}}{dx}\frac{d}{dx}(x\tilde{\psi})\, dx + \int_a^b \tilde{\phi}\tilde{\psi}\, dx - \frac{d\tilde{\phi}}{dx}x\tilde{\psi}\bigg|_a^b. \tag{e}$$

Integration by parts once again for the first term yields

$$(L_2\tilde{\phi}, \tilde{\psi}) = -\int_a^b \tilde{\phi}\frac{d^2}{dx^2}(x\tilde{\psi})\, dx + \int_a^b \tilde{\phi}\tilde{\psi}\, dx - \frac{d\tilde{\phi}}{dx}x\tilde{\psi}\bigg|_a^b + \tilde{\phi}\frac{d}{dx}(x\tilde{\psi})\bigg|_a^b \tag{f}$$

or

$$(L_2\tilde{\phi}, \tilde{\psi}) = (\tilde{\phi}, L_2^*\tilde{\psi}) + \text{boundary-condition terms,} \tag{g}$$

where

$$L_2^*\psi = -\frac{d^2}{dx^2}(x\psi) + \psi. \tag{h}$$

Equation (h) is the adjoint equation of $L_2\phi = 0$. Since L_2^* is not equal to L_2, the operator L is not self-adjoint.

We mentioned in the preceding section that variational principles of differential equations containing first derivatives do not exist. Thus, some revision of the variational method must be made. It is for this reason that we introduced the adjoint equation. By proper incorporation of such an adjoint equation, the derivation of a pseudovariational principle for a differential equation containing a first derivative thus becomes possible [33]. We shall describe such a treatment by considering Eq. (1-1).

Following the similar procedure described in Example 1-10, we are able to show that the adjoint equation of

$$L\phi = -\frac{d^2\phi}{dx^2} - \frac{d\phi}{dx} + 2\phi = 0$$

is simply

$$L^*\phi^* = -\frac{d^2\phi^*}{dx^2} + \frac{d\phi^*}{dx} + 2\phi^* = 0. \tag{1-58}$$

Observe that the only difference between the primary equation and the adjoint counterpart Eq. (1-58) is the sign before the first derivative. We further assume that ϕ^* satisfies the same boundary conditions as ϕ, i.e., $\phi^*(0) = 0$ and $\phi^*(4) = 1$. Slightly departing from the procedure in deriving the variational principle, we multiply Eq. (1-1) by the variation of the adjoint function $\delta\phi^*$, instead of $\delta\phi$, to obtain an inner product as

$$(L\tilde{\phi}, \delta\tilde{\phi}^*) = -\int_0^4 \frac{d^2\tilde{\phi}}{dx^2}\,\delta\tilde{\phi}^*\,dx - \int_0^4 \frac{d\tilde{\phi}}{dx}\,\delta\tilde{\phi}^*\,dx + 2\int_0^4 \tilde{\phi}\,\delta\tilde{\phi}^*\,dx = 0. \tag{1-59a}$$

Integrating the first term by parts and knowing that the boundary-condition term vanishes, we obtain

$$(L\tilde{\phi}, \delta\tilde{\phi}^*) = \int_0^4 \frac{d\tilde{\phi}}{dx}\,\delta\left(\frac{d\tilde{\phi}^*}{dx}\right)dx - \int_0^4 \frac{d\tilde{\phi}}{dx}\,\delta\tilde{\phi}^*\,dx + 2\int_0^4 \tilde{\phi}\,\delta\tilde{\phi}^*\,dx = 0. \tag{1-59b}$$

Next, we multiply Eq. (1-58) by $\delta\phi$ to obtain an inner product as

$$(L^*\tilde{\phi}^*, \delta\tilde{\phi}) = -\int_0^4 \frac{d^2\tilde{\phi}^*}{dx^2}\,\delta\tilde{\phi}\,dx + \int_0^4 \frac{d\tilde{\phi}^*}{dx}\,\delta\tilde{\phi}\,dx + 2\int_0^4 \tilde{\phi}^*\,\delta\tilde{\phi}\,dx = 0. \tag{1-60a}$$

Applying integration by parts to the first two terms yields

$$(L^*\tilde{\phi}^*, \delta\tilde{\phi}) = \int_0^4 \frac{d\tilde{\phi}^*}{dx}\,\delta\left(\frac{d\tilde{\phi}}{dx}\right)dx - \int_0^4 \tilde{\phi}^*\,\delta\left(\frac{d\tilde{\phi}}{dx}\right)dx + 2\int_0^4 \tilde{\phi}^*\,\delta\tilde{\phi}\,dx = 0. \tag{1-60b}$$

Close examination of Eqs. (1-59*b*) and (1-60*b*) reveals that combining these two equations will permit us to move the variation operator δ out of the integral. Hence, it follows that

$$(L\tilde{\phi}, \delta\tilde{\phi}^*) + (L^*\tilde{\phi}^*, \delta\tilde{\phi}) = \delta\left(\int_0^4 \frac{d\tilde{\phi}}{dx}\frac{d\tilde{\phi}^*}{dx}\,dx - \int_0^4 \frac{d\tilde{\phi}}{dx}\,\tilde{\phi}^*\,dx + 2\int_0^4 \tilde{\phi}\tilde{\phi}^*\,dx\right) = 0. \tag{1-61}$$

If we introduce a variation of a functional as

$$\delta I^* = (L\tilde{\phi}, \delta\tilde{\phi}^*) + (L^*\tilde{\phi}^*, \delta\tilde{\phi}),$$

we can generate a pseudovariational principle

$$I^* = \int_0^4 \frac{d\tilde{\phi}}{dx}\frac{d\tilde{\phi}^*}{dx}\,dx - \int_0^4 \frac{d\tilde{\phi}}{dx}\,\tilde{\phi}^*\,dx + 2\int_0^4 \tilde{\phi}\tilde{\phi}^*\,dx, \tag{1-62}$$

which will be called the adjoint variational principle of Eq. (1-1). On the basis of the variational calculus concepts outlined between Eqs. (1-35) and (1-44), finding $\tilde{\phi}$ and $\tilde{\phi}^*$ such that I^* is minimized is equivalent to finding the solutions to Eqs. (1-1) and (1-58). Furthermore, if the approximate adjoint function $\tilde{\phi}^*(x)$ is expanded by

$$\tilde{\phi}^*(x) = \sum_{j=0}^{J} N_j(x)\phi_j^*, \tag{1-63}$$

then I^* becomes a function of $\phi_1^*, \phi_2^*, \ldots, \phi_{J-1}^*$. Minimizing I^* dictates that

$$\frac{\partial I^*}{\partial \phi_1^*} = \frac{\partial I^*}{\partial \phi_2^*} = \cdots = \frac{\partial I^*}{\partial \phi_{J-1}^*} = 0. \tag{1-64}$$

Among $J-1$ equations, a typical one for grid point j can be derived as

$$\int_{x_{j-1}}^{x_{j+1}} \frac{d\tilde{\phi}}{dx}\frac{dN_j}{dx}\,dx - \int_{x_{j-1}}^{x_{j+1}} \frac{d\tilde{\phi}}{dx}N_j\,dx + 2\int_{x_{j-1}}^{x_{j+1}} \tilde{\phi}N_j\,dx = 0. \tag{1-65}$$

It is interesting to observe that Eq. (1-65) is identical to Eq. (1-25), derived by using the Galerkin formulation. Here we state, without proof,† that for differential equations (even nonlinear ones) that possess variational principles or adjoint variational principles, the final weak forms derived from the Galerkin formulation

†For more rigorous analysis, see [10, pp. 223–229].

and the variational formulation generally are identical. This statement is also valid for the problem subject to mixed-type boundary conditions.

Example 1-11 Heat conduction inside a one-dimensional slab with coordinate-dependent thermal conductivity and a heat source proportional to an exponential function of the local temperature can be described by a nonlinear ordinary differential equation as

$$A\phi = -\frac{d}{dx}\left[k(x)\frac{d\phi}{dx}\right] - ae^{b\phi} = 0, \qquad 0 \leqslant x \leqslant L.$$

The boundary conditions are $\phi(0) = 1$ and $(d\phi/dx)_L = c\phi(L)$, where a, b, and c are constants. With the approximation

$$\tilde{\phi}(x) = \sum_{j=0}^{J} N_j(x)\phi_j,$$

show that a final weak form derived by the Galerkin approach is identical to that derived by the variational method.

Solution In light of the Galerkin method, the inner product is written, after integration by parts, as

$$(A\tilde{\phi}, N_j) = \int_0^L k(x)\frac{d\tilde{\phi}}{dx}\frac{dN_j}{dx}\,dx - k(x)N_j\frac{d\tilde{\phi}}{dx}\bigg|_0^L - a\int_0^L e^{b\tilde{\phi}}N_j\,dx = 0. \tag{a}$$

For $j = J$ and $N_J(L) = 1$, Eq. (a) becomes

$$(A\tilde{\phi}, N_J) = \int_{x_{J-1}}^L k(x)\frac{d\tilde{\phi}}{dx}\frac{dN_J}{dx}\,dx - ck(L)\phi_J - a\int_{x_{J-1}}^L e^{b\tilde{\phi}}N_J\,dx = 0. \tag{b}$$

Using the variational method, on the other hand, gives

$$(A\tilde{\phi}, \delta\tilde{\phi}) = \int_0^L k(x)\frac{d\tilde{\phi}}{dx}\,\delta\left(\frac{d\tilde{\phi}}{dx}\right)dx - k(x)\frac{d\tilde{\phi}}{dx}\,\delta\tilde{\phi}\bigg|_0^L - a\int_0^L e^{b\tilde{\phi}}\,\delta\tilde{\phi}\,dx = 0 \tag{c}$$

or

$$(A\tilde{\phi}, \delta\tilde{\phi}) = \frac{1}{2}\,\delta\int_0^L k(x)\left(\frac{d\tilde{\phi}}{dx}\right)^2 dx - \frac{1}{2}\,\delta\,[k(L)c\tilde{\phi}^2(L)] - \frac{a}{b}\,\delta\int_0^L e^{b\tilde{\phi}}\,dx = 0. \tag{d}$$

The variational principle of $A\phi = 0$ is thus seen to exist; it takes the form

$$I = \frac{1}{2}\int_0^L k(x)\left(\frac{d\tilde{\phi}}{dx}\right)^2 dx - \frac{c}{2}k(L)\tilde{\phi}^2(L) - \frac{a}{b}\int_0^L e^{b\tilde{\phi}}\,dx. \qquad (e)$$

The Jth necessary condition of minimizing I

$$\frac{\partial I(\phi_1, \phi_2, \ldots, \phi_J)}{\partial \phi_J} = 0 \qquad (f)$$

leads to

$$\frac{\partial I}{\partial \phi_J} = \int_{x_{J-1}}^{L} k(x)\left(\frac{d\tilde{\phi}}{dx}\right)\frac{\partial}{\partial \phi_J}\left(\frac{d\tilde{\phi}}{dx}\right)dx - ck(L)\phi_J - \frac{a}{b}\int_{x_{J-1}}^{L} e^{b\tilde{\phi}}\frac{\partial}{\partial \phi_J}(b\tilde{\phi})\,dx = 0, \qquad (g)$$

which can be readily reduced to Eq. (*b*). Therefore, we established that

$$(A\tilde{\phi}, N_J) = \frac{\partial I}{\partial \phi_J}. \qquad (h)$$

For $j \neq J$, the boundary condition term disappears since N_j is zero at $x = 0$ and $x = L$. Based on Eqs. (*b*) and (*g*), it immediately follows that

$$(A\tilde{\phi}, N_j) = \frac{\partial I}{\partial \phi_j}. \qquad (i)$$

We have shown that, for this particular example, the weak forms derived by using both methods are identical.

SYMBOLS

C^k	space in which functions are k-time differentiable
h	uniform mesh size $(= x_j - x_{j-1} = x_{j+1} - x_j = \cdots)$
h_c	heat transfer coefficient, $W/m^2 \cdot K$
j	index for grid points in x direction
I	a variational integral
J	largest number of j
L	linear differential operator, or length of tube
m_j	first-derivative unknown at grid point $j\,[=(d\phi/dx)_{x=x_j}]$
M_j	second-derivative unknown at grid point $j\,[=(d^2\phi/dx^2)_{x=x_j}]$
$N_j(x)$	linear piecewise basis functions [defined in Eq. (1-16)]
R	residual, or radius of tube, m
t	isoparametric coordinate $(0 \leqslant t \leqslant 1)$
u	uniform flow velocity in $\bar{x}$ direction, m/s
$v(x)$	weighting function or test function

x	normalized coordinate ($= \bar{x}/R$)
$\bar{x}$	physical coordinate, m
$W_j(x)$	quadratic piecewise basis functions (defined in Example 1-6)
α	thermal diffusivity ($= k/\rho c_p$), m^2/s
$\delta\phi$	variation of function ϕ
$\delta\phi_j$	central-difference approximation of $d\phi/dx$ at $x = x_j$
$\delta^2\phi_j$	finite-difference approximation of $d^2\phi/dx^2$ at $x = x_j$
ϵ	a small pure number
$\Lambda_{d(j-1),j}$	$\int_{x_{j-1}}^{x_j} (dN_j/dx)N_j\, dx$
ρ	density of fluid, kg/m^3
ϕ	normalized temperature $[=(T - T_0)/(T_L - T_0), 0 \leqslant \phi \leqslant 1]$

Superscripts

~	approximation of a function
*	adjoint of a function, equation, or operator
+	right-side element to grid point j
−	left-side element to grid point j

REFERENCES

1. W. F. Ames, *Numerical Methods for Partial Differential Equations,* p. 113, Academic, New York, 1977.
2. E. Isaacson and H. B. Keller, *Analysis of Numerical Methods,* pp. 70–72, Wiley, New York, 1966.
3. T. Saitoh, A Numerical Method for Two-dimensional Navier-Stokes Equation by Multi-Point Finite Differences, *Int. J. Numer. Methods Eng.*, vol. 11, pp. 1439–1454, 1977.
4. P. J. Roache, Marching Methods for Elliptic Problems: Part 2, *Numer. Heat Transfer,* vol. 1, pp. 163–181, 1978.
5. R. H. Gallagher, *Finite Element Analysis, Fundamentals,* Prentice-Hall, Englewood Cliffs, N.J., 1975.
6. K. H. Huebner, *The Finite Element Method for Engineers,* pp. 9–13, Wiley, New York, 1975.
7. D. H. Norrie and G. de Vries, *Finite Element Bibliography,* Plenum, New York, 1976.
8. O. C. Zienkiewicz, *The Finite Element Method,* pp. 20–21, McGraw-Hill, London, 1977.
9. B. G. Galerkin, Rods and Plates, Series in Some Problems of Elastic Equilibrium of Rods and Plates, *Vestn. Inzh. Tekh.*, vol. 19, pp. 897–908, 1915.
10. B. A. Finlayson, *The Method of Weighted Residuals and Variational Principles,* Academic, New York, 1972.
11. G. Strang and G. J. Fix, *An Analysis of the Finite Element Method,* Prentice-Hall, Englewood Cliffs, N.J., 1973.
12. P. M. Prenter, *Splines and Variational Methods,* Wiley-Interscience, New York, 1975.
13. J. T. Oden and J. N. Reddy, *Introduction to Mathematical Theory of Finite Elements,* Wiley-Interscience, New York, 1976.
14. A. R. Mitchell and R. Wait, *The Finite Element Method in Partial Differential Equations,* Wiley-Interscience, London, 1977.
15. P. G. Ciarlet, *The Finite Element Method for Elliptic Problems,* North-Holland, Amsterdam, 1978.
16. P. Linze, *Theoretical Numerical Analysis,* Wiley-Interscience, New York, 1979.

17. C. R. Wylie, *Advanced Engineering Mathematics*, pp. 316–318, McGraw-Hill, New York, 1966.
18. L. Rayleigh, On the Theory of Resonance, *Philos. Trans. R. Soc., London Ser. A*, vol. A161, pp. 77–118, 1871.
19. W. Ritz, Über eine Neue Methode zur Lösung Gewisser Variationsprobeme der Mathematischen Physik, *J. Reine Angew. Math.*, vol. 135, pp. 1–61, 1908.
20. R. Weinstock, *Calculus of Variations with Applications to Physics and Engineering*, pp. 20–22, McGraw-Hill, New York, 1952.
21. R. Courant and D. Hilbert, *Methods of Mathematical Physics*, vol. 1, pp. 184–185, Wiley-Interscience, New York, 1955.
22. M. M. Vainberg, *Variational Methods for the Study of Nonlinear Operators*, Holden-Day, San Francisco, 1964.
23. S. G. Mikhlin, *Variational Methods in Mathematical Physics*, Pergamon, New York, 1964.
24. S. Usuki, The Application of a Variational Finite Element Method to Problems in Fluid Dynamics, *Int. J. Numer. Methods Eng.*, vol. 11, pp. 563–577, 1977.
25. O. Ofi and H. J. Hetherington, Application of the Finite Element Method to Natural Convection Heat Transfer from the Open Vertical Channel, *Int. J. Heat Mass Transfer*, vol. 20, pp. 1195–1204, 1977.
26. L. T. Yeh and B. T. F. Chung, A Variational Analysis of Freezing or Melting in a Finite Medium Subject to Radiation and Convection, *ASME J. Heat Transfer*, vol. 101, pp. 592–597, 1979.
27. K. E. Barrett, A Variational Principle for the Streamfunction–Vorticity Formulation of the Navier-Stokes Equations Incorporating No-Slip Conditions, *J. Comput. Phys.*, vol. 26, pp. 153–161, 1978.
28. A. Bejan, A General Variational Principle for Thermal Insulation System Design, *Int. J. Heat Mass Transfer*, vol. 22, pp. 219–228, 1979.
29. G. Lebon and P. Mathieu, A Numerical Calculation of Nonlinear Transient Heat Conduction in the Fuel Elements of a Nuclear Reactor, *Int. J. Heat Mass Transfer*, vol. 22, pp. 1187–1198, 1979.
30. A. H. C. Phillips and L. M. Delves, The Global Element Method for Stationary Advective Problems, *Int. J. Numer. Methods Eng.*, vol. 15, pp. 167–175, 1980.
31. A. Ecer, Variational Formulation of Viscous Flows, *Int. J. Numer. Methods Eng.*, vol. 15, pp. 1355–1361, 1980.
32. Y. T. Glazunov, Generalized Variational Principle of Molar-Molecular Heat and Mass-Transfer Phenomena, *Int. J. Heat Mass Transfer*, vol. 23, pp. 759–763, 1980.
33. H. van Dam and J. E. Hoogenboom, The Adjoint Space in Heat Transport Theory, *Int. J. Heat Mass Transfer*, vol. 23, pp. 349–353, 1980.

PROBLEMS

1-1 Discretize the ordinary differential equation given in Eq. (1-1) by using the backward difference for $d\phi/dx$. Then take $h = 1$ and solve for three nodal unknowns. Is the accuracy of this numerical solution lower than that obtained by the central-difference scheme? Why?

1-2 Solve Eq. (1-5) under the following conditions: $h = 3$, $J = 4$, $\phi(0) = 10$, and $\phi(12) = 0$. Do we observe any peculiar behavior exhibited by this numerical solution?

1-3 Consider the 3×3 matrix

$$[\mathbf{A}] = \begin{bmatrix} 1 & 2 & 1 \\ 0 & 3 & -1 \\ 1.4 & 1 & 2 \end{bmatrix}.$$

Is it positive-definite? Then consider the equation

$$[\mathbf{A}]\{x\} = \{\mathbf{B}\},$$

where $\{x\}^T = \{x_1\ x_2\ x_3\}$ is an unknown vector and $\{\mathbf{B}\}^T = \{b_1\ b_2\ b_3\}$ a known vector. Are we able to solve for x_1, x_2, and x_3?

1-4 It is always a good habit to check the correctness of the result derived from lengthy algebra. To do so for Eqs. (1-8) and (1-10), let $\phi(x) = x^2$, $x_j = 0$, and $h = 1$; evaluate ϕ_k, m_k, and M_k; and see if Eqs. (1-8) and (1-10) hold. Repeat the procedure with $\phi(x) = e^x$. Explain why the latter case falsely indicates that Eqs. (1-8) and (1-10) are invalid.

1-5 Repeat Example 1-1 with $h = \frac{1}{6}$ and $J = 6$, and compare the nodal results with the analytical solution

$$\phi(x) = \frac{e^{4t} - e^{-8t}}{e^4 - e^{-8}}.$$

The result obtained should be more accurate than that listed in Table 1-3.

1-6 Are $(1-e^{-x})/(1-e^{-1})$, $(1-e^{-2x})/(1-e^{-2}), \ldots, (1-e^{-nx})/(1-e^{-n})$ legitimate basis functions on $[0,1]$? Express $\phi(x) = x^2$ in terms of the three leading terms.

1-7 Let the variation of the function $\phi(x) = x - x^2$ be $\delta\phi(x) = \eta(x)$, where $\eta(x)$ is arbitrarily assumed to be $0.001\phi(x)$. Show that (*a*) δ and d/dx are commutative, i.e., $\delta(d\phi/dx) = d(\delta\phi)/dx$, and (*b*) $\delta F[\phi(x)] = (dF/d\phi)\delta\phi$, where $\delta F(\phi) = F(\phi + \delta\phi) - F(\phi)$ and $F(\phi) = 1 + \phi^2$.

1-8 The Galerkin method can also be used to solve integral equations such as

$$L\phi = \phi(x) - \int_0^1 K(x, \xi)\phi(\xi)\, d\xi = f(x), \tag{Pl-8}$$

where L is a linear operator. Solve for ϕ_1, ϕ_2, and ϕ_3 if $f(x) = x^2 - 4x - 3$, $K(x, \xi) = 12(x + \xi)$, $h = \frac{1}{4}$, $\phi_0 = 0$, and $\phi_1 = 1$. The analytical solution $\phi(x) = x^2$ can be used for comparison.

1-9 Consider the variational functional

$$I = \int_0^1 (\phi'^2 + 2\phi^2)\, dx. \tag{Pl-9}$$

(*a*) Prove that the stationary value of I is a minimum. [Hint: differentiate Eq. (1-38) with respect to ϵ to show $d^2I/d\epsilon^2 > 0$.]

(*b*) Substitute $\phi(x) = e\sqrt{2x} + \epsilon(x - x^2)$ into Eq. (Pl-9) and plot $I(\epsilon)$ against ϵ for $\epsilon = 0, \pm 0.1, \pm 0.2$.

1-10 Suppose that the end conditions of $\phi(x)$ are $\phi(0) = 1$ and $(d\phi/dx)_{x=4} = a\phi(4)$. Find the variational principle that corresponds to the Euler-Lagrange equation

$$-\phi'' + 2\phi = 0.$$

Note that the emphasis of this problem is on the mixed-type boundary condition.

1-11 Using the example $L = -d^2/dx^2 + 2$ for convenience, prove that, for a self-adjoint operator, the variational principle of the Euler-Lagrange equation

$$L\phi - f(x) = 0$$

is

$$I = \tfrac{1}{2}(L\phi, \phi) - (f, \phi),$$

in which the first integral is sometimes called the energy inner product.

1-12 Consider one-dimensional steady-state heat conduction in an infinite slab with a heat source $b\phi$ (b has dimensions of W/cm$^3\cdot$K) and two boundary conditions $\phi(0) = \phi_0$ and

$\phi(L) = \phi_L$. Verify that the steady-state temperature of the slab must be distributed such that the integral over the domain $[0, L]$ of the difference between the squared temperature gradient multiplied by the thermal conductivity and the squared temperature multiplied by b, i.e.,

$$I = \int_0^L \left[k\left(\frac{d\phi}{dx}\right)^2 - b\phi^2 \right] dx,$$

is minimized. By analogy with a conservative spring and block system in classical mechanics, this difference can be regarded as the Lagrangian if ϕ, k, and b correspond to the displacement of the block, the mass of the block, and the spring constant, respectively. It is also interesting to note (and to prove) that the sum $k(d\phi/dx)^2 + b\phi^2$ is constant throughout the entire domain.

1-13 Repeat Problem 1-8 by using the variational method. Note that the discretized equation $\partial I/\partial \phi_j = 0$ should be derived to be equal to

$$(L\tilde{\phi} - f, N_j) = 0.$$

1-14 Two heat transfer systems A and B are governed, respectively, by

$$\phi_A'' + \phi_A' + \phi_A = 0$$

and

$$\phi_B'' - \phi_B' + \phi_B = 0.$$

The boundary conditions will be ignored. Prove that the sum of $\phi_A \phi_B$ and $\phi_A' \phi_B'$ is constant.

1-15 For easy discussion, it is sometimes desirable that the governing equation be subject to homogeneous boundary conditions. Transform Eq. (1-1) into an inhomogeneous differential equation subject to $\phi(0) = 0$ and $\phi(4) = 0$.

1-16 Assume $\tilde{\phi}(x) = N_0(x)\phi_0 + N_1(x)\phi_1 + \cdots + N_J(x)\phi_J$. Minimize Eq. (1-46) with respect to ϕ_j and find the nonlinear discretized equation at $x = x_j$.

CHAPTER
TWO
NUMERICAL METHODS USED IN HEAT TRANSFER (II)

In the preceding chapter, the Galerkin and variational finite-element methods were considered. They were given priority because they are probably the two most widely used finite-element methods, and they were presented together because their final discretized equations are usually identical. In the present chapter, we will continue to consider these two methods with basis functions distributed throughout the entire domain, as well as other numerical methods that are also frequently used in the area of heat transfer.

In Section 2-1, the Galerkin and variational methods with global basis functions are presented. The adjoint variational principle exists for certain differential equations that are not self-adjoint. Related to this subject is the discussion of mixed-type boundary conditions.

In Section 2-2, we consider the central integration method, which is also called the subdomain method or the control volume method. This method leads to the same result as is obtained by directly considering the conservation law over a finite control volume and therefore is especially important to engineers.

Section 2-3 describes the method of least squares. In addition to its use, we are interested in its corresponding weighting function and the Euler-Lagrange equation of the least-squares functional. Since the Euler-Lagrange equation is of higher order than the original differential equation, care must be taken to constrain the additional degrees of freedom correctly. The penalty-functional approach that removes the restriction of the boundary conditions on the basis functions is also described.

Finally, the collocation method is presented in Section 2-4. The connections between this method and the finite-element and finite-difference methods are revealed. The orthogonal collocation is also described.

2-1 GALERKIN AND VARIATIONAL METHODS WITH GLOBAL BASIS FUNCTIONS

In Section 1-2 we discussed the Galerkin and variational methods with finite-element discretization, in which the basis functions are piecewise linear and have local support. By "support" of a function we mean the set of points on which the function is nonzero. For example, the function $f(x) = 2x + 1$ if $x \epsilon [0,1]$ and $f(x) = 0$ elsewhere has finite support, [0,1]. Alternatively, it is also possible to use nonlinear basis functions that have support everywhere in the entire domain. The procedure that involves the use of these global basis functions is called the spectral method [1-8]. Back in Example 1-3, we mentioned that

$$v_j(x) = \frac{x}{4}\left(1 - \frac{x^j}{4^j}\right), \quad j = 1, 2, 3, \ldots \tag{2-1}$$

are legitimate basis functions. One way to find a polynomial basis such as Eq. (2-1) is to first approximate $\tilde{\phi}(x)$ by a power series with undetermined constants and then set $\tilde{\phi}(x)$ to satisfy the boundary conditions. The following example will illustrate this procedure.

Example 2-1 Find a polynomial basis that can express any bounded function $\phi(x)$ on the interval [0,4] with Dirichlet end conditions $\phi(0) = 0$ and $\phi(4) = 1$.

Solution Let

$$\tilde{\phi}(x) = c_0 + c_1 x + c_2 x^2 + \cdots + c_n x^n. \tag{a}$$

Applying the end conditions $\phi(0) = 0$ and $\phi(4) = 1$ yields

$$c_0 = 0, \tag{b}$$

$$1 = 4c_1 + 16c_2 + 64c_3 + \cdots + 4^n c_n. \tag{c}$$

Equation (c) can be rearranged as

$$c_1 = \frac{1}{4} - 4c_2 - 16c_3 - \cdots - c_n 4^{n-1}. \tag{d}$$

Substituting Eqs. (b) and (d) into Eq. (a), we obtain

$$\tilde{\phi}(x) = \frac{x}{4} + c_2 x(x-4) + c_3 x(x^2 - 16) + \cdots + c_n x(x^{n-1} - 4^{n-1}). \tag{e}$$

As n becomes large, however, the value 4^{n-1} increases without limit. Therefore, it is inadequate to adopt these polynomials as basis functions. However, we may modify Eq. (e) into

$$\tilde{\phi}(x) = \frac{x}{4} + 16c_2 \frac{x}{4}\left(\frac{x}{4} - 1\right) + 64c_3 \frac{x}{4}\left(\frac{x^2}{16} - 1\right) + \cdots + 4^n c_n \frac{x}{4}\left(\frac{x^{n-1}}{4^{n-1}} - 1\right)$$

or

$$\tilde{\phi}(x) = \frac{x}{4} + e_1 \frac{x}{4}\left(1 - \frac{x}{4}\right) + e_2 \frac{x}{4}\left(1 - \frac{x^2}{16}\right) + \cdots + e_n \frac{x}{4}\left(1 - \frac{x^n}{4^n}\right). \tag{f}$$

Now, referring to Fig. 1-4*c*, we find that the polynomial expressions in Eq. (*f*) are bounded and can adequately represent a basis.

An alternative to Eq. (*a*) in the preceding example is

$$\tilde{\phi}(x) = c_0 + c_1(4 - x) + c_2(4 - x)^2 + \cdots + c_n(4 - x)^n. \tag{2-2}$$

Following procedures similar to those used from Eq. (*b*) to Eq. (*f*), we can derive

$$\tilde{\phi}(x) = \frac{x}{4} + e_1\left(1 - \frac{x}{4}\right)\left[\left(1 - \frac{x}{4}\right) - 1\right] + e_2\left(1 - \frac{x}{4}\right)\left[\left(1 - \frac{x}{4}\right)^2 - 1\right] + \cdots + e_n\left(1 - \frac{x}{4}\right)\left[\left(1 - \frac{x}{4}\right)^n - 1\right]. \tag{2-3}$$

Both Eq. (2-1) and Eq. (2-3) are good approximations of $\phi(x)$ on [0,4]. Since Eq. (2-1) is algebraically simpler, we will adopt it in the following sections. Meanwhile, Eq. (1-1) will continue to be our model equation.

2-1*a* Galerkin Method

We mentioned a few times that, in the Galerkin method, the weighting function is chosen to be the basis function. For easy presentation here, we will take only three leading basis functions to approximate $\phi(x)$, i.e.,

$$\tilde{\phi}(x) = \frac{x}{4} + e_1 \frac{x}{4}\left(1 - \frac{x}{4}\right) + e_2 \frac{x}{4}\left(1 - \frac{x^2}{16}\right). \tag{2-4}$$

The inner products of Eq. (1-1) with respect to ν_1 and ν_2 can be written as

$$(L\tilde{\phi}, \nu_1) = -\int_0^4 \tilde{\phi}''\nu_1\, dx - \int_0^4 \tilde{\phi}'\nu_1\, dx + 2\int_0^4 \tilde{\phi}\nu_1\, dx = 0 \tag{2-5a}$$

and

$$(L\tilde{\phi}, \nu_2) = -\int_0^4 \tilde{\phi}''\nu_2\, dx - \int_0^4 \tilde{\phi}'\nu_2\, dx + 2\int_0^4 \tilde{\phi}\nu_2\, dx = 0. \tag{2-5b}$$

In the finite-element method, it is preferable to use integration by parts once to integrate the second derivatives, which are otherwise undefined at the nodal points. In the global basis function formulation, since the basis functions are at least twice differentiable, integration by parts becomes optional. We will leave Eqs. (2-5*a*) and (2-5*b*) as they are. After straightforward algebra, Eqs. (2-5*a*) and (2-5*b*) are reduced to

$$\alpha_{11} e_1 + \alpha_{12} e_2 + \beta_1 = 0 \tag{2-6a}$$

and

$$\alpha_{21} e_1 + \alpha_{22} e_2 + \beta_2 = 0, \tag{2-6b}$$

where

$$\alpha_{kk} = \int_0^4 v_k'' v_k \, dx + \int_0^4 v_k' v_k \, dx - 2\int_0^4 v_k^2 \, dx, \qquad k = 1, 2, \tag{2-7a, b}$$

$$\alpha_{12} = \int_0^4 v_2'' v_1 \, dx + \int_0^4 v_2' v_1 \, dx - 2\int_0^4 v_2 v_1 \, dx, \tag{2-7c}$$

$$\alpha_{21} = \int_0^4 v_1'' v_2 \, dx + \int_0^4 v_1' v_2 \, dx - 2\int_0^4 v_1 v_2 \, dx, \tag{2-7d}$$

and

$$\beta_k = \frac{1}{4}\int_0^4 v_k \, dx - \frac{1}{2}\int_0^4 x v_k \, dx. \tag{2-7e, f}$$

These integrals are evaluated and their values are listed in Table 2-1. Using these values, we obtain $\alpha_{11} = -7/20$, $\alpha_{12} = -61/120$, $\beta_1 = -1/2$, $\alpha_{21} = -13/24$, $\alpha_{22} = -17/21$, and $\beta_2 = -49/60$. Substituting these numerical values into Eqs. (2-6*a*) and (2-6*b*) gives $e_1 = 1.2994$ and $e_2 = -1.8783$. The approximate solution $\tilde{\phi}(x)$ thus can be written as

$$\tilde{\phi}(x) = \frac{x}{4} + 1.2994\left(\frac{x}{4}\right)\left(1 - \frac{x}{4}\right) - 1.8783\left(\frac{x}{4}\right)\left(1 - \frac{x^2}{16}\right). \tag{2-8}$$

This solution is plotted in Fig. 2-1. It is seen to be in fair agreement with the exact solution. Obviously, the more terms we take to approximate $\tilde{\phi}(x)$, the closer to the exact solution the approximate will become. In a limit, the finite-dimensional subspace spanned by $v_1, v_2, \ldots, v_n$ will expand and coincide with the infinite-dimensional space. Such sets that belong to $C^\infty[0, 4]$ are numerous.

Example 2-2 Use the Galerkin method to solve Eq. (1-1), of which the approximate solution

Table 2-1 Numerical values of the integral $\int_0^1 AB\,dt$, where $\nu_1 = t(1-t)$ and $\nu_2 = t(1-t^2)$

	A							
B	1	ν_1	ν_2	ν_1'	ν_2'	ν_1''	ν_2''	t
1	1	$\frac{1}{6}$	$\frac{1}{4}$	0	0	-2	-3	$\frac{1}{2}$
ν_1		$\frac{1}{30}$	$\frac{1}{20}$	0	$\frac{1}{60}$	$-\frac{1}{3}$	$-\frac{1}{2}$	$\frac{1}{12}$
ν_2			$\frac{8}{105}$	$-\frac{1}{60}$	0	$-\frac{1}{2}$	$-\frac{4}{5}$	$\frac{2}{15}$
ν_1'				$\frac{1}{3}$	$\frac{1}{2}$	0	1	$-\frac{1}{6}$
ν_2'					$\frac{4}{5}$	0	$\frac{3}{2}$	$-\frac{1}{4}$
ν_1''						4	6	-1
ν_2''							12	-2
t								$\frac{1}{3}$

$$\tilde{\phi}(x) = \frac{x}{4} + e_1 \sin\frac{x\pi}{4} + e_2 \sin\frac{x\pi}{2} \qquad (a)$$

belongs to a subspace spanned by Fourier functions.

Solution Let ν_1 and ν_2 denote $\sin(x\pi/4)$ and $\sin(x\pi/2)$, respectively. The inner products are Eqs. (2-5*a*) and (2-5*b*) and the final algebraic equations for the

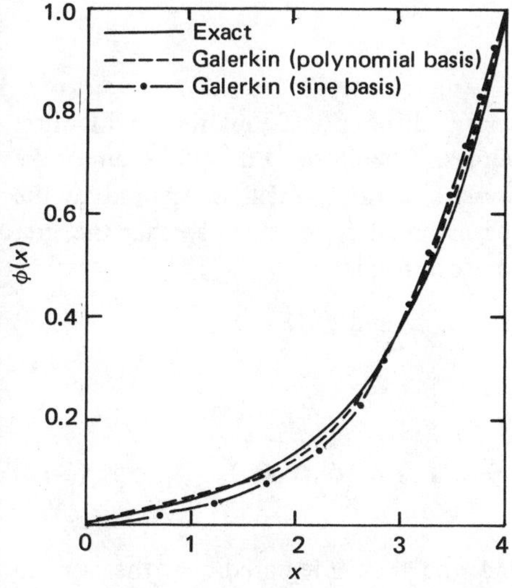

Fig. 2-1 Approximate solution of Eq. (1-1) obtained by the Galerkin method with the global basis functions given in Eq. (2-1) and the sine functions.

undetermined coefficients remain Eqs. (2-6) and (2-7). After the integrals are evaluated, we obtain

$\alpha_{11} = 5.2337, \quad \alpha_{12} = -1.3333, \quad \beta_1 = 1.9099, \quad \alpha_{21} = 1.3333,$

$\alpha_{22} = -8.9349, \quad \text{and } \beta_2 = 1.2732.$

Finally, Eqs. (2-6*a*) and (2-6*b*) can be solved to yield $e_1 = -0.3865$ and $e_2 = 0.0848$. The obtained approximation $\tilde{\phi}(x)$ is

$$\tilde{\phi}(x) = \frac{x}{4} - 0.3865 \sin\frac{x\pi}{4} + 0.0848 \sin\frac{x\pi}{2}, \tag{b}$$

which is also plotted in Fig. 2-1.

We will make two comments on the present method before moving to the next subject. (1) The fact that the approximation Eq. (2-8) appears more accurate than Eq. (*b*) in Example 2-2 does not suggest that the polynomial basis functions are generally better than the Fourier basis functions. A certain set of basis functions may be a good fit to one differential equation but not others. (2) Since we have studied the Galerkin method with finite-element discretization (Section 1-2) and with the global basis function formulation (this section), it may be desirable to compare the accuracies of these two schemes. Although the accuracies of numerical methods are usually important numerical properties, they heavily depend on the number of grid points or the number of basis functions adopted, respectively. If, for example, we compare the finite-element method with 100 grid points and the global basis function formulation with 3 trial functions, it is most likely that the former will be more accurate than the latter. Therefore, it is only fair to compare accuracies if the same amount of "effort" is devoted to both schemes. Readers who are interested in the advantages of the spectral method over the finite element method may consult [9].

2-1*b* Variational Principle

We mentioned in Section 1-2*c* that ordinary differential equations with first derivatives do not possess variational principles. Therefore, Eq. (1-1) cannot be solved by the variational method. In the present section, which is devoted to the variational principle, the flow velocity will be assumed to be small so that the first derivative can be dropped. Equation (1-1) therefore becomes

$$L\phi = -\phi'' + 2\phi = 0, \quad \phi(0) = 0, \quad \text{and } \phi(4) = 1. \tag{2-9}$$

The inner product takes the form

$$(L\phi, \delta\phi) = -\int_0^4 \phi''\,\delta\phi\,dx + 2\int_0^4 \phi\,\delta\phi\,dx = 0. \tag{2-10}$$

The reason why Eq. (2-9) is multiplied by $\delta\phi$ and then integrated over the domain was stated in Section 1-2*c*. Using integration by parts and recalling that the variation operator δ and the differentiation operator d/dx are commutative, we rewrite Eq. (2-10) as

$$\delta \int_0^4 (\phi'^2 + 2\phi^2)\, dx = 0. \tag{2-11}$$

If $\tilde{\phi}(x)$ is approximated according to Eq. (2-4), the functional

$$I(e_1, e_2) = \int_0^4 (\tilde{\phi}'^2 + 2\tilde{\phi}^2)\, dx \tag{2-12}$$

can be made stationary, i.e., be minimized, to yield

$$\frac{\partial I}{\partial e_1} = \int_0^4 (2\tilde{\phi}' v_1' + 4\tilde{\phi} v_1)\, dx = 0 \tag{2-13a}$$

and

$$\frac{\partial I}{\partial e_2} = \int_0^4 (2\tilde{\phi}' v_2' + 4\tilde{\phi} v_2)\, dx = 0. \tag{2-13b}$$

To utilize the result shown in Eqs. (2-5*a*) and (2-5*b*), we change Eqs. (2-13*a*) and (2-13*b*) with the help of integration by parts into

$$-\int_0^4 \tilde{\phi}'' v_k\, dx + 2\int_0^4 \tilde{\phi} v_k\, dx = 0, \qquad k = 1, 2. \tag{2-14a, b}$$

It can be seen that Eqs. (2-14*a*) and (2-14*b*) become the same as Eqs. (2-5*a*) and (2-5*b*) when the two middle terms in the latter are deleted. Hence, we may immediately write

$$\alpha_{k1} e_1 + \alpha_{k2} e_2 + \beta_k = 0, \tag{2-15a, b}$$

where α_{k1}, α_{k2}, and β_k are specified in Eqs. (2-7*a*)-(2-7*d*), excluding the second integral in Eqs. (2-7*a*)-(2-7*d*) and the first integral in Eqs. (2-7*e*) and (2-7*f*). Straightforward algebra with the aid of Table 2-1 yields $\alpha_{11} = -7/20$, $\alpha_{12} = -21/40$, $\beta_1 = -2/3$, $\alpha_{21} = -21/40$, $\alpha_{22} = -17/21$, and $\beta_2 = -16/15$. These numerical values subsequently give $e_1 = 2.6344$ and $e_2 = -3.0261$. Therefore, the approximate solution of Eq. (2-9) is written as

$$\tilde{\phi}(x) = \frac{x}{4} + 2.6344\left(\frac{x}{4}\right)\left(1 - \frac{x}{4}\right) - 3.026\left(\frac{x}{4}\right)\left(1 - \frac{x^2}{16}\right). \tag{2-16}$$

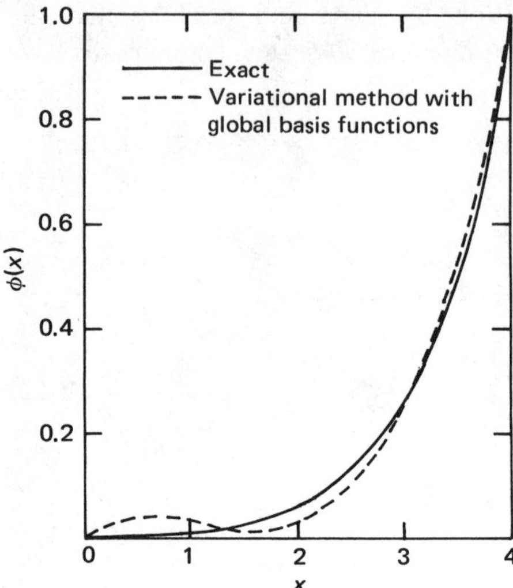

Fig 2-2 Solution of Eq. (2-9) obtained by the variational method.

This solution, along with the exact solution, is plotted in Fig. 2-2. It is expected that the wiggling behavior will disappear as more basis functions are retained.

Before departing from the subject of the variational principle, we should mention an important definition. So far, differential equations such as Eqs. (1-1) and (2-9) have been written with a negative sign assigned to the second derivative. This is done in order to make the differential operator positive-definite. An operator L is said to be positive-definite if

$$(L\phi, \phi) > 0 \qquad \text{for all } \phi \neq 0.^{\dagger} \tag{2-17}$$

On the basis of this definition, it can be shown that Eq. (1-1) has a positive-definite operator.

Example 2-3 Show that the operator $L = -d^2/dx^2 - d/dx + 2$ is positive-definite on $[0,4]$.

Solution The inner product

$$(L\phi, \phi) = -\int_0^4 \frac{d^2\phi}{dx^2}\,\phi\,dx - \int_0^4 \frac{d\phi}{dx}\,\phi\,dx + 2\int_0^4 \phi^2\,dx \tag{a}$$

can be rewritten as

$$(L\phi, \phi) = \int_0^4 \left[\left(\frac{d\phi}{dx}\right)^2 - \left(\frac{d\phi}{dx}\right)\phi + \left(\frac{1}{4} + \frac{7}{4}\right)\phi^2\right] dx \tag{b}$$

†A more precise definition of positive-definiteness that requires the concept of Sobolev spaces is described in Section 14-1q.

$$= \int_0^4 \left[\left(\frac{d\phi}{dx} - \frac{1}{2}\phi \right)^2 + \frac{7}{4}\phi^2 \right] dx > 0. \qquad (b)\ (cont.)$$

Therefore, we established that

$$(L\phi, \phi) > 0 \quad \text{for all } \phi \neq 0 \qquad (c)$$

and L is positive-definite.

We prefer to write the operator in positive-definite form for two reasons. First, if $L\phi = f$ and L is a positive-definite operator, then the weak solution

$$(L\tilde{\phi}, v) - (f, v) = 0 \qquad (2\text{-}18)$$

exists and is unique. This assertion is also known as the Lax-Milgram lemma [10-12]. Thus, once the operator is positive-definite, we feel confident that, in principle, there exists a unique approximate solution. Second, if ϕ satisfies $L\phi = f$, where L is a linear, symmetric, positive-definite operator, then it also minimizes the functional

$$I(\tilde{\phi}) = \tfrac{1}{2}\,(L\tilde{\phi}, \tilde{\phi}) - (f, \tilde{\phi}). \qquad (2\text{-}19)$$

Equation (2-19) is closely related to the variational principle, which will be of great interest throughout this book. Therefore, it is desirable to write an operator in positive-definite form.

Recalling the self-adjoint operator mentioned in the preceding chapter, we may wonder whether these two operators are related. Without rigorous proof, it is simply stated here that they are not. More precisely, a positive-definite operator is not necessarily a self-adjoint operator and vice versa.

Example 2-4 Find three operators that are, respectively, (*a*) positive-definite but not self-adjoint, (*b*) self-adjoint but not positive-definite, and (*c*) both positive-definite and self-adjoint.

Solution For case (*a*), we may consider the operator presented in Example 2-3. It was established that $L_a = -d^2/dx^2 - d/dx + 2$ is a positive-definite operator. Next, referring to Example 1-9, we found that L_a is not self-adjoint. Regarding case (*b*), let us examine $L_b = -d^2/dx^2 - 1$. It is self-adjoint since

$$(L_b\phi, \psi) = -\int \frac{d^2\phi}{dx^2}\psi\, dx - \int \phi\psi\, dx$$

$$= -\int \phi \frac{d^2\psi}{dx^2}\, dx - \int \phi\psi\, dx = (\phi, L_b\psi). \qquad (a)$$

However, it is not positive-definite because

$$(L_b\phi, \phi) = -\int \frac{d^2\phi}{dx^2}\,\phi\,dx - \int \phi^2\,dx = \int \left[\left(\frac{d\phi}{dx}\right)^2 - \phi^2\right] dx, \qquad (b)$$

where the sign of the integrand is indeterminate. Finally, for case (c) we consider $L_c = -d^2/dx^2 + 1$. As for Eq. (a), it can be shown that

$$(L_c\phi, \psi) = (\phi, L_c\psi) \qquad (c)$$

and that

$$(L_c\phi, \psi) = \int \left[\left(\frac{d\phi}{dx}\right)^2 + \phi^2\right] dx. \qquad (d)$$

Therefore, L_c is a positive-definite as well as self-adjoint operator.

2-1*c* Adjoint Variational Principle

In the preceding section we considered the variational method applied to a conservative system, e.g., a second-order ordinary differential equation without the first derivative. We also learned in Section 1-2 that a dissipative system governed by a differential equation having a first derivative does not possess a variational principle. It is the objective of the present section to introduce the adjoint variational method by using the global basis functions to solve Eq. (1-1).

To avoid repetition, we directly state the adjoint variational principle as

$$I = \int_0^4 \left(\frac{d\tilde{\phi}}{dx}\frac{d\tilde{\phi}^*}{dx} - \frac{d\tilde{\phi}}{dx}\,\tilde{\phi}^* + 2\tilde{\phi}\tilde{\phi}^*\right) dx, \qquad (2\text{-}20)$$

where ϕ is the solution of Eq. (1-1) and ϕ^* satisfies the adjoint equation

$$L^*\phi^* = -\frac{d^2\phi^*}{dx^2} + \frac{d\phi^*}{dx} + 2\phi^* = 0, \qquad \phi^*(0) = 0, \qquad \phi^*(4) = 1.$$

If ϕ and ϕ^* are the exact solutions that minimize I, then I is simply a pure number. If $\tilde{\phi}$ and $\tilde{\phi}^*$ belong to n-dimensional space and are approximated by

$$\tilde{\phi}(x) = \frac{x}{4} + e_1\frac{x}{4}\left(1 - \frac{x}{4}\right) + e_2\frac{x}{4}\left(1 - \frac{x^2}{16}\right) + \cdots + e_n\frac{x}{4}\left(1 - \frac{x^n}{4^n}\right) \qquad (2\text{-}21a)$$

and

$$\tilde{\phi}^*(x) = \frac{x}{4} + e_1^*\frac{x}{4}\left(1 - \frac{x}{4}\right) + e_2^*\frac{x}{4}\left(1 - \frac{x^2}{16}\right) + \cdots + e_n^*\frac{x}{4}\left(1 - \frac{x^n}{4^n}\right), \qquad (2\text{-}21b)$$

then I becomes a function of e_k and e_k^*, $k = 1, 2, \ldots, n$. Adjusting e_k and e_k^* such that $I(e_1, e_1^*, e_2, e_2^*, \ldots, e_n, e_n^*)$ is minimized dictates that

$$\frac{\partial I}{\partial e_1^*} = \frac{\partial I}{\partial e_2^*} = \cdots = \frac{\partial I}{\partial e_n^*} = 0 \quad \text{and} \quad \frac{\partial I}{\partial e_1} = \frac{\partial I}{\partial e_2} = \cdots = \frac{\partial I}{\partial e_n} = 0. \tag{2-22a, b, ..., 2n}$$

There are $2n$ algebraic equations for $2n$ unknowns $e_1, e_1^*, e_2, e_2^*, \ldots, e_n$, and e_n^*. When n is taken as 2, Eqs. (2-22*a*) and (2-22*b*) become

$$\frac{\partial I}{\partial e_1^*} = \int_0^4 \left(\frac{d\tilde{\phi}}{dx}\frac{dv_1}{dx} - \frac{d\tilde{\phi}}{dx} v_1 + 2\tilde{\phi} v_1 \right) dx = 0 \tag{2-23a}$$

and

$$\frac{\partial I}{\partial e_2^*} = \int_0^4 \left(\frac{d\tilde{\phi}}{dx}\frac{dv_2}{dx} - \frac{d\tilde{\phi}}{dx} v_2 + 2\tilde{\phi} v_2 \right) dx = 0. \tag{2-23b}$$

We observe that the adjoint function $\phi^*(x)$ does not appear in Eqs. (2-23*a*) and (2-23*b*). This suggests that equations containing $e_1, e_2, \ldots, e_n$ are decoupled from those containing $e_1^*, e_2^*, \ldots, e_n^*$. Since usually the adjoint problem is of little interest to us, it may be simply ignored after Eqs. (2-23*a*) and (2-23*b*) are obtained.

Upon integration by parts for the first terms, Eqs. (2-23*a*) and (2-23*b*) can be shown to become identical to Eqs. (2-5*a*) and (2-5*b*). This is similar to the situation in Chapter 1 where the Galerkin finite-element result was found to be the same as the adjoint-variational finite-element equation.

2-1*d* Mixed-Type Boundary Conditions

In many heat transfer problems, the boundary value of the field variable is prescribed implicitly by a mixed-type boundary condition such as

$$\left(\frac{d\phi}{dx}\right)_{x=4} = c\phi(4), \tag{2-24}$$

where c is a constant. Under such circumstances, the basis functions given in Eqs. (2-3), (2-4), and (2-21) are no longer appropriate since Eq. (2-24) cannot be satisfied regardless of any linear combination. We must therefore seek better basis functions. One procedure is to assume a polynomial with undetermined coefficients and then substitute it into the mixed-type boundary condition. Expressing one of the coefficients in terms of others and then grouping terms with the same coefficients, we obtain the desired basis functions. We illustrate the procedure in the following example.

Example 2-5 Find a complete set of basis functions that well approximate a function $\phi(x)$ satisfying two end conditions

$$\phi(0) = 0 \quad \text{and} \quad \left(\frac{d\phi}{dx}\right)_{x=4} = c\phi(4), \qquad (a,\ b)$$

where c is a constant not equal to $\frac{1}{4}$.

Solution We approximate $\phi(x)$ as

$$\tilde{\phi}(x) = b_0 + b_1 x + b_2 x^2 + b_3 x^3 + \cdots + b_n x^n. \qquad (c)$$

The first end condition gives

$$b_0 = 0. \qquad (d)$$

Substituting Eq. (c) into Eq. (b) leads to

$$b_1 + 8b_2 + 48b_3 + \cdots + nb_n 4^{n-1} = 4cb_1 + 16cb_2 + 64cb_3 + \cdots + 4^n cb_n \qquad (e)$$

or
$$b_1 = \left(\frac{16c - 8}{1 - 4c}\right) b_2 + \left(\frac{64c - 48}{1 - 4c}\right) b_3 + \cdots + \left(\frac{4^n c - 4^{n-1} n}{1 - 4c}\right) b_n. \qquad (f)$$

Substituting Eqs. (d) and (f) into Eq. (c), we obtain

$$\tilde{\phi}(x) = b_2 \left[x^2 - \left(\frac{16c - 8}{4c - 1}\right) x\right] + b_3 \left[x^3 - \left(\frac{64c - 48}{4c - 1}\right) x\right] + \cdots + b_n \left[x^n - \left(\frac{4^n c - n4^{n-1}}{4c - 1}\right) x\right]. \qquad (g)$$

Finally, it is observed that, as n increases, the last expression may grow without limit unless x is normalized on 4. Thus, Eq. (g) is rewritten as

$$\tilde{\phi}(x) = d_1 \frac{x}{4}\left[\left(\frac{x}{4}\right) - \left(\frac{4c - 2}{4c - 1}\right)\right] + d_2 \frac{x}{4}\left[\left(\frac{x}{4}\right)^2 - \left(\frac{4c - 3}{4c - 1}\right)\right] + \cdots + d_{n-1} \frac{x}{4}\left[\left(\frac{x}{4}\right)^{n-1} - \left(\frac{4c - n}{4c - 1}\right)\right] \qquad (h)$$

or
$$\tilde{\phi}(x) = \frac{x}{4} \sum_{j=1}^{n-1} d_j \left[\left(\frac{x}{4}\right)^j - 1 + \frac{j}{4c - 1}\right]. \qquad (i)$$

A few basis functions are plotted in Fig. 2-3 for $c = 1$.

In the following sections, we will turn our attention to other numerical methods.

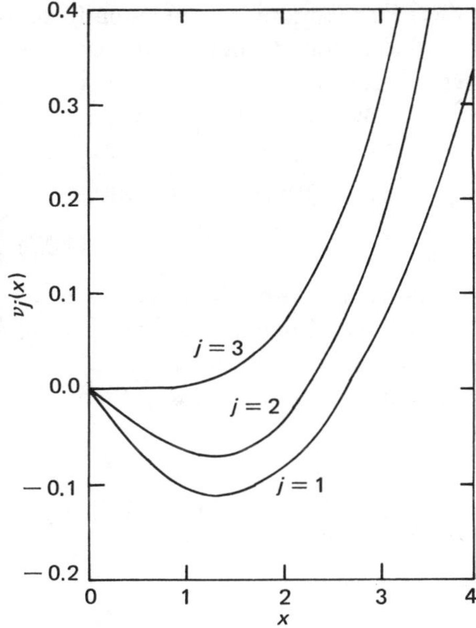

Fig. 2-3 Basis functions that do not vanish at the boundary point, as given in Eq. (*i*) of Example 2-5.

2-2 CENTRAL INTEGRATION METHOD

In the central integration method, the differential equation is integrated with respect to the coordinate variable over a typical interval $[x_{j-1/2}, x_{j+1/2}]$ with grid point j at the center. Integration once reduces the order of a differential equation by one and therefore permits us to use interpolants that require less interinterval continuity. The final discretized equations are similar to those obtained by the central finite-difference method. From the viewpoint of weighted-residuals methods, this integration procedure can be interpreted as equivalent to finding the inner product

$$(L\tilde{\phi}, W_j) = 0 \tag{2-25}$$

with the weighting function W_j defined as

$$W_j = \begin{cases} 1, & x \epsilon [x_{j-1/2}, x_{j+1/2}] \\ 0, & \text{elsewhere.} \end{cases} \tag{2-26}$$

The choice of such weighting functions leads to simple algebraic manipulation (in comparison with the Galerkin method) and tridiagonal matrix systems. These are attractive features for the numerical methods.

Before we illustrate this method by solving a specific problem, it is important to determine whether the set $(W_0, W_1, W_2, \ldots, W_J)$ is indeed a basis on the interval $[a,b]$. First, the linear combination $\alpha_0 W_0 + \alpha_1 W_1 + \cdots + \alpha_J W_J$ can be made zero if and only if $\alpha_0 = \alpha_1 = \cdots = \alpha_J = 0$. This criterion implies that all the members are linearly independent of one another. Furthermore, any function in the same space can be expressed in terms of a linear combination of these basis functions. The more terms we take, the better the approximation becomes. Figure 2-4 shows a fictitious function $f(x)$ on $[0,10]$. It is seen that this function is fairly approximated by

$$f(x) = 0.25W_0 + 0.5W_1 + 1.1W_2 + 2.4W_3 + \cdots + 6.8W_j + \cdots + 5.8W_{10}. \tag{2-27}$$

Since $(W_0, W_1, \ldots, W_J)$ is a set of basis functions, the residual $R = L\tilde{\phi}$ will be minimized if it is made orthogonal to all the members. Consequently, the central integration method can be classified as one of the weighted-residuals methods.

Now, let us proceed to solve the model equation (1-1) again by this method. The inner product in Eq. (2-25) can be written as

$$\int_{x_{j-1/2}}^{x_{j+1/2}} \frac{d^2\phi}{dx^2}\,dx + \int_{x_{j-1/2}}^{x_{j+1/2}} \frac{d\phi}{dx}\,dx - 2\int_{x_{j-1/2}}^{x_{j+1/2}} \phi\,dx = 0 \tag{2-28}$$

or

$$\left(\frac{d\phi}{dx}\right)_{x_{j+1/2}} - \left(\frac{d\phi}{dx}\right)_{x_{j-1/2}} + \phi(x_{j+1/2}) - \phi(x_{j-1/2}) - 2\int_{x_{j-1/2}}^{x_{j+1/2}} \phi\,dx = 0. \tag{2-29}$$

Referring to Fig. 2-5, we obtain the following approximations:

$$\left(\frac{d\phi}{dx}\right)_{x_{j+1/2}} = \frac{\phi_{j+1} - \phi_j}{h}, \qquad \phi(x_{j+1/2}) = \frac{\phi_{j+1} + \phi_j}{2}, \qquad \text{etc.} \tag{2-30a,b}$$

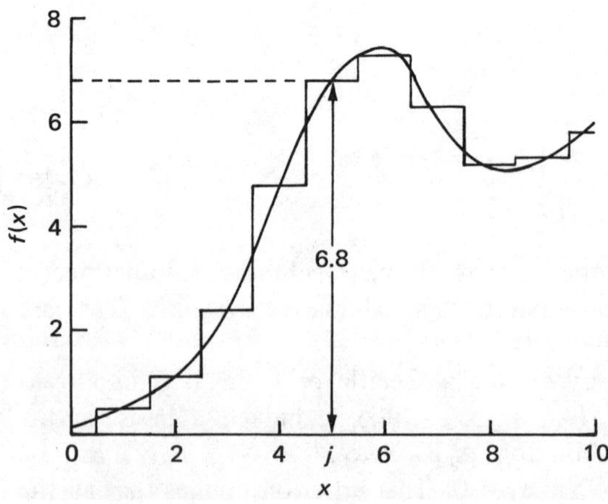

Fig. 2-4 A fictitious function $f(x)$ approximated by a linear combination of piecewise step basis functions $W_j(x)$.

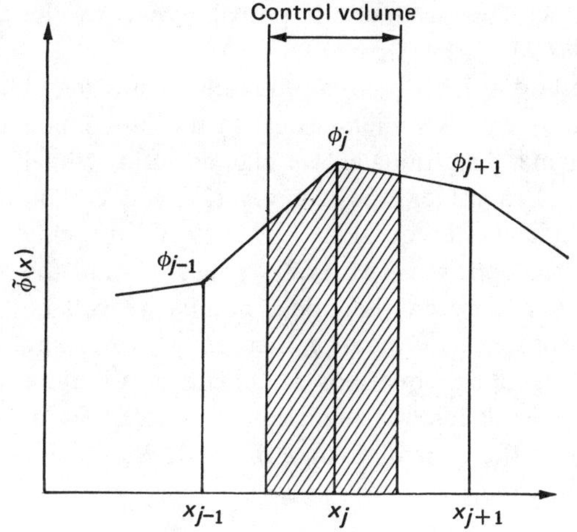

Fig. 2-5 Graphic representation of the integral $\int_{x_{j-1/2}}^{x_{j+1/2}} \phi\, dx$.

On the basis of these approximations, it is interesting to find that the discretizations

$$\int_{x_{j-1/2}}^{x_{j+1/2}} \frac{d^2\phi}{dx^2}\, dx = \frac{1}{h}\,(\phi_{j+1} - 2\phi_j + \phi_{j-1}) \quad \text{and} \quad \int_{x_{j-1/2}}^{x_{j+1/2}} \frac{d\phi}{dx}\, dx$$

$$= \frac{\phi_{j+1} - \phi_{j-1}}{2} \tag{2-30c,d}$$

yield algebraic expressions identical to those obtained by the central finite-difference method.

The only integral in Eq. (2-29), which is represented by the shaded area in Fig. 2-5, is evaluated by using the trapezoidal rule:

$$\int_{x_{j-1/2}}^{x_{j+1/2}} \phi\, dx = h\left(\frac{1}{8}\,\phi_{j-1} + \frac{3}{4}\,\phi_j + \frac{1}{8}\,\phi_{j+1}\right). \tag{2-31}$$

If the temperature is uniformly distributed inside the control volume such that $\phi_{j-1} \approx \phi_j \approx \phi_{j+1}$, the integral in Eq. (2-31) reduces to a lumped quantity $h\phi_j$, which would have been obtained if the finite-difference method had been used. In cases where large temperature gradients exist, however, Eq. (2-31) appears to be a better discretization result since it takes the nonuniformity of the temperature distribution into account. Substituting Eqs. (2-30) and (2-31) into Eq. (2-28) gives

$$\left(\frac{h^2}{2} + h - 2\right)\phi_{j-1} + (3h^2 + 4)\phi_j + \left(\frac{h^2}{2} - h - 2\right)\phi_{j+1} = 0. \tag{2-32}$$

This three-point relation is then repeatedly applied to the three grid points $j = 1, 2, 3$ with $h = 1$, $\phi_0 = 0$, and $\phi_4 = 1$, yielding $\phi_1 = 0.048$, $\phi_2 = 0.1344$, and $\phi_3 = 0.3667$. In comparison with the exact solution, these nodal values are more accurate than the finite-difference results. The accuracy is probably due to the adoption of the more accurate expression (2-31) rather than $h\phi_j$.

Since in the central integration method the differential equation is integrated over the subdomain $[x_{j-1/2}, x_{j+1/2}]$, the scheme is sometimes called the subdomain method. There is also another popular name. Traditionally, we take an infinitesimal control volume $[x - dx/2, x + dx/2]$ and consider the conservation of certain entities when deriving the governing differential equations. If, instead of being infinitesimal, the control volume is taken to be finite–i.e., $[x_{j-1/2}, x_{j+1/2}]$–and the variable distribution is linearly interpolated between the nodal points, we will be able to derive governing algebraic equations at grid point j, instead of governing differential equations. These governing algebraic equations are identical to those derived by direct central integration of the differential equation. For this reason, the central integration method is also called the control volume method. We present the following example to further clarify this remark.

Example 2-6 Consider the physical problem described at the beginning of the preceding chapter. Take the energy balance over the one-dimensional control volume $[x_{j-1/2}, x_{j+1/2}]$ and derive Eq. (2-32) directly. Ignore the availability of Eq. (1-1).

Solution Based on the linear interpolation, the rate of conductive heat transfer across the two cross-sectional areas into the control volume $[x_{j-1/2}, x_{j+1/2}]$ can be approximated as

$$q_{\text{cond}} \approx \pi R k \left(\frac{\phi_{j-1} - \phi_j}{h} + \frac{\phi_{j+1} - \phi_j}{h} \right) (T_L - T_0), \tag{a}$$

where the subscript "cond" denotes conduction and h is the dimensionless mesh size normalized on R. The net rate of convective heat transfer due to the flow into the control volume is

$$q_{\text{conv}} = \pi R^2 \rho u c_p (T_{j+1/2} - T_{j-1/2}) = \frac{\pi R^2 \rho u c_p (\phi_{j+1} - \phi_{j-1})(T_L - T_0)}{2}. \tag{b}$$

The rate of convective heat transfer across the circumferential area out into the tube wall at T_0 is

$$q_{\text{cir}} = 2\pi R h_c \int_{x_{j-1/2}}^{x_{j+1/2}} (T - T_0)\, dx \approx 2\pi R h_c \left\{ \int_{x_{j-1/2}}^{x_j} [N_{j-1}(x)\phi_{j-1} + N_j(x)\phi_j]\, dx \right.$$

$$\left. + \int_{x_j}^{x_{j+1/2}} [N_j(x)\phi_j + N_{j+1}(x)\phi_{j+1}]\, dx \right\} (T_L - T_0) \tag{c}$$

$$= 2\pi R^2 hh_c \left(\frac{1}{8} \phi_{j-1} + \frac{3}{4} \phi_j + \frac{1}{8} \phi_{j+1} \right) (T_L - T_0), \qquad (c)\ (cont)$$

where the subscript "cir" denotes circumferential. The overall energy balance on the control volume $[x_{j-1/2}, x_{j+1/2}]$ is thus

$$q_{\text{cond}} + q_{\text{conv}} - q_{\text{cir}} = 0. \qquad (d)$$

Substituting Eqs. (*a*)-(*c*) into Eq. (*d*) leads to

$$\left(\frac{-\xi h^2}{4} - \frac{uhR}{2\alpha} + 1 \right) \phi_{j-1} - \left(\frac{3\xi h^2}{2} + 2 \right) \phi_j + \left(\frac{-\xi h^2}{4} + \frac{uhR}{2\alpha} + 1 \right) \phi_{j+1} = 0, \qquad (e)$$

where $\xi = h_c R/k$ is the Nusselt number. To be consistent with the conditions according to which Eq. (1-1) was derived, we take $\xi = 1$ and $uR/\alpha = 1$. With these numerical values, Eq. (*e*) becomes the same as Eq. (2-32).

For comparison of the Galerkin method and the control volume method, see [13].

2-3 METHOD OF LEAST SQUARES

In the method of least squares, the functional

$$J = \int_a^b [L\phi - f(x)]^2 \, dx \qquad (2\text{-}33)$$

will be minimized, where $\phi(x)$ satisfies the differential equation

$$L\phi = f(x).$$

Being simpler, boundary conditions of the Dirichlet type will be considered first. The solution $\phi(x)$ can be approximated either by a set of piecewise functions (the finite-element method is so related) or by a series of global basis functions. The functional J is then differentiated partially with respect to the nodal unknowns or undetermined coefficients to yield a system of algebraic equations. The details of this method are illustrated in the following example.

Example 2-7 Use the method of least squares to solve Eq. (1-1), whose solution is approximated by

$$\tilde{\phi}(t) = t + e_1 \nu_1(t) + e_2 \nu_2(t), \qquad (a)$$

where $t = x/4$, $\nu_k = t(1 - t^k)$, and $k = 1, 2$.

Solution The functional in this particular problem is, according to Eq. (2-33),

$$J = \int_0^4 (-\phi'' - \phi' + 2\phi)^2 \, dx, \tag{b}$$

which is a pure number. If Eq. (*a*) is used, then $\tilde{\phi}(x)$ belongs to a two-dimensional subspace spanned by $v_1(t)$ and $v_2(t)$ and Eq. (*b*) becomes

$$J(e_1, e_2) = \int_0^4 (-\tilde{\phi}'' - \tilde{\phi}' + 2\tilde{\phi})^2 \, dx. \tag{c}$$

Minimization of $J(e_1, e_2)$ requires that

$$\frac{\partial J}{\partial e_1} = 2\int_0^4 (-\tilde{\phi}'' - \tilde{\phi}' + 2\tilde{\phi})(-v_1'' - v_1' + 2v_1)\, dx = 0 \tag{d}$$

and

$$\frac{\partial J}{\partial e_2} = 2\int_0^4 (-\tilde{\phi}'' - \tilde{\phi}' + 2\tilde{\phi})(-v_2'' - v_2' + 2v_2)\, dx = 0. \tag{e}$$

Equations (*d*) and (*e*) are then rearranged into

$$\beta_{11} e_1 + \beta_{12} e_2 + \gamma_1 = 0 \tag{f}$$

and

$$\beta_{21} e_1 + \beta_{22} e_2 + \gamma_2 = 0, \tag{g}$$

where

$$\beta_{kk} = \int_0^1 (v_k'' + 4v_k' - 32v_k)^2 \, dt,$$

$$\beta_{12} = \beta_{21} = \int_0^1 (v_1'' + 4v_1' - 32v_1)(v_2'' + 4v_2' - 32v_2)\, dt,$$

and

$$\gamma_k = \int_0^1 (v_k'' + 4v_k' - 32v_k)(4 - 32t)\, dt, \qquad k = 1, 2. \tag{h}$$

Note that the primes in Eq. (*h*) denote differentiation with respect to *t*. Utilizing Table 2-1, these coefficients can be evaluated as $\beta_{11} = 64.8$, $\beta_{12} = 101.2$, $\beta_{22} = 166.1$, $\gamma_1 = 109.3$, and $\gamma_2 = 180.3$, which lead to $e_1 = 0.1941$ and $e_2 = -1.204$. The final solution is plotted in Fig. 2-6, where the exact solution is also shown for comparison.

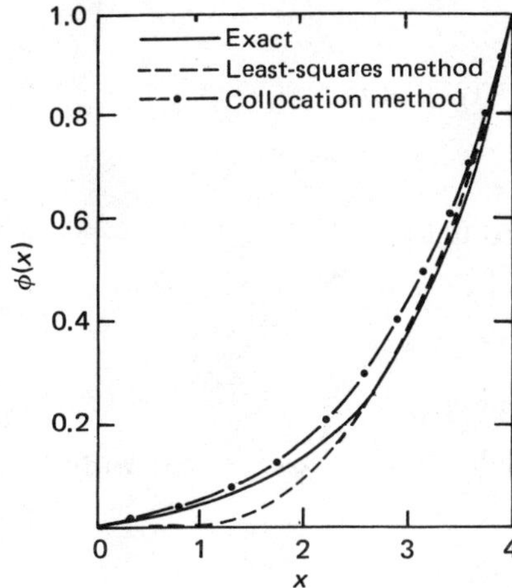

Fig. 2-6 Approximate solutions of Eq. (1-1) obtained by the least-squares method with the global basis functions given in Eq. (2-1), and by the collocation method.

We observe that the coefficient matrix in Eqs. (*f*) and (*g*) of the preceding example

$$[\beta] = \begin{bmatrix} \beta_{11} & \beta_{12} \\ \beta_{21} & \beta_{22} \end{bmatrix}$$

is positive-definite. This is because

$$\beta_{11} > 0 \qquad \text{and } \beta_{12} = \beta_{21},$$

and, on the basis of Schwarz's inequality

$$\int_a^b fg\, dx \leqslant \left(\int_a^b f^2\, dx\right)^{1/2} \left(\int_a^b g^2\, dx\right)^{1/2}$$

for all $f(x)$ and $g(x)$ that are squarely integrable on $[a,b]$, we also have

$$\beta_{11}\beta_{22} - \beta_{12}\beta_{21} = \left[\int_0^1 (L\nu_1)^2\, dt\right]\left[\int_0^1 (L\nu_2)^2\, dt\right] - \left[\int_0^1 (L\nu_1)(L\nu_2)\, dt\right]^2 > 0.$$

Since, with a positive-definite coefficient matrix, the algebraic system, when solved by an iterative scheme, has a unique and rapidly converging solution, the method of least squares appears attractive to some numerical researchers [14–18].

The method of least squares can also be related to the method of weighted residuals and the variational method.

2-3*a* Weighting Functions

It is possible to identify the weighting function for the least-squares formulation. Suppose that $\tilde{\phi}(x)$ is approximated by

$$\tilde{\phi}(x) = \sum_{j=1}^{J-1} e_j v_j(x), \qquad x \epsilon [a,b],$$

which is then substituted into $L\phi = f(x)$ to generate

$$L\tilde{\phi}(x,e_j) - f(x) = R(x,e_j), \tag{2-34}$$

where $R(x,e_j)$ is the residual resulting from the substitution. Now we may write the inner product

$$\left(L\tilde{\phi} - f, \frac{\partial R}{\partial e_j}\right) = \left(R, \frac{\partial R}{\partial e_j}\right). \tag{2-35}$$

If $\tilde{\phi}$ is a good approximation to ϕ, the residual should vanish in a certain fashion. Here we will set $(R, \partial R/\partial e_j)$ to zero to ensure that R is orthogonal to the members $\partial R/\partial e_j$. Hence,

$$\left(R, \frac{\partial R}{\partial e_j}\right) = \int_a^b R\frac{\partial R}{\partial e_j}\, dx = \frac{1}{2}\frac{\partial}{\partial e_j}\int_a^b R^2\, dx = 0 \tag{2-36}$$

leads to

$$\frac{\partial}{\partial e_j}\int_a^b [L\tilde{\phi} - f(x)]^2\, dx = 0. \tag{2-37}$$

Therefore, we established that the weighting functions that generate the functional defined in Eq. (2-33) are $\partial R/\partial e_j$. It is thus optional to consider the method of least squares as one of the weighted-residuals methods.

2-3*b* Euler-Lagrange Equation

Most functional integrals are associated with Euler-Lagrange equations, and Eq. (*b*) in Example 2-7 is not exceptional. To find its Euler-Lagrange equation, we make J stationary and obtain

$$\delta J = 2\int_0^4 (\phi'' + \phi' - 2\phi)[(\delta\phi)'' + (\delta\phi)' - 2\delta\phi]\, dx = 0, \tag{2-38}$$

which, for convenience, can be rearranged into

$$\tfrac{1}{2}\,\delta J = \delta J_1 + \delta J_2 - 2\delta J_3 = 0, \tag{2-39}$$

where

$$\delta J_1 = \int_0^4 (\phi'' + \phi' - 2\phi)(\delta\phi)''\,dx, \tag{2-40a}$$

$$\delta J_2 = \int_0^4 (\phi'' + \phi' - 2\phi)(\delta\phi)'\,dx, \tag{2-40b}$$

and

$$\delta J_3 = \int_0^4 (\phi'' + \phi' - 2\phi)\,\delta\phi\,dx. \tag{2-40c}$$

Using integration by parts twice, we rewrite Eq. (2-40*a*) as

$$\delta J_1 = \int_0^4 (\phi'''' + \phi''' - 2\phi'')\,\delta\phi\,dx$$

$$- (\phi''' + \phi'' - 2\phi')(\delta\phi)\Big|_0^4 + (\phi'' + \phi' - 2\phi)(\delta\phi)'\Big|_0^4. \tag{2-41a}$$

The boundary-condition terms should vanish since $L\phi$ and $(L\phi)'$ are zero everywhere on $[0,4]$. Similarly, Eq. (2-40*b*) becomes

$$\delta J_2 = -\int_0^4 (\phi''' + \phi'' - 2\phi')\,\delta\phi\,dx. \tag{2-41b}$$

Substitution of Eqs. (2-40*c*), (2-41*a*), and (2-41*b*) into Eq. (2-39) yields

$$\frac{1}{2}\,\delta J = \int_0^4 (\phi'''' - 5\phi'' + 4\phi)\,\delta\phi\,dx = 0, \tag{2-42}$$

Since $\delta\phi$ is arbitrary, it follows that

$$\phi'''' - 5\phi'' + 4\phi = 0. \tag{2-43}$$

Equation (2-43) is the Euler-Lagrange equation corresponding to Eq. (*b*) in Example 2-7. This equation can also be written as

$$\phi'''' - 5\phi'' + 4\phi = \left(-\frac{d^2}{dx^2} - \frac{d}{dx} + 2\right)\left(-\frac{d^2}{dx^2} + \frac{d}{dx} + 2\right)\phi = 0, \tag{2-44}$$

which suggests that, upon minimization of a least-squares functional, we generally obtain an Euler-Lagrange equation involving the operator LL^*, where L^* is the adjoint operator of L. This assertion will be further illustrated in the following example.

Example 2-8 Consider the following equation of the Sturm-Liouville type:

$$L\phi = -\frac{d}{dx}\left[\gamma(x)\frac{d\phi}{dx}\right] + \rho(x)\phi = 0, \qquad \phi(a) = \alpha, \qquad \text{and } \phi(b) = \beta. \tag{a}$$

Find the Euler-Lagrange equation corresponding to the least-squares functional.

Solution The least-squares functional takes the form

$$J = \int_a^b (-\gamma\phi'' - \gamma'\phi' + \rho\phi)^2\, dx, \tag{b}$$

where a slight variation around the minimal value of J can be written as

$$\delta J = 2\int_a^b (L\phi)[-\gamma(\delta\phi)'' - \gamma'(\delta\phi)' + \rho(\delta\phi)]\, dx = 0. \tag{c}$$

After integration by parts, Eq. (c) becomes

$$\frac{1}{2}\delta J = \int_a^b [-(\gamma L\phi)'' + (\gamma' L\phi)' + \rho L\phi]\, \delta\phi\, dx = 0. \tag{d}$$

The fact that $\delta\phi$ is arbitrary ensures that

$$-(\gamma L\phi)'' + (\gamma' L\phi)' + \rho L\phi = 0, \tag{e}$$

which is the Euler-Lagrange equation. By straightforward algebra, Eq. (e) can be shown to be equal to

$$L^2\phi = 0. \tag{f}$$

Since Eq. (a) is self-adjoint ($L = L^*$), we have established that $LL^*\phi = 0$ is the Euler-Lagrange equation corresponding to Eq. (b).

Example 2-9 Consider again Eq. (a) in the preceding example. Find the variational functional and the corresponding Euler-Lagrange equation.

Solution To avoid confusion in our notation here, we will use I to denote the variational functional. Recalling Section 1-2c, we obtain

$$\delta I = \int_a^b [-(\gamma\phi')' + \rho\phi]\, \delta\phi\, dx, \tag{a}$$

which, after integration by parts, can be readily transformed into

$$\delta I = \frac{\delta}{2}\int_a^b [\gamma(\phi')^2 + \rho\phi^2]\, dx. \tag{b}$$

Therefore, the variational functional is

$$I = \frac{1}{2}\int_a^b [\gamma(\phi')^2 + \rho\phi^2]\, dx. \tag{c}$$

Next, it is known that Eq. (*a*) in Example 2-8 is the Euler-Lagrange equation corresponding to Eq. (*c*). Yet, we can alternatively define a functional $F(x, \phi, \phi')$ as

$$F(x, \phi, \phi') = \gamma(\phi')^2 + \rho\phi^2 \tag{d}$$

and substitute it into

$$\frac{\partial F}{\partial \phi} - \frac{\partial}{\partial x}\left(\frac{\partial F}{\partial \phi'}\right) = 0 \tag{e}$$

to obtain exactly Eq. (*a*) in Example 2-8. It is worth mentioning here that, for a functional $F(x, \phi, \phi', \phi'')$ containing the second derivative, Eq. (*e*) is no longer the corresponding Euler-Lagrange equation. Instead, the counterpart of Eq. (*e*) was derived [19] as

$$\frac{\partial F}{\partial \phi} - \frac{\partial}{\partial x}\left(\frac{\partial F}{\partial \phi'}\right) + \frac{\partial^2}{\partial x^2}\left(\frac{\partial F}{\partial \phi''}\right) = 0. \tag{f}$$

A close examination of Examples 2-8 and 2-9 reveals that the Euler-Lagrange equations of the least-squares functional [such as Eq. (2-44) or Eq. (*e*) in Example 2-8] are of higher order than those of the variational functional (i.e., the original differential equation). If the original differential equation is of order m, then there are m additional degrees of freedom that must be constrained. These constraints should be imposed such that the least-squares functional is minimized. The following examples, although simple, make this point more precisely.

Example 2-10 The one-dimensional heat conduction inside an infinite slab having a coordinate-dependent heat sink, with one face maintained at a constant temperature and the other immersed in a convective fluid, can be described by the following nondimensionalized governing equation:

$$-\phi'' + x^2 = 0 \qquad \text{with } \phi(0) = 1 \qquad \text{and } \phi'(1) = 2\phi(1). \tag{a}$$

Find the Euler-Lagrange equation that corresponds to the least-squares functional and then solve it by arbitrarily constraining the additional two degrees of freedom.

Solution The least-squares functional

$$J = \int_0^1 (\phi'' - x^2)^2 \, dx \tag{b}$$

is made stationary to yield

$$\delta J = 2\int_0^1 (\phi'' - x^2)(\delta\phi)'' \, dx = 2\int_0^1 (\phi'''' - 2)\, \delta\phi \, dx = 0. \tag{c}$$

Since $\delta\phi$ is arbitrary, we obtain

$$\phi'''' - 2 = 0 \tag{d}$$

as the Euler-Lagrange equation, which can be analytically integrated to give

$$\tilde{\phi}(x) = \frac{x^4}{12} + d_3 x^3 + d_2 x^2 + d_1 x + d_0. \tag{e}$$

Equation (*e*) contains four undetermined constants, but it seems that only two constraints exist, namely $\phi(0) = 1$ and $\phi'(1) = 2\phi(1)$. It may be tempting to think that there are two degrees of freedom in specifying $\tilde{\phi}(x)$. Quite arbitrarily, let d_1 and d_2 be unity. Along with the two boundary constraints, Eq. (*e*) is evaluated as

$$\tilde{\phi}(x) = \frac{x^4}{12} + \frac{17}{6} x^3 + x^2 + x + 1, \tag{f}$$

whereas the exact solution to Eq. (*a*) is

$$\phi(x) = \frac{x^4}{12} - \frac{11}{6} x + 1. \tag{g}$$

Both the "approximate" solution and the exact solution are plotted in Fig. 2-7. It can be seen that the "approximate" solution is by no means an approximation to the exact solution. The inaccuracy is due to the assignment of arbitrary values to d_1 and d_2.

Example 2-11 Repeat Example 2-10, except that now the two degrees of freedom are restricted such that a minimal value of the least-squares functional is generated.

Solution In the preceding example, the solution to the Euler-Lagrange equation was found to be

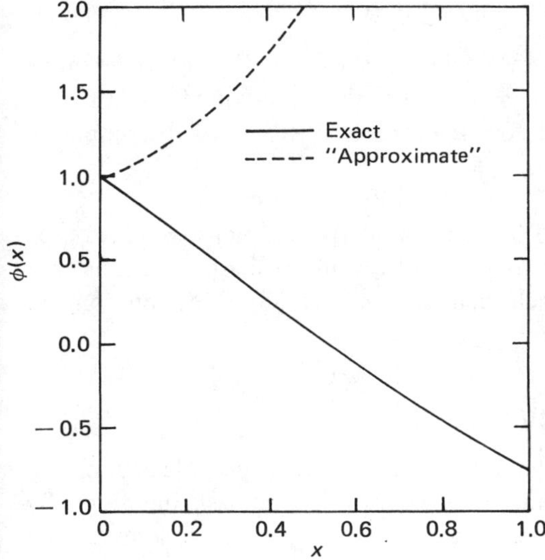

Fig. 2-7 Comparison of the exact solution of Eq. (*a*) in Example 2-10 and the "approximate" solution Eq. (*f*).

$$\tilde{\phi}(x) = \frac{x^4}{12} + d_3x^3 + d_2x^2 + d_1x + d_0. \tag{a}$$

If Eq. (*a*) is made to satisfy the boundary conditions, then it can be rewritten as

$$\tilde{\phi}(x) = \frac{x^4}{12} + d_3x^3 + d_2x^2 + \left(d_3 - \frac{11}{6}\right)x + 1. \tag{b}$$

Now, the two undetermined constants d_2 and d_3 will be specified in order that the functional

$$J(d_2, d_3) = \int_0^1 (\tilde{\phi}'' - x^2)^2 \, dx \tag{c}$$

is minimized. This minimization implies that

$$\frac{\partial J}{\partial d_2} = 2\int_0^1 (\tilde{\phi}'' - x^2)(2)\, dx = 8\int_0^1 (3d_3x + d_2)\, dx = 0 \tag{d}$$

and

$$\frac{\partial J}{\partial d_3} = 2\int_0^1 (\tilde{\phi}'' - x^2)(6x)\, dx = 24\int_0^1 (3d_3x + d_2)x\, dx = 0. \tag{e}$$

The simultaneous solutions of Eqs. (*d*) and (*e*) are found to be

$$d_2 = 0 \quad \text{and } d_3 = 0. \tag{f}$$

Therefore, Eq. (*b*) reduces to the exact solution shown in Eq. (*g*) of Example 2-10. However, it is only a coincidence that the approximate solution is identical to the exact solution. This coincidence arises because of the simplicity of the original equation $\phi'' - x^2 = 0$.

For problems with inhomogeneous Dirichlet boundary conditions or mixed-type boundary conditions, it is sometimes inconvenient to find the basis functions that fit the boundary data. Another approach that can remove all restrictions on the approximating subspace is thus desirable.

2-3*c* Penalty Functionals and Penalty Constants

It has been suggested [20, 21] that the inhomogeneous Dirichlet, Neumann, and mixed-type boundary conditions need not to be satisfied by the basis functions. By incorporating the boundary conditions into the so-called penalty functional and multiplying these functionals by penalty constants, it is possible to construct a new functional that relaxes the restrictions on the basis functions. For example, if the differential equation

$$A\phi = f(x), \qquad x \in [a, b] \tag{2-45}$$

is given (the operator A can be either linear or nonlinear), subject to

$$\phi(a) = g, \qquad \left(\frac{d\phi}{dx}\right)_{x=b} = c\phi(b),$$

where g and c are constants, we can generate a modified least-squares functional

$$J_p = \int_a^b (A\phi - f)^2 \, dx + \lambda_1 [\phi(a) - g]^2 + \lambda_2 [\phi'(b) - c\phi(b)]^2. \tag{2-46}$$

For clarity, the subscript p is used to denote penalty. The values of the penalty constants λ_1 and λ_2 can, in principle, be arbitrarily assigned. In practice, they are chosen such that the orders of magnitude of all the terms in Eq. (2-46) are the same.

Example 2-12 Consider again the problem

$$-\phi'' + x^2 = 0 \quad \text{with } \phi(0) = 1 \quad \text{and } \phi'(1) = 2\phi(1). \tag{a}$$

With the aid of the least-squares penalty functionals, find the solution of Eq. (*a*). What are the optimal values of the penalty constants?

Solution Following Eq. (2-46), we write

$$J_p = \int_0^1 (\phi'' - x^2)^2 \, dx + \lambda_1 [\phi(0) - 1]^2 + \lambda_2 [\phi'(1) - 2\phi(1)]^2. \qquad (b)$$

This functional is to be made stationary. Let us reuse Eq. (*a*) of Example 2-11

$$\tilde{\phi}(x) = \frac{x^4}{12} + d_3 x^3 + d_2 x^2 + d_1 x + d_0 \qquad (c)$$

as our approximate solution. The four degrees of freedom are constrained by the following four equations:

$$\frac{\partial J_p}{\partial d_0} = 2\lambda_1 (d_0 - 1) + 2\lambda_2 \left(\frac{1}{6} + d_3 - d_1 - 2d_0 \right) (-2) = 0, \qquad (d)$$

$$\frac{\partial J_p}{\partial d_1} = 2\lambda_2 \left(\frac{1}{6} + d_3 - d_1 - 2d_0 \right) (-1) = 0, \qquad (e)$$

$$\frac{\partial J_p}{\partial d_2} = 8 \int_0^1 (3d_3 x + d_2) \, dx = 0, \qquad (f)$$

and

$$\frac{\partial J_p}{\partial d_3} = 8 \int_0^1 (3d_3 x + d_2)(3x) \, dx + 2\lambda_2 \left(\frac{1}{6} + d_3 - d_1 - 2d_0 \right) = 0. \qquad (g)$$

Equations (*d*)-(*g*) are simplified and written in compact matrix form as

$$\begin{bmatrix} \lambda_1 + 4\lambda_2 & 2\lambda_2 & 0 & -2\lambda_2 \\ 2 & 1 & 0 & -1 \\ 0 & 0 & 1 & \frac{3}{2} \\ -2\lambda_2 & -\lambda_2 & 6 & 12 + \lambda_2 \end{bmatrix} \begin{Bmatrix} d_0 \\ d_1 \\ d_2 \\ d_3 \end{Bmatrix} = \begin{Bmatrix} \lambda_1 + \frac{\lambda_2}{3} \\ \frac{1}{6} \\ 0 \\ \frac{-\lambda_2}{6} \end{Bmatrix}. \qquad (h)$$

By straightforward algebra, Eq. (*h*) is solved to yield $d_0 = 1$, $d_1 = -11/6$, $d_2 = 0$, and $d_3 = 0$. Surprisingly, the values of d_0, d_1, d_2, and d_3 are free of λ_1 and λ_2. Therefore, in this particular example, the penalty constants can be chosen to have any real values and the least-squares solution always remains identical to the exact solution. Under normal circumstances, however, the coefficients d_k will be functions of the penalty constants and the values of these constants will be optimized from practical experience.

The advantage of using λ_1 and λ_2 in Eq. (*b*) can be seen from Eq. (*c*), where we use the power series $(1, x, x^2, x^3, x^4)$ as the set of our basis functions. The series has a very simple form and each member does not need to satisfy the boundary conditions at $x = 1$ (only x^2 does by coincidence). Had we not used λ_1 and λ_2, the basis functions would have been $x(x^j + j - 2)$, $j = 2, 3, \ldots, J$, which certainly cannot be obtained as easily as the power series.

Before we leave the subject of the penalty functional, it should also be made clear that adoption of the penalty functional is not a privilege of the method of least squares alone. In the variational method, for example, we may also add the penalty functional to the variational principle. See Section 8-2 for example.

2-3*d* Finite-Element Method

We now direct our attention to the method of least squares with the finite-element formulation. Again, the procedure will be described in terms of its application to our model equation (1-1).

The typical basis functions for the one-dimensional elements are pyramid functions. Since they have been successfully used in the previous methods, we may wish to use them here again. Unfortunately, in the least-squares functional

$$J = \int_0^4 (L\phi)^2 \, dx = \int_0^4 (-\phi'' - \phi' + 2\phi)^2 \, dx,$$

the presence of the second derivative requires at least C^1 continuity for the basis functions and, in contrast to the Galerkin and variational methods, the order of this continuity cannot be reduced through integration by parts. Therefore, we must seek certain polynomial basis functions of at least second degree. A standard set of these functions having local support was given in Example 1-6. An approximate solution expanded in their linear combination can be written as

$$\tilde{\phi}(x) = W_1(x)\phi_1 + W_2(x)\phi_2 + W_3(x)\phi_3 + W_4(x). \tag{2-47}$$

Strictly speaking, Eq. (2-47) is still not twice-differentiable at $x = x_2$, and the discontinuity in the first derivative at this point should cause difficulties in the evaluation of the functional integral J. At present, we simply give a warning about this pointwise discontinuity and ignore it. We will return to this subject in Chapter 3, where the nonconforming elements are considered. Equation (2-47) belongs to a three-dimensional subspace spanned by W_k, $k = 1, 2, 3$. The functional in this subspace

$$J(\phi_1, \phi_2, \phi_3) = \int_0^4 (-\tilde{\phi}'' - \tilde{\phi}' + 2\tilde{\phi})^2 \, dx \tag{2-48}$$

is now to be minimized to yield

$$\frac{\partial J}{\partial \phi_k} = 2\int_0^4 (\tilde{\phi}'' + \tilde{\phi}' - 2\tilde{\phi})(W_k'' + W_k' - 2W_k)\, dx = 0, \qquad k = 1, 2, 3. \tag{2-49a,b,c}$$

After straightforward algebra, Eqs. (2-49*a*)-(2-49*c*) are simplified into the matrix form

$$\begin{bmatrix} B_{11} & B_{12} & 0 \\ B_{21} & B_{22} & B_{23} \\ 0 & B_{32} & B_{33} \end{bmatrix} \begin{Bmatrix} \phi_1 \\ \phi_2 \\ \phi_3 \end{Bmatrix} = \begin{Bmatrix} 0 \\ -B_{24} \\ -B_{34} \end{Bmatrix}, \tag{2-50}$$

where

$$B_{11} = \int_0^2 (LW_1)^2\, dx, \qquad B_{22} = \int_0^4 (LW_2)^2\, dx, \qquad B_{33} = \int_2^4 (LW_3)^2\, dx,$$

$$B_{12} = B_{21} = \int_0^2 (LW_1)(LW_2^-)\, dx, \qquad B_{23} = B_{32} = \int_2^4 (LW_2^+)(LW_3)\, dx,$$

$$B_{24} = \int_2^4 (LW_2^+)(LW_4)\, dx \quad \text{and } B_{34} = \int_2^4 (LW_3)(LW_4)\, dx. \tag{2-51}$$

These coefficients can all be evaluated straightforwardly. The following example will illustrate the evaluation of a typical integral.

Example 2-13 Evaluate

$$\int_0^2 (LW_1)^2\, dx.$$

Solution In comparing the local element $[0, x_2]$ with the isoparametric element $[x_a, x_c]$, it is found that index 1 corresponds to index b. Therefore, we obtain

$$\int_0^2 (LW_1)^2\, dx = \int_{x_a}^{x_c} \left(\frac{d^2 W_b}{dx^2} + \frac{dW_b}{dx} - 2W_b\right)^2 dx$$

$$= \int_0^1 \left(\frac{1}{4}\frac{d^2 W_b}{dt^2} + \frac{1}{2}\frac{dW_b}{dt} - 2W_b\right)^2 2\, dt. \tag{a}$$

Table 2-2 Numerical values of the integral $\int_0^1 AB\,dt$, where $W_a = 1 - 3t + 2t^2$, $W_b = 4t - 4t^2$, $W_c = 2t^2 - t$, and $t = (x - x_a)/(x_c - x_a)$

	A						
B	1	W_a	W_b	W_c	W'_a	W'_b	W'_c
1	1	$\frac{1}{6}$	$\frac{2}{3}$	$\frac{1}{6}$	-1	0	1
W_a		$\frac{2}{15}$	$\frac{1}{15}$	$-\frac{1}{15}$	$-\frac{1}{2}$	$\frac{2}{3}$	$-\frac{1}{6}$
W_b			$\frac{8}{15}$	$\frac{1}{15}$	$-\frac{2}{3}$	0	$\frac{2}{3}$
W_c				$\frac{2}{15}$	$\frac{1}{6}$	$-\frac{2}{3}$	$\frac{1}{2}$
W'_a					$\frac{7}{3}$	$-\frac{8}{3}$	$\frac{1}{3}$
W'_b						$\frac{16}{3}$	$-\frac{8}{3}$
W'_c							$\frac{7}{3}$

The lengthy algebraic manipulation in evaluating the integral in Eq. (*a*) can be eliminated by consulting Table 2-2. The final numerical value is

$$\int_0^2 (LW_1)^2\,dx = \frac{128}{5}. \qquad (b)$$

2-4 COLLOCATION METHOD

In the collocation method, we first approximate the solution $\phi(x)$ by a linear combination of either the global basis functions [such as those given in Eq. (2-1)] or the splines with local support [such as those given in Eq. (7-9) or Eq. (9-15)]. Substituting this approximation into the differential equations, we then force the error to vanish at certain grid points. For the resulting set of algebraic equations to be closed, the number of grid points collocated should be the same as that of the undetermined coefficients.

Example 2-14 Use the collocation method with only two basis functions

$$v_1(t) = t(1 - t) \quad \text{and } v_2(t) = t(1 - t^2) \qquad (a)$$

to find an approximate solution to the model equation (1-1).

Solution To illustrate an alternative approach, we will transform Eq. (1-1) into an inhomogeneous equation subject to homogeneous boundary conditions as

$$\frac{d^2\psi}{dt^2} + 4\frac{d\psi}{dt} - 32\psi = 32t - 4, \qquad t\epsilon[0,1], \tag{b}$$

subject to

$$\psi(0) = 0 \quad \text{and } \psi(1) = 0, \tag{c}$$

where

$$t = \frac{x}{4} \quad \text{and } \psi(x) = \phi(x) - \frac{x}{4}.$$

To use the collocation method, we approximate

$$\tilde{\psi}(t) = e_1 v_1(t) + e_2 v_2(t). \tag{d}$$

Substituting Eq. (*d*) into Eq. (*b*) yields

$$(v_1'' + 4v_1' - 32v_1)e_1 + (v_2'' + 4v_2' - 32v_2)e_2 = 32t - 4. \tag{e}$$

Since there are two undetermined coefficients e_1 and e_2, we need two collocation grid points. Arbitrarily choosing $t = 0$ and $t = \frac{1}{2}$ enables Eq. (*e*) to generate a set of algebraic equations:

$$e_1 + 2e_2 = -2$$

and

$$5e_1 + 7e_2 = -6,$$

of which the solution is

$$e_1 = \tfrac{2}{3} \quad \text{and } e_2 = -\tfrac{4}{3}. \tag{f}$$

The collocation solution therefore becomes

$$\tilde{\phi}(x) = \frac{x}{4} + \frac{x}{6}\left(1 - \frac{x}{4}\right) + \frac{x}{3}\left(1 - \frac{x^2}{16}\right), \tag{g}$$

which is also shown in Fig. 2-6 for comparison.

For the application of the collocation method to heat transfer problems, see, for example, [22-33].

2-4*a* Connection between Collocation and Galerkin Methods

It is of interest to determine whether any connection exists between the collocation method and the Galerkin method. For convenience, we will use a practical example to do this. Let us start by rewriting Eq. (2-6*a*) as

$$e_1\left(\int_0^4 v_1''v_1\,dx + \int_0^4 v_1'v_1\,dx - 2\int_0^4 v_1^2\,dx\right) \tag{2-52}$$

$$+ e_2\left(\int_0^4 v_2''v_1\,dx + \int_0^4 v_2'v_1\,dx - 2\int_0^4 v_2v_1\,dx\right)$$

$$+\frac{1}{4}\int_0^4 v_1\,dx - \frac{1}{2}\int_0^4 xv_1\,dx = 0. \tag{2-52 (cont.)}$$

Suppose there exists an $x^*\epsilon[0,4]$ such that

$$\int_0^4 v_k''v_1\,dx = v_k''(x^*)\int_0^4 v_1\,dx, \tag{2-53a}$$

$$\int_0^4 v_k''v_1\,dx = v_k'(x^*)\int_0^4 v_1\,dx, \tag{2-53b}$$

$$\int_0^4 v_kv_1\,dx = v_k(x^*)\int_0^4 v_1\,dx, \qquad k = 1, 2 \tag{2-53c}$$

and

$$\int_0^4 xv_1dx = x^*\int_0^4 v_1\,dx. \tag{2-53d}$$

Then substitution of Eqs. (2-53*a*)-(2-53*d*) into Eq. (2-52) yields

$$e_1\,[v_1''(x^*) + v_1'(x^*) - 2v_1(x^*)] + e_2\,[v_2''(x^*) + v_2'(x^*) - 2v_2(x^*)] + \frac{1}{4} - \frac{x^*}{2} = 0. \tag{2-54}$$

If we substitute the approximate solution

$$\tilde{\phi}(x) = \frac{x}{4} + e_1\,\frac{x}{4}\left(1 - \frac{x}{4}\right) + e_2\,\frac{x}{4}\left(1 - \frac{x^2}{16}\right) = \frac{x}{4} + e_1v_1(x) + e_2v_2(x)$$

into Eq. (1-1) and set $x = x^*$, the identical equation will be derived. Therefore, we established that, when there exists an x^* such that Eqs. (2-53*a*)-(2-53*d*) are valid, the collocation method and the Galerkin method can yield the same set of algebraic equations. If there exists an x^* such that Eqs. (2-53*a*)-(2-53*d*) hold approximately, these two methods will yield similar approximate solutions. For more rigorous consideration, see [34, 35].

Example 2-15 Let

$$\tilde{\phi}(x) = \frac{x}{4} + e_1 \frac{x}{4}\left(1 - \frac{x}{4}\right) = \frac{x}{4} + e_1 v_1(x) \tag{a}$$

be a crude approximate solution to Eq. (1-1). Will the collocation method and the Galerkin method yield the same solution?

Solution We can simplify the inner product

$$(L\tilde{\phi}, v_1) = \int_0^4 (\tilde{\phi}'' + \tilde{\phi}' - 2\tilde{\phi})v_1 \, dx = 0 \tag{b}$$

to

$$e_1\left(\int_0^4 v_1''v_1 \, dx + \int_0^4 v_1'v_1 \, dx - 2\int_0^4 v_1^2 \, dx\right)$$

$$+\frac{1}{4}\int_0^4 v_1 \, dx - \frac{1}{2}\int_0^4 xv_1 \, dx = 0. \tag{c}$$

Now we seek an $x^* \epsilon [0,4]$ such that

$$\int_0^4 v_1'v_1 \, dx = v_1'(x^*)\int_0^4 v_1 \, dx, \tag{d}$$

$$\int_0^4 v_1^2 \, dx = v_1(x^*)\int_0^4 v_1 \, dx, \tag{e}$$

and

$$\int_0^4 xv_1 \, dx = x^*\int_0^4 v_1 \, dx. \tag{f}$$

With the aid of Table 2-1, Eqs. (*d*) and (*f*) yield

$$x^* = 2,$$

while Eq. (*e*) yields

$$x^* = 2.8944 \text{ or } 1.1056.$$

Hence there does not exist an x^* such that Eqs. (*d*)-(*f*) are simultaneously satisfied. It is most likely that the collocation method and the Galerkin method will not yield the same result regardless of the location of the collocation point.

2-4*b* Connection between Collocation and Finite-Difference Methods

By means of the C°-continuous quadratic basis functions (see Example 1-6), the collocation method may also lead to the same algebraic equations derived by the finite-difference method.

To establish this connection, we first assume

$$\tilde{\phi}(x) = \cdots + W_{j-1}(x)\phi_{j-1} + W_j(x)\phi_j + W_{j+1}(x)\phi_{j+1} + \cdots, \tag{2-55}$$

where, on the interval $[x_{j-1}, x_{j+1}]$,

$$W_{j-1}(t) = 1 - 3t + 2t^2,$$

$$W_j(t) = 4t - 4t^2,$$

and

$$W_{j+1}(t) = -t + 2t^2.$$

Here $t = (x - x_{j-1})/(x_{j+1} - x_{j-1})$. Therefore, at $x = x_j$, we obtain

$$\tilde{\phi}(x_j) = \phi_j, \tag{2-56a}$$

$$\left(\frac{d\tilde{\phi}}{dx}\right)_{x_j} = \frac{1}{x_{j+1} - x_{j-1}}(\phi_{j+1} - \phi_{j-1}), \tag{2-56b}$$

and

$$\left(\frac{d^2\tilde{\phi}}{dx^2}\right)_{x_j} = \frac{1}{(x_{j+1} - x_{j-1})^2}(\phi_{j-1} - 2\phi_j + \phi_{j+1}), \tag{2-56c}$$

which are the same as the equations obtained by the central finite-difference method. For the convergence analysis regarding this topic, see [24].

2-4*c* Orthogonal Collocation

So far, we have chosen the collocation points as evenly spaced grid nodes. This choice is merely a matter of convenience or intuition. There does exist a criterion for arranging the collocation points such that better accuracy can be achieved. This leads to the orthogonal collocation [27-30, 32].

In the orthogonal collocation method, the collocation points are taken as the roots of Legendre polynomials. On the interval [0,1], the first four polynomials are written as

$$P_0(x) = 1, \tag{2-57a}$$

$$P_1(x) = 2x - 1, \tag{2-57b}$$

$$P_2(x) = -6x^2 + 6x - 1, \tag{2-57c}$$

and

$$P_3(x) = 20x^3 - 30x^2 + 12x - 1. \tag{2-57d}$$

They satisfy the orthogonality criterion

$$\int_0^1 P_j(x)P_k(x) = 0 \quad \text{if } j \neq k. \tag{2-58}$$

Collocation at the roots of these polynomials, which happen to be the Gaussian quadrature knots, generally yields higher accuracy than collocation at evenly spaced grid points. In an elementary way, this can be shown by the following example. Suppose we wish to approximate a given function $f(x)$, $x \epsilon [0,1]$, by a linear combination of the leading three Legendre polynomials

$$\tilde{f}(x) = a + b(1 - 2x) + c(-6x^2 + 6x - 1). \tag{2-59}$$

If the collocation points are chosen to be $x = 0, \frac{1}{2}$, and 1, then a, b, and c can be determined from

$$\begin{Bmatrix} a \\ b \\ c \end{Bmatrix} = \begin{bmatrix} 1 & 1 & -1 \\ 1 & 0 & \frac{1}{2} \\ 1 & -1 & -1 \end{bmatrix}^{-1} \begin{Bmatrix} \tilde{f}(0) \\ \tilde{f}\left(\frac{1}{2}\right) \\ \tilde{f}(1) \end{Bmatrix}. \tag{2-60}$$

However, if the collocation points are chosen to be the three roots of $P_3(x)$, namely 0.112702, 0.5, and 0.887298, we are actually approximating

$$\tilde{f}(x) = a + b(1 - 2x) + c(-6x^2 + 6x - 1) + d(20x^3 - 30x^2 + 12x - 1) \tag{2-61}$$

and still have the opportunity of finding three undetermined constants from

$$\begin{Bmatrix} a \\ b \\ c \end{Bmatrix} = \begin{bmatrix} 1 & -0.7746 & -0.4 \\ 1 & 0 & 0.5 \\ 1 & 0.7746 & -0.4 \end{bmatrix}^{-1} \begin{Bmatrix} \tilde{f}(0.112702) \\ \tilde{f}(0.5) \\ \tilde{f}(0.887298) \end{Bmatrix} \tag{2-62}$$

because the fourth term in Eq. (2-61) vanishes at all three collocation points.

Example 2-15 Approximate $f(x) = \sin(\pi x/2)$ on [0,1] by orthogonal collocation with a linear combination of the first three Legendre polynomials.

Solution Utilizing Eq. (2-62), we write

$$\begin{Bmatrix} a \\ b \\ c \end{Bmatrix} = \begin{bmatrix} 1 & -0.7746 & -0.4 \\ 1 & 0 & 0.5 \\ 1 & 0.7746 & -0.4 \end{bmatrix}^{-1} \begin{Bmatrix} 0.1761 \\ 0.7071 \\ 0.9844 \end{Bmatrix} = \begin{Bmatrix} 0.6366 \\ 0.5217 \\ 0.1410 \end{Bmatrix}. \tag{a}$$

Therefore,

$$\tilde{f}(x) = 0.6366 + 0.5217(2x - 1) + 0.1410(-6x^2 + 6x - 1). \qquad (b)$$

Incidentally, if we compute a, b, and c from

$$a = \int_0^1 f(x)\, dx,$$

$$b = \frac{\int_0^1 f(x)(2x - 1)\, dx}{\int_0^1 (2x - 1)^2\, dx},$$

and

$$c = \frac{\int_0^1 f(x)(-6x^2 + 6x - 1)\, dx}{\int_0^1 (-6x^2 + 6x - 1)^2\, dx},$$

the values are found to be the same as those given in Eq. (a).

SYMBOLS

C^k	space in which functions are k-time differentiable
c_p	specific heat, J/g·K
e_j	undetermined coefficients in approximate expression
h	uniform mesh size ($= x_j - x_{j-1}$ for linear elements and $x_{j+1} - x_{j-1}$ for quadratic elements)
h_c	heat transfer coefficient, $W/m^2 \cdot K$
I	variational functional
J	least-squares functional, or largest number of j
$P_j(x)$	Legendre polynomials
R	residual, or radius of tube
t	isoparametric (natural) coordinate ($0 \leqslant t \leqslant 1$)
u	uniform flow velocity in x direction, m/s
W_j	constant
$W_j(x)$	quadratic piecewise basis function
$\delta\phi$	variation of function ϕ
$v_j(t)$	global basis function $[= t(1 - t^j)]$
ξ	dimensionless parameter ($= h_c R/k$)
ρ	density of fluid, kg/m^3
ϕ	principal field variable such as normalized temperature ($0 \leqslant \phi \leqslant 1$)

Superscripts

~ approximation of a function
* adjoint of a function, equation, or operator, or at a specific location [Eq. 2-53]
+ right-side element to grid point j
— left-side element to grid point j

REFERENCES

1. D. Gottlieb and S. A. Orszag, *Numerical Analysis of Spectral Methods: Theory and Applications,* Society for Industrial and Applied Mathematics, Philadelphia, 1978.
2. M. Kawahara and K. Hasegawa, Periodic Galerkin Finite Element Method of Tidal Flow, *Int. J. Numer. Methods Eng.,* vol. 12, pp. 115–127, 1978.
3. C.-M. Tang, Comparison of Spectral Methods for Flows on Spheres, *J. Comput. Phys.,* vol. 32, pp. 80–88, 1979.
4. D. B. Haidvogel and T. Zang, The Accurate Solution of Poisson's Equation by Expansion in Chebyshev Polynomials, *J. Comput. Phys.,* vol. 30, pp. 167–180, 1979.
5. S. A. Orszag, Spectral Methods for Problems in Complex Geometries, *J. Comput. Phys.,* vol. 37, pp. 70–92, 1980.
6. M. R. Malik, S. P. Wilkinson, and S. A. Orszag, Instability and Transition in Rotating Disk Flow, *AIAA J.,* vol. 19, pp. 1131–1138, 1981.
7. O. H. Hald, Convergence of Fourier Methods for Navier-Stokes Equations, *J. Comput. Phys.,* vol. 40, pp. 305–317, 1981.
8. T. D. Taylor and J. W. Murdock, Application of Spectral Methods to the Solution of Navier-Stokes Equations, *Comput. Fluids,* vol. 9, pp. 255–263, 1981.
9. B. Davies and J. A. Hendry, The Indirect Design of Fluid Channels Using a Global Variational Method with Trial Functions Not Satisfying the Prescribed Boundary Conditions, *Int. J. Numer Methods Eng.,* vol. 11, pp. 579–591, 1977.
10. J. P. Aubin, *Approximation of Elliptic Boundary Value Problems,* p. 34, Wiley-Interscience, New York, 1972.
11. J. T. Oden and J. N. Reddy, *Introduction to Mathematical Theory of Finite Elements,* Wiley-Interscience, New York, 1976.
12. P. G. Ciarlet, *The Finite Element Method for Elliptic Problems,* North-Holland, Amsterdam, 1978.
13. S. Ramadhyani and S. V. Patankar, Solution of the Poisson Equation: Comparison of the Galerkin and Control Volume Methods, *Int. J. Numer. Methods Eng.,* vol. 15, pp. 1395–1418, 1980.
14. G. V. Rao, K. K. Raju, N. Muthiyalu, and J. Venkataramana, Least Square and Galerkin Finite Element Solution of Flow Past a Flat Plate, *Int. J. Numer. Methods Eng.,* vol. 11, pp. 185–190, 1977.
15. W. L. Kwok, Y. K. Cheung, and C. Delcourt, Application of Least Square Collocation Technique in Finite Element and Finite Strip Formulation, *Int. J. Numer. Methods Eng.,* vol. 11, pp. 1391–1404, 1977.
16. J. N. Reddy and A. Satake, A Comparison of a Penalty Finite Element Model with the Stream Function-Vorticity Model of Natural Convection in Enclosures, *ASME J. Heat Transfer,* vol. 102, pp. 659–666, 1980.
17. G. R. Karr and T. J. Chung, Optimal Control Least-Squares Penalty Finite-Element Analysis in Convective Heat Transfer, *Numer. Heat Transfer,* vol. 3, pp. 35–46, 1980.
18. A. Kanarachos, The Discrete Least Square Approach to the Mixed Finite Element Solution of Viscous-Convective Type Flow, *Comput. Fluids,* vol. 8, pp. 189–198, 1980.

19. F. B. Hildebrand, *Methods of Applied Mathematics,* p. 203, Prentice-Hall, Englewood Cliffs, N.J., 1952.
20. J. H. Bramble and A. H. Schatz, Rayleigh-Ritz-Galerkin Methods for Dirichlet's Problem Using Subspaces with Boundary Conditions, *Comm. Pure Appl. Math.,* vol. 23, pp. 653–675, 1970.
21. I. Babuska, The Finite Element Method with Lagrangian Multipliers, *Numerische Math.,* vol. 20, pp. 179–192, 1973.
22. W. Zijl, Global Collocation Approximations of the Flow and Temperature Fields around a Gas and a Vapor Bubble, *Int. J. Heat Mass Transfer,* vol. 20, pp. 487–498, 1977.
23. S. H. Lin, Transient Heat Conduction in a Composite Slab with Variable Thermal Conductivity, *Int. J. Heat Mass Transfer,* vol. 22, pp. 1726–1731, 1979.
24. E. J. Doedel, Finite Difference Collocation Methods for Nonlinear Two-Point Boundary Value Problems, *SIAM (Soc. Ind. Appl. Math.) J. Numer. Anal.,* vol. 16, pp. 173–185, 1979.
25. H.-O. Kreiss and J. Oliger, Stability of the Fourier Method, *SIAM (Soc. Ind. Appl. Math.) J. Numer. Anal.,* vol. 16, pp. 421–433, 1979.
26. D. B. Haidvogel, Resolution of Downstream Boundary Layers in the Chebyshev Approximation to Viscous Flow Problems, *J. Comput. Phys.,* vol. 33, pp. 313–324, 1979.
27. Y. Mori, Y. Yamada, and K. Hijikata, Radiation Effects on Performances of Radiative Gas Heat Exchangers at High Temperatures, *Int. J. Heat Mass Transfer,* vol. 23, pp. 1079–1089, 1980.
28. P. Percell and M. F. Wheeler, A C^1 Finite Element Collocation Method for Elliptic Equations, *SIAM (Soc. Ind. Appl. Math.) J. Numer. Anal.,* vol. 17, pp. 605–622, 1980.
29. T. C. Chawla, W. J. Minkowycz, and G. Leaf, Spline-Collocation Solution of Integral Equations Occurring in Radiative Transfer and Laminar Boundary Layer Problems, *Numer. Heat Transfer,* vol. 3, pp. 133–148, 1980.
30. U. Ascher, Solving Boundary-Value Problems with a Spline-Collocation Code, *J. Comput. Phys.,* vol. 34, pp. 401–413, 1980.
31. B. M. Herbst, Collocation Methods and the Solution of Conduction-Convection Problems, *Int. J. Numer. Methods Eng.,* vol. 17, pp. 1093–1101, 1981.
32. A. Shapiro and G. F. Pinder, Analysis of an Upstream Weighted Collocation Approximation to the Transport Equation, *J. Comput. Phys.,* vol. 39, pp. 46–71, 1981.
33. J. Gary, The Multigrid Iteration Applied to the Collocation Method, *SIAM (Soc. Ind. Appl. Math.) J. Numer. Anal.,* vol. 18, pp. 211–224, 1981.
34. P. M. Prenter, *Splines and Variational Methods,* pp. 315–316, Wiley-Interscience, New York, 1975.
35. T. F. Hopkins and R. Wait, A Comparison of Galerkin, Collocation and the Method of Lines for Partial Differential Equations, *Int. J. Numer. Methods Eng.,* vol. 12, pp. 1981–2107, 1978.

PROBLEMS

2-1 (*a*) Express the function $g(t) = \sin(\pi t/2)$ in terms of the three basis functions $\nu_0(t) = t$, $\nu_1(t) = t(1-t)$, and $\nu_2(t) = t(1-t^2)$, $t \epsilon [0,1]$, by the following procedures: (1) let $g(t) = \nu_0(t) + a\nu_1(t) + b\nu_2(t)$; (2) evaluate the inner products in the two equations $(\nu_k, g) = (\nu_k, \nu_0) + a(\nu_k, \nu_1) + b(\nu_k, \nu_2)$, $k = 1, 2$; and (3) solve for a and b. After the exercise it may appear that the above algebra can be simplified if the basis functions ν_1 and ν_2 are orthogonal to each other. Therefore:

(*b*) Convert (ν_0, ν_1, ν_2) into an orthogonal set (ν_0, ν_1, μ_2) such that $(\nu_1, \mu_2) = 0$ by taking μ_2 as the linear combination of ν_1 and ν_2.

(*c*) Show that in general the nonorthogonal set of basis functions $(\nu_0, \nu_1, \nu_2, \nu_3, \ldots)$ can be converted into an orthogonal set $(\nu_0, \nu_1, \mu_2, \mu_3, \ldots)$.

2-2 If L is a self-adjoint, positive-definite linear operator and ϕ satisfies $L\phi - f(x) = 0$, prove that

$$(L\phi, \phi) - 2(f, \phi) < (L\bar{\phi}, \bar{\phi}) - 2(f, \bar{\phi}),$$

where $\bar{\phi} = \phi + \delta\phi$.

2-3 Follow a procedure similar to that in Example 2-6 to derive a discretized equation at $j = 0$ if $(d\phi/dx)_{x=0} = c\phi(0)$ and the control volume is taken to be $[0, \frac{1}{2}]$. Compare this equation with that directly obtained by using the control volume method.

2-4 Find the weighting functions $W_1(x)$ and $W_2(x)$ of the least-squares method applied to the governing equation $-\phi'' - \phi' + 2\phi = 0$ with the approximation $\phi(x) = e_0 v_0(x) + e_1 v_1(x) + e_2 v_2(x)$. Are they linearly independent?

2-5 Consider Example 2-8. Show that

(*a*) Equation (*e*) leads to Eq. (*f*).

(*b*) Differentiating Eq. (*a*) once and twice, along with some algebraic manipulation, leads to Eq. (*f*).

2-6 Find the Euler-Lagrange equation that corresponds to the least-squares functional

$$J = \int_1^3 (\phi' + \phi^2)^2 \, dx. \tag{P2-6}$$

What is the analytical solution (if it is unique) to the Euler-Lagrange equation?

2-7 Consider the fourth-order Euler-Lagrange equation

$$\phi'''' - 10\phi'' + 9\phi = 0, \tag{P2-7a}$$

with

$$\phi(0) = 0, \quad \phi'(0) = a_1, \quad \phi(1) = 1, \quad \text{and } \phi'(1) = a_2. \tag{P2-7b}$$

(*a*) What is the corresponding least-squares functional that has the negative sign in front of both the first derivative and the second derivative?

(*b*) Based on the functional obtained, the approximation $\tilde{\phi}(x) = t + e_1 t(1 - t) + e_2 t(1 - t^2)$, and Table 2-1, determine e_1 and e_2 such that the functional is minimized and thereby estimate a_1 and a_2.

(*c*) The analytical solution of Eq. (P2-7*a*) is

$$\phi(x) = c_1 e^x + c_2 e^{-x} + c_3 e^{3x} + c_4 e^{-3x}. \tag{P2-7c}$$

Determine c_1, c_2, c_3, and c_4 from Eq. (P2-7*b*) with a_1 and a_2 from (*b*). Then compare the analytical solution with the approximate solution.

2-8 Use the Rayleigh-Ritz method to find the approximate solution $\tilde{\phi}(x)$ to

$$-\phi'' - \phi' + 2\phi = 0, \quad \phi(0) = 1, \quad \left(\frac{d\phi}{dt}\right)_{t=1} = \phi(1).$$

(*a*) Assume $\tilde{\phi}(t) = d_1 t(t - \frac{2}{3}) + d_2 t(t^2 - \frac{1}{3})$.

(*b*) Assume $\tilde{\phi}(t) = e_1 t + e_2 t^2 + e_3 t^3$ and incorporate the penalty functional into the least-squares functional.

2-9 It is well known that numerical instability often arises when convection becomes dominant in convection-diffusion flow (see also Chapter 8). Consider the following nondimensionalized energy transport equation:

$$L\phi = -\frac{d^2\phi}{dt^2} + P\frac{d\phi}{dt} - 8\phi = 0 \tag{P2-9a}$$

subject to

$$\phi(0) = 0 \quad \text{and } \phi(1) = 1.$$

First find the exact solution for later comparisons. Discretize Eq. (P2-9*a*) and find ϕ_1, ϕ_2, and ϕ_3 ($h = \frac{1}{4}$) for $P = 2$, 8 using, respectively,

(*a*) The central finite-difference scheme. What happens at $P = 8$?

(*b*) The central integration (control volume) scheme.

(*c*) A quadratic finite-element scheme. In this scheme, we approximate $\tilde{\phi}(t)$ in the element $[t_{j-1}, t_j]$ by

$$\tilde{\phi}(t) = a + bt + ct^2 . \tag{P2-9b}$$

The constants *a*, *b*, and *c* are determined by the following three conditions:

$$\tilde{\phi}(t_{j-1}) = \phi_{j-1}, \quad \tilde{\phi}(t_j) = \phi_j, \quad \text{and } L\tilde{\phi}(t_{j-1/2}) = 0.$$

Note that the third condition is extremely important since the assumed profile will then preserve the convection-dominant characteristics of Eq. (P2-9*a*). After *a*, *b*, and *c* are determined, rearranging Eq. (P2-9*b*) leads to

$$\tilde{\phi}(t) = W_{j-1}(t)\phi_{j-1} + W_j(t)\phi_j, \quad t \in [t_{j-1}, t_j]. \tag{P2-9c}$$

Finally, find the inner product

$$(L\tilde{\phi}, N_j) = 0,$$

where $N_j(t)$ is the pyramid function.

CHAPTER

THREE

NUMERICAL METHODS USED IN HEAT TRANSFER (III)

In this chapter, we will continue to consider the finite-element method—applied to elements having polynomial basis functions of at least second degree—as well as some numerical methods that do not require evaluation of the integral inner products. By the end of this chapter, we will have covered more or less all the numerical methods commonly employed in the area of heat transfer.

In section 3-1, the higher-order elements that have C^0 and C^1 inter-elemental continuity will be considered. The Galerkin method is chosen to illustrate the use of these elements. Adoption of some nonconforming elements eases the computational task, but should be approved by the so-called patch test. Elements that pass or fail the test are introduced.

Section 3-2 presents the integral method and its general version—the method of moments. Although its application to ordinary differential equations may not be exciting, its ability to obtain approximate but quick solutions to partial differential equations deserves attention.

The perturbation method is described in Section 3-3. If the perturbed term is the highest derivative in a differential equation, the method is referred to as the singular perturbation method. Otherwise it is called the regular perturbation method.

In section 3-4, the methods for solving nonlinear two-point boundary-value problems are described. It is shown that all these techniques have been constructed on the basis of at least one of the following four fundamental strategies: (1) transformation of boundary-value problems into initial-value problems, (2) linearization, (3) discretization, and (4) iteration. A model differential equation in the nonlinear form subject to two-point boundary conditions is solved repeatedly by various techniques.

3-1 ONE–DIMENSIONAL HIGHER–ORDER ELEMENTS

In the preceding chapters, the interpolant in a finite element was chosen to consist of either linear or quadratic (in Section 2-3*d*) polynomials. Conventionally, elements in which the interpolants are linear are called first-order elements; those in which the interpolants are quadratic are called second-order elements. In the present section, we will further study some characteristics of elements that are at least of second order.

3-1*a* C^0 Continuity

Elements are said to have C^0 continuity if the interpolation functions in these elements are continuous on the elemental interfaces. For a one-dimensional linear C^0 element, the interpolant can simply be expressed in terms of the nodal functions at the two endpoints. For a one-dimensional quadratic C^0 element, we may introduce an additional nodal point, normally at the middle of the element so that the quadratic interpolant can be expressed in terms of three nodal functions. More precisely, the specification of this interpolation function can start with

$$\tilde{\phi}(x) = a_0 + a_1 x + a_2 x^2 \tag{3-1}$$

in an element $[x_a, x_c]$, with $x = x_b$ being the midpoint of the element as shown in Fig. 3-1. The three coefficients a_0, a_1, and a_2 are determined by the three nodal conditions

$$\tilde{\phi}(x_k) = \phi_k, \qquad k = a, b, c. \tag{3-2}$$

After algebra, we will be able to derive

$$\tilde{\phi}(x) = W_a(x)\phi_a + W_b(x)\phi_b + W_c(x)\phi_c, \tag{3-3}$$

where the basis functions (also called trial, expansion, or shape functions) $W_k(x)$ are preferably written in a transformed coordinate as

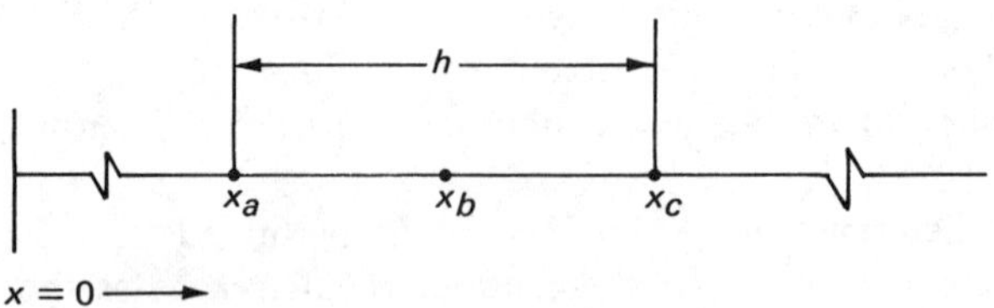

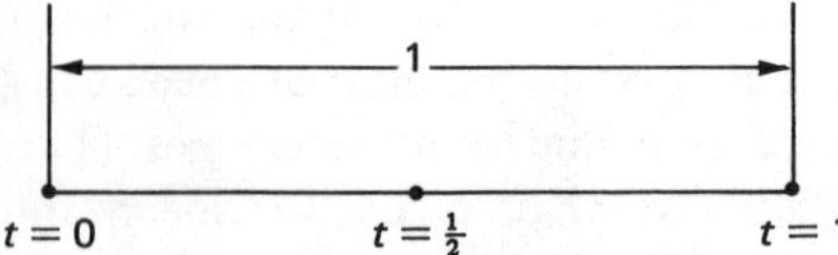

Figure 3-1 A typical element $[x_a, x_c]$ and the isoparametric element $[0,1]$, $t = (x - x_a)/h$.

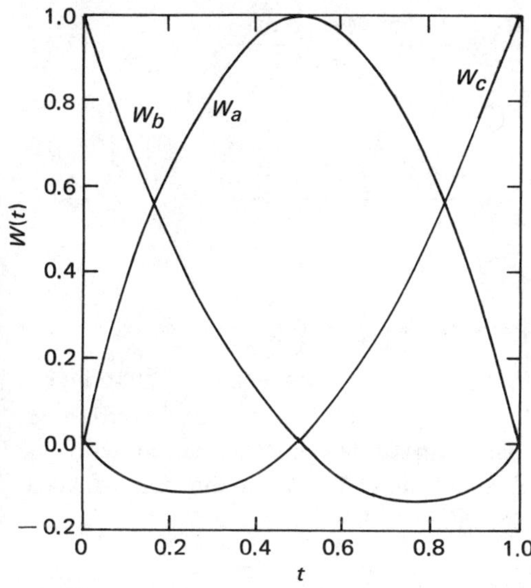

Figure 3-2 Three quadratic basis functions of C^0 elemental continuity. Note that $W_a + W_b + W_c = 1$ for all t.

$$W_a(t) = 1 - 3t + 2t^2, \tag{3-4a}$$

$$W_b(t) = 4(t - t^2), \tag{3-4b}$$

and

$$W_c(t) = 2t^2 - t, \tag{3-4c}$$

where $t = (x - x_a)/(x_c - x_a) = (x - x_a)/h$. For convenience, the element length h will be assumed to be uniform. These polynomial basis functions are shown in Fig. 3-2; their sum equals unity at any t. To demonstrate that such elements belong to C^0, we randomly choose two elements that are adjacent to each other as shown in Fig. 3-3. In the element $[x_{51}, x_{53}]$, the approximate function can be written according to Eq. (3-3) as

$$\tilde{\phi}(x) = (1 - 3t_1 + 2t_1^2)\phi_{51} + 4(t_1 - t_1^2)\phi_{52} + (2t_1^2 - t_1)\phi_{53}, \tag{3-5a}$$

where $t_1 = (x - x_{51})/(x_{53} - x_{51})$. In the element $[x_{53}, x_{55}]$, the interpolant is

$$\tilde{\phi}(x) = (1 - 3t_2 + 2t_2^2)\phi_{53} + 4(t_2 - t_2^2)\phi_{54} + (2t_2^2 - t_2)\phi_{55}, \tag{3-5b}$$

where $t_2 = (x - x_{53})/(x_{55} - x_{53})$. At the inter-element point $x = x_{53}$, we find that

$$\tilde{\phi}(x_{53}) \text{ on } [x_{51}, x_{53}] = \tilde{\phi}(x_{53}) \text{ on } [x_{53}, x_{55}] = \phi_{53}.$$

Therefore, these quadratic elements are of at least C^0 continuity. To further check whether they belong to C^1–i.e., whether the derivatives of the approximate functions are continuous at the inter-element point–we evaluate, on $[x_{51}, x_{53}]$,

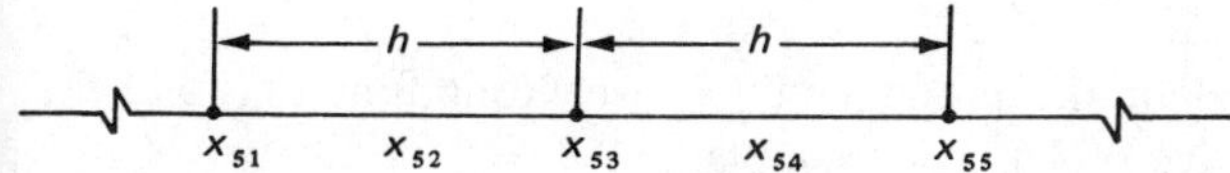

Figure 3-3 Two adjacent quadratic elements.

$$\left(\frac{d\tilde{\phi}}{dx}\right)_{x=x_{53}} = \left(\frac{d\tilde{\phi}}{dt_1}\right)_{t_1=1} \frac{dt_1}{dx} = \frac{\phi_{51} - 4\phi_{52} + 3\phi_{53}}{h}, \tag{3-6a}$$

and on $[x_{53}, x_{55}]$

$$\left(\frac{d\tilde{\phi}}{dx}\right)_{x=x_{53}} = \left(\frac{d\tilde{\phi}}{dt_2}\right)_{t_2=0} \frac{dt_2}{dx} = \frac{-3\phi_{53} + 4\phi_{54} - \phi_{55}}{h}. \tag{3-6b}$$

Therefore, unless the relation

$$\phi_{51} - 4\phi_{52} + 6\phi_{53} - 4\phi_{54} + \phi_{55} = 0$$

accidentally holds, the derivative given in Eq. (3-6*a*) is not equal to that in Eq. (3-6*b*). We conclude that the elements with Eqs. (3-4*a*)-(3-4*c*) are not C^1 continuous.

We now turn our attention to the Galerkin finite-element method with the quadratic elements, applied to the model equation (1-1). As in the case of linear elements, the solution to Eq. (1-1) will be assumed to be

$$\tilde{\phi}(x) = W_1(x)\phi_1 + W_2(x)\phi_2 + W_3(x)\phi_3 + W_4(x). \tag{3-7}$$

The quadratic basis functions, according to Eqs. (3-4*a*)-(3-4*c*), have the following local support:

$$W_1(t) = 4(t - t^2) \quad \text{if } t = \frac{x}{h} \text{ and } x\epsilon[0, x_2], \tag{3-8a}$$

$$W_2(t) = \begin{cases} W_2^-(t) = 2t^2 - t & \text{if } t = \dfrac{x}{h} \text{ and } x\epsilon[0, x_2] \\[2ex] W_2^+(t) = 1 - 3t + 2t^2 & \text{if } t = \dfrac{x - x_2}{h} \text{ and } x\epsilon[x_2, 4] \end{cases} \tag{3-8b}$$

$$W_3(t) = 4(t - t^2) \quad \text{if } t = \frac{x - x_2}{h} \text{ and } x\epsilon[x_2, 4]. \tag{3-8c}$$

It is noted that there are only two elements in the present problem and therefore the interval size h is twice that used with linear elements described in Chapter 1. Further, the basis function W_2 has finite support on both elements, whereas W_1 or W_3 has finite support on only one element.

Since Eq. (3-7) is similar in form to Eq. (1-27), we can immediately write the final algebraic equation for ϕ_j:

$$\begin{aligned}(\Lambda_{d(j-1),dj} - \Lambda_{d(j-1),j} + 2\Lambda_{j-1,j})\phi_{j-1} + (\Lambda_{dj,dj} - \Lambda_{dj,j} + 2\Lambda_{j,j})\phi_j \\ + (\Lambda_{d(j+1),d} - \Lambda_{d(j+1),j} + 2\Lambda_{j+1,j})\phi_{j+1} = 0, \quad j = 1, 3\end{aligned} \tag{3-9a, b}$$

where Λ stands for the integral of the product of two weighting functions and the subscript dj denotes the derivative of W_j. For example,

$$\Lambda_{d(j-1),j} = \int_{x_{j-1}}^{x_{j+1}} \frac{dW_{j-1}}{dx} W_j \, dx, \qquad \Lambda_{j,j} = \int_{x_{j-1}}^{x_{j+1}} W_j^2 \, dx, \qquad \text{etc.} \tag{3-10}$$

These integrals were evaluated and are listed in Table 2-2. Evaluation of these integrals can be facilitated by transforming the physical coordinate x into the isoparametric coordinate t. After the transformation, it is clear that some of the coefficients are equal to each other.

Example 3-1 Use the isoparametric element shown in Fig. 3-1 to evaluate $\Lambda_{d(j-1),j}$.

Solution In comparing a typical local element $[x_{j-1}, x_{j+1}]$ with the isoparametric element $[x_a, x_c]$, we establish the correspondence $j-1 \rightarrow a$, $j \rightarrow b$, and $j+1 \rightarrow c$. Therefore, the integral

$$\Lambda_{d(j-1),j} = \int_{x_{j-1}}^{x_{j+1}} \frac{dW_{j-1}(x)}{dx} W_j(x) \, dx \tag{a}$$

can be readily changed to

$$\Lambda_{d(j-1),j} = \int_{x_a}^{x_c} \frac{dW_a(x)}{dx} W_b(x) \, dx = \int_0^1 \frac{dW_a(t)}{dt} W_b(t) \, dt, \tag{b}$$

which can be evaluated as

$$\Lambda_{d(j-1),j} = \int_0^1 (-3 + 4t)(4t - 4t^2) \, dt = -\frac{2}{3}. \tag{c}$$

The derivation of the algebraic equation for the inter-elemental nodal point $j = 2$ is slightly lengthier than for those at the interior points $j = 1,3$ because the weighting function $W_2(x)$ has support on both elements. To derive this equation, we write the inner product

$$(L\tilde{\phi}, W_2) = -\int_0^4 \frac{d^2\tilde{\phi}}{dx^2} W_2 \, dx - \int_0^4 \frac{d\tilde{\phi}}{dx} W_2 \, dx + 2\int_0^4 \tilde{\phi} W_2 \, dx = 0. \tag{3-11}$$

Using integration by parts and dividing the integration range into [0,2] and [2,4], we obtain

$$(L\bar{\phi}, W_2) = \int_0^2 \frac{d\bar{\phi}}{dx}\left(\frac{dW_2^-}{dx} - W_2^-\right)dx + 2\int_0^2 \bar{\phi}W_2^- \, dx$$

$$+ \int_2^4 \frac{d\bar{\phi}}{dx}\left(\frac{dW_2^+}{dx} - W_2^+\right)dx + 2\int_2^4 \bar{\phi}W_2^+ \, dx, \tag{3-12}$$

which can be reduced to

$$a_0 \cdot 0 + a_1\phi_1 + a_2\phi_2 + a_3\phi_3 + a_4 \cdot 1 = 0, \tag{3-13}$$

where $$a_k = \Lambda_{dk,d2} - \Lambda_{dk,2} + 2\Lambda_{k,2}, \quad k = 0, 1, 3, 4 \tag{3-14a}$$

and $$a_2 = \Lambda^-_{d2,d2} + \Lambda^+_{d2,d2} - \Lambda^-_{d2,2} - \Lambda^+_{d2,2} + 2\Lambda^-_{2,2} + 2\Lambda^+_{2,2}. \tag{3-14b}$$

The superscripts − and + denote left element and right element, respectively. For completeness, it is our intention that Eq. (3-13) be written to include the terms $a_0 \cdot 0$. Consulting Table 2-2, we can evaluate Eqs. (3-9) and (3-13) as

$$\frac{24}{5}\phi_1 - \frac{26}{15}\phi_2 = 0, \tag{3-15a}$$

$$-\frac{2}{5}\phi_1 + \frac{51}{15}\phi_2 - \frac{26}{15}\phi_3 = -\frac{1}{15}, \tag{3-15b}$$

and $$-\frac{2}{5}\phi_2 + \frac{24}{5}\phi_3 = \frac{26}{15}, \tag{3-15c}$$

which are solved simultaneously to yield

$$\phi_1 = 0.0649, \quad \phi_2 = 0.1797, \quad \text{and } \phi_3 = 0.3761.$$

Table 3-1 shows the result obtained by the present scheme as well as those obtained by the linear element finite-element method (Section 1-2*a*) and the five-point finite-difference method (Section 1-1*a*). Also, it is interesting to compare the five-point discretization equations of $d^2\phi/dx^2$ and $d\phi/dx$ obtained by the present method with those obtained by the higher-order finite difference scheme.

Example 3-2 Use the quadratic-element formulation to obtain five-point relations for $(d^2\phi/dx^2)_{x=x_2}$ and $(d\phi/dx)_{x=x_2}$.

Table 3-1 Comparison of nodal unknowns computed by various schemes; only two elements were used in the quadratic finite element scheme

Scheme	ϕ_0	ϕ_1	ϕ_2	ϕ_3	ϕ_4
Exact	0	0.0473	0.1350	0.3679	1
Quadratic finite element	0	0.0649	0.1797	0.3761	1
Linear finite element	0	0.0444	0.1269	0.3563	1
Five-point finite difference	0	0.0505	0.1348	0.3918	1

Solution We will evaluate the following two terms:

$$H_1 = \frac{\int_0^4 (d^2\tilde{\phi}/dx^2)W_2\, dx}{\int_0^4 W_2\, dx} \tag{a}$$

and

$$H_2 = \frac{\int_0^4 (d\tilde{\phi}/dx)W_2\, dx}{\int_0^4 W_2\, dx.} . \tag{b}$$

But

$$\int_0^4 W_2(x)\, dx = \int_0^2 W_2^-(x)\, dx + \int_2^4 W_2^+(x)\, dx$$

$$= \frac{h}{6} + \frac{h}{6} = \frac{h}{3}, \tag{c}$$

$$-\int_0^4 \frac{d^2\tilde{\phi}}{dx^2} W_2\, dx = \Lambda_{d0,d2}\phi_0 + \Lambda_{d1,d2}\phi_1$$

$$+ (\Lambda_{\bar{d}2,d2} + \Lambda_{\dot{d}2,d2})\phi_2 + \Lambda_{d3,d2}\phi_3 + \Lambda_{d4,d2}\phi_4, \tag{d}$$

and

$$\int_0^4 \frac{d\tilde{\phi}}{dx} W_2\, dx = \Lambda_{d0,2}\phi_0 + \Lambda_{d1,2}\phi_1 + (\Lambda_{\bar{d}2,2} + \Lambda_{\dot{d}2,2})\phi_2$$

$$+ \Lambda_{d3,2}\phi_3 + \Lambda_{d4,2}\phi_4. \tag{e}$$

Noting that both local elements $[x_0, x_2]$ and $[x_2, x_4]$ are matched with the isoparametric element $[x_a, x_c]$, we obtain, with the aid of Table 2-2,

$$H_1 = -\frac{[(1/3h)\phi_0 - (8/3h)\phi_1 + (7/3h + 7/3h)\phi_2 - (8/3h)\phi_3 + (1/3h)\phi_4]}{h/3}$$

$$= -\frac{(\phi_0 - 8\phi_1 + 14\phi_2 - 8\phi_3 + \phi_4)}{h^2} \tag{f}$$

and

$$H_2 = \frac{\phi_0/6 - 2\phi_1/3 + (1/2 - 1/2)\phi_2 + 2\phi_3/3 - \phi_4/6}{h/3}$$

$$= \frac{\phi_0/2 - 2\phi_1 + 2\phi_3 - \phi_4/2}{h}. \tag{g}$$

There are two quick ways to partially check whether Eqs. (*f*) and (*g*) are derived correctly. First, if the function $\phi(x)$ is uniform in the vicinity of nodal point 2, we expect both $(d^2\phi/dx^2)_{x=x_2}$ and $(d\phi/dx)_{x=x_2}$ to vanish. Indeed, letting $\phi_0 = \phi_1 = \phi_2 = \phi_3 = \phi_4$ leads to $H_1 = H_2 = 0$. Second, we can assume $\phi(x)$ to be a simple function, such as x^2, and the grid point $x = x_2$ to be at $x = 2$. Then it immediately follows that $\phi_0 = 0$, $\phi_1 = 1$, $\phi_2 = 4$, $\phi_3 = 9$, $\phi_4 = 16$, $(d^2\phi/dx^2)_{x=2} = 2$ and $(d\phi/dx)_{x=2} = 4$. These nodal values are then used to compute Eqs. (*f*) and (*g*) to yield

$$H_1 = -\frac{(0 - 8 + 56 - 72 + 16)}{4} = 2 \tag{h}$$

and

$$H_2 = \frac{0 - 2 + 18 - 8}{2} = 4, \tag{i}$$

which are identical to the values from exact differentiation. After performing these two exercises, we therefore feel confident in our algebraic procedures that lead to Eqs. (*f*) and (*g*). Table 3-2 shows the values of the first and second derivatives for various functions x, e^x, and $\sin(\pi x/2)$ at $x = 2$ obtained by exact differentiation, five-point finite-difference Eqs. (1-8*a*) and (1-8*b*), and five-point finite-element Eqs. (*f*) and (*g*), respectively.

For elements $[x_a, x_d]$ having two interior nodal points $x = x_b$, x_c, the interpolation function $\tilde{\phi}(x)$ can be written as

$$\tilde{\phi}(x) = N_a(x)\phi_a + N_b(x)\phi_b + N_c(x)\phi_c + N_d(x)\phi_d,$$

where, for example,

$$N_a(x) = \frac{(x - x_b)(x - x_c)(x - x_d)}{(x_a - x_b)(x_a - x_c)(x_a - x_d)}$$

has unit value at $x = x_a$ and vanishes at $x = x_b$, x_c, x_d. These basis functions are also called the Lagrange polynomials and are very convenient in constructing higher-order interpolants. When multiplied by an exponential term, they can also be used for problems of infinite domain [1].

3-1*b* C^1 Continuity

In some problems [2, 3], we want the first derivatives of the functions, in addition to the functions themselves, to be continuous on the elemental boundaries. This

Table 3-2 Values of First and Second Derivatives of Various Functions at $x = 2$ with $h = \frac{1}{4}$

Scheme	Derivatives	x^2	e^x	$\sin\left(\frac{\pi x}{2}\right)$
Exact	$\frac{d^2\phi}{dx^2}$	2	7.3891	0
	$\frac{d\phi}{dx}$	4	7.3891	-1.5708
Five-point finite element	H_1	2	7.3109	0
	H_2	4	7.2317	-1.6473
Five-point finite difference	Eq. (1-8*a*)	2	7.3888	0
	Eq. (1-8*b*)	4	7.3880	-1.5696

requirement yields smoother approximate functions and higher accuracy than the C^0 continuity described in the preceding section. Since the number of nodal unknowns for one-dimensional problems doubles, the discretization procedure becomes cumbersome.

We start our consideration with a typical one-dimensional element $[x_a, x_b]$. The function distribution inside the element can be approximated by

$$\tilde{\phi}(x) = N_a(x)\phi_a + P_a(x)m_a + N_b(x)\phi_b + P_b(x)m_b, \tag{3-16}$$

where m_a and m_b are the first derivatives at grid points a and b. The basis functions $N_k(x)$ and $P_k(x)$, $k = a, b$, are to be determined. Assume that

$$\tilde{\phi}(t) = a_0 + a_1 t + a_2 t^2 + a_3 t^3, \quad t\epsilon[0,1], \tag{3-17}$$

where $t = (x - x_a)/(x_b - x_a) = (x - x_a)/h$. The four coefficients a_0, a_1, a_2, and a_3 can be determined by the following four conditions:

$$\phi_a = \phi(0) = a_0, \tag{3-18a}$$

$$m_a = \left(\frac{d\phi}{dt}\right)_{t=0} = a_1, \tag{3-18b}$$

$$\phi_b = \phi(1) = a_0 + a_1 + a_2 + a_3, \tag{3-18c}$$

and

$$m_b = \left(\frac{d\phi}{dt}\right)_{t=1} = a_1 + 2a_2 + 3a_3, \tag{3-18d}$$

which are simultaneously solved to yield $a_0 = \phi_a$, $a_1 = m_a$, $a_2 = 3(\phi_b - \phi_a) - 2m_a - m_b$, and $a_3 = 2(\phi_a - \phi_b) + m_a + m_b$. Therefore, Eq. (3-17) can be rewritten as

$$\tilde{\phi}(t) = (1 - 3t^2 + 2t^3)\phi_a + (t - 2t^2 + t^3)m_a + (3t^2 - 2t^3)\phi_b + (-t^2 + t^3)m_b. \tag{3-19}$$

Upon comparison of Eq. (3-16) with Eq. (3-19), we identify

$$N_a(t) = 1 - 3t^2 + 2t^3, \quad P_a(t) = t - 2t^2 + t^3,$$
$$N_b(t) = 3t^2 - 2t^3, \quad \text{and } P_b(t) = -t^2 + t^3. \tag{3-20}$$

These basis functions are also known as the cubic Hermite polynomials and they do not have support outside the interval $[x_a, x_b]$. Having constructed the basis functions for the isoparametric element, we are ready to continue on the Galerkin formulation. Again, the model equation (1-1) will be solved inside the domain consisting of four elements, [0,1] [1,2], [2,3], and [3,4]. Since two unknowns ϕ_j and m_j exist at nodal point j, the inner products we must evaluate there are two:

$$(L\tilde{\phi}, N_j) = \int_{x_{j-1}}^{x_{j+1}} \left(-\frac{d^2\tilde{\phi}}{dx^2} - \frac{d\tilde{\phi}}{dx} + 2\tilde{\phi} \right) N_j \, dx = 0 \tag{3-21a}$$

and

$$(L\tilde{\phi}, P_j) = \int_{x_{j-1}}^{x_{j+1}} \left(-\frac{d^2\tilde{\phi}}{dx^2} - \frac{d\tilde{\phi}}{dx} + 2\tilde{\phi} \right) P_j \, dx = 0. \tag{3-21b}$$

After straightforward algebra, Eqs. (3-21*a*) and (3-21*b*) can be reduced to

$$A_{j-1}\phi_{j-1} + B_{j-1}m_{j-1} + A_j\phi_j + B_jm_j + A_{j+1}\phi_{j+1} + B_{j+1}m_{j+1} = 0 \tag{3-22a}$$

and $$C_{j-1}\phi_{j-1} + D_{j-1}m_{j-1} + C_j\phi_j + D_jm_j + C_{j+1}\phi_{j+1} + D_{j+1}m_{j+1} = 0, \tag{3-22b}$$

where the coefficients A_k, B_k, C_k, and D_k, $k = j-1, j, j+1$, are listed in Table 3-3 for bookkeeping. Evaluation of these coefficients can be facilitated by transforming the physical coordinate $x \epsilon [x_{j-1}, x_j]$ into the isoparametric coordinate $t \epsilon [0,1]$.

Example 3-3 Use the concept of isoparametric elements to evaluate

$$\int_{x_{j-1}}^{x_j} \frac{d^2N_{j-1}}{dx^2} N_j \, dx.$$

Solution Let us match $[x_{j-1}, x_j]$ with $[x_a, x_b]$ and use the coordinate transformation $t = (x - x_{j-1})/h$. Thus,

$$\int_{x_{j-1}}^{x_j} \frac{d^2N_{j-1}(x)}{dx^2} N_j(x) \, dx = \int_{x_a}^{x_b} \frac{d^2N_a(x)}{dx^2} N_b(x) \, dx \tag{a}$$

Table 3-3 Integrals in Eqs. (3-22*a*) and (3-22*b*), $k = j - 1, j + 1$; for $k = j + 1$, the integration bounds are reversed

Symbol	Expression
A_k	$\int_{x_k}^{x_j} N_j(N_k'' + N_k' - 2N_k)\,dx$
B_k	$\int_{x_k}^{x_j} N_j L P_k\,dx$
C_k	$\int_{x_k}^{x_j} P_j L N_k\,dx$
D_k	$\int_{x_k}^{x_j} P_j L P_k\,dx$
A_j	$\int_{x_{j-1}}^{x_{j+1}} N_j L N_j\,dx$
B_j	$\int_{x_{j-1}}^{x_{j+1}} N_j L P_j\,dx$
C_j	$\int_{x_{j-1}}^{x_{j+1}} P_j L N_j\,dx$
D_j	$\int_{x_{j-1}}^{x_{j+1}} P_j L P_j\,dx$

$$= \frac{1}{h}\int_0^1 \frac{d^2N_a(t)}{dt^2}\,\mathrm{N}_b(t)\,dt \qquad (a)$$

(cont.)

which, after the use of Eq. (3-20), becomes

$$\frac{1}{h}\int_0^1 (-6 + 12t)(3t^2 - 2t^3)\,dt = \frac{6}{5h}. \qquad (b)$$

As an additional remark, we note that applying integration by parts to Eq. (*a*) yields

$$\int_{x_{j-1}}^{x_j} \frac{d^2N_{j-1}}{dx^2} N_j(x)\,dx = -\int_{x_{j-1}}^{x_j} \frac{dN_{j-1}}{dx}\frac{dN_j}{dx}\,dx + \frac{dN_{j-1}}{dx} N_j \bigg|_{x_{j-1}}^{x_j}. \tag{c}$$

Since dN_{j-1}/dx or N_j must vanish at nodal point x_{j-1} or x_j, the end-condition terms vanish. The relations

$$\int_0^1 \frac{d^2N_a}{dt^2} N_b(t)\,dt = -\int_0^1 \frac{dN_a}{dt}\frac{dN_b}{dt}\,dt, \text{ etc.} \tag{d}$$

are convenient to the construction of Table 3-4, which lists the numerical values of the integrals.

Table 3-4 Numerical values of the integral $\int_0^1 AB\,dt$, where $N_a = 1 - 3t^2 + 2t^3$, $P_a = t - 2t^2 + t^3$, $N_b = 3t^2 - 2t^3$, and $P_b = -t^2 + t^3$

	A			
B	N_a	N_b	P_a	P_b
1	$\frac{1}{2}$	$\frac{1}{2}$	$\frac{1}{12}$	$-\frac{1}{12}$
N_a''	$-\frac{6}{5}$	$\frac{6}{5}$	$-\frac{1}{10}$	$-\frac{1}{10}$
N_a'	$-\frac{1}{2}$	$-\frac{1}{2}$	$-\frac{1}{10}$	$\frac{1}{10}$
N_a	$\frac{13}{35}$	$\frac{9}{70}$	$\frac{11}{210}$	$-\frac{13}{420}$
N_b''	$\frac{6}{5}$	$-\frac{6}{5}$	$\frac{1}{10}$	$\frac{1}{10}$
N_b'	$\frac{1}{2}$	$\frac{1}{2}$	$\frac{1}{10}$	$-\frac{1}{10}$
N_b	$\frac{9}{70}$	$\frac{13}{35}$	$\frac{13}{420}$	$-\frac{11}{210}$
P_a''	$-\frac{11}{10}$	$\frac{1}{10}$	$-\frac{2}{15}$	$\frac{1}{30}$
P_a'	$\frac{1}{10}$	$-\frac{1}{10}$	0	$\frac{1}{60}$
P_a	$\frac{11}{210}$	$\frac{13}{420}$	$\frac{1}{105}$	$-\frac{1}{140}$
P_b''	$-\frac{1}{10}$	$\frac{11}{10}$	$\frac{1}{30}$	$-\frac{2}{15}$
P_b'	$-\frac{1}{10}$	$\frac{1}{10}$	$-\frac{1}{60}$	0
P_b	$-\frac{13}{420}$	$-\frac{11}{210}$	$-\frac{1}{140}$	$\frac{1}{105}$

With the aid of Tables 3-3 and 3-4, Eqs. (3-22*a*) and (3-22*b*) become

$$\begin{bmatrix} \frac{31}{70} & -\frac{13}{210} \\ -\frac{13}{210} & \frac{9}{140} \end{bmatrix} \begin{Bmatrix} \phi_{j-1} \\ m_{j-1} \end{Bmatrix} + \begin{bmatrix} -\frac{136}{35} & \frac{1}{5} \\ -\frac{1}{5} & -\frac{32}{105} \end{bmatrix} \begin{Bmatrix} \phi_j \\ m_j \end{Bmatrix} + \begin{bmatrix} \frac{101}{70} & -\frac{29}{210} \\ \frac{29}{210} & \frac{13}{420} \end{bmatrix} \begin{Bmatrix} \phi_{j+1} \\ m_{j+1} \end{Bmatrix} = 0, \tag{3-23}$$

which can be repeatedly used at $j = 1, 2, 3$. The algebraic system, however, is not closed, since m_0 and m_4 are not prescribed a priori. To remove this difficulty, we have at least three tactics available. The first and also the easiest is to assume

$$m_0 \approx \frac{\phi_1 - \phi_0}{h} = \phi_1 \qquad \text{and } m_4 \approx \frac{\phi_4 - \phi_3}{h} = 1 - \phi_3, \tag{3-24}$$

which indicate that the approximation solution is linear in the elements [0,1] and [3,4]. The second alternative is to adopt the three-point relation (1-10*a*). At $j = 1$ and $j = 3$, we obtain

$$m_0 = \frac{3}{h}(\phi_2 - \phi_0) - 4m_1 - m_2 = 3\phi_2 - 4m_1 - m_2 \tag{3-25a}$$

and

$$m_4 = \frac{3}{h}(\phi_4 - \phi_2) - 4m_3 - m_2 = 3(1 - \phi_2) - 4m_3 - m_2. \tag{3-25b}$$

The third choice concerns the modification of Eqs. (3-16), (3-17), and (3-19). For element $[x_a, x_b]$ we write

$$\tilde{\phi}(t) = \bar{N}_a(t)\phi_a + \bar{N}_b(t)\phi_b + \bar{P}_b(t)m_b, \qquad t = \frac{x - x_a}{h}, \tag{3-26a}$$

from which the missing boundary condition m_a is excluded. Following a procedure similar to that mentioned between Eqs. (3-16) and (3-19), we are able to derive

$$\bar{N}_a(t) = 1 - 2t + t^2, \qquad \bar{N}_b(t) = 2t - t^2, \qquad \bar{P}_b(t) = -t + t^2. \tag{3-26b}$$

The overbar is to distinguish the basis functions for the boundary elements from those for the interior elements. Similarly, we write

$$\tilde{\phi}(t) = \bar{N}_a(t)\phi_a + \bar{P}_a(t)m_a + \bar{N}_b(t)\phi_b, \tag{3-26c}$$

from which the missing boundary condition m_b is excluded. Clearly, Eqs. (3-26*a*) and (3-26*c*) are used as the interpolants in the elements $[0, x_1]$ and $[x_3, 1]$, respectively. The basis functions are derived as

$$\bar{N}_a(t) = 1 - t^2, \qquad \bar{P}_a(t) = t - t^2, \qquad \text{and } \bar{N}_b(t) = t^2. \tag{3-26d}$$

Although Eqs. (3-26*a*) and (3-26*c*) are polynomials of the second degree, which is lower than that of Eq. (3-19) by one, the discretization procedure with the former remains basically the same as that with the latter.

Example 3-4 Find the discretized form of Eq. (1-1) for grid points $j = 1$ by using Hermite polynomials.

Solution The two inner products at $j = 1$ are

$$(L\tilde{\phi}, N_1) = \int_0^2 \left(\frac{d^2\tilde{\phi}}{dx^2} + \frac{d\tilde{\phi}}{dx} - 2\tilde{\phi} \right) N_1(x)\, dx = 0 \tag{a}$$

and

$$(L\tilde{\phi}, P_1) = \int_0^2 \left(\frac{d^2\tilde{\phi}}{dx^2} + \frac{d\tilde{\phi}}{dx} - 2\tilde{\phi} \right) P_1(x)\, dx = 0, \tag{b}$$

where

$$\left.\begin{aligned} N_1 &= \bar{N}_b = 2t - t^2 \\ P_1 &= \bar{P}_b = -t + t^2 \end{aligned}\right\} \quad \text{if } t = x \text{ and } x \epsilon [0,1],$$

$$\left.\begin{aligned} N_1 &= N_a = 1 - 3t^2 + 2t^3 \\ P_1 &= P_a = t - 2t^2 + t^3 \end{aligned}\right\} \quad \text{if } t = x - 1 \text{ and } x \epsilon [1,2]. \tag{c}$$

After algebra, Eqs. (a) and (b) become

$$\bar{A}_1\phi_1 + \bar{B}_1 m_1 + A_2\phi_2 + B_2 m_2 = 0 \tag{d}$$

and

$$\bar{C}_1\phi_1 + \bar{D}_1 m_1 + C_2\phi_2 + D_2 m_2 = 0, \tag{e}$$

where

$$\bar{A}_1 = \int_0^1 (\bar{N}_b''\bar{N}_b + N_a''N_a + \bar{N}_b'\bar{N}_b + N_a'N_a - 2\bar{N}_b^2 - 2N_a^2)\, dt,$$

$$\bar{B}_1 = \int_0^1 (\bar{P}_b''\bar{N}_b + P_a''N_a + \bar{P}_b'\bar{N}_b + P_a'N_a - 2\bar{P}_b\bar{N}_b - 2P_aN_a)\, dt,$$

$$\bar{C}_1 = \int_0^1 (\bar{N}_b''\bar{P}_b + N_a''P_a + \bar{N}_b'\bar{P}_b + N_a'P_a - 2\bar{N}_b\bar{P}_b - 2N_aP_a)\, dt,$$

and

$$\bar{D}_1 = \int_0^1 (\bar{P}_b''\bar{P}_b + P_a''P_a + \bar{P}_b'\bar{P}_b + P_a'P_a - 2\bar{P}_b^2 - 2P_a^2)\, dt.$$

whereas the coefficients A_2, B_2, C_2, and D_2 can be directly obtained from Tables 3-3 and 3-4 by letting $j = 2$.

After substitution of the numerical values for these coefficients, Eqs. (*d*) and (*e*) can be written in the following matrix form (with the aid of Table 3-5*a*):

$$\begin{bmatrix} -\frac{152}{35} & \frac{22}{35} \\ \frac{2}{21} & -\frac{58}{105} \end{bmatrix} \begin{Bmatrix} \phi_1 \\ m_1 \end{Bmatrix} + \begin{bmatrix} \frac{101}{70} & -\frac{29}{210} \\ \frac{29}{210} & \frac{13}{420} \end{bmatrix} \begin{Bmatrix} \phi_2 \\ m_2 \end{Bmatrix} = \begin{Bmatrix} 0 \\ 0 \end{Bmatrix}. \tag{3-27}$$

Table 3-5*a* Numerical Values of the Integral $\int_0^1 AB\, dt$, where $\bar{N}_a(t) = 1 - 2t + t^2$, $\bar{N}_b = 2t - t^2$, and $\bar{P}_b = -t + t^2$

	A	
B	$\bar{N}_b$	$\bar{P}_b$
$\bar{N}_b''$	$-\frac{4}{3}$	$\frac{1}{3}$
$\bar{N}_b'$	$\frac{1}{2}$	$-\frac{1}{6}$
$\bar{N}_b$	$\frac{8}{15}$	$-\frac{7}{60}$
$\bar{P}_b''$	$\frac{4}{3}$	$-\frac{1}{3}$
$\bar{P}_b'$	$\frac{1}{6}$	0
$\bar{P}_b$	$-\frac{7}{60}$	$\frac{1}{30}$
$\bar{N}_a''$	$\frac{4}{3}$	$-\frac{1}{3}$
$\bar{N}_a'$	$-\frac{1}{2}$	$\frac{1}{6}$
$\bar{N}_a$	$\frac{2}{15}$	$-\frac{1}{20}$

Following the procedure described in the preceding example, we can also derive the discretized equations at $j = 3$ as (with the aid of Table 3-5b):

$$\begin{bmatrix} \frac{31}{70} & \frac{-13}{210} \\ \frac{-13}{210} & \frac{9}{140} \end{bmatrix} \begin{Bmatrix} \phi_1 \\ m_2 \end{Bmatrix} + \begin{bmatrix} \frac{-152}{35} & \frac{-2}{21} \\ \frac{-22}{35} & \frac{-58}{105} \end{bmatrix} \begin{Bmatrix} \phi_3 \\ m_3 \end{Bmatrix} = \begin{Bmatrix} \frac{-47}{30} \\ \frac{-2}{5} \end{Bmatrix}. \tag{3-28}$$

Therefore, Eq. (3-23) with $j = 2$ and Eqs. (3-27) and (3-28) constitute a 2×2 block matrix system

$$\begin{bmatrix} \begin{bmatrix} \frac{-152}{35} & \frac{22}{35} \\ \frac{2}{21} & \frac{-58}{105} \end{bmatrix} & \begin{bmatrix} \frac{101}{70} & \frac{-29}{210} \\ \frac{29}{210} & \frac{13}{420} \end{bmatrix} & \begin{bmatrix} 0 & 0 \\ 0 & 0 \end{bmatrix} \\ \begin{bmatrix} \frac{31}{70} & \frac{-13}{210} \\ \frac{-13}{210} & \frac{9}{140} \end{bmatrix} & \begin{bmatrix} \frac{-136}{35} & \frac{1}{5} \\ \frac{-1}{5} & \frac{-32}{105} \end{bmatrix} & \begin{bmatrix} \frac{101}{70} & -\frac{29}{210} \\ \frac{29}{210} & \frac{13}{420} \end{bmatrix} \\ \begin{bmatrix} 0 & 0 \\ 0 & 0 \end{bmatrix} & \begin{bmatrix} \frac{31}{70} & \frac{-13}{210} \\ \frac{-13}{210} & \frac{9}{140} \end{bmatrix} & \begin{bmatrix} \frac{-152}{35} & \frac{-2}{21} \\ \frac{-22}{35} & \frac{-58}{105} \end{bmatrix} \end{bmatrix} \begin{Bmatrix} \begin{Bmatrix} \phi_1 \\ m_1 \end{Bmatrix} \\ \begin{Bmatrix} \phi_2 \\ m_2 \end{Bmatrix} \\ \begin{Bmatrix} \phi_3 \\ m_3 \end{Bmatrix} \end{Bmatrix} = \begin{Bmatrix} \begin{Bmatrix} 0 \\ 0 \end{Bmatrix} \\ \begin{Bmatrix} 0 \\ 0 \end{Bmatrix} \\ \begin{Bmatrix} \frac{-47}{30} \\ \frac{-2}{5} \end{Bmatrix} \end{Bmatrix}. \tag{3-29}$$

Equation (3-29) can be solved by Gaussian elimination modified for a block matrix system. This modified version is similar to the procedure for solving an element matrix system. For example, to change the coefficient of the vector $\{\phi_1 m_1\}^T$ (T denotes transpose) into a unity block, we multiply the inverse of the coefficient block to the first row of Eq. (3-29) to yield

$$\begin{bmatrix} 1 & 0 \\ 0 & 1 \end{bmatrix} \begin{Bmatrix} \phi_1 \\ m_1 \end{Bmatrix} + \begin{bmatrix} \frac{-152}{35} & \frac{22}{35} \\ \frac{2}{21} & \frac{-58}{105} \end{bmatrix}^{-1} \begin{bmatrix} \frac{101}{70} & \frac{-29}{210} \\ \frac{29}{210} & \frac{13}{420} \end{bmatrix} \begin{Bmatrix} \phi_2 \\ m_2 \end{Bmatrix} = \begin{Bmatrix} 0 \\ 0 \end{Bmatrix}. \tag{3-30}$$

Likewise in the Gaussian elimination procedure for an element matrix system that has a first equation such as $a_{11}v_1 + a_{12}v_2 = 0$, we first multiply the equation by a_{11}^{-1} and obtain $v_1 + (a_{12}/a_{11})v_2 = 0$. The analogy between these two procedures can thus be clearly seen. We will describe Gaussian elimination for block matrix systems in greater detail in Chapter 7. Here we simply list in Table 3-6 the final numerical result which is seen to be more accurate than the C° solution.

Table 3-5b Numerical Values of the integral $\int_0^1 AB\,dt$, where $\bar{N}_b = t^2, \bar{N}_a = 1 - t^2$, and $\bar{P}_a = t - t^2$

B	A: $\bar{N}_a$	A: $\bar{P}_a$
$\bar{N}_a''$	$-\frac{4}{3}$	$-\frac{1}{3}$
$\bar{N}_a'$	$-\frac{1}{2}$	$-\frac{1}{6}$
$\bar{N}_a$	$\frac{8}{15}$	$\frac{7}{60}$
$\bar{P}_a''$	$-\frac{4}{3}$	$-\frac{1}{3}$
$\bar{P}_a'$	$\frac{1}{6}$	0
$\bar{P}_a$	$\frac{7}{60}$	$\frac{1}{30}$
$\bar{N}_b''$	$\frac{4}{3}$	$\frac{1}{3}$
$\bar{N}_b'$	$\frac{1}{2}$	$\frac{1}{6}$
$\bar{N}_b$	$\frac{2}{15}$	$\frac{1}{20}$

Clearly, Eq. (3-29) does not necessarily have to be written in a block matrix form. It may simply be constructed as a 6×6 element matrix system. For a network of only three grid points, there is little difference between these two choices. When the number of grid points becomes large, however, the algorithm for solving the block matrix system proves to be more efficient, since it requires less computer time and memory storage than those for the element matrix systems.

3-1*c* Conforming Elements and Patch Test

In Chapters 1 and 2 we considered C^0 elements in which the interpolation functions are continuous at the inter-elemental points. Even though the derivatives of the

Table 3-6 Nodal solution of the C^1 element

Scheme	ϕ_0	ϕ_1	m_1	ϕ_2	m_2	ϕ_3	m_3	ϕ_4
Exact	0	0.0473	0.0547	0.1350	0.1360	0.3679	0.3680	1
Finite element	0	0.0486	0.0483	0.1355	0.1083	0.3663	0.3046	1

approximate solutions do not exist there, they are integrable over the entire domain. Typical C^0 elements are those having piecewise linear basis functions. A finite-element solution of Eq. (1-1) is $\tilde{\phi}(x) = 0.0444N_1(x) + 0.1269N_2(x) + 0.3563 N_3(x) + N_4(x)$, obtained in Section 1-2*a*. Naturally, the solution itself is C^0 continuous. The nodal derivatives $(d\tilde{\phi}/dx)_{x=1}$, $(d\tilde{\phi}/dx)_{x=2}$, and $(d\tilde{\phi}/dx)_{x=3}$ do not exist, but the integral of the derivative, which is exactly the function $\tilde{\phi}(x)$ itself, exists over the entire domain [0,4]. This example may help us to better understand the definition of the so-called conforming elements.

Elements are said to be conforming if, for a differential equation of order $2k$, they are of C^{k-1} continuity for the Galerkin and variational methods, or of C^{2k-1} for the method of least squares. In many heat transfer problems, $k = 1$. An integral inner product of the Galerkin version will then take form as

$$\left(-\frac{d^2\tilde{\phi}}{dx^2}, v\right) = \int_a^b \frac{d\tilde{\phi}}{dx}\frac{dv}{dx}\,dx + \text{boundary condition terms}, \qquad v \epsilon C^0[a, b]. \tag{3-31a}$$

Since the integral contains only first-order derivatives, the C^0 elements are sufficiently conforming according to the definition. With regard to the method of least squares, its functional to be minimized is of the form

$$J = \int_a^b \left(\frac{d^2\tilde{\phi}}{dx^2}\right)^2 dx, \tag{3-31b}$$

containing a second derivative that cannot be integrated by parts to become first order. Therefore, the elements must be at least C^1 continuous.

For one-dimensional problems that require up to C^1 continuity or two-dimensional systems in which C^0 continuity is already satisfactory, it is relatively simple to construct the conforming elements. For example, the C^0 interpolant in a typical triangular element (as shown in Fig. 3-4)

$$\tilde{\phi}(x, y) = a_0 + a_1 x + a_2 y$$

has three degrees of freedom, which can be determined by the three nodal functions ϕ_a, ϕ_b, and ϕ_c. As the requirement of inter-elemental continuity becomes more strict, the task of constructing the conforming elements becomes more difficult. To be ensured of C^1 continuity, for example, the interpolant in the triangular element shown in Fig. 3-4 must have at least 21 degrees of freedom

$$\tilde{\phi}(x, y) = a_1 + a_2 x + a_3 y + a_4 x^2 + a_5 xy + a_6 y^2 + a_7 x^3 + a_8 y^3 + \cdots + a_{21} y^5$$

so that it can be expressed in terms of the 21 nodal variables ϕ_k, $(\partial\phi/\partial x)_k$, $(\partial\phi/\partial y)_k$, $(\partial^2\phi/\partial x^2)_k$, $(\partial^2\phi/\partial y^2)_k$, $(\partial^2\phi/\partial x\partial y)_k$, $k = a, b, c$, and normal derivatives $\partial\phi/\partial n$ specified at three side nodes. For three-dimensional tetrahedral elements of C^1 continuity, the degree of the interpolating polynomials must be raised as high as

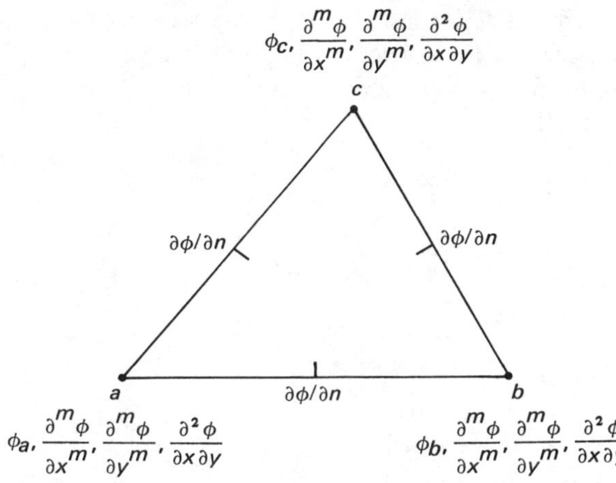

Figure 3-4 A typical C^1 triangular element and its 21 nodal unknowns, $m = 1, 2$.

nine! Therefore, from a computational viewpoint, it is desirable to use nonconforming elements if they work as well as their conforming counterparts.

Here, by "work well," we mean "converge, for the finite-element solution, to the true solution." In other words, nonconforming elements can be used only when the finite-element solution approaches the exact solution as $h \to 0$. Some nonconforming elements will yield convergent solutions, while some will not. It is thus important to understand the behavior of the elements and to use them with care.

One method for determining whether the nonconforming elements yield convergent solutions is the so-called patch test [4, 5]. In the patch test, the functional is minimized with respect to an inter-elemental nodal unknown ϕ_j. A polynomial $\psi(x)$ of the same degree r as the highest derivative in the functional is then chosen to a particular solution within two adjacent elements (called the patch). This polynomial $\psi(x)$ must satisfy the original governing differential equation, but not necessarily the boundary conditions. The approximate solution $\tilde{\phi}(x)$ is also chosen such that its nodal values are the same as those computed from the polynomial except for ϕ_j. If, after discretization of the functional, ϕ_j is found to be identical to $\psi(x_j)$, the nonconforming elements are considered to pass the patch test and the finite-element solution with these elements will converge to the true solution.

Clearly, elements that are conforming should pass the patch test. The following example is presented to confirm this assertion. Since we seldom need to use nonconforming elements for one-dimensional second-order differential equations, we will intentionally design a problem with fourth order regardless of the fact that heat transfer equations are rarely of such high order.

Example 3-5 Consider the linear differential equation

$$L\phi = \frac{d^4\phi}{dx^4} - \frac{d^2\phi}{dx^2} + 2 = 0, \qquad x \epsilon [-10,10]. \tag{a}$$

Boundary conditions will not be prescribed since they are irrelevant to the present problem. Let the patch be [0,2], consisting of two adjacent elements [0,1] and [1,2], as shown in Fig. 3-5. Do nonconforming [with respect to Eq. (*a*)] elements $[x_a, x_b]$ having local basis functions as given in Eq. (3-20) pass the patch test for the Galerkin method?

Solution The inner product

$$(L\tilde{\phi}, N_1) = \int_0^2 \frac{d^4\tilde{\phi}}{dx^4} N_1\, dx - \int_0^2 \frac{d^2\tilde{\phi}}{dx^2} N_1\, dx + 2\int_0^2 N_1\, dx = 0 \tag{b}$$

can be integrated by parts to become

$$(L\tilde{\phi}, N_1) = \int_0^2 \left(\frac{d^2\tilde{\phi}}{dx^2}\frac{d^2N_1}{dx^2} + \frac{d\tilde{\phi}}{dx}\frac{dN_1}{dx} + 2N_1 \right) dx = 0. \tag{c}$$

All the boundary condition terms vanish because the basis function $N_1(x)$ is zero at $x = 0{,}2$. It is noteworthy that Eq. (*c*) can also be derived by minimizing the variational principle

$$I = \int_{-10}^{10} \left[\frac{1}{2}\left(\frac{d^2\tilde{\phi}}{dx^2}\right)^2 + \frac{1}{2}\left(\frac{d\tilde{\phi}}{dx}\right)^2 + 2\tilde{\phi} \right] dx + \text{boundary condition terms} \tag{d}$$

with respect to ϕ_j. Matching patch elements [0,1] and [1,2], respectively, with $[x_a, x_b]$, we change Eq. (*c*) to

$$(L\tilde{\phi}, N_1) = \int_0^1 \left[\frac{d^2\phi^{(A)}}{dx^2}\frac{d^2N_b}{dx^2} + \frac{d\phi^{(A)}}{dx}\frac{dN_b}{dx} + 2N_b \right] dx + \int_1^2 \left[\frac{d^2\phi^{(B)}}{dx^2}\frac{d^2N_a}{dx^2} + \frac{d\phi^{(B)}}{dx}\frac{dN_a}{dx} + 2N_a \right] dx, \tag{e}$$

where the interpolants $\phi^{(A)}(x)$ and $\phi^{(B)}(x)$ are expressed in terms of the C^1-continuity basis functions as

$$\phi^{(A)}(x) = N_a(x)\phi_0 + P_a(x)m_0 + N_b(x)\phi_1 + P_b(x)m_1 \tag{f}$$

and

$$\phi^{(B)}(x) = N_a(x)\phi_1 + P_a(x)m_1 + N_b(x)\phi_2 + P_b(x)m_2. \tag{g}$$

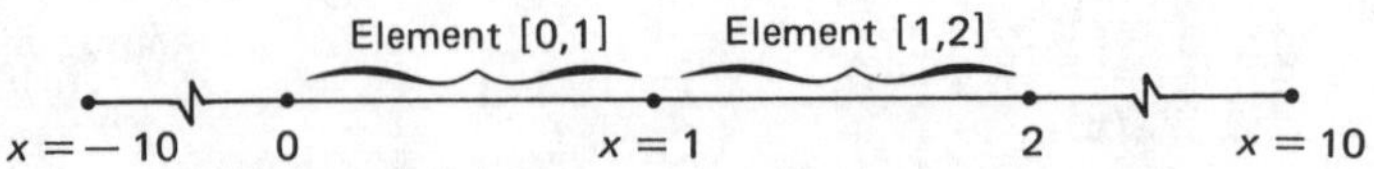

Element [0,1] Element [1,2]

$x = -10$ 0 $x = 1$ 2 $x = 10$

Figure 3-5 A typical patch consisting of two conforming C^1 elements.

Substituting Eqs. (*f*) and (*g*) into Eq. (*e*), we obtain the algebraic equation

$$a_0\phi_0 + b_0 m_0 + a_1\phi_1 + b_1 m_1 + a_2\phi_2 + b_2 m_2 = -2c, \tag{h}$$

where

$$a_0 = \int_0^1 (N_a''N_b'' + N_a'N_b')\, dt,$$

$$b_0 = \int_0^1 (P_a''N_b'' + P_a'N_b')\, dt,$$

$$a_1 = \int_0^1 [(N_a'')^2 + (N_b'')^2 + (N_a')^2 + (N_b')^2]\, dt,$$

$$b_1 = \int_0^1 (P_a''N_a'' + P_b''N_b'' + P_a'N_a' + P_b'N_b')\, dt,$$

$$a_2 = a_0,$$

$$b_2 = \int_0^1 (P_b''N_a'' + P_b'N_a')\, dt,$$

and

$$c = \int_0^1 (N_a + N_b)\, dt.$$

The prime in these integrals denotes differentiation with respect to $t \epsilon [0,1]$. Utilizing Table 3-4, we can readily obtain the following numerical values: $a_0 = -66/5$, $b_0 = -61/10$, $a_1 = 132/5$, $b_1 = 0$, $a_2 = -66/5$, $b_2 = 61/10$, and $c = 1$.

So far, we have not accomplished much other than deriving a discretized equation (*h*) at grid point 1. Incorporating this equation into the patch test is our next task. According to the test, a second-degree polynomial $\psi(x)$ that satisfies Eq. (*a*) is needed. With little effort, we find that

$$\psi(x) = x^2 \tag{i}$$

is a qualified candidate. Thus, let

$$\phi_0 = \psi(0) = 0, \qquad m_0 = \left(\frac{d\psi}{dx}\right)_{x=0} = 0,$$

$$m_1 = \left(\frac{d\psi}{dx}\right)_{x=1} = 2, \quad \phi_2 = \psi(2) = 4,$$

and

$$m_2 = \left(\frac{d\psi}{dx}\right)_{x=2} = 4. \tag{j}$$

Substitution of these nodal values into Eq. (*h*) yields

$$\phi_1 = \frac{5}{132}\left[-2 + \frac{66}{5} \times 4 - \frac{61}{10} \times 4\right] = 1, \tag{k}$$

which is identical to $\psi(1)$. Therefore, the line segment $[x_a, x_b]$ passes the patch test. It should be noted that passing the test for this specific problem does not guarantee success for other fourth-order problems. The patch test must be performed whenever a new problem is tackled.

In the preceding example, it is expected that the element should pass the patch test since it is conforming. Now, we naturally grow curious about the chance for certain nonconforming elements to pass the test.

Example 3-6 Consider the nonconforming elements $[x_a, x_c]$, inside which the interpolants are

$$\tilde{\phi}(x) = W_a(x)\phi_a + W_b(x)\phi_b + W_c(x)\phi_c \tag{a}$$

where

$$W_a(x) = 1 - 3t + 2t^2,$$

$$W_b(x) = 4t - 4t^2,$$

and

$$W_c(x) = 2t^2 - t, \qquad t = \frac{x - x_a}{x_c - x_a} = \frac{x - x_a}{h}$$

Determine whether these elements will pass the patch test for the fourth-order problem described in the preceding example.

Solution An arbitrary patch consisting of two adjacent elements [0,2] and [2,4] is shown in Fig. 3-6. These two elements are nonconforming because the first derivatives of the two interpolants are not continuous at the inter-elemental point $x = x_2 = 2$ and the highest derivative present in the functional integral is of second order. Avoiding repetition, we directly write the inner product [similar to that in Eq. (*e*) of the preceding example] as

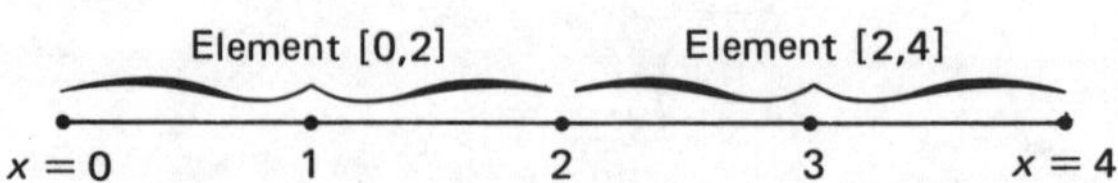

Figure 3-6 A tested patch consisting of two C^0 elements that fail the patch test in Example 3-6 but pass it in Example 3-7.

$$(L\tilde{\phi}, W_2) = \int_0^2 \left[\frac{d^2\phi^{(A)}}{dx^2}\frac{d^2W_c}{dx^2} + \frac{d\phi^{(A)}}{dx}\frac{dW_c}{dx} + 2W_c\right]dx$$

$$+ \int_2^4 \left[\frac{d^2\phi^{(B)}}{dx^2}\frac{d^2W_a}{dx^2} + \frac{d\phi^{(B)}}{dx}\frac{dW_a}{dx} + 2W_a\right]dx = 0, \tag{b}$$

where the elemental interpolants are approximated as

$$\phi^{(A)}(x) = W_a(x)\phi_0 + W_b(x)\phi_1 + W_c(x)\phi_2 \tag{c}$$

and
$$\phi^{(B)}(x) = W_a(x)\phi_2 + W_b(x)\phi_3 + W_c(x)\phi_4. \tag{d}$$

Upon substitution of Eqs. (*c*) and (*d*) into Eq. (*b*), a discretized equation

$$a_0\phi_0 + a_1\phi_1 + a_2\phi_2 + a_3\phi_3 + a_4\phi_4 = -32a_5 \tag{e}$$

can be derived, where

$$a_0 = \int_0^1 (W_a''W_c'' + 4W_a'W_c')\,dt, \qquad a_1 = \int_0^1 (W_b''W_c'' + 4W_b'W_c')\,dt,$$

$$a_2 = \int_0^1 [(W_a'')^2 + (W_c'')^2 + 4(W_a')^2 + 4(W_c')^2]\,dt,$$

$$a_3 = \int_0^1 (W_a''W_b'' + 4W_a'W_b')\,dt, \qquad a_4 = a_0,$$

and

$$a_5 = \int_0^1 (W_a + W_c)\,dt.$$

The prime again stands for differentiation with respect to the isoparametric coordinate $t\epsilon[0,1]$. Consulting Table 2-2, we obtain the following numerical values: $a_0 = a_4 = 52/3$, $a_1 = a_3 = -128/3$, $a_2 = 152/3$, and $a_5 = 1/3$. Equation (*e*) is then rearranged as

$$\phi_2 = \frac{3}{152}\left[-\frac{32}{3} - \frac{52}{3}(\phi_0 + \phi_4) + \frac{128}{3}(\phi_1 + \phi_3)\right]. \tag{f}$$

After the nodal values ϕ_0, ϕ_1, ϕ_3, and ϕ_4 are chosen, respectively, to be the same as the polynomial values $\psi(0) = 0$, $\psi(1) = 1$, $\psi(3) = 9$, and $\psi(4) = 16$, Eq. (*f*) is reduced to

$$\phi_2 = \frac{52}{19}. \tag{g}$$

But $\psi(2)$ is equal to 4. Therefore, these nonconforming elements fail the patch test.

We recall that, in Chapter 2, the least-squares functional

$$J = \int_0^4 (\phi'' + \phi' - 2\phi)^2 \, dx$$

was minimized by using the nonconforming elements $[x_a, x_c]$ with the second-degree basis functions $W_a(x)$, $W_b(x)$, and $W_c(x)$. We did not justify the legitimacy of our procedure there. Now that we have learned the patch test, it may be appropriate at this point to investigate whether these nonconforming elements pass the test. However, there is one difficulty: a polynomial of second degree that satisfies $L\phi = \phi'' + \phi' - 2\phi = 0$ cannot be found. Under such circumstances, it is difficult to perform the patch test. Therefore, for convenience, let us slightly modify the governing equation to

$$L\phi = \phi'' + \phi' - 2(1 + x) = 0 \tag{3-32}$$

so that a particular solution of a second-degree polynomial $\psi(x) = x^2$ becomes available. Therefore, in the following example Eq. (3-32) instead of Eq. (1-1) will be tested against the patch test. We make this accommodation because the main objective of this section is to describe the procedure of the patch test and then present a few examples in which elements either pass or fail the test. To strive for rigorous rules for one-dimensional elements may not be rewarding because the patch test becomes useful only for higher-dimensional patches.

Example 3-7 Consider Eq. (3-32), to which the polynomial $\psi(x) = x^2$ is a particular solution. Use the patch test to determine whether the nonconforming elements $[x_a, x_c]$ with the basis functions $W_a(x)$, $W_b(x)$, and $W_c(x)$ are acceptable for minimization of the least-squares functional

$$J = \int_0^4 \left[\frac{d^2\phi}{dx^2} + \frac{d\phi}{dx} - 2(1 + x) \right]^2 dx. \tag{a}$$

Solution The discontinuity in $d\phi/dx$ takes place at $x = x_2 = 2$ within the patch [0,4]. To be consistent with the patch test, the functional J is minimized with respect to ϕ_2 to yield

$$\frac{\partial J}{\partial \phi_2} = 2 \int_0^4 \left[\frac{d^2\tilde{\phi}}{dx^2} + \frac{d\tilde{\phi}}{dx} - 2(1 + x) \right] (W_2'' + W_2') \, dx = 0. \tag{b}$$

Since within the elements [0,2] and [2,4] we assume

$$\tilde{\phi}(x) = W_b(x)\psi(1) + W_c(x)\phi_2, \qquad x\epsilon[0,2], \tag{c}$$

and

$$\tilde{\phi}(x) = W_a(x)\phi_2 + W_b(x)\psi(3) + W_c(x)\psi(4), \qquad x\epsilon[2,4], \tag{d}$$

Eq. (b) can be reduced, after algebra and with the aid of Table 2-2, to

$$a_1\psi(1) + a_2\phi_2 + a_3\psi(3) + a_4\psi(4) = a_5, \tag{e}$$

where

$$a_1 = \int_0^1 \left(\frac{1}{4} W_b' W_c' + \frac{1}{2} W_b' - W_c' - 2\right) dt = -\frac{11}{3},$$

$$a_2 = \int_0^1 \left[\frac{1}{4} (W_a')^2 + \frac{1}{4} (W_c')^2 + W_a' + W_c' + 2\right] dt = \frac{19}{6},$$

$$a_3 = \int_0^1 \left(\frac{1}{4} W_a' W_b' + \frac{1}{2} W_b' - W_a' - 2\right) dt = -\frac{5}{3},$$

$$a_4 = \int_0^1 \left(\frac{1}{4} W_a' W_c' + \frac{1}{2} W_a' + \frac{1}{2} W_c' + 1\right) dt = \frac{13}{12},$$

and

$$a_5 = \int_0^1 \left[(2 + 4t)\left(1 + \frac{1}{2} W_c'\right) + (6 + 4t)\left(1 + \frac{1}{2} W_a'\right)\right] dt = \frac{34}{3}.$$

Substituting these numerical values, along with $\psi(1) = 1$, $\psi(3) = 9$, and $\psi(4) = 16$, into Eq. (e) leads to

$$\phi_2 = 4 = \psi(2). \tag{f}$$

We conclude that these second-order nonconforming elements pass the patch test for the problem governed by Eq. (a).

So far we have shown three cases, in which (1) conforming elements pass (of course); (2) nonconforming elements fail, and (3) nonconforming elements pass. It is hoped that these examples demonstrate the procedure for the patch test. More examples can be found in [5, 6] for two-dimensional patches. The limitations of the patch test are described in [7], and the relation between the variational principle and the patch test is described in [8, 9].

3-2 INTEGRAL METHOD: A SPECIAL CASE OF THE METHOD OF MOMENTS

In this section, we will describe another method of weighted residuals, which employs members in the power series $(1, t, t^2, \ldots, t^n)$, $t \in [0,1]$, as the weighting functions. Since successively higher moments of the residual are forced to vanish, this technique is called the method of moments [10,11]. Consider again the model equation (1-1):

$$L\phi = -\phi'' - \phi' + 2\phi = 0, \qquad \phi(0) = 0, \qquad \phi(4) = 1. \tag{1-1}$$

If the approximate solution $\tilde{\phi}(x)$ is expressed by

$$\tilde{\phi}(x) = \frac{x}{4} + e_1 \frac{x}{4}\left(1 - \frac{x}{4}\right) + e_2 \left(\frac{x}{4}\right)^2 \left(1 - \frac{x}{4}\right) + e_3 \left(\frac{x}{4}\right)^3 \left(1 - \frac{x}{4}\right) + \cdots,$$

its substitution into Eq. (1-1) should generate a residual, that is,

$$L\tilde{\phi} = -\tilde{\phi}'' - \tilde{\phi}' + 2\tilde{\phi} = R(x).$$

This residual is then diminished by being made orthogonal to the power basis functions:

$$[R(x), 1] = 0, \tag{3-33a}$$

$$[R(x), t] = 0, \tag{3-33b}$$

$$[R(x), t^2] = 0, \tag{3-33c}$$

and so forth. When n undetermined constants appear in the approximate solution, correspondingly n inner products are evaluated. We will omit the details of the procedure here since the method is no longer in widespread use and its mathematical formulation is fairly straightforward. Our intention in this section is to present a special case of the method of moments–evaluation of $[R(x), 1]$ alone. This is called the integral method [12–15]. Although its accuracy is limited, this method is able to obtain a reliable solution faster than other methods, especially for physical systems having similarity distributions of the field variables. This method is also a very popular topic in heat transfer courses. For these reasons, we will describe the procedure in the following example, keeping in mind that its main application lies in solving partial differential equations.

Example 3-8 Use the integral method to solve Eq. (1-1). Assume that

$$\tilde{\phi}(x) = \frac{x}{4} + e_1 \frac{x}{4}\left(1 - \frac{x}{4}\right). \tag{a}$$

Solution In light of Eq. (3-33*a*), we obtain

$$\int_0^4 (\tilde{\phi}'' + \tilde{\phi}' - 2\tilde{\phi})\, dx = 0. \tag{b}$$

Substitution of Eq. (*a*) into Eq. (*b*) yields

$$4\int_0^1 [-2e_1 + (1 + e_1 - 2e_1 t) - 2(t + e_1 t - e_1 t^2)]\, dt = 0, \tag{c}$$

where $t = x/4$. Therefore, the only undetermined constant is evaluated as

$$e_1 = -\frac{18}{11} \tag{d}$$

and the integral solution becomes

$$\tilde{\phi}(x) = \frac{x}{4} - \frac{18}{11}\left(\frac{x}{4}\right)\left(1 - \frac{x}{4}\right). \tag{e}$$

So far, we have considered more or less all the numerical methods associated with the evaluation of integral inner products. Although the illustration was limited to solving the simple ordinary differential equation (1-1), these methods are generally applicable to more complicated problems as well as to partial differential equations. At this point, however, we prefer not to continue with the application of these methods to more advanced problems, but will devote our efforts to other numerical techniques.

3-3 PERTURBATION METHOD

In the beginning of Chapter 1 we considered an ordinary differential equation that governed heat conduction and heat convection in a one-dimensional system. We also assigned certain specific numerical values to the dimensionless parameters such as uR/α and $h_c R/k$ so that the three terms in Eq. (1-1) are of the same order of magnitude. In some physical problems, however, either the velocity of the flow may be very low or the thermal conductivity of the fluid may be very poor, leading to, respectively, small or large values of the parameter uR/α (also called the Peclet number). For low flow velocities, the flow can be viewed as a stationary solid given a minute displacement. For fluids having poor thermal conductivity, on the other hand, the system can be treated as a thermally insulated one perturbed by a minor heat conduction effect. In either circumstance, it is appropriate to focus on the undisturbed system and then gradually add the disturbing effect to it. The analytical or numerical technique associated with this approach is the perturbation method.

We will describe this method by applying it to a one-dimensional heat transfer problem similar to that considered in the preceding chapters. Consider the following differential equation governing energy conservation over an infinitesimal control volume $[\bar{x}, \bar{x} + d\bar{x}]$

$$\alpha \frac{d^2T}{d\bar{x}^2} + u \frac{dT}{d\bar{x}} + a(T - T_0) = 0, \tag{3-34a}$$

subject to two one-point boundary conditions (or two initial conditions)

$$T(0) = T_0 \quad \text{and} \quad \left(\frac{\partial T}{\partial \bar{x}}\right)_{\bar{x}=0} = \frac{\dot{q}_0''}{k}, \tag{3-34b}$$

where a is defined as $2h_c\alpha/k$ and has dimension s^{-1}. The physical system and the derivation of the governing equation are similar to those described at the beginning of Chapteı 1. Depending on the relative strength of u and α, Eq. (3-34*a*) is classified as either a regular perturbation equation or a singular perturbation equation. The former can be solved rather easily and therefore will be considered first.

3-3*a* Regular Perturbation

When a small dimensionless parameter $\epsilon \ll \mathbf{O}(1)$ is embedded in the low-order derivatives, the equation is to be solved by the regular perturbation method. In reference to Eq. (3-34*a*), we assume that u is much smaller than $(\alpha a)^{1/2}$. For convenience, it is preferable to nondimensionalize Eqs. (3-34*a*) and (3-34*b*) into the following forms:

$$\frac{d^2\theta}{dx^2} + 2\epsilon \frac{d\theta}{dx} + \theta = 0, \tag{3-35}$$

subject to

$$\theta(0) = 0 \quad \text{and} \quad \left(\frac{d\theta}{dx}\right)_{x=0} = 1,$$

where

$$\theta = \frac{(T - T_0)\rho c_p(\alpha a)^{1/2}}{\dot{q}_0''}, \qquad x = \bar{x}\left(\frac{a}{\alpha}\right)^{1/2}$$

and

$$\epsilon = \frac{u}{2(\alpha a)^{1/2}} \ll \mathbf{O}(1).$$

We will assume that the function $\theta(x)$ can be expressed in terms of an asymptotic series† as

†A series is asymptotic if $\epsilon^n\theta(x) \gg \epsilon^{n+1}\theta_{n+1}(x) + \epsilon^{n+2}\theta_{n+2}(x) + \cdots$. It is noted that the choice of $\epsilon, \epsilon^2, \ldots, \epsilon^n$ is not arbitrary. The rationale for the choice can be found in [16, 17].

$$\theta(x) = \theta_0(x) + \epsilon\theta_1(x) + \epsilon^2\theta_2(x) + \cdots. \tag{3-36}$$

Substituting Eq. (3-36) into Eq. (3-35) leads to

$$(\theta_0'' + \epsilon\theta_1'' + \epsilon^2\theta_2'' + \cdots) + 2\epsilon(\theta_0' + \epsilon\theta_1' + \epsilon^2\theta_2'' + \cdots) + (\theta_0 + \epsilon\theta_1 + \epsilon^2\theta_2 + \cdots) = 0.$$

Collection of terms with the same power of ϵ yields

$$\theta_0'' + \theta_0 = 0 \quad \text{with } \theta_0(0) = 0 \quad \text{and } \theta_0'(0) = 1, \tag{3-37a}$$

$$\theta_1'' + \theta_1 = -2\theta_0' \quad \text{with } \theta_1(0) = 0 \quad \text{and } \theta_1'(0) = 0, \tag{3-37b}$$

and so forth. Equation (3-37*a*) is homogeneous with inhomogeneous initial conditions; Eq. (3-37*b*) and the remaining higher-order equations are inhomogeneous with homogeneous initital conditions. They can be analytically integrated to give

$$\theta_0(x) = \sin x \tag{3-38a}$$

and

$$\theta_1(x) = -x \sin x. \tag{3-38b}$$

Therefore, the solution of Eq. (3-35), with only two leading terms evaluated, can be written as

$$\theta(x) = \sin x - \epsilon x \sin x + O(\epsilon^2). \tag{3-39}$$

Both Eq. (3-39) and the exact solution

$$\theta_{\text{exact}}(x) = (1 - \epsilon^2)^{-1/2} \exp(-\epsilon x) \sin[(1 - \epsilon^2)^{1/2} x] \tag{3-40}$$

are plotted in Fig. 3-7 for comparison. Clearly, the asymptotic solution will become closer to the exact solution if more terms in the asymptotic series are used.

3-3*b* Singular Perturbation

When the small parameter appears in the highest-order derivative, the equation is to be solved by the singular perturbation method. For example, in reference to Eq. (3-34*a*) again, if the thermal conductivity of the fluid is very poor so that $\alpha \ll u^2/a$, then the nondimensionalized form of Eq. (3-34*a*)

$$\omega \frac{d^2\phi}{dx^2} + \frac{d\phi}{dx} + \phi = 0 \tag{3-41}$$

becomes a singular perturbation equation, subject to

$$\phi(0) = 0 \quad \text{and} \quad \left(\frac{d\phi}{dx}\right)_{x=0} = \frac{1}{\omega}, \tag{3-42a,b}$$

where

$$\phi = \frac{(T - T_0)\rho c_p u}{\dot{q}_0''}, \qquad x = \frac{a\bar{x}}{u},$$

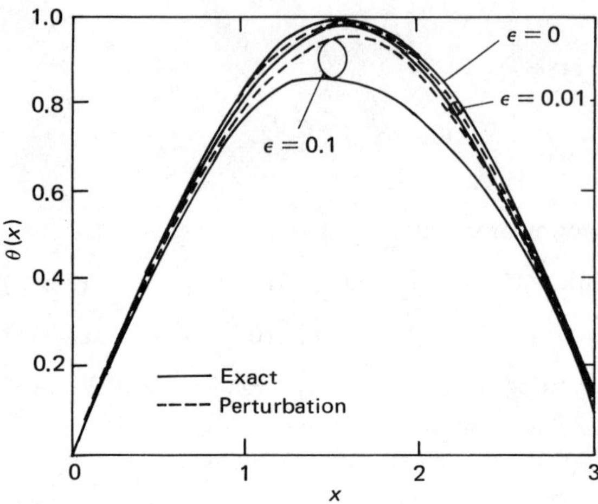

Figure 3-7 The regular perturbation solution and the exact solution of Eq. (3-35) for $\epsilon = 0, 0.01$, and 0.1.

and

$$\omega = \frac{\alpha a}{u^2} \ll \mathbf{O}(1).$$

Equation (3-41) is more difficult to solve than the regular perturbation equation (3-35) for the following reason. Since ω is a small number, the solution of Eq. (3-41) should be, according to our intuition, close to that of

$$\frac{d\phi}{dx} + \phi = 0 \qquad \text{as } \omega \to 0. \tag{3-43}$$

However, Eq. (3-43) is a differential equation of first order, which clearly must be different from the second-order differential equation (3-41). Furthermore, although the order of Eq. (3-43) is first, the number of initial conditions that it must satisfy is two. Mathematically, this inconsistency cannot be properly handled. Therefore, Eq. (3-41) must be solved with a special treatment in order to remove this difficulty.

To avoid the inconsistency that a first-order differential equation is subject to two initial conditions, we temporarily ignore these two conditions and proceed to solve Eq. (3-43), realizing that the solution obtained will not be valid in the vicinity of the boundary $x = 0$. The solution is called the outer-expansion (short for outer-region asymptotic expansion) solution, and involves some undetermined constants. We then modify Eq. (3-41) by normalizing the coordinate x on ω, so that the normalized domain is stretched and the parameter ω is shifted to the lower derivatives. Now the solution is made to satisfy the two initial conditions and is called the inner-expansion solution. By matching the inner and outer solutions around the interface, we thus obtain a composite expansion that is valid throughout the domain.

We will consider the outer-expansion solution first. Again, let

$$\phi(x) = \phi_0(x) + \omega\phi_1(x) + \omega^2\phi_2(x) + \cdots. \tag{3-44}$$

Substituting Eq. (3-44) into Eq. (3-41) and collecting terms with the same power of ω, we obtain

$$\phi_0' + \phi_0 = 0, \tag{3-45a}$$

$$\phi_1' + \phi_1 = -\phi_0'', \tag{3-45b}$$

and so forth. The analytical solutions of Eqs. (3-45*a*) and (3-45*b*) are found to be, respectively,

$$\phi_0 = c_0 e^{-x} \tag{3-46a}$$

and

$$\phi_1 = c_1 e^{-x} - c_0 x e^{-x}. \tag{3-46b}$$

The outer-expansion solution, with two leading terms, can be written as

$$\phi_{\text{outer}}(x) = c_0 e^{-x} + \omega(c_1 e^{-x} - c_0 x e^{-x}) + \mathbf{O}(\omega^2). \tag{3-47}$$

We will leave the constants c_0 and c_1 temporarily undetermined and proceed to seek the inner-expansion solution.

As mentioned earlier, the coordinate of the inner expansion should be made to be of $\mathbf{O}(1)$. This can be achieved by introducing

$$z = \frac{x}{\omega}$$

as the normalized coordinate. Consequently, Eq. (3-41) is transformed to

$$\frac{d^2\phi}{dz^2} + \frac{d\phi}{dz} + \omega\phi = 0 \tag{3-48}$$

subject to

$$\phi(0) = 0 \quad \text{and} \left(\frac{d\phi}{dz}\right)_{z=0} = 1. \tag{3-49a,b}$$

Equation (3-48) is recognized as a regular perturbation problem and therefore will be solved similarly to Eq. (3-35). Once again, let

$$\phi(z) = \phi_0(z) + \omega\phi_1(z) + \omega^2\phi_2(z) + \cdots \tag{3-50}$$

be the asymptotic series in the inner-expansion region. Substitution of Eq. (3-50) into Eq. (3-48) and collection of terms having the same power of ω lead to

$$\phi_0'' + \phi_0' = 0 \quad \text{with } \phi_0(0) = 0 \quad \text{and } \phi_0'(0) = 1 \tag{3-51a}$$

and

$$\phi_1'' + \phi_1' = -\phi_0 \quad \text{with } \phi_1(0) = 0 \quad \text{and } \phi_1'(0) = 0. \tag{3-51b}$$

The analytical solutions of Eqs. (3-51*a*) and (3-51*b*) are, respectively,

$$\phi_0 = 1 - e^{-z} \tag{3-52a}$$

and

$$\phi_1 = 2 - z - (2 + z)e^{-z}. \tag{3-52b}$$

Supposing that only these two leading terms are sufficiently accurate for the construction of the asymptotic series, we obtain the final inner-expansion solution

$$\phi_{\text{inner}}(z) = 1 - e^{-z} + \omega[2 - z - (2 + z)e^{-z}] + O(\omega^2). \tag{3-53}$$

Equation (3-53) can be used to determine the unknown constants c_0 and c_1 in Eqs. (3-46*a*) and (3-46*b*). Since the inner-expansion and outer-expansion solutions should match in the interface region, it follows that

$$\lim_{x\to 0} \phi_{\text{outer}}(x) = \lim_{z\to\infty} \phi_{\text{inner}}(z).$$

But, using the approximation $e^{-t} \approx 1 - t$ when $t \ll 1$, we obtain

$$\begin{aligned}\lim_{x\to 0} \phi_{\text{outer}}(x) &\approx c_0(1 - x) + \omega[c_1(1 - x) - c_0 x(1 - x)] \\ &= c_0 + \omega(-c_0 z + c_1) + O(\omega^2).\end{aligned} \tag{3-54a}$$

On the other hand, Eq. (3-53) can be approximated by

$$\lim_{z\to\infty} \phi_{\text{inner}}(z) \approx 1 + \omega(-z + 2) + O(\omega^2). \tag{3-54b}$$

Comparison of Eq. (3-54*a*) with Eq. (3-54*b*) gives $c_0 = 1$ and $c_1 = 2$. Therefore, the outer-expansion solution becomes

$$\phi_{\text{outer}}(x) = e^{-x} + \omega(2e^{-x} - xe^{-x}) + O(\omega^2). \tag{3-55}$$

Equation (3-55) is valid only for large x, while Eq. (3-53) represents the solution only for small x. It will be desirable to construct a composite solution that is uniformly applicable throughout the domain. To achieve this, we notice that the overlapping portion of Eq. (3-53) and Eq. (3-55) is

$$\phi_{\text{overlapping}}(z) = 1 + \omega(-z + 2). \tag{3-56}$$

If Eq. (3-56) is subtracted from the combination of the inner-expansion and outer-expansion solutions, the resulting function

$$\phi^*(z) = \phi_{\text{inner}}(z) + \phi_{\text{outer}}(\omega z) - \phi_{\text{overlapping}}(z) \tag{3-57}$$

should constitute a series solution that, although not necessarily asymptotic, satisfies Eq. (3-41) with an error of $O(\omega^2)$. Such a solution $\phi^*(x)$ is called the "composite-expansion solution" [18] and takes the final form

$$\phi^*(x) = (1 + 2\omega)(e^{-x} - e^{-x/\omega}) - \omega x e^{-x} + x e^{-x/\omega} + O(\omega^2). \tag{3-58}$$

The exact solution of Eq. (3-41) is found to be

$$\phi_{\text{exact}}(x) = \frac{2}{\sqrt{1 - 4\omega}} \exp\left(-\frac{x}{2\omega}\right) \sinh\left(\frac{x}{2\omega}\sqrt{1 - 4\omega}\right). \tag{3-59}$$

Both Eq. (3-58) and Eq. (3-59) are plotted in Fig. 3-8 for $\omega = 0.01$ and 0.1.

3-4 NONLINEAR TWO-POINT BOUNDARY-VALUE PROBLEMS

Up to now we have dealt with differential equations only in the linear form. These are useful for basic numerical studies but are slightly idealized. In countless practical

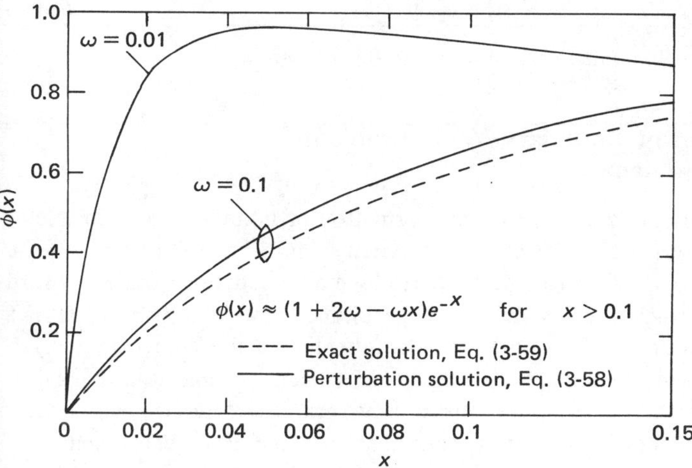

Figure 3-8 The singular perturbation solutions and the exact solution of Eqs. (3-41) and (3-42) for $\omega = 0.01$ and 0.1. For $\omega = 0.01$, the perturbation and exact solutions are indistinguishable.

heat transfer problems the governing equations are nonlinear. An example was given at the beginning of Chapter 1, when the thermal conductivity became temperature-dependent or the flow velocity and temperature were coupled. Nonlinear problems generally require iteration and testing for the convergence of the numerical scheme, and therefore are more difficult to handle. Since nonlinear problems are both common and difficult, numerical analysts have directed much effort toward seeking satisfactory methods.

Although there are probably a dozen numerical techniques, all the techniques are usually constructed on the basis of one or more of four fundamental strategies. These techniques may have various names, yet it is usually possible to decompose the procedures into a few elemental strategies. In the following sections, we will apply them to the model equation.

$$-A\phi = \frac{d^2\phi}{dx^2} + 2\phi\frac{d\phi}{dx} = 0 \tag{3-60}$$

subject to

$$\phi(0) = 1 \quad \text{and } \phi(4) = \tfrac{1}{5}. \tag{3-61a,b}$$

Equations (3-60) and (3-61) are derived from the heat transport problem given at the beginning of Chapter 1 with zero heat transfer coefficient h_c and with the flow velocity being proportional to the temperature. The exact solution of Eqs. (3-60) and (3-61) is

$$\phi_{\text{exact}}(x) = \frac{1}{1+x}, \tag{3-62}$$

which gives the exact missing initial condition (hereafter abbreviated as MIC) as

$$\phi'(0) = -1. \tag{3-63}$$

Finally, we will assume that Eq. (3-62) is the unique solution.

3-4*a* Transformation of Boundary-Value Problems into Initial-Value Problems

The first strategy to be considered is the transformation of boundary-value problems into initial-value counterparts. Once all $m-1$ initial conditions are prescribed, the differential equation order of m can be numerically integrated in a straightforward manner. These integration schemes, which range in accuracy from $\mathbf{O}(h)$ to $\mathbf{O}(h^5)$, where h is the integration step size, are well documented in the numerical analysis literature [19]. Some of them have already been written as subroutine packages, available to be called into a computer main program. Therefore, we will briefly present only two elementary integration schemes here, realizing that schemes of higher-order accuracy are based on similar principles and that the details of these schemes can be easily acquired in the literature.

Consider the Taylor's series of $\phi(x+h)$ expaned about $\phi(x)$ as

$$\phi(x+h) = \phi(x) + \phi'(x)h + \mathbf{O}(h^2), \tag{3-64a}$$

which leads to a forward finite-difference approximation (also known as the Euler method)

$$\phi'(x) = \frac{\phi(x+h) - \phi(x)}{h} + \mathbf{O}(h). \tag{3-64b}$$

Likewise,

$$\left(\frac{d^k\phi}{dx^k}\right)_{x+h} = \left(\frac{d^k\phi}{dx^k}\right)_x + \left(\frac{d^{k+1}\phi}{dx^{k+1}}\right)_x h + \mathbf{O}(h^2) \tag{3-64c}$$

or

$$\frac{d^{k+1}\phi}{dx^{k+1}} = \frac{(d^k\phi/dx^k)_{x+h} - (d^k\phi/dx^k)_x}{h} + \mathbf{O}(h), \qquad k = 1, 2, 3, \ldots. \tag{3-64d}$$

If a differential equation of order m, with $m-1$ initial conditions prescribed at $x=a$, is to be integrated, we may first evaluate $\phi^{(m)}(a)$ from the governing differential equation, find $\phi(a+h)$, $\phi'(a+h)$, $\phi''(a+h), \ldots, \phi^{(m-1)}(a+h)$ from Eqs. (3-64*a*) and (3-64*c*), evaluate $\phi^{(m)}(a+h)$ again from the differential equation, and then repeat the same procedure for $x = a + 2h$, $a + 3h$, etc.†

Example 3-9 Use the forward finite-difference scheme to integrate the initial-value problem

†Note also that in using an integration scheme it is customary to reduce an mth order differential equation into a set of m first-order differential equations.

$$\phi'' + 2\phi\phi' = 0 \tag{a}$$

subject to

$$\phi(0) = 1, \qquad \phi'(0) = -1. \tag{b}$$

Let the step size h be 0.1.

Solution See Table 3-7. Solutions at $x \geqslant 0.4$ can be obtained in a similar fashion. The exact solution to Eqs. (*a*) and (*b*) is also tabulated for comparison.

The forward finite-difference integration scheme is probably the most elementary of all algorithms. For those of higher-order accuracy such as the Runge-Kutta method [20], the Gear method [21-23], and the predictor-corrector method [19, pp. 386-393] the procedures appear more sophisticated, but the strategy remains the same in that the prediction of the forward-step values is based on the current values. We will now briefly describe the predictor-corrector method.

In the predictor-corrector method, we first expand ϕ_{n+1} and ϕ_{n-1} about ϕ_n in the Taylor's series and rearrange to obtain

$$\phi_{n+1} = \phi_{n-1} + 2h\phi'_n + \frac{h^3}{3}\phi'''_n, \tag{3-65a}$$

in which the subscript n denotes "at $x = nh$." The predicted value of ϕ_{n+1} is calculated by dropping the $\mathbf{O}(h^3)$ term. Next, we differentiate

$$\phi_{n+1} = \phi_n + h\phi'_n + \frac{h^2}{2}\phi''_n + \frac{h^3}{6}\phi'''_n \tag{3-65b}$$

with respect to x once, rearrange the result, and obtain

$$\phi'_n = \frac{1}{2}\phi'_n + \frac{1}{2}\phi'_{n+1} - \frac{h}{2}\phi''_n - \frac{h^2}{4}\phi'''_n - \frac{h^3}{12}\phi''''_n,$$

which is, again, substituted into Eq. (3-65*b*) to yield

Table 3-7 Finite-difference (FD) solution, predictor-corrector (PC) solution, and exact solution to the initial-value problem in Example 3-9

	ϕ		$\dfrac{d\phi}{dx}$		$\dfrac{d^2\phi}{dx^2}$	
x	FD	Exact (or PC)	FD	Exact (or PC)	FD	Exact (or PC)
0	1	1	−1	−1	2	2
0.1	0.90	0.9090	−0.80	−0.8264	1.440	1.5026
0.2	0.820	0.8333	−0.6560	−0.6944	1.0758	1.1574
0.3	0.7544	0.7692	−0.5484	−0.5917	0.8275	0.9103
0.4	0.6996	0.7143	−0.4657	−0.5102	0.6516	0.7288

$$\phi_{n+1} = \phi_n + \frac{h}{2}(\phi'_n + \phi'_{n+1}) - \frac{h^3}{12}\phi'''_n. \tag{3-65c}$$

The corrected value of ϕ_{n+1} is then calculated by dropping the $\mathbf{O}(h^3)$ term. Comparing Eq. (3-65*c*) with Eq. (3-65*a*), we find that the two values of ϕ_{n+1} thus computed differ by $5(h^3\phi'''_n/12)$ and that the true value of ϕ_{n+1} lies between the two values and differs from the corrected value by $h^3\phi'''_n/12$. Consider, for example, the following problem:

$$\phi' = \phi^2 + 1, \qquad \phi(0) = 0, \qquad x \geqslant 0.$$

Take $h = 0.2$. In order to compute ϕ_2 from Eq. (3-65*a*), we must be given ϕ_0 and ϕ'_1. By Taylor's series expansion, it follows that

$$\phi_1 = \phi_0 + h\phi'_0 + \frac{h^2}{2}\phi''_0 + \frac{h^3}{6}\phi'''_0 + \ldots = \phi_0 + h(\phi_0^2 + 1)$$

$$+ \frac{h^2}{2}(2\phi_0\phi'_0) + \frac{h^3}{6}(2\phi_0\phi''_0 + 2\phi_0'^2) + \ldots = h + \frac{h^3}{3} + \ldots$$

$$= 0.2027$$

and

$$\phi'_1 = \phi_1^2 + 1 = \left(h + \frac{h^3}{3}\right)^2 + 1 = 1.0411.$$

Using Eq. (3-65*a*) with the $\mathbf{O}(h^3)$ term, we thus predict

$$p_2 = \phi_0 + 2h\phi'_1 = 0.4164$$

and

$$p'_2 = p_2^2 + 1 = 1.1734.$$

Also, using Eq. (3.65*c*), we obtain the corrected value as

$$c_2 = \phi_1 + \frac{h}{2}(\phi'_1 + p'_2) = 0.4242.$$

The difference between p_2 and c_2 is then

$$c_2 - p_2 = 5\left(\frac{h^3\phi'''_1}{12}\right) = 0.0078.$$

Since the true value differs from c_2 by one-fifth of $c_2 - p_2$, it follows that

$$\phi_2 = c_2 - \frac{0.0078}{5} = 0.4226.$$

Using the predictor-corrector method, we solve Eqs. (3-60) and (3-61) and find that the solution is almost indistinguishable from the exact solution listed in Table 3-7. For application of the predictor-corrector method, see, for example, [24].

Knowing why it is relatively easy to solve initial-value problems, we now describe a few techniques for transforming boundary-value problems into initial-value problems. The most intuitive way is simply to guess the MICs. Using these randomly guessed MICs, we solve the initial-value problem, whose solution is in general different from that of the original boundary-value problem. This incorrect solution is then improved by some systematic operations such as interpolation and Taylor's series expansion. The method can be well illustrated by solving Eq. (3-60) and (3-61) in the following example.

Example 3-11 Transform Eqs. (3-60) and (3-61) into two initial-value problems by randomly guessing the MICs twice. Interpolate the computed terminal-point values to find the third improved MIC.

Solution Let the MIC be first chosen as zero. The incorrect solution is obtained as

$$\phi(x) = 1, \quad \phi'(x) = 0, \quad \phi''(x) = 0, \qquad \text{for all } x \in [0,4].$$

Next, let the MIC arbitrarily be -1.5. The forward finite-difference integration of Eq. (3-60) with $\phi(0) = 1$ and $\phi'(0) = -1.5$ is shown on Table 3-8. The terminal-point value is computed to be

$$\phi(4) = -1.5632. \tag{a}$$

These MICs and their corresponding terminal-point values can be used for the following interpolation:

Table 3-8 Solution of Eq. (3-60) subject to $\phi(0) = 1$ and $\phi'(0) = -1.5$ (by finite difference) and $\phi(0) = 1$ and $\phi'(0) = -0.4682$† (by predictor corrector)

x	$\phi(x)$	$\phi'(x)$	$\phi''(x)$	$\phi(x)$	$\phi'(x)$	$\phi''(x)$
0	1	−1.5	3	1.0000	−.4682	0.9364
0.2	0.7000	−0.9000	1.2600	0.8320	−.1562	0.2599
0.4	0.5200	−0.6480	0.6739	0.8115	−.1224	0.1986
0.6	0.3904	−0.5132	0.4007	0.7899	−.0877	0.1386
0.8	0.2878	−0.4331	0.2492	0.7748	−.0642	0.0995
1.0	0.2011	−0.3832	0.1542	0.7637	−.0471	0.0720
1.2	0.1245	−0.3524	0.0877	0.7555	−.0347	0.0525
1.4	0.0540	−0.3348	0.0362	0.7495	−.0257	0.0385
1.6	−0.0129	−0.3276	−0.0085	0.7451	−.0190	0.0284
1.8	−0.0785	−0.3293	−0.0517	0.7418	−.0141	0.0209
2.0	−0.1443	−0.3396	−0.0980	0.7393	−.0105	0.0155
2.2	−0.2123	−0.3592	−0.1525	0.7375	−.0078	0.0115
2.4	−0.2841	−0.3898	−0.2215	0.7361	−.0058	0.0085
2.6	−0.3621	−0.4340	−0.3143	0.7351	−.0043	0.0064
2.8	−0.4489	−0.4969	−0.4461	0.7344	−.0032	0.0047
3.0	−0.5482	−0.5861	−0.6427	0.7338	−.0024	0.0035
3.2	−0.6655	−0.7147	−0.9512	0.7334	−.0018	0.0026
3.4	−0.8084	−0.9049	−1.4630	0.7331	−.0013	0.0020
3.6	−0.9894	−1.1975	−2.3696	0.7329	−.0010	0.0015
3.8	−1.2289	−1.6714	−4.1079	0.7327	−.0007	0.0011
4.0	−1.5632	−2.4930	−7.7939	0.7326	−.0006	0.0008

†For $x \geqslant 6$, this wrong MIC yields $\phi(x) = 0.7322$, $\phi'(x) = 0$, and $\phi''(x) = 0$. Therefore, Eq. (3-60) with this MIC accidentally resembles a similarity equation governing a certain type of boundary-layer flows.

$$\frac{0-(-1.5)}{1-(-1.5632)} = \frac{0-\kappa}{1-0.2}, \tag{b}$$

where 0.2 is the true terminal-point value and κ is the third improved MIC, which is found to be

$$\kappa = -0.4682. \tag{c}$$

This MIC subsequently yields an improved terminal-point value 0.7326 as also shown in Table 3-8.

For nonlinear problems such as the one given here, iteration is generally needed to finally "shoot' the desired terminal target. Hence the technique is sometimes called the shooting method [25-27]. Also, if more than two MICs are present in a boundary-value problem, an interpolation incorporated with the Taylor's series expansion is needed. Such a shooting method will be illustrated in Chapter 11 by solving the similarity equation governing the free-convection boundary-layer flow over a vertical plate. For application of the shooting method to two-dimensional problems, see [28].

If the differential equations are originally linear or have somehow (see Section 3-4*b*) been linearized, we may use the principle of superposition to transform the boundary-value problem into a system of initial-value problems. Suppose

$$L\phi(x) = f(x), \quad \phi(a) = c_1, \quad \phi(b) = c_2, \quad a \leqslant x \leqslant b, \tag{3-66}$$

where L is a linear second-order differential operator. If $\phi_1(x)$ and $\phi_2(x)$ are, respectively, the solutions of two initial-value problems

$$L\phi_1(x) = 0, \quad \phi_1(a) = 0, \quad \phi_1'(a) = 1 \tag{3-67a}$$

and

$$L\phi_2(x) = f(x), \quad \phi_2(a) = c_1, \quad \phi_2'(a) = 0, \tag{3-67b}$$

then the superposition

$$\phi(x) = \phi_1(x)\phi'(a) + \phi_2(x) \tag{3-68}$$

can be established because of the linearity of Eq.(3-66). The unknown coefficient $\phi'(a)$, which is the true MIC of Eq. (3-66), can be obtained as

$$\phi'(a) = \frac{c_2 - \phi_2(b)}{\phi_1(b)}, \tag{3-69}$$

where $\phi_1(b)$ and $\phi_2(b)$ are the terminal-point values computed by a certain integration scheme. Once $\phi'(a)$ is obtained from Eq. (3-69), it is seen that Eq. (3-66) is already transformed into an initial-value problem.

Example 3-11 Consider Eq. (1-1), rewritten here for convenience:

$$\phi'' + \phi' - 2\phi = 0, \quad \phi(0) = 0, \quad \phi(4) = 1.$$

Find the missing initial condition $\phi'(0)$ by the principle of superposition.

Solution The two superposition functions $\phi_1(x)$ and $\phi_2(x)$ should satisfy, respectively,

$$\phi_1'' + \phi_1' - 2\phi_1 = 0, \quad \phi_1(0) = 0, \quad \phi_1'(0) = 1 \tag{a}$$

and

$$\phi_2'' + \phi_2' - 2\phi_2 = 0, \quad \phi_2(0) = 0, \quad \phi_2'(0) = 0. \tag{b}$$

Equation (*a*) is computed by using the forward finite-difference scheme, and its solution is listed in Table 3-9. Here we add that the use of such an inaccurate scheme for constructing this table is by no means a proper choice. Since our emphasis in Part I of this book is on numerical analysis and not computer programming, we intentionally present problems that are solvable with a desk calculator alone. In actual practice, it is recommended that the reader use more accurate integration schemes.

With regard to Eq. (*b*), it is a homogeneous equation with homogeneous initial conditions and therefore the solution vanishes everywhere. According to Eq. (3-69), we obtain

$$\phi'(0) = \frac{1}{\phi_1(4)} = \frac{1}{5}. \tag{c}$$

Although the differential equation considered in the example above is of second order, the superposition principle can be extended to linear problems of higher order.

3-4*b* Linearization

When all $m-1$ initial conditions of an mth order differential equation are specified a priori, the equation can be immediately integrated step by step regardless of the nonlinearity. When some initial conditions are missing but the equation is linear, the principle of superposition can be used to transform the boundary-value problem into a system of initial-value problems with little difficulty. When there exist some MICs and the equation is nonlinear, however, the situation becomes difficult. As mentioned in the preceding section, the shooting method can be used to solve a nonlinear boundary-value problem. Although the method is advantageous in that it

Table 3-9 Forward finite-difference solution for the superposition function

x	$\phi_1(x)$	$\phi_1'(x)$	$\phi_1''(x)$
0	0	1	−1
1	1	0	2
2	1	2	0
3	3	2	4
4	5	6	4

requires little mathematical analysis and programming preparation, its solution sometimes fails to converge to the true one when the problems are sensitive to the initial conditions. In some cases even slight changes in the initial conditions grow into machine overflow. Therefore, other techniques are called for that do not require integration of the nonlinear differential equation from the initial point.

Linearization of the differential equation is an alternative to the shooting method. The most conventional linearization scheme is the Taylor's series expansion. In this scheme, we start with a certain guessed solution that satisfies the boundary conditions, and expand the nonlinear terms about this guessed solution.

Example 3-12 Linearize Eq. (3-60) by Taylor's series expansion.

Solution Let $f(\phi, \phi')$ be a many-times differentiable function of ϕ and ϕ'. Then a Taylor's series expansion gives

$$f(\phi, \phi') = f(\bar{\phi}, \bar{\phi}') + \left(\frac{\partial f}{\partial \phi}\right)_{\bar{\phi},\bar{\phi}'}(\phi - \bar{\phi}) + \left(\frac{\partial f}{\partial \phi'}\right)_{\bar{\phi},\bar{\phi}'}(\phi' - \bar{\phi}') + \cdots, \tag{a}$$

where the overbar denotes the initially guessed solution or, in general, the previously iterated solution. With the aid of Eq. (*a*), it follows that

$$\phi\phi' \approx \bar{\phi}\bar{\phi}' + \bar{\phi}'(\phi - \bar{\phi}) + \bar{\phi}(\phi' - \bar{\phi}') = -\bar{\phi}\bar{\phi}' + \bar{\phi}\phi' + \bar{\phi}'\phi. \tag{b}$$

With Eq. (*b*), immediately Eq. (3-60) becomes a standard linear equation

$$L\phi = f, \tag{c}$$

where

$$L = \frac{d^2}{dx^2} + 2\bar{\phi}\frac{d}{dx} + 2\bar{\phi}'$$

and

$$f = 2\bar{\phi}\bar{\phi}'. \tag{d}$$

Once a differential equation has been linearized, we can use the superposition (described in Section 3-4*a*) that transforms the boundary-value problem into an initial-value problem or the discretization (to be presented in Section 3-4*c*) that transforms the boundary-value problem into a set of nonlinear algebraic equations.

Another linearization technique is called parametric differentiation [29-31]. Associated with this scheme, an artificial parameter λ must be embedded in the differential equation. The purpose of embedding λ is to find a starting solution at $\lambda = 0$ and then relate this solution to the genuine solution at $\lambda = 1$ [32]. Consider the following nonlinear differential equation:

$$\phi^{(m)} + B[\phi^{(m-1)}, \phi^{(m-2)}, \ldots, \phi', \phi] = 0, \qquad m = 2, 3, \ldots \tag{3-70}$$

where B is a nonlinear function of low-order derivatives, whose boundary conditions will be assumed to be prescribed either at the initial point or at the terminal point. Placing the artificial parameter λ in front of B, we obtain

$$\phi^{(m)} + \lambda B = 0. \tag{3-71}$$

At $\lambda = 0$, Eq. (3-71) reduces to

$$\phi^{(m)} = 0, \tag{3-72}$$

which can be analytically integrated with ease. Consequently, $\phi(x, \lambda = 0)$ can serve as the known starting solution. We then attempt to relate $\phi(x, \lambda = 0)$ to $\phi(x, \lambda = 1)$, which is the true solution that we seek. First, we differentiate Eq. (3-71) with respect to λ (not x). The resulting expression

$$\psi^{(m)} + B + \lambda \frac{\partial B}{\partial \lambda} = 0 \tag{3-73}$$

becomes a linear differential equation in ψ, where $\psi = \partial\phi/\partial\lambda$. The boundary conditions are homogeneous if the original ones of Eq. (3-70) are free of λ. The linearity in Eq. (3-73) arises because the values of ϕ and its lower-order derivatives are known [e.g., at $\lambda = 0$ these values are obtained by analytically integrating Eq. (3-72)]. The procedure for differentiating Eq. (3-71) into Eq. (3-73) is called parametric differentiation. The following examples will further illustrate this procedure.

Example 3-13 Linearize

$$\phi'' + 2\lambda\phi\phi' = 0 \tag{a}$$

by parametric differentiation.

Solution Let $\psi(x, \lambda)$ denote $\partial\phi(x, \lambda)/\partial\lambda$. Differentiating Eq. (*a*) with respect to λ leads to

$$\psi'' + 2\phi\phi' + 2\lambda(\phi\psi' + \phi'\psi) = 0 \tag{b}$$

or

$$L\psi = f, \tag{c}$$

where

$$L = \frac{d^2}{dx^2} + 2\lambda\phi\frac{d}{dx} + 2\lambda\phi' \quad \text{and } f = -2\phi\phi'.$$

The procedure that relates $\phi(x, 0)$ to $\phi(x, 1)$ remains to be explained. The function $\phi(x, \lambda + \Delta\lambda)$ can be expanded about $\phi(x, \lambda)$ in Taylor's series as

$$\phi(x, \lambda + \Delta\lambda) = \phi(x, \lambda) + \frac{\partial\phi}{\partial\lambda}\Delta\lambda + \frac{\partial^2\phi}{\partial\lambda^2}\frac{(\Delta\lambda)^2}{2} + \cdots. \tag{3-74}$$

If only the two leading terms in Eq. (3-74) are taken (retaining higher-order terms is described in Chapter 6), we obtain

$$\phi(x, \lambda + \Delta\lambda) = \phi(x, \lambda) + \psi(x, \lambda)\Delta\lambda, \tag{3-75}$$

which is essentially the result of forward finite difference. Clearly, Eq. (3-75) can be repeatedly used for $\lambda = 0, \Delta\lambda, 2\Delta\lambda, \ldots, 1$ to eventually yield $\phi(x, 1)$, provided the linear equation (3-73) is somehow solved. Since the superposition scheme was described in Section 3-4*a*, we will use it here.

Example 3-14 Use both parametric differentiation and the principle of superposition to solve the following parameterized version of Eq. (3-60):

$$\phi'' + 2\lambda\phi\phi' = 0, \quad \phi(0) = 1, \quad \phi(4) = 0.2 \tag{a}$$

for $\lambda = 0.2$ and $\Delta\lambda = 0.1$.

Solution The starting solution at $\lambda = 0$ is

$$\phi(x, 0) = 1 - 0.2x. \tag{b}$$

Then at $\lambda = 0$ Eq. (*b*) in Example 3-13 becomes

$$\psi'' - 0.4(1 - 0.2x) = 0 \tag{c}$$

subject to the homogeneous boundary conditions

$$\psi(0,0) = \psi(4,0) = 0. \tag{d}$$

Therefore, solving Eqs. (*c*) and (*d*) yields

$$\psi(x, 0) = -\frac{0.04}{3}x^3 + 0.2x^2 - \frac{1.76}{3}x. \tag{e}$$

Then, using Eqs. (3-75), (*b*), and (*e*), we obtain

$$\phi(x, 0.1) = 1 - \frac{0.776}{3}x + 0.02x^2 - \frac{0.004}{3}x^3. \tag{f}$$

At $\lambda = 0.1$, Eq. (*c*) in Example 3-13 now becomes

$$L\psi = \psi'' + 0.2\phi(x, 0.1)\psi' + 0.2\phi'(x, 0.1)\psi = -2\phi(x, 0.1)\phi'(x, 0.1) \tag{g}$$

subject to $\qquad \psi(0, 0.1) = \psi(4, 0.1) = 0.$

From the superposition principle given in Eqs. (3-67*a*) and (3-67*b*), we obtain

$$L\psi_1 = 0, \quad \psi_1(0, 0.1) = 0, \quad \psi_1'(0, 0.1) = 1 \tag{h}$$

and $\quad L\psi_2 = -2\phi(x, 0.1)\phi'(x, 0.1), \quad \psi_2(0, 0.1) = \psi_2'(0, 0.1) = 0. \qquad (i)$

Being initial-value problems, Eqs. (*h*) and (*i*) can be directly integrated by using an integration scheme (we use the first-order-accurate forward-difference scheme with $\Delta x = 1$ here for clarity of presentation) as

x	ψ_1	ψ_1'	ψ_1''	ψ_2	ψ_2'	ψ_2''	ψ
0	0	1	−0.2	0	0	0.5173	0
1	1	0.8	−0.0771	0	0.5173	0.2599	−0.6863
2	1.8	0.7229	−0.0097	0.5173	0.7772	0.1493	−0.7184
3	2.5229	0.7132	0.0356	1.2945	0.9265	0.1054	−0.4376
4	3.2361	0.7485	0.0753	2.2210	1.0319	0.0758	0

From Eq. (3-69), we calculate

$$\psi'(0, 0.1) = -\frac{\psi_2(4, 0.1)}{\psi_1(4, 0.1)} = -0.6863. \tag{j}$$

Consequently, Eq. (3-68) becomes

$$\psi(x, 0.1) = -0.6863\psi_1(x, 0.1) + \psi_2(x, 0.1) \tag{k}$$

and thus, at $\lambda = 0.2$, Eq. (3-75) can be rewritten as

$$\phi(x, 0.2) = \phi(x, 0.1) + 0.1\psi(x, 0.1), \tag{l}$$

which yields

$$\phi(1, 0.2) = 0.6914, \quad \phi(2, 0.2) = 0.4802$$

$$\phi(3, 0.2) = 0.3242, \quad \text{and } \phi(4, 0.2) = 0.2.$$

The procedure in the example above can be repeated until $\lambda = 1$, at which point Eq. (3-60) is recovered and the solution $\phi(x, 1)$ is exactly the true solution.

3-4*c* Transformation of Nonlinear Differential Equations into Nonlinear Algebraic Equations

It is not absolutely necessary that differential equations be linearized first. Instead, we may directly transform the differential equations into a set of algebraic equations. Such a transformation can be achieved by using the finite-difference method or the Galerkin method. These procedures are similar to those employed for linear differential equations in the preceding chapters.

Example 3-15 Use the central finite-difference method to transform Eq. (3-60) into a set of nonlinear algebraic equations.

Solution We substitute the approximations

$$\frac{d^2\phi}{dx^2} = \frac{\phi_{j-1} - 2\phi_j + \phi_{j+1}}{h^2} \tag{a}$$

and

$$\frac{d\phi}{dx} = \frac{\phi_{j+1} - \phi_{j-1}}{2h} \tag{b}$$

into Eq. (3-60) to obtain

$$\phi_{j-1} - 2\phi_j + \phi_{j+1} + \phi_j\phi_{j+1} - \phi_j\phi_{j-1} = 0, \quad j = 1, 2, 3 \tag{c}$$

where h is assumed to be equal to 1.

Example 3-16 Use the Galerkin finite-element method to transform Eq. (3-60) into a set of nonlinear algebraic equations. Let the domain be subdivided into four linear elements.

Solution The inner product

$$(A\tilde{\phi}, N_j) = \int_{x_{j-1}}^{x_{j+1}} (\tilde{\phi}' N_j' - 2\tilde{\phi}\tilde{\phi}' N_j)\, dx \qquad (a)$$

can be simplified, after lengthy but straightforward algebra, into

$$-\phi_{j-1} + 2\phi_j - \phi_{j+1} + \frac{1}{3}\phi_{j-1}^2 - \frac{1}{3}\phi_{j+1}^2 + \frac{1}{3}\phi_j\phi_{j-1} - \frac{1}{3}\phi_j\phi_{j+1} = 0,$$

$$j = 1, 2, 3. \qquad (b)$$

Whenever lengthy algebra has been performed, it is advisable to find some quick way, if possible, to confirm the final result. Here we will attempt to check the correctness of Eq. (*b*) by a simple test. Let

$$\phi(x) = x, \quad x_{j-1} = 0, \quad \text{and } x_{j+1} = 2. \qquad (c)$$

Then, substituting Eq. (*c*) into Eq. (*a*), we obtain

$$(A\tilde{\phi}, N_j) = \int_0^1 (1 - 2x^2)\, dx + \int_1^2 [-1 - 2x(2 - x)]\, dx = -2. \qquad (d)$$

On the other hand, according to Eq. (*c*), the nodal values are

$$\phi_{j-1} = 0, \quad \phi_j = 1, \text{ and } \phi_{j+1} = 2. \qquad (e)$$

Substituting Eq. (*e*) into Eq. (*b*) leads to

$$-0 + 2 - 2 + 0 - \frac{1}{3} \times 4 + 0 - \frac{1}{3} \times 2 = -2, \qquad (f)$$

which has the same value as Eq. (*d*). We thus gain confidence in the correctness of Eq. (*b*).

The two examples above demonstrate the transformation of a differential equation into a system of algebraic equations by discretization. There is still another transformation scheme that does not discretize the differential equation. We recall that in Section 2-1*a* the approximate solution was expanded in terms of a set of global basis functions. Clearly, for nonlinear differential equations, the Galerkin method with global basis functions remains applicable. Using this method, we are able to transform the differential equation into a set of algebraic equations containing undetermined coefficients.

Example 3-17 The approximate solution $\tilde{\phi}(x)$ can be written as

$$\tilde{\phi}(x) = 1 - \frac{x}{4} + e_1 v_1(x) + e_2 v_2(x), \quad x \epsilon [0,4] \qquad (a)$$

where $$v_0 = 1 - \frac{x}{4}$$

and

$$v_k = \frac{x}{4}\left[1 - \left(\frac{x}{4}\right)^k\right], \qquad k = 1, 2. \tag{b}$$

Use the Galerkin method to transform Eq. (3-60) into two nonlinear algebraic equations.

Solution The two inner products are

$$(A\tilde{\phi}, v_1) = \int_0^4 (\tilde{\phi}' v_1' - 2\tilde{\phi}\tilde{\phi}' v_1)\, dx = 0 \tag{c}$$

and

$$(A\tilde{\phi}, v_2) = \int_0^4 (\tilde{\phi}' v_2' - 2\tilde{\phi}\tilde{\phi}' v_2)\, dx = 0. \tag{d}$$

Substituting Eq. (a) into Eqs. (c) and (d) gives two algebraic equations:

$$a_1 e_1 + a_2 e_2 - 2a_{12} e_1 e_2 - 2a_{11} e_1^2 - 2a_{22} e_2^2 + \tfrac{1}{4}\, b = 0, \tag{e, f}$$

where

$$a_1 = \int_0^4 (v_1' v_k' - 2v_0 v_1' v_k + \tfrac{1}{2}\, v_1 v_k)\, dx, \qquad k = 1, 2,$$

$$a_2 = \int_0^4 (v_2' v_k' - 2v_0 v_2' v_k + \tfrac{1}{2}\, v_2 v_k)\, dx,$$

$$a_{12} = \int_0^4 (v_1 v_k v_2' + v_2 v_k v_1')\, dx,$$

$$a_{11} = \int_0^4 v_1 v_1' v_k\, dx,$$

$$a_{22} = \int_0^4 v_1 v_2' v_k\, dx,$$

and

$$b = \int_0^4 (2v_0 v_k - v_k')\, dx.$$

The main difference between Eq. (e) in this example and Eq. (b) in the preceding example is that the unknowns in the former are undetermined coefficients, whereas those in the latter are nodal functions. The two sets of algebraic equations can be linearized by the techniques described in Section 3-4b. These linearized equations are then arranged into matrix systems and are solved by iteration techniques.

3-4*d* Iteration of Algebraic Systems

In this section we will begin by stating that an algebraic system reduced by linearization first and then discretization is the same as that reduced by discretization first and the linearization. While a rigorous proof will not be attempted here, we can present an example to illustrate this statement.

Example 3-18 Consider the nonlinear term $\phi\phi'$: (a) Linearize it by using Taylor's series expansion and then discretize the linearized result by using the Galerkin method. (b) Discretize it by using the Galerkin method and then linearize the discretized result by using Taylor's series expansion.

Solution (a) For convenience, we copy Eq. (b) of Example 3-12:

$$\phi\phi' = -\bar{\phi}\bar{\phi}' + \bar{\phi}\phi' + \bar{\phi}'\phi. \tag{a}$$

To discretize Eq. (a) by the Galerkin method, we evaluate the inner products

$$(\bar{\phi}\bar{\phi}', N_j) = \tfrac{1}{6}\,(-\bar{\phi}_{j-1}^2 + \bar{\phi}_{j+1}^2 - \bar{\phi}_j\bar{\phi}_{j-1} + \bar{\phi}_j\bar{\phi}_{j+1}), \tag{b}$$

$$(\bar{\phi}\phi', N_j) = \tfrac{1}{6}\,(-\bar{\phi}_{j-1}\phi_{j-1} + \bar{\phi}_{j+1}\phi_{j+1} + \bar{\phi}_{j-1}\phi_j - 2\bar{\phi}_j\phi_{j-1} - \bar{\phi}_{j+1}\phi_j + 2\bar{\phi}_j\phi_{j+1}), \tag{c}$$

and

$$(\bar{\phi}'\phi, N_j) = \tfrac{1}{6}\,(-\bar{\phi}_{j-1}\phi_{j-1} + \bar{\phi}_{j+1}\phi_{j+1} + \bar{\phi}_j\phi_{j-1} - 2\bar{\phi}_{j-1}\phi_j - \bar{\phi}_j\phi_{j+1} + 2\bar{\phi}_{j+1}\phi_j). \tag{d}$$

Combining Eqs. (b)-(d) thus gives

$$6(\phi\phi', N_j) = (-2\bar{\phi}_{j-1} - \bar{\phi}_j)\phi_{j-1} + (-\bar{\phi}_{j-1} + \bar{\phi}_{j+1})\phi_j + (2\bar{\phi}_{j+1} + \bar{\phi}_j)\phi_{j+1} + \bar{\phi}_{j-1}^2 - \bar{\phi}_{j+1}^2 + \bar{\phi}_j(\bar{\phi}_{j-1} - \bar{\phi}_{j+1}). \tag{e}$$

(b) Equation (b) of Example 3-16 can be utilized. We write

$$(\phi\phi', N_j) = \tfrac{1}{6}\,(-\phi_{j-1}^2 + \phi_{j+1}^2 - \phi_j\phi_{j-1} + \phi_j\phi_{j+1}). \tag{f}$$

Next, the nonlinear terms in Eq. (f) can be linearized, respectively, into

$$\phi_{j-1}^2 = -\bar{\phi}_{j-1}^2 + 2\bar{\phi}_{j-1}\phi_{j-1}, \tag{g}$$

$$\phi_{j+1}^2 = -\bar{\phi}_{j+1}^2 + 2\bar{\phi}_{j+1}\phi_{j+1}, \tag{h}$$

$$\phi_j\phi_{j-1} = -\bar{\phi}_j\bar{\phi}_{j-1} + \bar{\phi}_j\phi_{j-1} + \bar{\phi}_{j-1}\phi_j, \tag{i}$$

and

$$\phi_j\phi_{j+1} = -\bar{\phi}_j\bar{\phi}_{j+1} + \bar{\phi}_j\phi_{j+1} + \bar{\phi}_{j+1}\phi_j. \tag{j}$$

Substituting Eqs. (*g*)-(*j*) into Eq. (*f*), we obtain the same result as Eq. (*e*). Although the sequence of linearization and discretization is immaterial, we prefer, in most problems, to follow the sequence in case (*b*), namely discretization first and then linearization, since it is algebraically simpler.

Now we will briefly describe the procedure for solving the set of linearized (to be distinguished from linear) algebraic equations reduced from Eq. (3-60). After the discretization by central finite difference and the linearization by Taylor's series expansion with $h = 1$, the final result can be written in matrix form as

$$[\mathbf{A}]\,\{\phi\} = \{\mathbf{B}\}, \tag{3-76}$$

where

$$\mathbf{A} = \begin{bmatrix} -2 + \bar{\phi}_2 - \bar{\phi}_0 & 1 + \bar{\phi}_1 & 0 \\ 1 + \bar{\phi}_2 & -2 + \bar{\phi}_3 - \bar{\phi}_1 & 1 + \bar{\phi}_2 \\ 0 & 1 - \bar{\phi}_3 & -2 + \bar{\phi}_4 - \bar{\phi}_2 \end{bmatrix},$$

$$\{\phi\}^T = \{\phi_1 \quad \phi_2 \quad \phi_3\},$$

and

$$\{\mathbf{B}\}^T = \{\bar{\phi}_1\bar{\phi}_2 - \phi_0 \quad \phi_2(\bar{\phi}_3 - \bar{\phi}_1) \quad -\bar{\phi}_2\bar{\phi}_3 - \phi_4\}.$$

To solve Eq. (3-76), we guess an initial solution for $\phi(x)$ (preferably one that satisfies the boundary conditions), calculate [**A**] and {**B**}, solve for {ϕ}, and use these updated values of {ϕ} to calculate [**A**] and {**B**} for iteration. The procedure is applied to momentum and energy transport problems in Chapters 8 and 11.

For matrix systems of moderate size, the accepted solution algorithms can generally be categorized into direct methods and iterative methods. Among the former, the Gauss elimination (or its modified version Gauss-Jordan elimination which does not require back substitution) and the triangular decomposition (or its special case Cholesky decomposition for symmetric and positive-definite matrices) are most widely used. Among the latter, the Gauss-Seidel method (or its modified version successive over-relaxation) is most popular. See Chapters 4 and 5 and [33-35] for details.

Conventionally systems of nonlinear equations are solved by the Newton-Raphson method (or its modified version wherein the Jacobian is held constant for a fixed number of iterations). This method converges rapidly if the initial guess lies within the vicinity of the solution. It will diverge, however, if the starting solution cannot be well guessed. Alternatively, the method of steepest descent which converges (but slowly) from any starting guess can be used initially. Then it is replaced by the Newton-Raphson method when the true solution is approached. For details, see [36, 37].

SYMBOLS

a	a convenient term, s^{-1} $(= 2h_c\alpha/k)$
C^k	space in which functions are k-time differentiable
h	uniform mesh size $[= x_j - x_{j-1}$ for linear and cubic (Hermitian) elements and $x_{j+1} - x_{j-1}$ for quadratic (Lagrangian) elements]
h_c	heat transfer coefficient, $W/m^2 \cdot K$
k	thermal conductivity, $W/m \cdot K$
m_j	first derivative at grid point $j\,[= (d\phi/dx)_{x=x_j}]$
$N_j(x)$	linear or cubic piecewise basis functions
$P_a(x), P_b(x)$	cubic piecewise basis functions corresponding to m_a and m_b
$\dot{q}''_0$	heat flux at $x = 0$, W/m^2
t	isoparametric coordinate $\{= (x - x_a)/(x_c - x_a),\ x \epsilon [x_a, x_c]$, or $(x - x_a)/(x_b - x_a),\ x \epsilon [x_a, x_b]\}$
u	uniform flow velocity in x direction, m/s
$\bar{x}$	physical coordinate, m
x	dimensionless coordinate in regular perturbation $[= \bar{x}(a/\alpha)^{1/2}]$ or coordinate in singular perturbation $(= a\bar{x}/u)$
$W_j(x)$	quadratic piecewise basis function
z	inner-expansion coordinate $(= x/\omega)$
ϕ	field variable or normalized temperature $[= (T - T_0)\rho c_p u/\dot{q}''_0]$
$\Lambda_{d(j-1),j}$	convenient expression for integrals $[\int_{x_{j-1}}^{x_j} (dW_{j-1}/dx) W_j\, dx]$
α	thermal diffusivity, m^2/s $(= k/\rho c_p)$
ϵ	regular perturbation parameter $[= u/z(\alpha a)^{1/2}]$
θ	normalized temperature $[= (T - T_0)\rho c_p(\alpha a)^{1/2}/\dot{q}''_0]$
λ	artificially embedded parameter varying from 0 to 1
$\psi(x)$	a particular solution in polynomial form
ω	singular perturbation parameter $(= \alpha a/u^2)$

Subscripts

a,b	at grid points $x = x_a, x_b$

Superscripts

$(A), (B)$	in element (A) and element (B)
~	approximation of a function
+	right-side element to grid point j
—	left-side element to grid point j
–	overbar, in the element immediately adjacent to the boundary points or previously iterated values

REFERENCES

1. P. Bettess, Infinite Elements, *Int. J. Numer. Methods Eng.*, vol. 11, pp. 53–64, 1977.
2. T. C. Chawla and S. H. Chan, Spline Collocation Solution of Combined Radiation-Convec-

tion in Thermally Developing Flows with Scattering, *Numer. Heat Transfer,* vol. 3, pp. 47–65, 1980.
3. B. M. Herbst, Collocation Methods and the Solution of Conduction-Convection Problems, *Int. J. Numer. Methods Eng.*, vol. 17, pp. 1093–1101, 1981.
4. B. M. Irons and A. Razzaque, Experience with the Patch Test for Convergence of Finite Element, in A. K. Aziz (ed.), *The Mathematical Foundations of the Finite Element Method with Applications to Partial Differential Equations,* pp. 557–587, Academic, New York, 1972.
5. G. Strang and G. J. Fix, *An Analysis of the Finite Element Method,* Prentice-Hall, Englewood Cliffs, N.J., pp. 174–181, 1973.
6. A. R. Mitchell and R. Wait, *The Finite Element Method in Partial Differential Equations,* Wiley-Interscience, London, pp. 166–173, 1977.
7. F. Strummel, The Limitations of the Patch Test, *Int. J. Numer. Methods Eng.*, vol. 15, pp. 177–188, 1980.
8. B. F. de Veubeke, Variational Principles and the Patch Test, *Int. J. Numer. Methods Eng.*, vol. 8, pp. 783–801, 1974.
9. G. Sander and P. Beckers, The Influence of the Choice of Connectors in the Finite Element Method, *Int. J. Numer. Methods Eng.*, vol. 11, pp. 1491–1505, 1977.
10. H. Yamada, An Approximate Method of Integration of Laminar Boundary-Layer Equations (in Japanese), *Rep. Res. Inst. Fluid Eng.*, vol. 4, pp. 27–42, 1950.
11. B. A. Kader and A. A. Gukhman, Turbulent Mass Transfer with a First-Order Chemical Reaction on a Wall at Pr $\gg$ 1, *Int. J. Heat Mass Transfer,* vol. 20, pp. 1339–1350, 1977.
12. K. Pohlhausen, The Approximate Integration of the Differential Equation for the Laminar Boundary Layer, *Z. Angew. Math. Mech.*, vol. 1, pp. 252–268, 1921.
13. T. von Karman, Über laminare und turbulente reibung, *Z. Angew. Math. Mec.*, vol. 1, pp. 233–252, 1946.
14. J. Karnis and V. Pechoc, The Thermal Laminar Boundary Layer on a Continuous Cylinder, *Int. J. Heat Mass Transfer,* vol. 21, pp. 43–47, 1978.
15. J. Sucec, Extension of a Modified Integral Method to Boundary Conditions of Prescribed Surface Heat Flux, *Int. J. Heat Mass Transfer,* vol. 22, pp. 771–774, 1979.
16. M. van Dyke, *Perturbation Method in Fluid Mechanics,* pp. 5 and 28–30, Academic, New York, 1964.
17. A. H. Nayfeh, *Perturbation Methods,* pp. 16–19, Wiley-Interscience, New York, 1973.
18. G. E. Lath, Singular Perturbation Problems, Ph.D. dissertation, California Inst. of Technology, Pasadena, Calif., 1951.
19. E. Isaacson and B. H. Keller, *Analysis of Numerical Methods,* p. 402, Wiley, New York, 1966.
20. C. Runge and W. Kutta, Beitrag zur näherungsweisen Integration totaler Differentialgleichungen, *Zeit. Math. Phys.*, vol. 46, pp. 435–453, 1901.
21. C. W. Gear, *Numerical Initial Problems in Ordinary Differential Equations,* Prentice-Hall, Englewood Cliffs, N.J., 1971.
22. A. C. Hindmarsh, Gear: O.D.E. System Solver, UCID-30001, rev. 2, Lawrence Livermore Lab., Livermore, Calif., 1972.
23. D. H. Cho and M. Epstein, Laminar Film Condensation of Flowing Vapor on a Horizontal Melting Surface, *Int. J. Heat Mass Transfer,* vol. 20, pp. 23–30, 1977.
24. O. A. Plumb and J. C. Huenefeld, Non-Darcy Natural Convection from Heated Surfaces in Saturated Porous Media, *Int. J. Heat Mass Transfer,* vol. 24, pp. 765–768, 1981.
25. S. M. Roberts and J. S. Shipman, *Two-Point Boundary Value Problems: Shooting Methods,* American Elsevier, New York, 1972.
26. Z. Aktas and H. J. Stetter, A Classification and Survey of Numerical Methods for Boundary Value Problems in Ordinary Differential Equations, *Int. J. Numer. Methods Eng.*, vol. 11, pp. 771–796, 1977.
27. A. Granas, R. B. Guenther, and J. W. Lee, The Shooting Method for the Numerical Solution of a Class of Nonlinear Boundary Value Problems, *SIAM (Soc. Ind. Appl. Math.) J. Numer. Anal.*, vol. 16, pp. 828–836, 1979.

28. P. J. Roache, Marching Methods for Elliptic Problems: Part I, *Numer. Heat Transfer,* vol. 1, pp. 1-25, 1978.
29. P. E. Rubbert and M. T. Landahl, Solution of Nonlinear Flow Problems through Parametric Differentiation, *Phys. Fluids,* vol. 10, pp. 831-835, 1967.
30. C. W. Tan and R. DiBano, A Parametric Study of Falkner-Skan Problem with Mass Transfer, *AIAA J.,* vol. 10, pp. 923-925, 1972.
31. T. Y. Na and I. S. Habib, Solution of the Natural Convection Problem by Parametric Differentiation, *Int. J. Heat Mass Transfer,* vol. 17, pp. 457-459, 1974.
32. T. M. Shih, A Method to Solve Two-Point Boundary-Value Problems in Boundary-Layer Flows or Flames, *Numer. Heat Transfer,* vol. 2, pp. 177-191, 1979.
33. R. S. Varga, *Matrix Iterative Analysis,* pp. 61-68, Prentice-Hall, Englewood Califfs, N.J., 1962.
34. L. Fox, An Introduction to Numerical Linear Algebra, Oxford University Press, Belfast, 1964.
35. L. A. Hageman and D. M. Young, Applied Iterative Methods, Academic, New York, 1981.
36. J. M. Ortega and W. C. Rheinboldt, Iterative Solution of Nonlinear Equations in Several Variables, Academic, New York, 1970.
37. A. Ralston and P. Rabinowitz, A First Course in Numerical Analysis, McGraw-Hill, New York, 1978.

PROBLEMS

3-1 Find the quadratic interpolant on [0,1] that is expressed in terms of ϕ_a, m_b, and ϕ_c, where x_a, x_b, and x_c are matched with $t = 0$, α, and 1, respectively; $0 \leqslant \alpha \leqslant 1$. What happens if α is set to 0, $\frac{1}{2}$, or 1?

3-2 Suppose $\phi(t) = e^t$ is an analytical solution to a certain differential equation on [0, 1]. The linear interpolation and the quadratic interpolation can be written, respectively, as

$$\tilde{\phi}_l(t) = (1 - t)\phi(0) + t\phi(1) \tag{P3-2a}$$

and

$$\tilde{\phi}_q(t) = (1 - 3t + 2t^2)\phi(0) + 4(t - t^2)\phi(\tfrac{1}{2}) + (2t^2 - t)\phi(1). \tag{P3-2b}$$

Evaluate the error bounds $\|\sigma_l\|$ and $\|\sigma_q\|$ defined as

$$\sigma(t) = \phi(t) - \tilde{\phi}(t), \qquad \|\sigma\| = \left\{\int_0^1 \left[\sigma^2 + \left(\frac{d\sigma}{dt}\right)^2\right] dt\right\}^{1/2}.$$

3-3 Use the central-integration method with the quadratic interpolants given in Eq. (3-4) to discretize Eq. (1-1). Is the solution more accurate than that obtained with linear interpolants?

3-4 Consider the fourth-order linear differential equation

$$\phi'''' - \phi'' + 2 = 0.$$

Let the patch be [0,2], consisting of two adjacent elements [0,1] and [1,2]. Do these quadratic nonconforming elements $[x_a, x_c]$ given in Problem 3-1 pass the patch test (with respect to the variational principle)?

3-5 The quadratic elements given in Problem 3-1 are nonconforming with respect to minimization of the least-squares functional

$$J = \int_0^4 [\phi'' + \phi' - 2(1 + x)]^2 \, dx.$$

Determine whether they can pass the patch test.

3-6 Use the method of moments to solve Eq. (1-1) by taking the three leading moments (1, x, x^2).

3-7 Use the Euler method to solve the singular perturbation equation (initial-value problem)

$$\omega\phi'' + \phi' + \phi = 0$$

for $\omega = 0.1$, subject to $\phi(0) = 0$ and $\phi'(0) = 10$. Use the step size $h = 0.2$ and integrate the equation from 0 to 1. Repeat the problem, using the predictor-corrector method.

3-8 It is believed that one of the main differences between finite-difference and finite-element schemes is in their ability to deal with a nonuniform grid. Consider Eq. (1-1) with nonuniform grid sizes $h_1 = 1.75$, $h_2 = 1.25$, $h_3 = 0.75$, and $h_4 = 0.25$, as shown below. Find ϕ_1, ϕ_2, and ϕ_2, using (*a*) the central-difference scheme and (*b*) the Galerkin scheme.

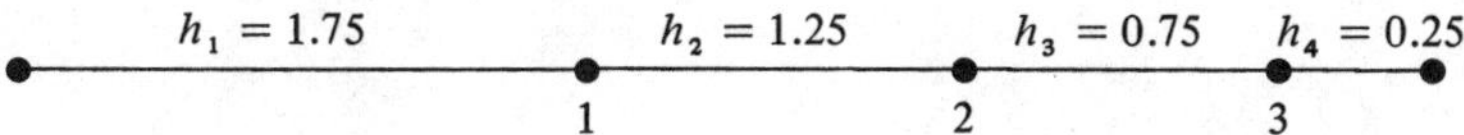

Compare the approximate solutions with the exact solution. If the finite-element solution thus obtained is more accurate than the uniform-grid solution, comment on this.

3-9 Solve Eqs. (3-60) and (3-61) according to Eq. (3-76) until convergence is attained within $\pm$ 0.1%. Compare your result with the exact solution.

CHAPTER

FOUR

NUMERICAL PROPERTIES OF VARIOUS DISCRETIZATION SCHEMES

In the preceding chapters we described the procedure for various numerical techniques and intentionally ignored the effect of interval size on the numerical solution. It was assumed that we may achieve higher accuracy simply by reducing the interval size and that the only price we must pay for this is to put aside the desk calculator and use the computer. Unfortunately, this assumption is not always true. For certain schemes, reduction of the interval size will yield unrealistic results. Even worse situations arise when some numerical results appear to be realistic, but actually are incorrect. Furthermore, even when the numerical result is reliable, questions remain in choosing the optimal interval size ratios to achieve the highest accuracy for transient problems. It is thus important to familiarize ourselves with the characteristics of the numerical schemes, in addition to their use.

In Section 4-1, numerical stability is defined and discussed. The stability criteria for both steady and transient systems are described with emphasis on the matrix stability criterion and the von Neumann stability criterion. The effect of the boundary conditions on stability is also considered.

Section 4-2 presents the second numerical property—consistency. Examples of schemes that are stable but inconsistent, unstable but consistent, and both stable and consistent follow.

The most important numerical property, convergence, is then described in Section 4-3. Its definition varies slightly, depending on the group of numerical techniques being discussed. For finite-difference methods applied to initial-value problems, stability and consistency imply convergence, and vice versa. More details concerning convergence for methods employing inner-product integrals are presented in Chapter 14. Finally, we discuss convergence of the iterative techniques.

Sections 4-4 and 4-5 cover accuracy and efficiency, respectively. It is clear that only if a numerical solution is reliable–i.e., convergent–is it significant to further investigate the accuracy of the solution and the efficiency of the scheme.

4-1 NUMERICAL STABILITY

A discretization scheme is said to be unstable (oscillatory, or wiggling) if its nodal solution to a steady-state equation exhibits oscillatory behavior or if its nodal solution to a transient equation starts being oscillatory and gradually, through amplification, grows unbounded. For second-order linear differential equations, numerical instability is generally caused by the discretization of the first derivative with improper interval size. Since only linear equations have been intensively studied with some conclusive results [1-3] and since most heat transfer problems are of second order, we will restrict our interest to the numerical instability associated with the discretization of the first derivative. Furthermore, it should be realized that numerical instability is tested against the discretized algebraic equation, not the original differential equation. In other words, it does not make sense to ask, "Is Eq. (1-1) stable?" A meaningful question would be, "Is Eq. (1-5) stable?" In the following section, we will present some examples to show how numerical instability arises.

4-1*a* Examples in which Numerical Instability Arises

In a steady-state system, the discretized equation may become unstable (wiggling) because of the choice of a large grid size or the existence of large convection.

Example 4-1 Let the interval size h be 3 in Eq. (1-5). Find the nodal solution ϕ_1, ϕ_2, and ϕ_3 that is subject to, respectively, (*a*) $\phi(0) = 0$ and $\phi(12) = 1$, and (*b*) $\phi(0) = 1$ and $\phi(12) = 0$.

Solution For case (*a*), substituting $h = 3$ into Eq. (1-5) leads to

$$40\phi_1 - 5\phi_2 = 0, \tag{a}$$

$$\phi_1 + 40\phi_2 - 5\phi_3 = 0, \tag{b}$$

$$\phi_2 + 40\phi_3 = 5, \tag{c}$$

which are simultaneously solved to yield

$$\phi_1 = 0.0019, \quad \phi_2 = 0.0155, \quad \text{and } \phi_3 = 0.1246. \tag{d}$$

In comparison with the exact solution $\phi(3) = 0.00012$, $\phi(6) = 0.00248$, and $\phi(9) = 0.04979$, this solution is inaccurate, but physically realistic (work out Problem 4-4 to see why the solution is stable).

For case (*b*), the system of equations becomes

$$40\phi_1 - 5\phi_2 = -1, \tag{e}$$

$$\phi_1 + 40\phi_2 - 5\phi_3 = 0, \tag{f}$$

$$\phi_2 + 40\phi_3 = 0, \tag{g}$$

which is simultaneously solved to yield

$$\phi_1 = -0.02492, \quad \phi_2 = 0.00062, \quad \text{and } \phi_3 = -0.000016.$$

The oscillatory behavior of this unstable solution is depicted in Fig. 4-1. The solution is not only inaccurate but also physically impossible.

The discretized equation governing a transient system may become unstable if the ratio of the time interval to the spatial grid size is too large. The following classic example is presented to show this undesirable behavior.

Example 4-2 By a so-called explicit discretization scheme (to be described in Chapter 5), the one-dimensional transient heat conduction equation

$$\frac{\partial \phi}{\partial t} = \alpha \frac{\partial^2 \phi}{\partial x^2}, \quad x \epsilon [0,4], \quad t \epsilon [0,\infty), \tag{a}$$

subject to

$$\phi(x, 0) = 1, \quad \phi(0, t) = \phi(4, t) = 0, \tag{b}$$

can be discretized into

$$\phi_j^{(n+1)} = r\phi_{j-1}^{(n)} + (1 - 2r)\phi_j^{(n)} + r\phi_{j+1}^{(n)}, \tag{c}$$

where $r = \alpha \Delta t / h^2$; the superscript (n) denotes "at $t = n\Delta t$" and is to be distinguished from the power n. If $r = 2$, find $\phi_1^{(n+1)}$, $\phi_2^{(n+1)}$, and $\phi_3^{(n+1)}$ for $n = 0, 1, 2, 3, 4$.

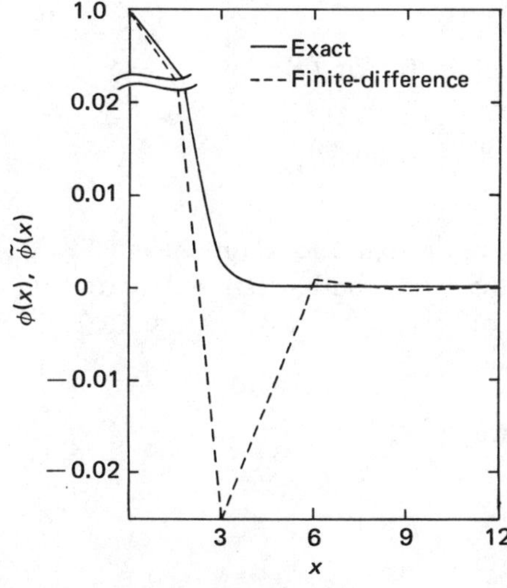

Figure 4-1 Wiggling solution of Eq. (1-1) subject to $\phi(0) = 1$ and $\phi(12) = 0$.

Solution Initially, we have

$$\phi_0^{(0)} = \phi_1^{(0)} = \phi_2^{(0)} = \phi_3^{(0)} = \phi_4^{(0)} = 1. \tag{d}$$

With $r = 2$, Eq. (c) becomes

$$\phi_j^{(n+1)} = 2\phi_{j-1}^{(n)} - 3\phi_j^{(n)} + 2\phi_{j+1}^{(n)}, \tag{e}$$

which is repeatedly used to generate the following tabulated solution:

t	ϕ_0	ϕ_1	ϕ_2	ϕ_3	ϕ_4
0	1	1	1	1	1
Δt	0	1	1	1	0
$2\Delta t$	0	-1	1	-1	0
$3\Delta t$	0	5	-7	5	0
$4\Delta t$	0	-29	41	-29	0

Clearly, this numerical solution is unrealistic because it becomes oscillatory and sometimes even negative. If the computation continues for $t > 4\Delta t$, the errors will be amplified and eventually grow unbounded. Therefore, we find that Eq. (c) with $r = 2$ is unstable.

For a transient two-dimensional system, a similar numerical instability arises.

Example 4-3 Consider the two-dimensional transient heat conduction equation

$$\frac{\partial \phi}{\partial t} = \alpha \left(\frac{\partial^2 \phi}{\partial x^2} + \frac{\partial^2 \phi}{\partial y^2} \right), \quad 0 \leqslant x, y \leqslant 3, \quad t \geqslant 0, \tag{a}$$

subject to

$$\phi(x, y, 0) = 0, \quad \phi(0, y, t) = \phi(3, y, t) = \phi(x, 0, t) = 0,$$

and

$$\phi(x, 3, t) = 1. \tag{b}$$

A certain discretization scheme is used to discretize Eq. (a) into

$$\phi_j^{(n+1)} = r \sum_{p=W}^{N} \phi_p^{(n)} + (1 - 4r)\phi_j^{(n)}, \tag{c}$$

where $r = \alpha\Delta t/h^2$, $\Delta x = \Delta y = h$, the subscript p stands for position, and $\Sigma_{p=W}^{N}$ stands for the summation of values at the west, south, east, and north grid points surrounding j. Let $r = 1$ and find $\phi_1^{(n+1)}$, $\phi_2^{(n+1)}$, $\phi_3^{(n+1)}$, and $\phi_4^{(n+1)}$ for $n = 0, 1, 2, 3$. The network is shown in Fig. 4-2.

Solution With $r = 1$, Eq. (c) can be rewritten as

$$\phi_j^{(n+1)} = \sum_{p=W}^{N} \phi_p^{(n)} - 3\phi_j^{(n)}, \quad j = 1, 2, 3, 4, \tag{d}$$

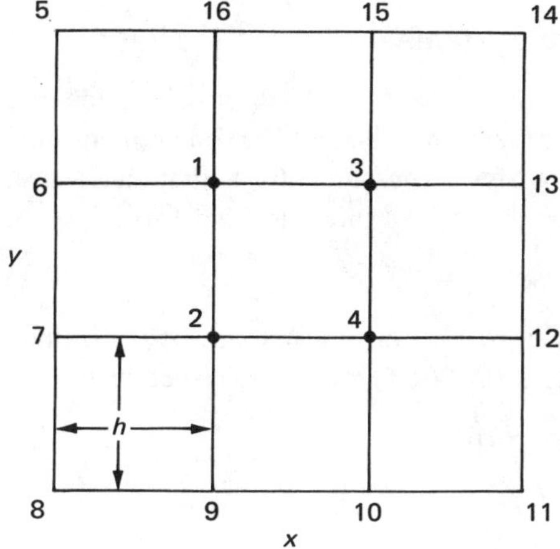

Figure 4-2 Network of a transient two-dimensional system with $\Delta x = \Delta y = h = 1$.

or in matrix form as

$$\begin{Bmatrix} \phi_1^{(n+1)} \\ \phi_2^{(n+1)} \\ \phi_3^{(n+1)} \\ \phi_4^{(n+1)} \end{Bmatrix} = \begin{bmatrix} -3 & 1 & 1 & 0 \\ 1 & -3 & 0 & 1 \\ 1 & 0 & -3 & 1 \\ 0 & 1 & 1 & -3 \end{bmatrix} \begin{Bmatrix} \phi_1^{(n)} \\ \phi_2^{(n)} \\ \phi_3^{(n)} \\ \phi_4^{(n)} \end{Bmatrix} + \begin{Bmatrix} \phi_6^{(n)} + \phi_{16}^{(n)} \\ \phi_7^{(n)} + \phi_9^{(n)} \\ \phi_{13}^{(n)} + \phi_{15}^{(n)} \\ \phi_{10}^{(n)} + \phi_{12}^{(n)} \end{Bmatrix}. \tag{e}$$

Equation (*e*) is solved successively for $n = 0, 1, 2, 3$ to yield the nodal values $\phi_j^{(n+1)}$, $j = 1, 2, 3, 4$, listed in Table 4-1. Since at each time step the information on the right side of Eq. (*e*) is available, the evaluation procedure is fairly easy. For example,

$$\phi_1^{(1)} = -3\phi_1^{(0)} + \phi_2^{(0)} + \phi_3^{(0)} + 0 \times \phi_4^{(0)} + \phi_6^{(0)} + \phi_{16}^{(0)} = \phi_{16}^{(0)} = 1.$$

The oscillatory behavior of these nodal values and the trend of the amplified oscillation are seen from Table 4-1.

Table 4-1 Historical nodal temperatures of a two-dimensional transient system

t	ϕ_1	ϕ_2	ϕ_3	ϕ_4
0	0	0	0	0
Δt	1	0	1	0
$2\Delta t$	-1	1	-1	1
$3\Delta t$	4	-3	4	-3
$4\Delta t$	-10	10	-10	10

4-1*b* Stability Criteria for Steady-State Systems

Since an unstable solution is erroneous, it is desirable to establish certain guidelines to ensure the numerical stability of a discretization scheme. For one-dimensional steady-state systems with convective heat transfer represented by a first derivative, the homogeneous differential equation can be discretized into the final form

$$A\phi_{j-1} + B\phi_j + C\phi_{j+1} = 0. \tag{4-1}$$

We now seek the stability criteria for Eq. (4-1). For convenience, it is assumed that ϕ_0 and ϕ_J are prescribed. Thus, at $j = 1, 2$, Eq. (4-1) becomes, respectively,

$$B\phi_1 + C\phi_2 = -A\phi_0$$

and

$$A\phi_1 + B\phi_2 + C\phi_3 = 0.$$

Following the procedure of Gaussian elimination, we obtain

$$E_2\phi_2 - F\phi_3 = G_2,$$

where

$$E_2 = \frac{B^2 - AC}{-AB}, \quad F = \frac{C}{A}, \text{ and } G_2 = -\frac{A\phi_0}{B}.$$

Repeating the procedure leads to the following general expressions:

$$E_j\phi_j - F\phi_{j+1} = G_j, \quad j = 3, 4, \ldots, J-2, \tag{4-2a}$$

where

$$E_j = \frac{BE_{j-1} + C}{-AE_{j-1}} \quad \text{and } G_j = \frac{G_{j-1}}{E_{j-1}}. \tag{4-2b,c}$$

Finally,

$$G_{J-1} = \frac{G_{J-2}}{E_{J-2}} + \frac{C\phi_J}{A}, \quad E_{J-1}\phi_{J-1} = G_{J-1}. \tag{4-2d,e}$$

Now let Eq. (4-1) be arranged such that $B > 0$. Under this condition, we will examine the following three cases:

1. $A < 0$ and $C > 0$.

Since G_{J-2} and E_{J-2} can be calculated without knowledge of ϕ_J value, G_{J-1} may become negative if ϕ_J is large, according to Eq. (4-2*d*). To ensure that ϕ_{J-1} is positive, we then must have, based on Eq. (4-2*e*), $E_{J-1} < 0$. Rearranging Eq. (4-2*b*) and letting $j = J - 1$, we obtain

$$E_{J-2} = \frac{-C}{AE_{J-1} + B} < 0.$$

Rearranging Eq. (4-2*a*) and letting $j = J - 2$, on the other hand, yields

$$\phi_{J-2} = \frac{G_{J-2}}{E_{J-2}} + \frac{F\phi_{J-1}}{E_{J-2}} = G_{J-1} + \frac{F\phi_{J-1}}{E_{J-2}},$$

which may become negative if $|G_{J-1}| > |F\phi_{J-1}/E_{J-2}|$. However, a negative value of the temperature is physically unrealistic. Therefore, the conditions $A < 0$ and $C > 0$ should be excluded.

2. $A > 0$ and $C < 0$.

In this case,

$$G_2 = -\frac{A\phi_0}{B} \leqslant 0$$

and

$$E_2 = -\frac{[B^2 + A(-C)]}{AB} < 0.$$

In light of Eqs. (4-2*b*) and (4-2*c*) for $j = 3$, we obtain

$$E_3 = -\left[\frac{B}{A} + \frac{(-C)}{A(-E_2)}\right] < 0$$

and

$$G_3 = \frac{G_2}{E_2} \geqslant 0.$$

As the same procedure is carried out for $j > 3$, we find that E_j is always negative, and the sign of G_j is alternating (except for $G_j = 0$). Eventually, however, we must obtain a negative value for G_{J-1} in order for ϕ_{J-1} to be positive, as shown in Eq. (4-2*e*). If J is an even number, then G_{J-1} is positive and consequently ϕ_{J-1} becomes negative. Since there is no reason why J cannot be even, the conditions $A > 0$ and $C < 0$ should also be ruled out.

3. $A > 0$ and $C > 0$.

The three-point relation

$$A\phi_{j-1} + B\phi_j + C\phi_{j+1} = 0$$

cannot hold unless at least one of the three nodal values is negative.

On the basis of the arguments in cases 1-3, we established that the conditions for the nodal functions in Eq. (4-1) to have physically significant values are

$$B > 0, \quad A < 0, \quad \text{and } C < 0. \tag{4-3}$$

It is interesting to note that, if a differential equation and the boundary conditions are mathematically well defined but physically unmeaningful, the corresponding discretized equation satisfying Eq. (4-3) will also yield a numerically stable but physically unrealistic solution. This may happen when a source term exists in a conduction system governed by, e.g.,

$$\frac{d^2\phi}{dx^2} + a^2\phi = 0,$$

where a is a constant (see problem 4-15).

Applying these criteria to Eq. (1-5), we find that if

$$h < 2, \tag{4-4}$$

then Eq. (1-5) is stable. In Example 4-1, h is taken as 3 and consequently the nodal solution becomes oscillatory.

Now suppose the flow velocity is so small that the first derivative in Eq. (1-1) can be neglected. The resulting discretized equation can be derived as

$$-2\phi_{j-1} + (4 + 4h^2)\phi_j - 2\phi_{j+1} = 0. \tag{4-5}$$

The coefficients in Eq. (4-5) satisfy Eq. (4-3) regardless of the choice of interval size h. It therefore seems that the discretization of the first derivative in a differential equation is generally the potential source of numerical instability. If the central-difference scheme or the Galerkin method with the symmetric pyramid function is to be used for discretizing the first derivative, the interval size must be chosen with care. Alternatively, the restriction on interval size can be relaxed by the upwind technique [4-7], where the one-sided finite difference or the asymmetric weighting function is adopted for the discretization of the first derivative. We will discuss this topic in Chapter 8.

4-1*c* Stability Criteria for Transient Systems

In this section, we will introduce two stability criteria that are used to test the numerical stability of the discretized equation governing transient systems. The first criterion is called the matrix method [2, 3]. In summary, this method asserts that if the nodal solution at $t = (n + 1)\Delta t$ can eventually be expressed in terms of that at $t = n\Delta t$ as

$$\{\phi^{(n+1)}\} = [\mathbf{D}]\ \{\phi^{(n)}\} + \{\text{boundary conditions}\}, \qquad n = 0, 1, 2, \ldots, \tag{4-6}$$

where $\{\phi^{(n+1)}\}^T = \{\phi_1^{(n+1)}\phi_2^{(n+1)} \cdots \phi_{J-1}^{(n+1)}\}$, $[\mathbf{D}]$ is a $J-1$ by $J-1$ matrix, $J = L/\Delta x$, and L is the domain of the one-dimensional system, then Eq. (4-6) is stable provided the absolute eigenvalues of the matrix $[\mathbf{D}]$ are not greater than 1. Being important, this assertion will be proved at some length as follows. If Eq. (4-6) holds, it follows that

$$\{\phi^{(1)}\} = [\mathbf{D}]\ \{\phi^{(0)}\} + \{\text{b.c.}\} \tag{4-7a}$$

and
$$\{\phi^{(2)}\} = [\mathbf{D}]\ \{\phi^{(1)}\} + \{\text{b.c.}\} = [\mathbf{D}]^2\ \{\phi^{(0)}\} + ([\mathbf{D}] + [\mathbf{I}])\ \{\text{b.c.}\}. \tag{4-7b}$$

Performance of the induction leads to

$$\{\phi^{(n+1)}\} = [\mathbf{D}]^{n+1}\ \{\phi^{(0)}\} + \cdots, \tag{4-7c}$$

where, for clarity, the matrices multiplied by the boundary-condition terms are not written because they play minor roles in the present discussion. Now, a few theories and examples of matrix applications will be presented before we move to the topic of matrix stability analysis. By definition, the eigenvalues λ_j of a $J \times J$ matrix

$$[\mathbf{A}] = \begin{bmatrix} a_{11} & a_{12} & \cdots & a_{1J} \\ a_{21} & a_{22} & \cdots & a_{2J} \\ \cdots & \cdots & \cdots & \cdots \\ a_{J1} & a_{J2} & \cdots & a_{JJ} \end{bmatrix} \tag{4-8}$$

are the roots of the equation

$$\begin{vmatrix} a_{11}-\lambda & a_{12} & \cdots & a_{1J} \\ a_{21} & a_{22}-\lambda & \cdots & \cdots \\ \cdots & \cdots & \cdots & \cdots \\ a_{J1} & a_{J2} & \cdots & a_{JJ}-\lambda \end{vmatrix} = 0. \tag{4-9}$$

Example 4-4 Find the eigenvalues of the matrix

$$[\mathbf{D}] = \begin{bmatrix} -1 & 1 & 0 \\ 1 & -1 & 1 \\ 0 & 1 & -1 \end{bmatrix}. \tag{a}$$

Solution The eigenvalues of [**D**] are the roots of

$$\begin{vmatrix} -1-\lambda & 1 & 0 \\ 1 & -1-\lambda & 1 \\ 0 & 1 & -1-\lambda \end{vmatrix} = 0 \tag{b}$$

or

$$-(1+\lambda)^3 + 2(1+\lambda) = 0. \tag{c}$$

They are found to be $\lambda_1 = -1$, $\lambda_2 = -1 + \sqrt{2}$, and $\lambda_3 = -1 - \sqrt{2}$.

The J eigenvectors $\mathbf{V}^{(j)}$ of a $J \times J$ matrix given in Eq. (4-8) satisfy the relation

$$[\mathbf{A}]\ \{\mathbf{V}^{(j)}\} = \lambda_j \{\mathbf{V}^{(j)}\}. \tag{4-10}$$

Example 4-5 Find the three eigenvectors of the matrix

$$[\mathbf{D}] = \begin{bmatrix} -1 & 1 & 0 \\ 1 & -1 & 1 \\ 0 & 1 & -1 \end{bmatrix}. \tag{a}$$

Solution According to Eq. (4-10), the eigenvector $\mathbf{V}^{(j)}$ corresponding to the eigenvalue λ_j is specified such that

$$\begin{bmatrix} -1 & 1 & 0 \\ 1 & -1 & 1 \\ 0 & 1 & -1 \end{bmatrix} \begin{Bmatrix} x_1 \\ x_2 \\ x_3 \end{Bmatrix} = \lambda_j \begin{Bmatrix} x_1 \\ x_2 \\ x_3 \end{Bmatrix} \tag{b}$$

or

$$-(1+\lambda_j)x_1 + x_2 = 0, \tag{c}$$

$$x_1 - (1+\lambda_j)x_2 + x_3 = 0, \tag{d}$$

$$x_2 - (1+\lambda_j)x_3 = 0. \tag{e}$$

Equations (*c*)-(*e*) are simultaneously solved to yield x_1, x_2, and x_3. For $\lambda_1 = -1$, Eqs. (*c*)-(*e*) become

$$x_2 = 0, \tag{f}$$

$$x_1 + x_3 = 0, \tag{g}$$

$$x_2 = 0. \tag{h}$$

Since the system of Eqs. (*f*)-(*h*) is indeterminate, it is our option to assign some convenient values to x_1 and x_3.† Letting x_1 and x_3 be 1 and -1, respectively, we obtain

$$\{\mathbf{V}^{(1)}\} = \begin{Bmatrix} 1 \\ 0 \\ -1 \end{Bmatrix} \tag{i}$$

as the first eigenvector. Likewise, the second and third eigenvectors can be obtained as

$$\{\mathbf{V}^{(2)}\} = \begin{Bmatrix} 1 \\ \sqrt{2} \\ 1 \end{Bmatrix} \quad \text{and} \quad \{\mathbf{V}^{(3)}\} = \begin{Bmatrix} 1 \\ -\sqrt{2} \\ 1 \end{Bmatrix}. \tag{j, k}$$

According to matrix theory (the proof will be omitted here), any J-component vector can be expressed as a linear combination of the J eigenvectors of a $J \times J$ matrix if these eigenvectors are all distinct [8].

Example 4-6 Express the vector $\{\mathbf{U}\}^T = \{1 \;\; 1 \;\; 1\}$ in terms of the three eigenvectors $\{\mathbf{V}^{(j)}\}$, $j = 1, 2, 3$, given in Example 4-5.

† An excellent option is to normalize the eigenvectors such that $\{\mathbf{V}_j\}^T\{\mathbf{V}_j\} = 1$.

Solution Let {U} be expressed by

$$\begin{Bmatrix} 1 \\ 1 \\ 1 \end{Bmatrix} = a_1 \begin{Bmatrix} 1 \\ 0 \\ -1 \end{Bmatrix} + a_2 \begin{Bmatrix} 1 \\ \sqrt{2} \\ 1 \end{Bmatrix} + a_3 \begin{Bmatrix} 1 \\ -\sqrt{2} \\ 1 \end{Bmatrix}, \tag{a}$$

To determine the coefficients a_1, a_2, and a_3, we rearrange Eq. (*a*) into

$$\begin{bmatrix} 1 & 1 & 1 \\ 0 & \sqrt{2} & -\sqrt{2} \\ -1 & 1 & 1 \end{bmatrix} \begin{Bmatrix} a_1 \\ a_2 \\ a_3 \end{Bmatrix} = \begin{Bmatrix} 1 \\ 1 \\ 1 \end{Bmatrix}. \tag{b}$$

Since the coefficient matrix is positive-definite, i.e.,

$$|1| > 0, \quad \begin{vmatrix} 1 & 1 \\ 0 & \sqrt{2} \end{vmatrix} > 0, \quad \text{and} \begin{vmatrix} 1 & 1 & 1 \\ 0 & \sqrt{2} & -\sqrt{2} \\ -1 & 1 & 1 \end{vmatrix} > 0,$$

it is guaranteed that Eq. (*b*) has a unique solution. After arithmetic, we obtain

$$a_1 = 0, \quad a_2 = \frac{\sqrt{2}+1}{2\sqrt{2}}, \quad \text{and } a_3 = \frac{\sqrt{2}-1}{2\sqrt{2}}. \tag{c}$$

Having studied Examples 4-4 to 4-6, we now return to the description of the matrix method. In light of Example 4-6, the vector $\{\phi^{(0)}\}$ in Eq. (4-7*c*) can be expressed as

$$\{\phi^{(0)}\} = a_1\{\mathbf{V}^{(1)}\} + a_2\{\mathbf{V}^{(2)}\} + \dots + a_j\{\mathbf{V}^{(j)}\} + \dots + a_{J-1}\{\mathbf{V}^{(J-1)}\}, \tag{4-11}$$

where $\{\mathbf{V}^{(j)}\}$, $j = 1, 2, \dots, J-1$, are eigenvectors of the matrix [**D**] given in Eq. (4-6). Substituting Eq. (4-11) into Eq. (4-7*c*) leads to

$$\{\phi^{(n+1)}\} = \sum_{j=1}^{J-1} a_j[\mathbf{D}]^{n+1}\{\mathbf{V}^{(j)}\} + \dots. \tag{4-12}$$

Referring now to Eq. (4-10), we may write

$$[\mathbf{D}]\{\mathbf{V}^{(j)}\} = \lambda_j\{\mathbf{V}^{(j)}\}, \quad [\mathbf{D}]^2\{\mathbf{V}^{(j)}\} = [\mathbf{D}]\lambda_j\{\mathbf{V}^{(j)}\} = \lambda_j^2\{\mathbf{V}^{(j)}\} \tag{4-13}$$

or eventually

$$[\mathbf{D}]^{n+1}\{\mathbf{V}^{(j)}\} = \lambda_j^{n+1}\{\mathbf{V}^{(j)}\}. \tag{4-14}$$

Substituting Eq. (4-14) into Eq. (4-12), we obtain the desired final result

$$\{\phi^{(n+1)}\} = \sum_{j=1}^{J-1} a_j \lambda_j^{n+1} \{\mathbf{V}^{(j)}\} + \cdots. \tag{4-15}$$

Clearly, if the absolute value of any of the eigenvalues λ_j is larger than 1, the nodal solution $\{\phi^{(n+1)}\}$ will grow unbounded as $n \rightarrow \infty$ and Eq. (4-6) will be unstable. The assertion of the matrix method is thus established.

Let us now use the matrix method to examine the numerical stability of Eq. (*c*) in Example (4-2). This equation can be written in matrix form as

$$\{\phi^{(n+1)}\} = [\mathbf{D}]\,\{\phi^{(n)}\} + \{\text{b.c.}\}, \tag{4-16}$$

where

$$[\mathbf{D}] = \begin{bmatrix} 1-2r & r & 0 \\ r & 1-2r & r \\ 0 & r & 1-2r \end{bmatrix}. \tag{4-17}$$

For the sake of easy illustration, the number of nodal unknowns is taken as only 3. The following presentation, however, is valid for general cases. For convenience in finding the eigenvalues of $[\mathbf{D}]$, we split $[\mathbf{D}]$ into

$$[\mathbf{D}] = \begin{bmatrix} 1 & 0 & 0 \\ 0 & 1 & 0 \\ 0 & 0 & 1 \end{bmatrix} + r \begin{bmatrix} -2 & 1 & 0 \\ 1 & -2 & 1 \\ 0 & 1 & -2 \end{bmatrix}. \tag{4-18}$$

Instead of finding the eigenvalues of $[\mathbf{D}]$, we seek those of

$$[\mathbf{S}] = \begin{bmatrix} -2 & 1 & 0 \\ 1 & -2 & 1 \\ 0 & 1 & -2 \end{bmatrix}. \tag{4-19}$$

In reference to Example 4-4, the eigenvalues of $[\mathbf{S}]$ are the three roots of

$$\begin{vmatrix} -2-\mu & 1 & 0 \\ 1 & -2-\mu & 1 \\ 0 & 1 & -2-\mu \end{vmatrix} = 0, \tag{4-20}$$

which can be solved to yield

$$\mu_1 = -2 + \sqrt{2}, \quad \mu_2 = -2, \quad \text{and } \mu_3 = -2 - \sqrt{2}. \tag{4-21}$$

In fact, these three roots can be expressed by a single general equation as

$$\mu_j = -4 \sin^2\left(\frac{j\pi}{2J}\right), \qquad j = 1, 2, \ldots, J-1. \tag{4-22}$$

In this problem, $J = 4$. Since normally it is tedious to find eigenvalues of a large matrix, we customarily express the matrix, such as Eq. (4-17), in terms of some standard matrices whose eigenvalues were previously found.

Now, let Eq. (4-18) be rewritten as

$$[\mathbf{D}] = [\mathbf{I}] + r[\mathbf{S}] \tag{4-23}$$

and eigenvalues of $[\mathbf{D}]$ be denoted as λ_1, λ_2, and λ_3. From Eq. (4-10), we obtain

$$([\mathbf{D}] - \lambda_j[\mathbf{I}])\{\mathbf{V}^{(j)}\} = 0, \qquad j = 1, 2, 3. \tag{4-24}$$

Upon substitution of Eq. (4-23) into Eq. (4-24) and rearrangement of the result, we derive

$$\left([\mathbf{S}] - \frac{\lambda_j - 1}{r}[\mathbf{I}]\right)\{\mathbf{V}^{(j)}\} = 0, \tag{4-25}$$

which suggests that the $\{\mathbf{V}^{(j)}\}$ are also the eigenvectors of $[\mathbf{S}]$. Furthermore, Eq. (4-25) can be compared with

$$([\mathbf{S}] - \mu_j[\mathbf{I}])\{\mathbf{V}^{(j)}\} = 0$$

to yield

$$\mu_j = \frac{\lambda_j - 1}{r}. \tag{4-26}$$

Since the absolute value of λ_j must not be greater than 1, we derive, with the aid of Eq. (4-22), a restriction on r, i.e.,

$$r \leqslant \frac{1}{2\sin^2(j\pi/2J)}. \tag{4-27}$$

The possible maximum value of the sine function is unity and thus r should not be greater than the possible minimum value of the right-hand side of Eq. (4-27). Consequently, the criterion for numerical stability of the discretization equation (*c*) in Example 4-2 is

$$r \leqslant \tfrac{1}{2}. \tag{4-28}$$

Leaving one more example of the matrix method to Chapter 5, we will now turn to the criterion for numerical stability introduced by von Neumann [9, 10].

In von Neumann stability analysis, the time-dependent nodal unknowns are assumed to take a form similar to the analytical solution. For example, let us again consider the transient one-dimensional heat conduction equations (*a*) and (*b*) in Example 4-2. The analytical solution satisfying the differential equation (*a*) and the two boundary conditions (but not the initial condition) given in Eq. (*b*) takes the form

$$\phi_m(x, t) = A_m \exp\left(-\frac{\alpha m^2 \pi^2 t}{L}\right) \sin\left(\frac{m\pi x}{L}\right), \qquad m = 1, 2, 3, \ldots \tag{4-29a}$$

which can be obtained by separation of variables. These solutions are then superimposed in such a manner that the linear combination at $t = 0$ approximates the initial condition $\phi(x, 0) = 1$. We will not proceed to evaluate A_m here since this evaluation is somewhat irrelevant to our stability analysis (A_m is given in Example 4-15). Furthermore, for the sake of clarity, let us consider a particular solution at $m = 1$

$$\phi(x, t) = A_1 \exp\left(\frac{-\alpha \pi^2 t}{L}\right) \sin\left(\frac{\pi x}{L}\right). \tag{4-29b}$$

Observe that as $t \to \infty$ this analytical solution approaches zero as it should. Analogously, the numerical approximation

$$\phi_j^{(n)} = A_1 \exp\left(\frac{-\alpha \pi^2 n \Delta t}{L}\right) \sin\left(\frac{\pi j h}{L}\right) \tag{4-30}$$

should also be bounded as $n \to \infty$. The von Neumann stability criterion thus states that a discretization scheme is stable if its nodal solution can be assumed to take the form

$$\phi_j^{(n)} = A_1 \, \xi^n \sin\left(\frac{\pi j h}{L}\right) \tag{4-31}$$

and if

$$|\xi| \leqslant 1, \tag{4-32}$$

where ξ is called the amplification factor. To further demonstrate the usefulness of this criterion, Eqs. (4-31) and (4-32), we normally substitute Eq. (4-31) into the discretization equation that is being investigated. For instance, substitution of Eq. (4-31) into Eq. (*c*) in Example 4-2 yields

$$\xi \sin\left(\frac{j\pi}{J}\right) = r \sin\left[\frac{(j-1)\pi}{J}\right] + (1 - 2r) \sin\left(\frac{j\pi}{J}\right) + r \sin\left[\frac{(j+1)\pi}{J}\right],$$

where the relation $h/L = 1/J$ has been used. After simplification, we obtain

$$\xi = 1 - 2r + 2r \cos\left(\frac{\pi}{J}\right). \tag{4-33}$$

Since Eq. (4-32) is required for stability, we deduce

$$r \leqslant \frac{1}{1 - \cos(\pi/J)} = \frac{1}{2 \sin^2(\pi/2J)}, \tag{4-34}$$

which is identical to Eq. (4-27) with $j = 1$.

Sometimes a complex exponential form

$$\phi_m(x, t) = B_m \exp\left(\frac{-\alpha m^2 \pi^2 t}{L}\right) \exp\left(\frac{im\pi x}{L}\right), \qquad m = \pm 1, \pm 2, \ldots \tag{4-35}$$

is used instead of that given in Eq. (4-29*a*), where i is the imaginary number. It can be shown that these complex exponential functions are orthogonal to one another and that they all satisfy the two homogeneous Dirichlet boundary conditions at $x = L, -L$. Following a procedure similar to that described above, we will also be able to derive Eq. (4-34). This is left to the reader as an exercise (Problem 4-7).

In comparing these two stability analyses, we found that, for the transient one-dimensional heat conduction equation with Dirichlet boundary conditions, the criteria for numerical stability are the same. For problems with other types of boundary conditions, however, the matrix method is sometimes capable of taking the effects of boundary conditions into account, while the von Neumann analysis fails to do so. In the following examples, we will present a problem subject to certain Neumann and mixed-type boundary conditions and a corresponding difference scheme that is supposed to be unstable. The matrix method succeeds in predicting the instability but the von Neumann method fails.

Example 4-7 Consider the transient equation

$$\frac{\partial \phi}{\partial t} = \alpha \frac{\partial^2 \phi}{\partial x^2} \tag{a}$$

subject to the boundary conditions

$$\frac{\partial \phi}{\partial x}(0, t) = \phi(0, t), \qquad \frac{\partial \phi}{\partial x}(1, t) = 0 \tag{b, c}$$

and the initial condition

$$\phi(x, 0) = 1.$$

Use the matrix method to determine whether the discretized equation [let $r = \frac{1}{2}$ in Eq. (*c*) of Example 4-2]

$$\phi_j^{(n+1)} = \frac{1}{2}\phi_{j-1}^{(n)} + \frac{1}{2}\phi_{j+1}^{(n)} \tag{d}$$

is stable.

Solution To be consistent with the treatment in [11], in which this problem was initially designed, let us approximate Eq. (*b*) as

$$\frac{\phi_1^{(n)} - \phi_{-1}^{(n)}}{2h} = \phi_0^{(n)}, \tag{e}$$

where $\phi_{-1}^{(n)}$ is a fictitious nodal unknown beyond the domain and $h = 1/J$. This unknown $\phi_{-1}^{(n)}$ can be eliminated from both Eq. (*d*) with $j = 0$ and Eq. (*e*) to yield

$$\phi_0^{(n+1)} = -h\phi_0^{(n)} + \phi_1^{(n)}. \qquad (f)$$

Similarly, we approximate Eq. (*c*) by

$$\phi_{J-1}^{(n)} = \phi_{J+1}^{(n)}, \qquad (g)$$

which is substituted into Eq. (*d*) with $j = J$ to give

$$\phi_J^{(n+1)} = \phi_{J-1}^{(n)}. \qquad (h)$$

Therefore, Eq. (*d*) along with Eqs. (*f*) and (*h*) can be written in matrix form as

$$\begin{Bmatrix} \phi_0^{(n+1)} \\ \phi_1^{(n+1)} \\ \cdots \\ \phi_{J-1}^{(n+1)} \\ \phi_J^{(n+1)} \end{Bmatrix} = \begin{bmatrix} -h & 1 & 0 & \cdots & 0 \\ \frac{1}{2} & 0 & \frac{1}{2} & \cdots & 0 \\ \cdots & \cdots & \cdots & \cdots & \cdots \\ 0 & 0 & \frac{1}{2} & 0 & \frac{1}{2} \\ 0 & 0 & 0 & 1 & 0 \end{bmatrix} \begin{Bmatrix} \phi_0^{(n)} \\ \phi_1^{(n)} \\ \cdots \\ \phi_{J-1}^{(n)} \\ \phi_J^{(n)} \end{Bmatrix} \qquad (i)$$

or

$$\{\phi^{(n+1)}\} = [\mathbf{D}]\,\{\phi^{(n)}\}. \qquad (j)$$

To seek the eigenvalues of [**D**] is a separate laborious task. To avoid this additional difficulty, let us take $J = 2$. The matrix [**D**] is reduced to

$$[\mathbf{D}] = \begin{bmatrix} -\frac{1}{2} & 1 & 0 \\ \frac{1}{2} & 0 & \frac{1}{2} \\ 0 & 1 & 0 \end{bmatrix} \qquad (k)$$

of which the eigenvalues are the roots of

$$f(\lambda) = \begin{vmatrix} -\frac{1}{2} - \lambda & 1 & 0 \\ \frac{1}{2} & -\lambda & \frac{1}{2} \\ 0 & 1 & -\lambda \end{vmatrix} = 0 \qquad (l)$$

or

$$-4f(\lambda) = 4\lambda^3 + 2\lambda^2 - 4\lambda - 1 = 0. \qquad (m)$$

To assess the numerical instability of Eqs. (*j*) and (*k*) according to the matrix method, we only need to know whether any absolute eigenvalue of [**D**] is greater than 1. That is, we do not need to determine the exact values of the roots. We find that

$$f(-1) = -\frac{1}{4} \quad \text{and } f(-2) = \frac{17}{4}. \tag{n}$$

This suggests that there is at least one real root on $[-2, -1]$. Consequently,

$$|\lambda| > 1. \tag{o}$$

In a nonrigorous way, we have established that Eq. (*d*) is unstable. The numerical solution listed below

t	ϕ_{-1}	ϕ_0	ϕ_1	ϕ_2	ϕ_3
0	0	1	1	1	1
Δt	$\frac{1}{2}$	$\frac{1}{2}$	1	1	1
$2\Delta t$	0	$\frac{3}{4}$	$\frac{3}{4}$	1	$\frac{3}{4}$
$3\Delta t$	$\frac{1}{2}$	$\frac{3}{8}$	$\frac{7}{8}$	$\frac{3}{4}$	$\frac{7}{8}$
...	...	...	...	...	...

indeed exhibits the oscillatory behavior.

Example 4-8 Use von Neumann stability analysis to repeat the preceding example.

Solution The analytical solutions to Eqs. (*a*)-(*c*) in the preceding example are

$$\phi_m(x, t) = A_m \exp(-\alpha t \gamma_m^2)(\gamma_m \cos \gamma_m x + \sin \gamma_m x),$$
$$0 \leqslant x \leqslant 1, \quad m = 1, 2, \ldots \tag{a}$$

where γ_m are the eigenvalues that satisfy the transcendental equation

$$\gamma_m \tan \gamma_m - 1 = 0. \tag{b}$$

Analogously, we will assume that the finite-difference solution takes the form

$$\phi_j^n = A\xi^n \left[\gamma \cos\left(\frac{\gamma j}{J}\right) + \sin\left(\frac{\gamma j}{J}\right)\right], \tag{c}$$

where the subscript m is dropped for simplicity and the relation $x = j/J$ has been used. Substituting Eq. (*c*) into

$$\phi_j^{(n+1)} = \frac{1}{2}\phi_{j-1}^{(n)} + \frac{1}{2}\phi_{j+1}^{(n)}, \tag{d}$$

we obtain

$$\xi = \frac{1}{2}\frac{\{\gamma \cos[\gamma(j+1)/J] + \sin[\gamma(j+1)/J] + \gamma \cos[\gamma(j-1)/J] + \sin[\gamma(j-1)/J]\}}{\gamma \cos(\gamma j/J) + \sin(\gamma j/J)} \tag{e}$$

which, after simplification, becomes

$$\xi = \cos\left(\frac{\gamma}{J}\right). \tag{f}$$

It thus follows that

$$|\xi| \leqslant 1. \tag{g}$$

Based on Eq. (*g*), we arrive at the false conclusion that the discretization equation (*d*) is stable. This example demonstrates that the von Neumann method is insensitive to the effect of boundary conditions.

Both the matrix and von Neumann stability analyses can also be applied to two-dimensional transient problems (see [7, pp. 51-53; 12] for details). We will now leave the topic of stability and proceed to consider the second important numerical property.

4-2 NUMERICAL CONSISTENCY

4-2*a* Definition

The discretized equation is said to be consistent with the original differential equation if the truncation errors approach zero as h, $\Delta t \to 0$. Since we are considering truncation errors that arise upon use of the Taylor's series expansion, numerical consistency is primarily a property of finite-difference schemes. Table 4-2 lists various differencing schemes for the first and second derivatives and the corresponding truncation errors. The variable z may stand for either the space coordinates x and y or the time t. The most noteworthy term is the truncation error, whose derivation is demonstrated in the following example.

Example 4-9 Use the Taylor's series expansion to derive the truncated leading terms for the approximation $\delta^2 \phi_j^{(n+1)}$ in Table 4-2.

Solution About the point (x, t) the functions $\phi(x+h, t+\Delta t)$, $\phi(x-h, t+\Delta t)$, and $\phi(x, t+\Delta t)$ can be expanded in Taylor's series as, respectively,

$$\phi(x+h, t+\Delta t) = \left[1 + \left(h\frac{\partial}{\partial x} + \Delta t\frac{\partial}{\partial t}\right) + \frac{1}{2}\left(h\frac{\partial}{\partial x} + \Delta t\frac{\partial}{\partial t}\right)^2 + \frac{1}{6}\left(h\frac{\partial}{\partial x} + \Delta t\frac{\partial}{\partial t}\right)^3 + \frac{1}{24}\left(h\frac{\partial}{\partial x} + \Delta t\frac{\partial}{\partial t}\right)^4 + \cdots\right]\phi(x, t), \tag{a}$$

$$\phi(x-h,\, t+\Delta t) = \left[1 + \left(-h\frac{\partial}{\partial x} + \Delta t\frac{\partial}{\partial t}\right) + \frac{1}{2}\left(-h\frac{\partial}{\partial x} + \Delta t\frac{\partial}{\partial t}\right)^2 + \frac{1}{6}\left(-h\frac{\partial}{\partial x} + \Delta t\frac{\partial}{\partial t}\right)^3 + \frac{1}{24}\left(-h\frac{\partial}{\partial x} + \Delta t\frac{\partial}{\partial t}\right)^4 + \cdots\right]\phi(x, t), \qquad (b)$$

and

$$\phi(x,\, t+\Delta t) = \left[1 + \Delta t\frac{\partial}{\partial t} + \frac{(\Delta t)^2}{2}\frac{\partial^2}{\partial t^2} + \frac{(\Delta t)^3}{6}\frac{\partial^3}{\partial t^3} + \frac{(\Delta t)^4}{24}\frac{\partial^4}{\partial t^4}\right]\phi(x, t). \qquad (c)$$

If we define

$$\delta^2\phi_j^{(n+1)} = \frac{\phi(x+h,\, t+\Delta t) - 2\phi(x,\, t+\Delta t) + \phi(x-h,\, t+\Delta t)}{h^2}, \qquad (d)$$

then substitution of Eqs. (*a*)-(*c*) into Eq. (*d*) yields

$$\delta^2\phi_j^{(n+1)} = \frac{\partial^2\phi}{\partial x^2} + \Delta t\,\frac{\partial^3\phi}{\partial x^2\,\partial t} + \frac{h^2}{12}\frac{\partial^4\phi}{\partial x^4} + \frac{(\Delta t)^2}{2}\,\frac{\partial^4\phi}{\partial x^2\,\partial t^2}, \qquad (e)$$

which can be readily changed to the form in Table 4-2.

Table 4-2 Various differencing schemes and their corresponding truncated terms[a]

Derivative	Differencing approximation	Truncated leading terms
$\frac{\partial\phi}{\partial z}$	$\frac{\phi(z^+)-\phi(z)}{\Delta z}$ (forward)	$-\frac{\Delta z}{2}\frac{\partial^2\phi}{\partial z^2} - \frac{(\Delta z)^2}{6}\frac{\partial^3\phi}{\partial z^3}$
	$\frac{\phi(z)-\phi(z^-)}{\Delta z}$ (backward)	$+\frac{\Delta z}{2}\frac{\partial^2\phi}{\partial z^2} - \frac{(\Delta z)^2}{6}\frac{\partial^3\phi}{\partial z^3}$
	$\frac{\phi(z^+)-\phi(z^-)}{2\Delta z}$ (central)	$-\frac{(\Delta z)^2}{6}\frac{\partial^3\phi}{\partial z^3}$
$\frac{\partial^2\phi}{\partial x^2}$	$\frac{\phi(x^+, t^+) - 2\phi(x, t^+) + \phi(x^-, t^+)}{(\Delta x)^2}$	$-\frac{(\Delta x)^2}{12}\frac{\partial^4\phi}{\partial x^4} - (\Delta t)\frac{\partial^3\phi}{\partial x^2\,\partial t}$ $-\frac{(\Delta t)^2}{2}\frac{\partial^4\phi}{\partial x^2\,\partial t^2}$
	$\frac{\phi(x^+, t) - 2\phi(x, t) + \phi(x^-, t)}{(\Delta x)^2}$	$-\frac{(\Delta x)^2}{12}\frac{\partial^4\phi}{\partial x^4}$

[a] $z^+ = z + \Delta z$, $z^- = z - \Delta z$, and z may denote variables such as x, y, or t.

Table 4-2 is a very convenient reference when the numerical consistency of a finite-difference equation is to be considered.

Example 4-10 Use Table 4-2 to determine whether Eq. (1-5) is consistent with Eq. (1-1).

Solution Using Table 4-2, we rewrite Eq. (1-1) as

$$\begin{aligned} L\phi &= \frac{d^2\phi}{dx^2} + \frac{d\phi}{dx} - 2\phi \\ &= \delta^2\phi_j - \frac{h^2}{12}\frac{d^4\phi}{dx^4} + \frac{\phi_{j+1} - \phi_{j-1}}{2h} - \frac{h^2}{6}\frac{d^3\phi}{dx^3} - 2\phi_j \qquad (a) \\ &= \tilde{L}\phi_j + \tau, \end{aligned}$$

where

$$\tilde{L}\phi_j = \delta^2\phi_j + \frac{\phi_{j+1} - \phi_{j-1}}{2h} - 2\phi_j = 0 \qquad (b)$$

and

$$\tau = -\frac{h^2}{12}\frac{d^4\phi}{dx^4} - \frac{h^2}{6}\frac{d^3\phi}{dx^3} \qquad (c)$$

is the truncation error of $\mathbf{O}(h^2)$. Clearly, τ approaches zero as $h \to 0$. Therefore, Eq. (b) is consistent with Eq. (1-1) and it can be readily changed to Eq. (1-5).

Generally, one-dimensional steady-state finite-difference equations are consistent with the differential equations and the task of examining their consistency is trivial. For transient systems, however, the time derivative can be approximated by central, forward, or backward differencing schemes. Various combinations with the spatial difference are possible. Without the guidance of consistency analysis, an inconsistent scheme can be generated. Let us first consider an example that is consistent.

Example 4-11 Determine whether the implicit scheme

$$\tilde{L}\phi_j = \frac{\phi_j^{(n)} - \phi_j^{(n-1)}}{\Delta t} - \alpha\delta^2\phi_j^{(n)} \qquad (a)$$

is consistent with

$$L\phi = \frac{\partial\phi}{\partial t} - \alpha\frac{\partial^2\phi}{\partial x^2}. \qquad (b)$$

Solution Employing a backward-difference scheme for $\partial\phi/\partial t$, we discretize Eq. (b) to

$$L\phi = \frac{\phi_j^{(n)} - \phi_j^{(n-1)}}{\Delta t} + \frac{\Delta t}{2}\frac{\partial^2\phi}{\partial t^2} - \alpha\left[\delta^2\phi_j^{(n)} - \frac{h^2}{12}\frac{\partial^4\phi}{\partial x^4}\right]. \qquad (c)$$

It follows that

$$\tau = L\phi - \tilde{L}\phi_j = \frac{\Delta t}{2}\frac{\partial^2 \phi}{\partial t^2} + \frac{\alpha h^2}{12}\frac{\partial^4 \phi}{\partial x^4}\,. \qquad (d)$$

Clearly, $$\tau \to 0 \quad \text{as } \Delta t, h \to 0\,. \qquad (e)$$

Therefore, Eq. (*a*) is consistent with Eq. (*b*). Note that these high-order derivatives are all evaluated at $x = j\Delta x$ and $t = n\Delta t$.

4-2*b* Stable but Inconsistent Schemes

Next, we will introduce a well-known finite-difference scheme, proposed by Du Fort and Frankel [13], that could be inconsistent with the original differential equation.

Example 4-12 Consider the Du Fort-Frankel scheme

$$\tilde{L}_{DF}\phi_j = \frac{\phi_j^{(n+1)} - \phi^{(n-1)}}{2\Delta t} - \frac{\alpha}{h^2}\,[\phi_{j-1}^{(n)} - \phi_j^{(n-1)} - \phi_j^{(n+1)} + \phi_{j+1}^{(n)}\,], \qquad (a)$$

which is derived by replacing $2\phi_j^{(n)}$ with $\phi_j^{(n-1)} + \phi_j^{(n+1)}$ and therefore involves three time steps. Examine its numerical consistency.

Solution Using the central-difference scheme, we obtain

$$L\phi = \frac{\partial \phi}{\partial t} - \alpha\frac{\partial^2 \phi}{\partial x^2} = \frac{\phi_j^{(n+1)} - \phi_j^{(n-1)}}{2\Delta t} - \frac{(\Delta t)^2}{6}\frac{\partial^3 \phi}{\partial t^3}$$

$$- \alpha\left[\frac{\phi_{j-1}^{(n)} - 2\phi_j^{(n)} + \phi_{j+1}^{(n)}}{h^2} - \frac{h^2}{12}\frac{\partial^4 \phi}{\partial x^4}\right]. \qquad (b)$$

From the forward and backward approximations of Table 4-2, we derive

$$-2\phi_j^{(n)} = -\phi_j^{(n-1)} - \phi_j^{(n+1)} + (\Delta t)^2\frac{\partial^2 \phi}{\partial t^2}, \qquad (c)$$

which is substituted into Eq. (*b*) to yield

$$L\phi = \tilde{L}_{DF}\phi_j - \frac{(\Delta t)^2}{6}\frac{\partial^3 \phi}{\partial t^3} + \frac{\alpha h^2}{12}\frac{\partial^4 \phi}{\partial x^4} - \frac{\alpha(\Delta t)^2}{h^2}\frac{\partial^2 \phi}{\partial t^2}\,. \qquad (d)$$

The last term on the right-hand side of Eq. (*d*) contains the ratio $(\Delta t/h)^2$. When Δt and h approach zero with values of the same order of magnitude, the ratio remains finite and Eq. (*d*) becomes

$$\beta\frac{\partial^2 \phi}{\partial t^2} + \frac{\partial \phi}{\partial t} - \alpha\frac{\partial^2 \phi}{\partial x^2} = \tilde{L}_{DF}\phi_j, \qquad (e)$$

where $\beta = \alpha(\Delta t)^2/h^2$. Therefore, instead of solving the parabolic equation, we are actually solving a hyperbolic equation. To make the situation worse, Eq. (*a*) can be proved to be unconditionally stable. Based on the stable result, we may be led to a false conclusion that our computation is successful. Such hidden mistakes must be detected by consistency analysis.

The Du Fort-Frankel scheme is a typical three-level (or three-time-step) scheme for which the numerical stability criterion is derived slightly differently than for the two-level scheme. In the following example, we will use the von Neumann stability method to examine the stability of this scheme.

Example 4-13 Is the three-level Du Fort-Frankel scheme

$$(2r+1)\phi_j^{(n+1)} - 2r[\phi_{j-1}^{(n)} + \phi_{j+1}^{(n)}] + (2r-1)\phi_j^{(n-1)} = 0 \tag{a}$$

stable?

Solution As usual, we assume that the finite-difference solution

$$\phi_j^{(n)} = A_1 \xi^n \sin\left(\frac{j\pi}{J}\right), \qquad j = 1, 2, \ldots, J-1 \tag{b}$$

can represent a particular solution of Eq. (*a*) subject to two homogeneous Dirichlet boundary conditions.† Substituting Eq. (*b*) into Eq. (*a*) yields

$$(2r+1)\xi^2 - 4r\xi \cos\left(\frac{\pi}{J}\right) + 2r - 1 = 0. \tag{c}$$

Two roots of this quadratic equation are

$$\xi = \frac{2r\cos(\pi/J) \pm \sqrt{1 - 4r^2 \sin^2(\pi/J)}}{1+2r}. \tag{d}$$

It is inconvenient to prove $|\xi| \leqslant 1$ directly from Eq. (*d*). Instead, let us make an a priori assumption that $|\xi| \leqslant 1$ is true. If the derived result based on this assumption can be proved to be true and if the derivation is reversible, then this a priori assumption is proved to be valid. Hence we first suppose that

$$\xi \leqslant 1. \tag{e}$$

It then follows from Eq. (*d*) that

$$\pm\sqrt{1 - 4r^2 \sin^2\left(\frac{\pi}{J}\right)} \leqslant 1 + 2r\left[1 - \cos\left(\frac{\pi}{J}\right)\right]. \tag{f}$$

If the negative sign is taken, Eq. (*f*) is seen immediately to be true. If the positive sign is considered, Eq. (*f*) can be squared without changing the direction of the inequality. After a few steps, we obtain

†If boundary conditions of other types are present, we may have to replace $\sin(j\pi/J)$ with $\sin(mj\pi/J)$, $m \neq 1$, or with other harmonic functions.

$$0 \leqslant \left[1 - \cos\left(\frac{\pi}{J}\right)\right](8r^2 + 4r), \tag{g}$$

which, of course, is true. Next, suppose that

$$\xi \geqslant -1. \tag{h}$$

Equation (d) can be arranged to

$$\pm\sqrt{1 - 4r^2 \sin^2\left(\frac{\pi}{J}\right)} \geqslant -1 - 2r\left[1 + \cos\left(\frac{\pi}{J}\right)\right]. \tag{i}$$

If the positive sign is chosen, Eq. (i) is automatically valid. If the negative sign is taken, then the inequality must be reversed when Eq. (i) is squared. Keeping this in mind, we can derive

$$0 \leqslant \left[1 + \cos\left(\frac{\pi}{J}\right)\right](8r^2 + 4r), \tag{j}$$

which is always true. In light of Eqs. (g) and (j), we conclude that

$$|\xi| \leqslant 1. \tag{k}$$

Consequently, $\tilde{L}_{DF}\phi_j$ is proved to be a stable scheme.

For three-level schemes such as Eq. (a), it is also customary to replace the discretized equation with an equivalent two-level system [2, p. 43]. For example, Eq. (a) was replaced by

$$(2r + 1)\phi_j^{(n+1)} - 2r[\phi_{j-1}^{(n)} + \phi_{j+1}^{(n)}] + (2r - 1)\psi_j^{(n)} = 0 \tag{l}$$

and

$$\psi_j^{(n+1)} = \phi_j^{(n)}. \tag{m}$$

For the numerical stability of Eqs. (l) and (m) to be guaranteed, the two eigenvalues of the 2×2 amplification matrix must not be greater than zero. See [2, p. 72, chap. 4] for the proof.

Despite the danger of becoming inconsistent, the Du Fort-Frankel scheme remains attractive to many researchers [14–18] partly because it has second-order accuracy.

4-2*c* Unstable but Consistent Schemes

For completeness, we will present an example of an unstable but consistent scheme. It is recommended, however, that a scheme be abandoned regardless of other advantageous numerical properties if it is unstable.

Example 4-14 Consider the explicit scheme

$$\tilde{L}\phi_j^{(n)} = \frac{\phi_j^{(n+1)} - \phi_j^{(n)}}{\Delta t} - \alpha\delta^2\phi_j^{(n)} \tag{a}$$

as the difference equation of

$$L\phi = \frac{\partial \phi}{\partial t} - \alpha \frac{\partial^2 \phi}{\partial x^2}. \tag{b}$$

Prove that Eq. (*a*) is consistent with Eq. (*b*) and is unstable if r is taken to be, say, 1. Take $J = 4$.

Solution With the aid of Table 4-2, the truncation error τ can be derived as

$$\tau = L\phi - \tilde{L}\phi_j = -\frac{\Delta t}{2}\frac{\partial^2 \phi}{\partial t^2} + \frac{h^2 \alpha}{12}\frac{\partial^4 \phi}{\partial x^4}, \tag{c}$$

which approaches zero as $\Delta t,\ h \to 0$. Hence $\tilde{L}\phi_j$ is consistent with $L\phi$. Next, with $r = 1$ and $J = 4$, Eq. (*a*) can be written as

$$\begin{Bmatrix} \phi_1^{(n+1)} \\ \phi_2^{(n+1)} \\ \phi_3^{(n+1)} \end{Bmatrix} = \begin{bmatrix} -1 & 1 & 0 \\ 1 & -1 & 1 \\ 0 & 1 & -1 \end{bmatrix} \begin{Bmatrix} \phi_1^{(n)} \\ \phi_2^{(n)} \\ \phi_3^{(n)} \end{Bmatrix} + \begin{Bmatrix} \phi_0 \\ 0 \\ \phi_4 \end{Bmatrix}. \tag{d}$$

The three eigenvalues of the matrix were found in Examples 4-4 to be $\lambda = -1$, $-1 + \sqrt{2}$, and $-1 - \sqrt{2}$. Since the absolute magnitude of the last eigenvalue is greater than 1, Eq. (*d*) is unstable.

4-3 CONVERGENCE

Convergence is probably the most important numerical property of a discretized scheme. The reason for this will become clear in the present section. While the two numerical properties just discussed are primarily associated with finite-difference schemes involving Taylor's series expansion, convergence is related to most numerical methods. The definition of convergence is stated according to the category of the numerical technique.

4-3*a* Definition

Let us first consider finite-difference schemes. Let $\phi_{j,k}^{(n)}$ denote the nodal solution and $\phi(j\Delta x, k\Delta y, n\Delta t)$ the exact solution at that nodal point. The finite-difference solution or scheme is said to be convergent if

$$|\phi(j\Delta x, k\Delta y, n\Delta t) - \phi_{j,k}^{(n)}| \to 0 \quad \text{as } \Delta x, \Delta y, \Delta t \to 0. \tag{4-36}$$

For the class of methods employing inner-product integrals such as Galerkin and variational methods, the conventional definition of convergence is that

$$\|\phi(x, y, t) - \tilde{\phi}(x, y, t)\| \to 0 \quad \text{as } \Delta x, \Delta y, \Delta t \to 0, \tag{4-37}$$

which will be further explained as follows. First, ϕ and $\bar{\phi}$ denote the exact solution and the approximate solution, respectively. The symbol $\| \quad \|$ denotes a certain type of norm, usually the Sobolev norm defined as

$$\|e\| = \left\{ \iint_{\Omega} \left[e^2 + \left(\frac{\partial e}{\partial x} \right)^2 + \left(\frac{\partial e}{\partial y} \right)^2 \right] dx\, dy \right\}^{1/2}, \tag{4-38}$$

where $e \epsilon C^0$, Ω is the domain of interest, and the partial differential equation is of second order in x, y and of first order in t. The Sobolev norm and space are described in more detail in Chapter 14.

When there is a nonlinearity in the differential or algebraic equations, or when a large linear matrix system is to be solved efficiently, iterative schemes are usually used. An iterative solution or scheme is said to be convergent if

$$\frac{|\phi^{[k+1]} - \phi^{[k]}|}{\phi^{[k]}} \to 0 \qquad \text{as } k \to \infty, \tag{4-39}$$

where the superscript $[k]$ denotes the kth iteration. We note that the convergence criterion shown in Eq. (4-39) does not guarantee the accuracy of the numerical solution for iterative schemes.

In the following sections we will discuss convergence criteria for these three classes of numerical methods.

4-3*b* Finite-Difference Formulations

According to the Lax theorem [19], the convergence of a finite-difference scheme is ensured if and only if the scheme is both stable and consistent. While the rigorous proof is omitted here, we will present three simple examples to support the theorem by showing that the scheme (or the numerical solution) (1) diverges if it is unstable but consistent; (2) diverges if it is stable but inconsistent, and (3) converges if it is both stable and consistent.

Example 4-15 Consider the finite-difference scheme

$$\phi_j^{(n+1)} = \phi_{j-1}^{(n)} - \phi_j^{(n)} + \phi_{j+1}^{(n)} \tag{a}$$

whose instability and consistency were established in Example 4-14. Is Eq. (*a*) divergent? Take $J = 10$, $\phi_0^{(n)} = \phi_J^{(n)} = 0$ if $n \geqslant 1$, and $\phi_0^{(0)} = \phi_1^{(0)} = \dots = \phi_{10}^{(0)} = 1$.

Solution The analytical solution to the heat conduction equation

$$\frac{\partial \phi}{\partial t} = \alpha \frac{\partial^2 \phi}{\partial x^2}, \qquad \phi(0, t) = \phi(L, t) = 0, \qquad \phi(x, 0) = 1 \tag{b}$$

can be obtained by separation of variables as

$$\phi(x, t) = \sum_{m=1}^{\infty} \frac{2}{m\pi}(1 - \cos m\pi) \exp\left(-\frac{\alpha m^2 \pi^2 t}{L^2}\right) \sin\left(\frac{m\pi x}{L}\right). \tag{c}$$

For the purpose of comparison with the approximate solution later, we choose

$$\frac{x}{L} = 0.2 \quad \text{and} \quad \frac{\alpha t}{L^2} = 0.03. \tag{d}$$

Therefore,

$$\phi\left(0.2L, \frac{0.03L^2}{\alpha}\right) \approx \frac{4}{\pi} \exp(-0.03\pi^2) \sin(0.2\pi)$$

$$+ \frac{4}{3\pi} \exp(-0.27\pi^2) \sin(0.6\pi) = 0.5847. \tag{e}$$

Corresponding to the numerical values given in Eq. (*d*) are

$$j = \left(\frac{x}{L}\right) J = 2 \tag{f}$$

and

$$n = \left(\frac{\alpha t}{L^2}\right) \frac{J^2}{r} = 3. \tag{g}$$

The numerical solution obtained with Eq. (*a*) is tabulated as

t	ϕ_0	ϕ_1	ϕ_2	ϕ_3	ϕ_4	ϕ_5	ϕ_6	$\cdots$
0	1	1	1	1	1	1	1	$\cdots$
Δt	0	1	1	1	1	1	1	$\cdots$
$2\Delta t$	0	0	1	1	1	1	1	$\cdots$
$3\Delta t$	0	1	0	1	1	1	1	$\cdots$

It follows that

$$\left|\phi\left(0.2L, \frac{0.03L^2}{\alpha}\right) - \phi_2^{(3)}\right| = |0.5847 - 0| = 0.5847. \tag{h}$$

Could this error be due to the large value of h or Δt? Let us increase J to 20 such that r remains 1. Correspondingly,

$$j = 4 \quad \text{and } n = 12. \tag{i}$$

We continue to tabulate the numerical solution as

t	ϕ_0	ϕ_1	ϕ_2	ϕ_3	ϕ_4	ϕ_5
...	...	...	...	...	...	...
$4\Delta t$	0	−1	2	0	1	1
$5\Delta t$	0	3	−3	3	0	1
$6\Delta t$	0	−6	9	−6	4	0
...	...	...	...	...	...	...
$12\Delta t$	0	−1585	2628	−2872	2441	...

t	ϕ_6	ϕ_7	ϕ_8	ϕ_9	ϕ_{10}	ϕ_{11}
...	...	...	...	...	...	...
$4\Delta t$	1	1	1	1	1	1
$5\Delta t$	1	1	1	1	1	1
$6\Delta t$	1	1	1	1	1	1
...	...	...	...	...	...	...
$12\Delta t$	...	...	...	...	...	...

Therefore,

$$\left|\phi\left(0.2L, \frac{0.03L^2}{\alpha}\right) - \phi_4^{(12)}\right| = |0.5847 - 2441| = 2440.6. \qquad (j)$$

In comparing Eq. (*j*) with Eq. (*h*), it is seen that the error increases even when h and Δt are reduced. It is most likely that the scheme is divergent.

Example 4-16 The Du Fort-Frankel scheme

$$(1 + 2r)\phi_j^{(n+1)} - 2r\phi_{j-1}^{(n)} - 2r\phi_{j+1}^{(n)} + (2r - 1)\phi_j^{(n-1)} = 0 \qquad (a)$$

is stable but inconsistent. Take $r = 6$ and $J = 20$. Is Eq. (*a*) convergent or divergent?

Solution Since Eq. (*a*) has three time levels, we lack information about the initial values $\phi_j^{(1)}$ at $n = 1$. We may rely on the implicit scheme

$$-r\phi_{j-1}^{(n)} + (2r + 1)\phi_j^{(n)} - r\phi_{j+1}^{(n)} = \phi_j^{(n-1)} \qquad (b)$$

or, with $r = 6$ and $n = 1$,

$$-6\phi_{j-1}^{(1)} + 13\phi_j^{(1)} - 6\phi_{j+1}^{(1)} = \phi_j^{(0)} \qquad (c)$$

to provide these initial values. They are shown, in the second row of the following table. Values for $n \geqslant 2$ are then computed by using Eq. (*a*) with $r = 6$, i.e.,

$$13\phi_j^{(n+1)} - 12\phi_{j-1}^{(n)} - 12\phi_{j+1}^{(n)} + 11\phi_j^{(n-1)} = 0. \qquad (d)$$

The analytical solution at $x = 0.2L$ and $t = 0.06L^2/\alpha$ that corresponds to $n = 4$ and $j = 4$ is

t	ϕ_0	ϕ_1	ϕ_2	ϕ_3	ϕ_4	ϕ_5
0	1	1	1	1	1	1
Δt	0.0	0.3331	0.5550	0.7028	0.8010	0.8661
$2\Delta t$	0.0	−0.3338	0.110	0.4056	0.6020	0.7321
$3\Delta t$	0.0	−0.1803	−0.4034	0.0626	0.3724	0.5776
$4\Delta t$	0.0	−0.0899	−0.2017	−0.3718	0.0816	0.3819
t	ϕ_6	ϕ_7	ϕ_8	ϕ_9	ϕ_{10}	
0	1	1	1	1	1	
Δt	0.9088	0.9364	0.9533	0.9624	0.9653	
$2\Delta t$	0.8176	0.8727	0.9066	0.9249	0.9307	
$3\Delta t$	0.7124	0.7993	0.8527	0.8815	0.8907	
$4\Delta t$	0.5791	0.7063	0.7844	0.8266	0.840	

$$\phi\left(0.2L, \frac{0.06L^2}{\alpha}\right) \approx \frac{4}{\pi} \exp(-0.06\pi^2) \sin(0.2\pi) + \frac{4}{3\pi} \exp(-0.54\pi^2) \sin(0.6\pi) = 0.4160. \qquad (e)$$

Therefore

$$\left|\phi\left(0.2L, \frac{0.06L^2}{\alpha}\right) - \phi_4^{(4)}\right| = |0.4160 - 0.0816| = 0.3344, \qquad (f)$$

which suggests that Eq. (*a*) is divergent from the heat conduction equation (but convergent to a certain hyperbolic equation).

Example 4-17 According to Eq. (*c*) in Example 4-2 with $r = \frac{1}{4}$, the stability criterion Eq. (4-28), and Example 4-14, the explicit scheme

$$\phi_j^{(n+1)} = \frac{1}{4} \phi_{j-1}^{(n)} + \frac{1}{2} \phi_j^{(n)} + \frac{1}{4} \phi_{j+1}^{(n)} \qquad (a)$$

is both stable and consistent. Is it also convergent? Assume $\phi_0^{(n)} = \phi_5^{(n)} = 0$ if $n \geqslant 1$ and $\phi_0^{(0)} = \phi_1^{(0)} = \quad .. = \phi_5^{(0)} = 1$.

Solution Using Eq. (*a*), we obtain the following tabulated solution:

t	ϕ_0	ϕ_1	ϕ_2	ϕ_3	ϕ_4	ϕ_5
0	1	1	1	1	1	1
Δt	0	1	1	1	1	0
$2\Delta t$	0	$\frac{3}{4}$	1	1	$\frac{3}{4}$	0
$3\Delta t$	0	$\frac{5}{8}$	$\frac{15}{16}$	$\frac{15}{16}$	$\frac{5}{8}$	0

The analytical solution $\phi(0.2L, 0.03L^2/\alpha)$ was given in Eq. (*e*) of Example 4-15 as 0.585, while the corresponding discrete solution $\phi_1^{(3)}$ is found to be $\frac{5}{8}$. The fact that they agree within 7% suggests that Eq. (*a*) is convergent.

4-3*c* Galerkin and Variational Finite-Element Formulations

We mentioned the definition of convergence for the finite-element method in Section 4-3*a*. Here, on the basis of this definition, we will examine the sufficient conditions for the finite-element solution (computed from the inner product) to converge to the true solution. To establish these conditions, we may divide the logical path that leads to the convergence criteria into two steps. The first step is to show that, if ϕ is the solution of the differential equation

$$L\phi = f,$$

where L is a linear, positive-definite, self-adjoint operator, then

$$\|\phi - \tilde{\phi}\| \leqslant \|\phi - \bar{\phi}\|, \tag{4-40}$$

where $\tilde{\phi}$ is the approximate finite-element solution and $\bar{\phi}$ is a polynomial interpolant. At the nodal points $\bar{\phi}$ coincides with ϕ, whereas $\tilde{\phi}$ may not. These three functions ϕ, $\tilde{\phi}$, and $\bar{\phi}$ are depicted in Fig. 4-3 for a fictitious exact solution ϕ. The proof of Eq. (4-40), which requires some mathematical background in functional analysis, will be delayed until Chapter 14. The second logical step following Eq. (4-40) is then to show

$$\|\phi - \bar{\phi}\| \leqslant Mh^{\nu}, \tag{4-41}$$

where h is the interval size, M is a positive number free of h, and $\nu \geqslant 1$. We will prove Eq. (4-41) here for the special case that $\bar{\phi}$ is a piecewise linear interpolant. The proof for the case where $\bar{\phi}$ is quadratic is presented in Chapter 14.

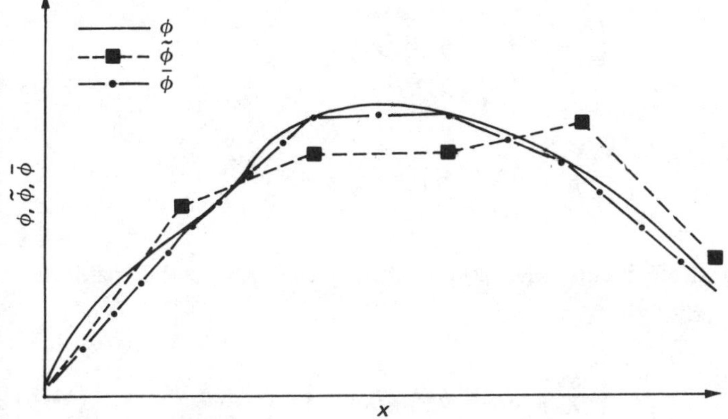

Figure 4-3 Fictitious exact solution ϕ, its finite-element solution $\tilde{\phi}$, and its linear interpolation function $\bar{\phi}$.

Define the error e as

$$e = \phi - \bar{\phi}. \tag{4-42}$$

Then if $\bar{\phi}$ is piecewise linear we obtain

$$\frac{d^2 e}{dx^2} = \frac{d^2 \phi}{dx^2}, \quad x \epsilon [x_{j-1}, x_j]. \tag{4-43}$$

Since $\bar{\phi}$ coincides with ϕ pointwise at $x = x_{j-1}$, x_j, it follows that

$$e(x_{j-1}) = 0 \quad \text{and } e(x_j) = 0. \tag{4-44}$$

Recalling Rolle's theorem in calculus, we know that there exists a $\xi_1 \epsilon (x_{j-1}, x_j)$†

$$\left(\frac{de}{dx}\right)_{x=\xi_1} = 0. \tag{4-45}$$

Consequently,

$$\frac{de}{dx} = \frac{de}{dx} - \left(\frac{de}{dx}\right)_{x=\xi_1} = \int_{\xi_1}^{x} \frac{d^2 e}{dt^2}\, dt \leqslant (x - \xi_1) \max \left|\frac{d^2 e}{dx^2}\right| \leqslant h \max \left|\frac{d^2 e}{dx^2}\right|$$

or

$$\left(\frac{de}{dx}\right)^2 \leqslant h^2 M_j, \tag{4-46}$$

where $M_j = \{\max|d^2 e/dx^2|\}^2$, $x \epsilon [x_{j-1}, x_j]$. Integrating Eq. (4-46) over the interval $[x_{j-1}, x_j]$, we obtain

$$\int_{x_{j-1}}^{x_j} \left(\frac{de}{dx}\right)^2 dx \leqslant h^2 M_j (x_j - x_{j-1}). \tag{4-47}$$

Similarly, for other intervals, we can also derive

$$\int_0^{x_1} \left(\frac{de}{dx}\right)^2 dx \leqslant h^2 M_1 x_1,$$

$$\int_{x_1}^{x_2} \left(\frac{de}{dx}\right)^2 dx \leqslant h^2 M_2 (x_2 - x_1),$$

and so forth. Summing these inequalities and assuming that M_k is the maximum among $M_1, M_2, \ldots, M_J$, we obtain

$$\int_0^{L} \left(\frac{de}{dx}\right)^2 dx \leqslant h^2 M_k L. \tag{4-48}$$

†The parentheses here denote an open interval.

Before proceeding to establish the upper bound for $\int_0^L e^2\,dx$, we will present an example to illustrate Eq. (4-48).

Example 4-18 Let the error $e(x)$ be represented by $\sin(\pi x/h)$, which satisfies Eq. (4-44) at $x=0, h, 2h, \ldots, jh, \ldots, Jh$, where $J=L/h$. Show that Eq. (4-48) holds.

Solution We evaluate

$$\int_0^L \left(\frac{de}{dx}\right)^2 dx = \frac{L\pi^2}{2h^2} \tag{a}$$

and

$$\left|\frac{d^2e}{dx^2}\right| = \frac{\pi^2}{h^2}\sin\left(\frac{\pi x}{h}\right). \tag{b}$$

Since the maximum of $|\sin(\pi x/h)|$ is unity, it follows that

$$M_k = \frac{\pi^4}{h^4}{}^\dagger \tag{c}$$

or

$$h^2 M_k L = \frac{L\pi^4}{h^2}. \tag{d}$$

In comparing Eqs. (*a*) and (*d*), it is established that Eq. (4-48) is valid.

To seek the upper bound of $\int_0^L e^2\,dx$, we write

$$e(x) = e(x) - e(x_{j-1}) = \int_{x_{j-1}}^{x} \frac{de}{dx}\,dx, \qquad x\epsilon[x_{j-1}, x_j]. \tag{4-49}$$

Recalling the mean-value theorem in calculus, we derive

$$e(x) = \left(\frac{de}{dx}\right)_{x=\xi_2}(x - x_{j-1}), \qquad \xi_2\epsilon[x_{j-1}, x]. \tag{4-50}$$

Further, since Eq. (4-46) holds for all $x\epsilon[x_{j-1}, x_j]$, it follows that

$$e(x) \leqslant h\left(\frac{de}{dx}\right)_{x=\xi_2} \leqslant h^2 M_j^{1/2}. \tag{4-51}$$

By the same argument between Eqs. (4-47) and (4-48), we obtain

$$\int_0^L e^2\,dx \leqslant h^4 M_k L. \tag{4-52}$$

†In this example we can ignore the fact that M_k is a function of h.

Combining Eqs. (4-48) and (4-52), we finally derive

$$\|e\| = \left\{\int_0^L \left[e^2 + \left(\frac{de}{dx}\right)^2\right] dx\right\}^{1/2} \leqslant h[M_k L(1 + h^2)]^{1/2}. \tag{4-53}$$

In comparing Eqs. (4-41) and (4-53), it is found that the power ν is unity for the piecewise linear interpolation function.

The two logical steps expressed by Eqs. (4-40) and (4-41) immediately lead to

$$\|\phi - \tilde{\phi}\| \leqslant Mh^{\nu}. \tag{4-54}$$

Thus, as $h \to 0$, this Sobolev norm approaches zero and, according to Eq. (4-37), the finite-element solution $\tilde{\phi}$ converges to the exact solution ϕ. A numerical scheme must be able to provide convergent solutions in order to be considered for adoption. However, if it converges slowly, the scheme will not be attractive regardless of its convergent behavior. In Eq. (4-54), the power ν determines the rate of convergence for a given interpolation function. The higher the power becomes, the faster the finite-element solution converges. For quadratic C^0 elements presented in Section 3-1*a*, the power ν is 2. This result will be derived in Chapter 14.

4-3*d* Iteration Schemes

In this section we will concentrate on establishing convergence criteria for the iteration schemes used to solve the linear matrix system

$$[\mathbf{A}]\{\phi\} = \{\mathbf{B}\}, \tag{4-55}$$

where [A] is usually not tridiagonally banded so that the Thomas algorithm (see Section 5-1*b*) is not applicable. Also, it is of large dimension so that matrix inversion becomes inefficient. These iteration schemes should not be confused with those applied to nonlinear problems. For the nonlinear problems described in Section 3-4 the iteration techniques is also needed, but it is used to deal with the nonlinearity and not the linear matrix system. The convergence theory for the iteration techniques that are applied to nonlinear problems has not been as well developed as that for the linear systems [20] and it will be omitted here.

Instead of solving Eq. (4-55), most iteration schemes actually solve the following rearranged form:

$$\{\Phi^{[k]}\} = [\mathbf{G}]\{\phi^{[k-1]}\} + [\mathbf{H}]\{\mathbf{B}\}, \tag{4-56}$$

where the superscript $[k]$ denotes the kth iteration and the expressions for [G] and [H] depend on the scheme. The matrix [G] is the essential one that is associated with the convergent behavior of Eq. (4-56). If Eq. (4-56) is compared with Eq. (4-6), it is seen that these two equations are very similar mathematically. Using the argument described between Eqs. (4-6) and (4-7*c*), we may also derive

$$\{\phi^{[k+1]}\} = [\mathbf{G}]^{k+1}\{\phi^{[0]}\} + \cdots. \tag{4-57}$$

In light of Eq. (4-57), we assert that the iteration scheme converges if all the absolute eigenvalues of [G] are not greater than 1. In this section, we will discuss the Gauss-Seidel method [21] exclusively. Other iterative schemes will be described in Chapter 5.

In the Gauss-Seidel method, the calculation of $\phi_1^{[k]}$ is based on the previously iterated values $\phi_2^{[k-1]}, \ldots, \phi_3^{[k-1]}, \ldots, \phi_{J-1}^{[k-1]}$. After $\phi_1^{[k]}$ is calculated, the subsequent calculation of $\phi_2^{[k]}$ is then based on this updated $\phi_1^{[k]}$ as well as the previously iterated values $\phi_3^{[k-1]}$, $\phi_4^{[k-1]}, \ldots, \phi_{J-1}^{[k-1]}$. The same algorithm is repeated, and finally the calculation of $\phi_{J-1}^{[k]}$ is based on all the updated values of other unknowns.

Example 4-19 Use the Gauss-Seidel method to solve the following linear system:

$$\begin{aligned} 2x_1 + x_2 + x_3 &= 7, \\ x_1 + 2x_2 + x_3 &= 8, \\ x_1 + x_2 + 2x_3 &= 9. \end{aligned} \tag{a}$$

Solution Let the initial guesses arbitrarily be

$$x_1^{[0]} = x_2^{[0]} = x_3^{[0]} = 1. \tag{b}$$

Then Eq. (*a*) can be rearranged to

$$\begin{aligned} 2x_1^{[1]} &= -x_2^{[0]} - x_3^{[0]} + 7 = 5, \\ x_1^{[1]} + 2x_2^{[1]} &= 0 - x_3^{[0]} + 8 = 7, \\ x_1^{[1]} + x_2^{[1]} + 2x_3^{[1]} &= 0 + 0 + 9 = 9. \end{aligned} \tag{c}$$

Successively (not simultaneously), we solve Eq. (*c*) to obtain

$$x_1^{[1]} = \frac{5}{2}, \quad x_2^{[1]} = \frac{9}{4}, \quad \text{and } x_3^{[1]} = \frac{17}{8}. \tag{d}$$

For the second-round iteration, we write

$$\begin{aligned} 2x_1^{[2]} &= -x_2^{[1]} - x_3^{[1]} + 7 = \frac{21}{8} \\ x_1^{[2]} + 2x_2^{[2]} &= 0 - x_3^{[1]} + 8 = \frac{47}{8}, \\ x_1^{[2]} + x_2^{[2]} + 2x_3^{[2]} &= 0 + 0 + 9 = 9, \end{aligned} \tag{e}$$

which is again successively solved to yield

$$x_1^{[2]} = \frac{21}{16}, \quad x_2^{[2]} = \frac{73}{32}, \quad \text{and } x_3^{[2]} = \frac{173}{64}. \tag{f}$$

After a few more iterations, the iterated solution converges to the exact solution, namely

$$x_1 = 1, \quad x_2 = 2, \quad \text{and } x_3 = 3. \tag{g}$$

Using the convergence criterion given in Eq. (4-57), we can also show that, for the system (*a*) of Example 4-19, the Gauss-Seidel method converges.

Example 4-20 Consider again the system (*a*) of Example 4-19. Based on Eq. (4-57), prove that the Gauss-Seidel method is convergent.

Solution In reference to Eqs. (*c*) and (*e*), we can write the general equation

$$\begin{bmatrix} 2 & 0 & 0 \\ 1 & 2 & 0 \\ 1 & 1 & 2 \end{bmatrix} \begin{Bmatrix} x_1^{[k+1]} \\ x_2^{[k+1]} \\ x_3^{[k+1]} \end{Bmatrix} = \begin{bmatrix} 0 & -1 & -1 \\ 0 & 0 & -1 \\ 0 & 0 & 0 \end{bmatrix} \begin{Bmatrix} x_1^{[k]} \\ x_2^{[k]} \\ x_3^{[k]} \end{Bmatrix} + \begin{Bmatrix} 7 \\ 8 \\ 9 \end{Bmatrix}. \tag{a}$$

Performing the inversion of the matrix yields, after lengthy arithmetic,

$$\begin{Bmatrix} x_1^{[k+1]} \\ x_2^{[k+1]} \\ x_3^{[k+1]} \end{Bmatrix} = \begin{bmatrix} 0 & -\frac{1}{2} & -\frac{1}{2} \\ 0 & \frac{1}{4} & -\frac{1}{4} \\ 0 & \frac{1}{8} & \frac{3}{8} \end{bmatrix} \begin{Bmatrix} x_1^{[k]} \\ x_2^{[k]} \\ x_3^{[k]} \end{Bmatrix} + \begin{Bmatrix} \frac{7}{2} \\ \frac{9}{4} \\ \frac{13}{8} \end{Bmatrix}. \tag{b}$$

Next, the task is to prove that the absolute eigenvalues of the matrix $[\mathbf{G}]$, defined as the coefficient matrix in Eq. (*b*), or the roots of the determinant

$$\begin{vmatrix} -\lambda & -\frac{1}{2} & -\frac{1}{2} \\ 0 & \frac{1}{4}-\lambda & -\frac{1}{4} \\ 0 & \frac{1}{8} & \frac{3}{8}-\lambda \end{vmatrix} = 0, \tag{c}$$

are not greater than 1. We reduce Eq. (*c*) to

$$\lambda(8\lambda^2 - 5\lambda + 1) = 0, \tag{d}$$

which has the three roots

$$\lambda = 0, \quad \frac{5+\sqrt{7}i}{16}, \quad \text{and } \frac{5-\sqrt{7}i}{16}.$$

Since the absolute values of these complex numbers are less than 1, we conclude that, for matrix system (*a*), the Gauss-Seidel method is convergent.

4-4 ACCURACY AND ERROR BOUNDS

When the exact solution of a governing differential equation is available, the accuracy of a numerical solution can be easily checked; the closer the numerical solution is to the exact solution, the more accurate the numerical scheme is. In numerical analysis, however, the accuracy that will be examined is the one when the exact solution is unavailable.

4-4*a* Accuracy for Finite-Difference Methods

Finite-difference schemes are said to be accurate if the truncation errors τ of the difference equations, defined as

$$\tau = L\phi - \tilde{L}\phi_j^{(n)},$$

are small. There are at least two ways to reduce the truncation error. First, we may simply decrease the mesh size. While it is easy to do so, this reduction will sometimes undesirably increase the round-off error and will definitely increase the computing time. The second alternative, especially applicable to transient equations, is to choose a proper mesh-size ratio such that the higher derivatives cancel each other. One well-known example [22] is the choice of $\alpha\Delta t/h^2$ to be equal to $\frac{1}{6}$ in the explicit scheme

$$\tilde{L}\phi_j^{(n)} = \frac{\phi^{(n+1)} - \phi^{(n)}}{\Delta t} + \frac{\alpha}{h^2}\left[\phi_{j-1}^{(n)} - 2\phi_j^{(n)} + \phi_{j+1}^{(n)}\right] = 0, \tag{4-58}$$

which is the discretization equation of

$$L\phi = \frac{\partial\phi}{\partial t} - \alpha\frac{\partial^2\phi}{\partial x^2} = 0. \tag{4-59}$$

With the aid of Table 4-2, we obtain

$$L\phi = \tilde{L}\phi_j + \frac{\alpha h^2}{12}\frac{\partial^4\phi}{\partial x^4} - \frac{\Delta t}{2}\frac{\partial^2\phi}{\partial t^2} + \mathrm{O}(h^4) + \mathrm{O}[(\Delta t)^2]. \tag{4-60}$$

But since

$$\frac{\partial}{\partial t}\left(\frac{\partial\phi}{\partial t}\right) = \frac{\partial}{\partial t}\left(\alpha\frac{\partial^2\phi}{\partial x^2}\right) = \alpha\frac{\partial^2}{\partial x^2}\left(\frac{\partial\phi}{\partial t}\right) = \alpha^2\frac{\partial^4\phi}{\partial x^4}, \tag{4-61}$$

Eq. (4-60) can be rearranged to

$$L\phi = \tilde{L}\phi_j^{(n)} + \frac{\alpha h^2}{2}\frac{\partial^4\phi}{\partial x^4}\left(\frac{1}{6} - r\right) + \frac{\alpha h^4}{6}\frac{\partial^6\phi}{\partial x^6}\left(\frac{1}{60} - r^2\right) + \mathrm{O}[(\Delta t)^2]. \tag{4-62}$$

If r is chosen to be $\frac{1}{6}$, it is seen that the accuracy of $\tilde{L}\phi_j^{(n)}$ is increased to $\mathbf{O}[h^4, (\Delta t)^2]$.

4-4*b* Error Bounds for Finite-Element Methods

With regard to the finite-element method, since Taylor's series expansion is not employed, we generally can no longer speak of truncation errors (unless these errors are defined in a broader sense [23]). Instead, the accuracy of a finite-element scheme is measured by the error bound represented by a certain type of norm [see Eq. (4-38) and Chapter 14]

$$\|e\| = \|\phi - \tilde{\phi}\| \leqslant Mh^{\nu}. \tag{4-63}$$

Finite-element schemes are said to be accurate if the power ν is high. In one-dimensional problems, ν is equal to 1 for piecewise linear basis functions and 2 for piecewise quadratic basis functions. Equation (4-63) is also the criterion for the convergence rate of a finite-element scheme, and it is not always possible to obtain Eq. (4-63) for a given differential equation. We will discuss these topics in greater detail in Chapters 14 and 15.

4-5 EFFICIENCY

A computation algorithm can be vaguely said to be efficient if it requires both little memory storage [24, 25] and short central processor unit (CPU) time. The memory storage includes specifications of the data and instructions; the CPU time is proportional to the number and the length of instructions. Instead of attempting to describe a general efficiency criterion, which may not exist at all, let us present a simple example to show how the memory storage and CPU time of two algorithms can be estimated and thereby show how their efficiencies are evaluated.

Example 4-21 Consider the following simple matrix system:

$$\begin{aligned} 2x + y &= 4 \\ x + 2y &= 5 \qquad \text{or } [\mathbf{A}]\{\mathbf{X}\} = \{\mathbf{B}\}. \end{aligned} \tag{a}$$

In order to solve for x and y, how many data must we store in the computer memory and how many instructions (assuming that addition, subtraction, multiplication, division, and equating all have the same speed) must we perform when (*a*) the Gauss-Seidel method and (*b*) matrix inversion are used?

Solution (*a*) For the sake of bookkeeping, we construct Table 4-3 as follows. To obtain $x^{[1]}$ in the second column, we must multiply a_{12} by $y^{[0]}$ (one instruction), subtract the product from b_1 (one instruction), divide the result

Table 4-3 Assessment of the efficiency of the Gauss-Seidel method

k	$x^{[k]}$	$y^{[k]}$	Data that need to be stored	Number of instructions
0	0	0	x, y, k	0
1	$\dfrac{b_1 - a_{12}y^{[0]}}{a_{11}} = 2$	$\dfrac{b_2 - a_{21}x^{[1]}}{a_{22}} = \dfrac{5}{2}$	$b_1, b_2, a_{11}, a_{12}, a_{21}, a_{22}$	8
2	$\dfrac{b_1 - a_{12}y^{[1]}}{a_{11}} = \dfrac{3}{4}$	$\dfrac{b_2 - a_{21}x^{[2]}}{a_{22}} = \dfrac{17}{8}$	0	8
3	$\dfrac{b_1 - a_{12}y^{[2]}}{a_{11}} = \dfrac{15}{16}$	$\dfrac{b_2 - a_{21}x^{[3]}}{a_{22}} = \dfrac{65}{32}$	0	8

by a_{11} (one instruction), and then equate the result to $x^{[1]}$ (one instruction). Likewise $y^{[1]}$ is computed by 4 operations. Therefore, we need a total of 8 operations to construct the second row. If we terminate the iteration procedure at $k = 3$ (the exact solution is $x = 1$ and $y = 2$), then the number of data that must be stored in the memory is 9 and the number of instructions is 24.

(*b*) Now consider matrix inversion. The procedures are list in Table 4-4. From Table 4-4 we find that the number of stored data is 13 and the number of instructions is 20. We should not be impressed by the smaller number of instructions required for matrix inversion. The fact is that as the dimension of

Table 4-4 Assessment of the efficiency of the matrix inversion scheme

Procedures	Data that need to be stored	Number of instructions
Find $\lvert A\rvert = a_{11}a_{22} - a_{12}a_{21}$	$\lvert A\rvert, a_{11}, a_{12}, a_{21}, a_{22}$	4
Find $[\mathbf{C}] = [\mathbf{A}]^{-1}$, i.e., $c_{11} = \dfrac{a_{22}}{\lvert A\rvert}, c_{12} = \dfrac{-a_{12}}{\lvert A\rvert}$, $c_{21} = -\dfrac{a_{21}}{\lvert A\rvert}, c_{22} = \dfrac{a_{22}}{\lvert A\rvert}$	$c_{11}, c_{12}, c_{21}, c_{22}$	8
Calculate $\{\mathbf{X}\} = [\mathbf{C}]\{\mathbf{B}\}$, i.e., $x = c_{11}b_1 + c_{12}b_2 = 1$, $y = c_{21}b_1 + c_{22}b_2 = 2$	x, y, b_1, b_2	8

matrix [A] becomes large, the matrix inversion, especially the computation of matrix [C], will eventually require much more memory and cpu time than the iterative scheme. Therefore, on the basis of this example, we may conclude that the iterative scheme is generally more efficient than matrix inversion.

SYMBOLS

e	error of interpolation function $(= \phi - \bar{\phi})$, or of approximate function $(= \phi - \tilde{\phi})$
[I]	identity matrix
J	largest number of index $j(= L/h)$
$\tilde{L}$	finite-difference operator derived from differential operator L
M_j	convenient term $(= \{\max \lvert d^2e/dx^2 \rvert\}^2)$
r	mesh-size ratio $(= \alpha\Delta t/h^2)$
[S]	standard matrix [Eq. (4-19)]
λ	eigenvalues of a matrix
μ_j	eigenvalues of matrix [S]
ξ	amplification factor [Eq. (4-31)]
τ	truncated high-order terms

Subscript

DF	Du Fort-Frankel scheme

Superscripts

(k)	at $t = k\Delta t$
$[k]$	at kth iteration

REFERENCES

1. G. E. Forsythe and W. R. Wasow, *Finite Difference Methods for Partial Differential Equations,* pp. 92–95, Wiley, New York, 1960.
2. R. D. Richtmyer and K. W. Morton, *Difference Methods for Initial-Value Problems,* chap. 4, Wiley-Interscience, New York, 1967.
3. W. F. Ames, *Numerical Methods for Partial Differential Equations,* chap. 2, Academic, New York, 1977.
4. R. Courant, E. Isaacson, and M. Rees, On the Solution of Nonlinear Hyperbolic Differential Equations by Finite Differences, *Commun. Pure Appl. Math.*, vol. 5, pp. 243–255, 1952.
5. K. E. Torrance and J. A. Rockett, Numerical Study of Natural Convection in an Enclosure with Localized Heating from Below, *J. Fluid Mech.*, vol. 36, pp. 33–54, 1969.
6. D. B. Spalding, A Novel Finite Difference Formulation for Differential Equations Involving both First and Second Derivatives, *Int. J. Numer. Methods Eng.*, vol. 4, pp. 551–559, 1972.
7. P. J. Roache, *Computational Fluid Dynamics,* p. 64, Hermosa, Albuquerque, N.M., 1976.
8. C. R. Wiley, *Advanced Engineering Mathematics,* p. 481, McGraw-Hill, New York, 1966.
9. J. von Neumann, Proposal and Analysis of a Numerical Method for the Treatment of Hydrodynamical Shock Problems, *Natl. Def. Res. Committee Rept. AM-551,* March 1944.

10. G. G. O'Brien, M. A. Hyman, and S. Kaplan, A Study of the Numerical Solution of Partial Differential Equations, *J. Math. Phys.*, vol. 29, pp. 223-251, 1951.
11. S. H. Crandall, *Engineering Analysis*, pp. 376-380, McGraw-Hill, New York, 1956.
12. H. O. Kreiss, On Difference Approximation of the Dissipative Type for Hyperbolic Differential Equations, *Commun. Pure Appl. Math.*, vol. 17, pp. 335-353, 1964.
13. E. C. Du Fort and S. P. Frankel, Stability Conditions in the Numerical Treatment of Parabolic Differential Equations, *Math. Tables Other Aids Comput.*, vol. 7, pp. 135-152, 1953.
14. I. K. Madni and R. H. Pletcher, Buoyant Jets Discharging Nonvertically into a Uniform, Quiescent Ambient–A Finite Difference Analysis and Turbulence Modelling, *ASME J. Heat Transfer*, vol. 99, pp. 641-646, 1977.
15. J. W. Ou and K. C. Cheng, Natural Convection Effects on Graetz Problem in Horizontal Isothermal Tubes, *Int. J. Heat Mass Transfer*, vol. 20, pp. 953-960, 1977.
16. C. R. Gane and P. L. Stephenson, An Explicit Numerical Method for Solving Transient Combined Heat Conduction and Convection Problems, *Int. J. Numer. Methods Eng.*, vol. 14, pp. 1141-1163, 1979.
17. A. F. Emery, P. K. Neighbors, and F. B. Gessner, The Numerical Prediction of Developing Turbulent Flow and Heat Transfer in a Square Duct, *ASME J. Heat Transfer*, vol. 102, pp. 51-57, 1980.
18. M. R. Malik and R. H. Pletcher, Calculation of Variable Property Heat Transfer in Ducts of Annular Cross Section, *Numer. Heat Transfer*, vol. 3, pp. 241-257, 1980.
19. P. D. Lax, Nonlinear Hyperbolic Equations, *Commun. Pure Appl. Math.*, vol. 6, p. 231, 1953.
20. L. A. Hageman and D. M. Young, Applied Iterative Methods, Academic, New York, 1981.
21. P. L. Seidel, Über ein Verfahren, die Gleichungen, auf welche die Methode der kleinsten Quadrate führt, sowie lineäre Gleichungen überhaupt, durch successive Annäherung aufzulösen, *Abh. Bayer. Akad.*, vol. 11, pp. 81-108, 1874.
22. W. E. Milne, *Numerical Solution of Differential Equations*, p. 122, Wiley-Interscience, New York, 1953.
23. M. J. P. Cullen and K. W. Morton, Analysis of Evolutionary Error in Finite Element and Other Methods, *J. Comput. Phys.*, vol. 34, pp. 245-267, 1980.
24. J. H. Williamson, Low-Storage Runge-Kutta Schemes, *J. Comput. Phys.*, vol. 35, pp. 48-56, 1980.
25. O. Axelsson and I. Gustafsson, On the Use of Preconditioned Conjugate Gradient Methods for Red-Black Ordered Five-Point Differences Schemes, *J. Comput. Phys.*, vol. 35, pp. 284-289, 1980.

PROBLEMS

4-1 Consider Example 4-1. The coefficients A, B, and C in Eqs. (a)-(c) satisfy case (2) of the stability criteria for steady-state systems. Find E_2, G_2, E_3, and G_3. Is E_j always negative? Is the sign of G_j alternating?

4-2 Repeat Example 4-3 by using the implicit scheme. Is the solution stable?

4-3 Show that at least one absolute eigenvalue of the matrix

$$[D] = \begin{bmatrix} -3 & 1 & 1 & 0 \\ 1 & -3 & 0 & 1 \\ 1 & 0 & -3 & 1 \\ 0 & 1 & 1 & -3 \end{bmatrix}$$

is greater than one. This implies that the solution of Eq. (e) in Example 4-3 is unstable.

4-4 Following the argument given in Section 4-1*b* (case 2), show that Eq. (4-1) with $A > 0$ and $C < 0$ can be stable if $\phi_0 = 0$. This proof implies that the solution of Eqs. (*a*)-(*c*) of Example 4-1 is stable.

4-5 Is the matrix

$$[\mathbf{D}] = \begin{bmatrix} 3 & 1 & 0 \\ 1 & 3 & 1 \\ 0 & 1 & 3 \end{bmatrix}$$

positive-definite? What are its eigenvalues and corresponding eigenvectors? Are these eigenvectors orthogonal to one another?

4-6 A transient one-dimensional heat conduction system is subject to the initial condition $\phi(x, 0) = \sin(\pi x/L)$ and the boundary conditions $\phi(0, t) = \phi(L, t) = 0$. The exact solution is given in Eq. (4-29*b*) with $A_1 = 1$. Use Eqs. (4-31) and (4-33) to find $\phi_2^{(5)}$ if $r = \frac{1}{6}$ and $J = 4$. Compare $\phi_2^{(5)}$ with the corresponding exact value.

4-7 Starting with Eq. (4-35), derive Eq. (4-34).

4-8 Use Newton's method described by

$$x^{[k+1]} = x^{[k]} - \frac{f(x^{[k]})}{f'(x^{[k]})}$$

to find the root within the interval $[-2, -1]$ of Eq. (*m*) of Example 4-7. Let $x^{[0]}$ be -1.

4-9 Apply the matrix method to prove that the three-level Du Fort Frankel scheme is stable. Hint: rewrite $[\mathbf{A}]\{\phi^{(n+1)}\} = [\mathbf{B}]\{\phi^{(n)}\} + [\mathbf{C}]\{\phi^{(n-1)}\}$ as $\{\psi^{(n+1)}\} = [\mathbf{P}]\{\psi^{(n)}\}$, where

$$\{\psi^{(n+1)}\} = \begin{Bmatrix} \phi^{(n+1)} \\ \phi^{(n)} \end{Bmatrix} \quad \text{and} \quad [\mathbf{P}] = \begin{bmatrix} [\mathbf{A}]^{-1}[\mathbf{B}] & [\mathbf{A}]^{-1}[\mathbf{C}] \\ [\mathbf{I}] & 0 \end{bmatrix}.$$

Then find the eigenvalues of $[\mathbf{P}]$.

4-10 Consider Eq. (*d*) of Example 4-14. Let $\phi_1^{(0)} = \phi_2^{(0)} = \phi_3^{(0)} = 1$. Express $\{\phi^{(3)}\} = \{\phi_1^{(3)}\ \phi_2^{(3)}\ \phi_3^{(3)}\}^T$ in terms of the two eigenvectors $\{\mathbf{V}_2\} = \{1\ \sqrt{2}\ 1\}$ and $\{\mathbf{V}_3\} = \{1\ -\sqrt{2}\ 1\}$ and the two corresponding eigenvalues $\lambda_2 = -1 + \sqrt{2}$ and $\lambda_3 = -1 - \sqrt{2}$. Do we see the unstable trend of the solution $\{\phi^{(3)}\}$?

4-11 We are given a set of two nonlinear equations

$$f(x, y) = x^2 - x + 2y^2 - 3y - 11 = 0 \tag{P4-11a}$$

and

$$g(x, y) = 2x^2 - 3x + y^2 - 4y + 1 = 0. \tag{P4-11b}$$

Solve Eqs. (P4-11*a*) and (P4-11*b*) by using Newton-Raphson method described by

$$\begin{Bmatrix} x^{[k+1]} \\ y^{[k+1]} \end{Bmatrix} = \begin{Bmatrix} x^{[k]} \\ y^{[k]} \end{Bmatrix} - [\mathbf{J}]^{-1} \begin{Bmatrix} f(x^{[k]}, y^{[k]}) \\ g(x^{[k]}, y^{[k]}) \end{Bmatrix},$$

where $[J]$ is the so-called Jacobian matrix, defined as

$$[\mathbf{J}] = \begin{bmatrix} \dfrac{\partial f}{\partial x} & \dfrac{\partial f}{\partial y} \\ \dfrac{\partial g}{\partial x} & \dfrac{\partial g}{\partial y} \end{bmatrix}.$$

Take $x^{[0]} = y^{[0]} = 1$.

4-12 Suppose that an exact solution is found to be

$$\phi(x) = \sin \pi x. \qquad \text{(P4-12}a\text{)}$$

An interpolating function $\bar{\phi}(x)$ is given by

$$\bar{\phi}(x) = \begin{cases} 2x, & x \in \left[0, \frac{1}{2}\right] \\ 2 - 2x, & x \in \left[\frac{1}{2}, 1\right] \end{cases}. \qquad \text{(P4-12}b\text{)}$$

With Eqs. (P4-12a) and (P4-12b) assess the validity of Eq. (4-48).

4-13 Use the Gauss-Seidel method to solve the linear system

$$\begin{aligned} -x_1 + 2x_2 + x_3 &= 6, \\ x_1 + 2x_2 + x_3 &= 8, \\ x_1 + x_2 + 2x_3 &= 9. \end{aligned} \qquad \text{(P4-13}a\text{)}$$

Does the method converge? What is the major difference between the coefficient matrix of Eq. (P4-13a) and that of Eq. (a) in Example 4-19?

4-14 Repeat Example 4-2 with $r = \frac{1}{6}$. Compare the finite-difference solution obtained with the exact solution.

4-15 Consider a heat-conduction system governed by

$$\frac{d^2\phi}{dx^2} + a^2\phi = 0, \quad \phi(0) = 0 \quad \text{and } \phi(1) = 100, \qquad \text{(P4-14)}$$

whose exact solution is $\phi(x) = 100 \sin ax/\sin a$. Use the finite difference method (with $h = \frac{1}{8}$) to solve Eq. (P4-14) for two cases: (a) $a = 3.1$ and (b) $a = 3.2$. Comment on the drastic difference between the solutions of these two cases.

PART TWO

FUNDAMENTAL HEAT TRANSFER MODES

CHAPTER

FIVE

HEAT CONDUCTION

In the preceding four chapters, we have described the uses as well as the numerical properties of most numerical methods employed in the computation of heat transfer problems. Beginning with the present chapter, we will apply this knowledge to analyses of the three basic heat transfer phenomena: conduction, convection, and radiation. Heat conduction will be considered first.

Transient heat conduction in a two-dimensional system with heat sources (or sinks) can be described by the partial differential equation

$$A\phi = \rho c_v \frac{\partial \phi}{\partial t} - \frac{\partial}{\partial x}\left(k \frac{\partial \phi}{\partial x}\right) - \frac{\partial}{\partial y}\left(k \frac{\partial \phi}{\partial y}\right) - S(x, y, t) = 0, \qquad (5\text{-}1)$$

where ρ, c_v, and k are, respectively, the density, heat capacity, and thermal conductivity of the medium; $S(x, y, t)$ represents the heat source, if positive. Equation (5-1) will be our model equation, and this equation and its special cases, subject to certain prescribed boundary conditions, will be considered throughout this chapter. When the heat conduction is one-dimensional and steady, the governing equation is in general analytically solvable and therefore will not be examined here.

In Section 5-1, the one-dimensional transient system is analyzed by the finite-difference method and the Galerkin finite-element method. The Thomas algorithm is briefly described. Finally, the system with temperature-dependent thermal conductivity is examined.

The two-dimensional steady-state system with a constant heat sink is considered in Section 5-2. We adopt finite-difference methods of various accuracies as well as the Galerkin finite-element method to discretize the governing equation. We also explain the inconvenience of the formulation using global basis functions. Finally, the procedures for a few iterative matrix solvers are presented.

Section 5-3 describes the two-dimensional transient system. The emphasis is placed on the stability criterion of the explicit scheme. In addition, the eigenvalue method that utilizes the eigenvalues of the conductance matrix is introduced.

5-1 ONE-DIMENSIONAL TRANSIENT SYSTEMS

For one-dimensional transient heat conduction systems with constant thermal conductivity and without heat generation, Eq. (5-1) reduces to

$$L\phi = \frac{\partial \phi}{\partial t} - \alpha \frac{\partial^2 \phi}{\partial x^2} = 0, \tag{5-2a}$$

where L is a linear operator to be distinguished from the nonlinear operator A. For convenience, we will consider the simple initial condition and boundary conditions

$$\phi(x, 0) = 1 \quad \text{and} \quad \phi(0, t) = \phi(L, t) = 0. \tag{5-2b}$$

The exact solution of Eqs. (5-2*a*) and (5-2*b*) is obtained by separation of variables as

$$\phi(x, t) = \sum_{m=1}^{\infty} \frac{2}{m\pi} (1 - \cos m\pi) \exp\left(\frac{-\alpha m^2 \pi^2 t}{L^2}\right) \sin\left(\frac{m\pi x}{L}\right), \tag{5-3}$$

which can be compared later with the numerical solution. We will now use several schemes to discretize Eq. (5-2*a*).

5-1*a* Finite-Difference Method (Explicit, Implicit, and Hybrid)

The classic finite-difference method is the explicit scheme [1, 2], in which the current nodal unknowns can be explicitly expressed in terms of the previous ones. With the aid of Table 4-2, Eq. (5-2*a*) is transformed into

$$\tilde{L}_{\text{ex}}\phi_j^{(n)} = \frac{\phi_j^{(n+1)} - \phi_j^{(n)}}{\Delta t} - \frac{\alpha}{h^2} [\phi_{j-1}^{(n)} - 2\phi_j^{(n)} + \phi_{j+1}^{(n)}] - \tau_{\text{ex}} = 0, \tag{5-4a}$$

where

$$n = \frac{t}{\Delta t}, \quad j = \frac{x}{h} = 1, 2, \ldots, J-1, \quad J = \frac{L}{h} \tag{5-4b}$$

and

$$\tau_{\text{ex}} = \frac{\Delta t}{2} \frac{\partial^2 \phi}{\partial t^2} - \frac{\alpha h^2}{12} \frac{\partial^4 \phi}{\partial x^4} + \mathbf{O}[(\Delta t)^2, h^4]. \tag{5-4c}$$

The subscript ex denotes explicit. Defining the mesh-size ratio r as

$$r = \frac{\alpha \Delta t}{h^2},$$

we rearrange Eq. (5-4*a*) into the familiar algebraic form

$$\phi_j^{(n+1)} = r\phi_{j-1}^{(n)} + (1 - 2r)\phi_j^{(n)} + r\phi_{j+1}^{(n)} + \mathbf{O}(\Delta t, h^2). \tag{5-5}$$

The stability of Eq. (5-5) is ensured if

$$r \leqslant \tfrac{1}{2}.$$

See Section 4-1 for the proof. Since information on the right-hand side of Eq. (5-5) is available, the solution procedure is relatively easy. Further, if r is chosen to be $\frac{1}{6}$, the accuracy of Eq. (5-5) is raised to $\mathbf{O}[(\Delta t)^2, h^4]$ (see Section 4-4).

The disadvantage of the explicit finite-difference scheme is that the size of the time step must be restricted to ensure numerical stability. Another scheme [3] that relaxes this restriction is to discretize Eq. (5-2*a*) to

$$\tilde{L}_{\text{im}}\phi_j^{(n)} = \frac{\phi_j^{(n)} - \phi_j^{(n-1)}}{\Delta t} - \frac{\alpha}{h^2}\left[\phi_{j-1}^{(n)} - 2\phi_j^{(n)} + \phi_{j+1}^{(n)}\right] - \tau_{\text{im}} = 0, \tag{5-6}$$

where

$$\tau_{\text{im}} = -\frac{\Delta t}{2}\frac{\partial^2 \phi}{\partial t^2} - \frac{\alpha h^2}{12}\frac{\partial^4 \phi}{\partial x^4} + \mathbf{O}[(\Delta t)^2, h^4].$$

The subscript im stands for implicit because Eq. (5-6) can be rearranged to

$$-r\phi_{j-1}^{(n)} + (2r + 1)\phi_j^{(n)} - r\phi_{j+1}^{(n)} = \phi_j^{(n-1)} + \mathbf{O}(\Delta t, h^2), \tag{5-7}$$

in which the current nodal unknowns are expressed implicitly in terms of the previous ones. Although a set of Eqs. (5-7) for $j = 1, 2, \ldots, J-1$ must now be solved simultaneously, the numerical stability is unconditionally ensured for any value of r. The proof is left as an exercise (Problem 5-2).

The accuracy of Eqs. (5-5) and (5-7) is only $\mathbf{O}(\Delta t)$. It is possible to raise this accuracy by using a so-called hybrid scheme [4, 5], which combines the explicit scheme and the implicit scheme:

$$\tilde{L}_{CN}\phi_j^{(n)} = \frac{1}{2}\tilde{L}_{\text{ex}}\phi_j^{(n-1)} + \frac{1}{2}\tilde{L}_{\text{im}}\phi_j^{(n)}, \tag{5-8}$$

where the subscript *CN* stands for the names of the initiators of the scheme, Crank and Nicolson. From Eqs. (5-4) and (5-6) it follows that

$$\tilde{L}_{CN}\phi_j^{(n)} = \frac{\phi_j^{(n)} - \phi_j^{(n-1)}}{\Delta t} - \frac{\alpha}{2}\left[\delta^2\phi_j^{(n-1)} + \delta^2\phi_j^{(n)}\right] - \tau_{CN} = 0, \tag{5-9}$$

where $$\delta^2\phi_j^{(m)} = \frac{1}{h^2}\left[\phi_{j-1}^{(m)} - 2\phi_j^{(m)} + \phi_{j+1}^{(m)}\right], \qquad m = n-1, n,$$

and $$\tau_{CN} = \frac{(\Delta t)^2}{6}\frac{\partial^3 \phi}{\partial t^3} - \frac{\alpha h^2}{12}\frac{\partial^4 \phi}{\partial x^4} + \mathbf{O}[(\Delta t)^3, h^4].$$

We observe that the accuracy of Eq. (5-9) becomes $\mathbf{O}[(\Delta t)^2]$ instead of $\mathbf{O}(\Delta t)$ as in Eqs. (5-5) and (5-7). Furthermore, the scheme remains unconditionally stable. We will now use the matrix method given in Section 4-1 to examine this stability.

To proceed, we rearrange Eq. (5-9) to

$$-\frac{r}{2}\,\phi_{j-1}^{(n)} + (1+r)\phi_j^{(n)} - \frac{r}{2}\,\phi_{j+1}^{(n)} = \frac{r}{2}\,\phi_{j-1}^{(n-1)} + (1-r)\phi_j^{(n-1)} + \frac{r}{2}\,\phi_{j+1}^{(n-1)}, \tag{5-10}$$

which can be written in matrix form as

$$[\mathbf{A}]\,\{\phi^{(n)}\} = [\mathbf{B}]\,\{\phi^{(n-1)}\} + \{\text{boundary conditions}\}, \tag{5-11}$$

where

$$[\mathbf{A}] = \begin{bmatrix} 1+r & -\frac{r}{2} & 0 & \cdots & 0 \\ -\frac{r}{2} & 1+r & -\frac{r}{2} & \cdots & 0 \\ \cdots & \cdots & \cdots & \cdots & \cdots \\ 0 & \cdots & -\frac{r}{2} & 1+r & -\frac{r}{2} \\ 0 & 0 & 0 & -\frac{r}{2} & 1+r \end{bmatrix},$$

$$[\mathbf{B}] = \begin{bmatrix} 1-r & \frac{r}{2} & 0 & \cdots & 0 \\ \frac{r}{2} & 1-r & \frac{r}{2} & \cdots & 0 \\ \cdots & \cdots & \cdots & \cdots & \cdots \\ 0 & \cdots & \frac{r}{2} & 1-r & \frac{r}{2} \\ 0 & 0 & 0 & \frac{r}{2} & 1-r \end{bmatrix},$$

and $$\{\mathbf{b.c.}\}^T = \{r\phi_0 \;\; 0 \;\; \cdots \;\; 0 \;\; r\phi_J\}.$$

For convenience, we have assumed here that ϕ_0 and ϕ_J are prescribed. The mixed-type boundary conditions were considered in Example 4-7. Equation (5-11) can be further rewritten as

$$\{\phi^{(n)}\} = [\mathbf{D}]\,\{\phi^{(n-1)}\} + [\mathbf{A}]^{-1}\,\{\mathbf{b.c.}\}, \tag{5-12}$$

where $$[\mathbf{D}] = [\mathbf{A}]^{-1}\,[\mathbf{B}]. \tag{5-13}$$

According to the matrix method, Eq. (5-10) is stable if the absolute eigenvalues $|\lambda_j|$ of [D] are not greater than 1. Therefore, our next task is to seek the eigenvalues λ_j of the matrix [D]. This task will be made easier if, rather than conducting a direct search, we split [A] and [B], respectively, into

$$[\mathbf{A}] = [\mathbf{I}] - \frac{r}{2}\,[\mathbf{S}] \tag{5-14a}$$

and

$$[\mathbf{B}] = [\mathbf{I}] + \frac{r}{2}\,[\mathbf{S}], \tag{5-14b}$$

where [I] is the identity matrix and

$$[\mathbf{S}] = \begin{bmatrix} -2 & 1 & 0 & \cdots & 0 \\ 1 & -2 & 1 & \cdots & 0 \\ \cdots & \cdots & \cdots & \cdots & \cdots \\ 0 & \cdots & 0 & 1 & -2 \end{bmatrix}.$$

Consequently,

$$\begin{aligned}[\mathbf{D}] - \lambda_j[\mathbf{I}] &= [\mathbf{A}]^{-1}[\mathbf{B}] - \lambda_j[\mathbf{I}] \\ &= [\mathbf{A}]^{-1}\left\{[\mathbf{S}] - \frac{2(\lambda_j - 1)}{r(\lambda_j + 1)}\,[\mathbf{I}]\right\}(1 + \lambda_j)\,\frac{r}{2} = 0.\end{aligned} \tag{5-15}$$

From Eq. (4-22), however, we found that the eigenvalues ϕ_j of the standard matrix [S] are given as

$$\mu_j = -4\sin^2\left(\frac{j\pi}{2J}\right), \qquad j = 1, 2, \ldots, J-1.$$

Furthermore, Eq. (5-15) implies that

$$[\mathbf{S}] - \mu_j[\mathbf{I}] = [\mathbf{S}] - \frac{2(\lambda_j - 1)}{r(\lambda_j + 1)}\,[\mathbf{I}] = 0. \tag{5-16}$$

Hence, we obtain the relation

$$-4\sin^2\left(\frac{j\pi}{2J}\right) = \frac{2(\lambda_j - 1)}{r(\lambda_j + 1)}.$$

The stability criterion of the matrix method dictates that

$$-1 \leqslant \lambda_j = \frac{2 - 4r\sin^2(j\pi/2J)}{2 + 4r\sin^2(j\pi/2J)} \leqslant 1, \tag{5-17}$$

which is always true regardless of the value of r (as long as r is positive). We therefore conclude that the Crank-Nicolson scheme is unconditionally stable. For its application to heat transfer problems, see [6-9].

5-1*b* Thomas Algorithm

So far, we have derived a number of algebraic equations, such as Eqs. (5-5), (5-7), and (5-10), that take the standard form

$$A_j\phi_{j-1} + B_j\phi_j + C_j\phi_{j+1} = D_j, \quad j = 1, 2, \ldots, J-1. \tag{5-18}$$

In fact, even the various discretized equations corresponding to Eq. (1-1) described in Chapters 1-3 can be represented by Eq. (5-18) with $D_j = 0$. Such three-point relations will constitute a tridiagonal matrix system, which can be solved very efficiently by the so-called Thomas algorithm [5, pp. 198-201; 10]. Briefly, this method resembles Gaussian elimination. For $j = 1$ and 2, Eq. (5-18) becomes, respectively,

$$\phi_1 + \frac{C_1}{B_1}\phi_2 = \frac{D_1 - A_1\phi_0}{B_1} \tag{5-19a}$$

and

$$\phi_1 + \frac{B_2}{A_2}\phi_2 + \frac{C_2}{A_2}\phi_3 = \frac{D_2}{A_2}. \tag{5-19b}$$

Subtracting Eq. (5-19*b*) from Eq. (5-19*a*) yields

$$E_2\phi_2 + F_2\phi_3 = G_2, \tag{5-20}$$

where the coefficients E_2, F_2, and G_2 are given in Table 5-1. Repeating the procedure, we obtain

$$E_j\phi_j + F_j\phi_{j+1} = G_j, \quad j = 3, 4, \ldots, J-2, \tag{5-21}$$

and

$$E_{J-1}\phi_{J-1} = G_{J-1}, \tag{5-22}$$

where the expressions for the coefficients are also listed in Table 5-1. A close examination of Table 5-1 reveals that these coefficients are free of nodal unknowns and are always expressed in terms of those at lower index numbers. This enables us to compute all the coefficients first. With these computed values, we then use Eqs.

Table 5-1 Expressions for the coefficients of Thomas algorithm

j	E_j	F_j	G_j
2	$\frac{C_1}{B_1} - \frac{B_2}{A_2}$	$-\frac{C_2}{A_2}$	$\frac{D_1 - A_1\phi_0}{B_1} - \frac{D_2}{A_2}$
$3 \sim J-2$	$\frac{F_{j-1}}{E_{j-1}} - \frac{B_j}{A_j}$	$-\frac{C_j}{A_j}$	$\frac{G_{j-1}}{E_{j-1}} - \frac{D_j}{A_j}$
$J-1$	$\frac{F_{J-2}}{E_{J-2}} - \frac{B_{J-1}}{A_{J-1}}$	0	$\frac{G_{J-2}}{E_{J-2}} - \frac{D_{J-1}}{A_{J-1}} + \frac{C_{J-1}\phi_J}{A_{J-1}}$

(5-22), (5-21), (5-20), and (5-19*a*) to evaluate the nodal unknowns, going from ϕ_{J-1} back to ϕ_1.

Example 5-1 Use the Thomas algorithm to solve the following set of linear equations:

$$\begin{aligned} 2\phi_1 + 3\phi_2 &= 8, \\ 3\phi_1 + 2\phi_2 + 3\phi_3 &= 4, \\ 3\phi_2 + 2\phi_3 + 3\phi_4 &= -2, \\ 3\phi_3 + 2\phi_4 &= -7, \end{aligned} \tag{a}$$

where ϕ_0 and ϕ_5 are assumed to be prescribed and are already combined into D_1 and D_4.

Solution: First, we list the given coefficients A_j, B_j, C_j, and D_j in Table 5-2. Then we use Table 5-1 to calculate E_j, F_j, and G_j, varying j from 2 to 4. Finally, varying j backward from 4 to 1, we calculate ϕ_4, ϕ_3, ϕ_2, and ϕ_1 according to Eqs. (5-22), (5-21), (5-20), and (5-19*a*).

5-1*c* Galerkin Finite-Element Method

The Galerkin finite-element method can also be used to discretize Eqs. (5-2*a*) and (5-2*b*). The inner product produced by the piecewise linear basis function $N_j(x)$ can be written, after integration by parts, as

$$(L\tilde{\phi}, N_j) = \int_0^L \frac{\partial \tilde{\phi}}{\partial t} N_j \, dx + \alpha \int_0^L \frac{\partial \tilde{\phi}}{\partial x} \frac{\partial N_j}{\partial x} \, dx = 0. \tag{5-23}$$

Consulting Table 1-4, we reduce Eq. (5-23) to

$$\frac{d\phi_{j-1}}{dt} + 4\frac{d\phi_j}{dt} + \frac{d\phi_{j+1}}{dt} - \frac{6\alpha}{h^2}(\phi_{j-1} - 2\phi_j + \phi_{j+1}) = 0, \quad j = 1, 2, 3, \ldots, J-1. \tag{5-24}$$

Equation (5-24) constitutes a set of linear first-order ordinary differential equations. If all initial values of $\phi_1, \phi_2, \ldots, \phi_{J-1}$ are specified, an integration scheme can be used to integrate this initial-value problem readily. This technique, in which the governing equation is solved discretely in one variable and continuously in the other, is sometimes called the Kantorovich method [11, 12], which is similar to the method of lines [13–17]. For comparisons of the method of lines with the Galerkin, collocation, and finite-difference methods, see [18, 19].

If the approximate solution $\tilde{\phi}(x, t)$ is expressed by

$$\tilde{\phi}(x, t) = \sum_{j=1}^{n} a_j(t) v_j(x),$$

Table 5-2 Solution procedure for the Thomas algorithm applied to Example 5-1

j	A_j	B_j	C_j	D_j	E_j	F_j	G_j	ϕ_j
1	0	2	3	8	–	–	–	1
2	3	2	3	4	$\frac{5}{6}$	-1	$\frac{8}{3}$	2
3	3	2	3	-2	$-\frac{28}{15}$	-1	$\frac{58}{15}$	-1
4	3	2	0	-7	$-\frac{11}{84}$	0	$\frac{11}{42}$	-2
		Given				$j = 2 \to 4$		$j = 4 \to 1$

where $v_j(x)$ are global basis functions having support over the entire domain $[0, L]$, then the Galerkin formulation is also called the spectral method (see also Section 2-1). In this formulation, we take the inner product of $L\tilde{\phi}$ and $v_j(x)$ and derive

$$(L\tilde{\phi}, v_j(x)) = \sum_{k=1}^{n} \frac{da_k}{dt} \int_0^L v_k v_j \, dx$$

$$- \alpha \sum_{k=1}^{n} a_k \int_0^L \frac{d^2 v_k}{dx^2} v_j \, dx = 0. \tag{5-25}$$

One of the main differences between Eq. (5-24) and Eq. (5-25) is that Eq. (5-24) involves only three unknown functions $\phi_{j-1}(t)$, $\phi_j(t)$, and $\phi_{j+1}(t)$, whereas Eq. (5-25) involves all the unknown functions $a_1(t), a_2(t), \ldots, a_n(t)$.

Alternatively, we may continue to discretize Eq. (5-24), using either the Galerkin formulation or the finite-difference schemes. For example, if the explicit finite-difference scheme is adopted, then Eq. (5-24) becomes

$$\phi_{j-1}^{(n+1)} + 4\phi_j^{(n+1)} + \phi_{j+1}^{(n+1)} = (6r + 1)\phi_{j-1}^{(n)} + (4 - 12r)\phi_j^{(n)} + (6r + 1)\phi_{j+1}^{(n)}. \tag{5-26}$$

Equation (5-26) should not be used unless it is stable. We will use the von Neumann stability criterion (see Section 4-1) to examine the stability of Eq. (5-26). Assume

$$\phi_j^{(n)} = B_m \xi^n \exp\left(\frac{i\pi mjh}{L}\right), \qquad m = \pm 1, \pm 2, \ldots \tag{5-27}$$

to be a solution of Eq. (5-26). Here i is the complex number $\sqrt{-1}$. Substituting Eq. (5-27) into Eq. (5-26) yields

$$\xi e^{-i\theta} + 4\xi + \xi e^{i\theta} = (6r + 1)e^{-i\theta} + 4 - 12r + (6r + 1)e^{i\theta},$$

where $\theta = m\pi h/L$. The von Neumann stability criterion dictates that

$$-1 \leqslant \xi = \frac{(6r+1)\cos\theta + 2 - 6r}{\cos\theta + 2} \leqslant 1 \tag{5-28}$$

or

$$r \leqslant \frac{2+\cos\theta}{3(1-\cos\theta)}. \tag{5-29}$$

The minimum value of the right-hand side of Eq. (5-29) is $\frac{1}{6}$ when $\cos\theta = -1$. Therefore

$$r \leqslant \tfrac{1}{6} \tag{5-30}$$

is the stability condition for Eq. (5-26).† It is thus found that Eq. (5-30) is more strict than the stability condition $r \leqslant \frac{1}{2}$ obtained by the explicit finite-difference scheme described in Section 4-1. In addition, Eq. (5-26) is not explicit at all. These two shortcomings suggest that it is preferable to discretize Eq. (5-24) in the implicit manner, i.e.,

$$(1-6r)\phi_{j-1}^{(n)} + (4+12r)\phi_j^{(n)} + (1-6r)\phi_{j+1}^{(n)} = \phi_{j-1}^{(n-1)} + 4\phi_j^{(n-1)} + \phi_{j+1}^{(n-1)}, \tag{5-31}$$

which, although remaining implicit, is unconditionally stable. Proof of unconditional stability for Eq. (5-31) is left as an exercise (Problem 5-5).

5-1*d* Temperature-Dependent Thermal Conductivity

For transient one-dimensional source-free systems with variable thermal conductivity, Eq. (5-1) reduces to

$$A\phi = \rho c_v \frac{\partial \phi}{\partial t} - \frac{\partial}{\partial x}\left[k(\phi)\frac{\partial \phi}{\partial x}\right] = 0. \tag{5-32}$$

For convenience, we will assume that $k(\phi) = c_1\phi$, where c_1 is a constant. This nonlinear problem can be numerically solved by using a discretization scheme and then a linearization scheme. Let us adopt the finite-difference scheme to discretize the nonlinear term and then the Taylor's series expansion to linearize the discretized result. Using the central difference, we obtain

$$\frac{-\partial}{\partial x}\left(\phi\frac{\partial\phi}{\partial x}\right) = \frac{1}{h}\left[\phi_{j-1/2}\left(\frac{\partial\phi}{\partial x}\right)_{j-1/2} - \phi_{j+1/2}\left(\frac{\partial\phi}{\partial x}\right)_{j+1/2}\right]$$

$$= \frac{1}{2h^2}(-\phi_{j-1}^2 + 2\phi_j^2 - \phi_{j+1}^2). \tag{5-33}$$

It is interesting to show (Problem 5-6) that an identical discretized result can be obtained with the Galerkin method. Following the linearization formula $\phi_k^2 \approx -\bar{\phi}_k^2 + 2\bar{\phi}_k\phi_k$, we linearize Eq. (5-33) to

†The initial conditions at the boundary points should also be made continuous with the boundary conditions for $t > 0$ to ensure stability.

$$\frac{-\partial}{\partial x}\left(\phi\frac{\partial\phi}{\partial x}\right)\approx\frac{1}{2h^2}(\bar{\phi}_{j-1}^2-2\bar{\phi}_j^2+\bar{\phi}_{j+1}^2-2\bar{\phi}_{j-1}\phi_{j-1}+4\bar{\phi}_j\phi_j-2\bar{\phi}_{j+1}\phi_{j+1}).$$

(5-34)

We have mentioned that it is always advisable to use a simple test, if available, to check the final algebraic result after lengthy algebra. Here let $\phi = x$, then

$$-\frac{\partial}{\partial x}\left(\phi\frac{\partial\phi}{\partial x}\right)=-\frac{\partial}{\partial x}(x)=-1. \tag{5-35a}$$

On the other hand, we take $x_j = 1$, $\bar{\phi} = \phi$, and $h = 1$, and the right-hand side of Eq. (5-34) becomes

$$\tfrac{1}{2}(0-2+4-0+4-8)=-1. \tag{5-35b}$$

The fact that Eq. (5-35*a*) is identical to Eq. (5-35*b*) partly demonstrates the validity of our algebra. We leave it to the reader to show that the reversed sequence–linearization first and then discretization–will also lead from the nonlinear term $-\partial(\phi\,\partial\phi/\partial x)/\partial x$ to Eq. (5-34), but the algebra is more cumbersome than that shown above (see Problem 5-7). Readers who are interested in the numerical computation of heat transfer problems with variable thermal conductivity may consult [6, 20–25].

5-2 TWO-DIMENSIONAL STEADY-STATE SYSTEMS

Steady-state heat conduction in a two-dimensional system with a heat sink is governed by the Poisson equation

$$-\frac{\partial^2\phi}{\partial x^2}-\frac{\partial^2\phi}{\partial y^2}+2=0, \tag{5-36a}$$

which is reduced from Eq. (5-1) with $\partial/\partial t = 0$, $S(x, y, t) = 2k$, and a constant k. If the boundary conditions are given as

$$\phi(0, y) = -y^2, \quad \phi(2, y) = 8 - y^2, \quad \phi(x, 0) = 2x^2,$$

and

$$\phi(x, 3) = 2x^2 - 9, \tag{5-36b}$$

the exact solution of Eqs. (5-36*a*) and (5-36*b*) exists and is given by

$$\phi(x, y) = 2x^2 - y^2. \tag{5-37}$$

Equation (5-37) can be used for comparison with the numerical solution obtained later.

5-2*a* Finite-Difference Method (Second-Order Accuracy)

As it is the simplest method, the finite-difference scheme with second-order accuracy will be used first to solve Eqs. (5-36*a*) and (5-36*b*). Figure 5-1 shows the

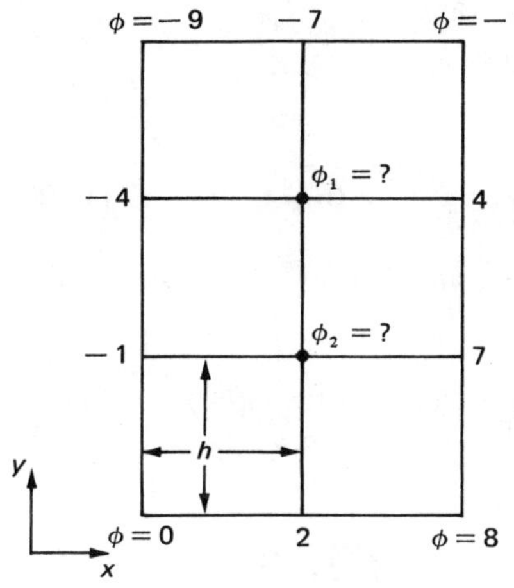

Figure 5-1 Mesh domain of a two-dimensional steady-state system, $\Delta x = \Delta y = h = 1$.

subdivided domain and the boundary conditions. For the purpose of easy presentation, only two grid nodes are present. Equation (5-36*a*) can be discretized into a five-point relation

$$\frac{1}{h^2}(-\phi_W + 2\phi_j - \phi_E) + \frac{1}{h^2}(-\phi_N + 2\phi_j - \phi_S) + 2 + O(h^2) = 0$$

or

$$-\sum_{p=W}^{N} \phi_p + 4\phi_j + 2h^2 = 0, \quad j = 1, 2, \tag{5-38}$$

where the subscript p denotes position (west, south, east, or north) relative to the grid point j. With $h = 1$ and the boundary conditions in Eq. (5-36*b*) incorporated, Eq. (5-38) is reduced to

$$4\phi_1 - \phi_2 = -9 \tag{5-39a}$$

and

$$-\phi_1 + 4\phi_2 = 6, \tag{5-39b}$$

which are solved simultaneously to yield

$$\phi_1 = -2 \quad \text{and } \phi_2 = 1,$$

identical to the exact solution. If Eqs. (5-36*a*) and (5-36*b*) are not so well designed, the numerical solution, in general, will not be identical to the exact solution and will have only second-order accuracy.

5-2*b* Finite-Difference Method (Higher-Order Accuracy)

In order to increase the accuracy, we may approximate

$$\frac{\partial^2 \phi}{\partial x^2} = \frac{1}{h^2}(\phi_W - 2\phi_j + \phi_E) - \frac{h^2}{12}\frac{\partial^2 M^x}{\partial x^2} + O(h^4) \tag{5-40a}$$

and

$$\frac{\partial^2 \phi}{\partial y^2} = \frac{1}{h^2}(\phi_N - 2\phi_j + \phi_S) - \frac{h^2}{12}\frac{\partial^2 M^y}{\partial y^2} + O(h^4), \tag{5-40b}$$

where M^x and M^y denote, respectively, $\partial^2\phi/\partial x^2$ and $\partial^2\phi/\partial y^2$. Rearranging Eqs. (5-40*a*) and (5-40*b*), we then write

$$\frac{1}{12}M_W^x + \frac{5}{6}M_j^x + \frac{1}{12}M_E^x = \frac{1}{h^2}(\phi_W - 2\phi_j + \phi_E) + O(h^4) \tag{5-41a}$$

and

$$\frac{1}{12}M_S^y + \frac{5}{6}M_j^y + \frac{1}{12}M_N^y = \frac{1}{h^2}(\phi_S - 2\phi_j + \phi_N) + O(h^4). \tag{5-41b}$$

These equations are similar to Eq. (1-10*b*) derived for the one-dimensional case. The governing equation can also be rewritten as

$$M_j^x + M_j^y - 2 = 0. \tag{5-42}$$

Equations (5-41*a*), (5-41*b*), and (5-42) are the three discretized equations at the interior point j. If the mesh system shown in Fig. 5-2 is to be considered, then the only interior point is $j = 5$. At this interior point, Eqs. (5-41) and (5-42) can be rewritten in matrix form, with $h = 1$, as

$$[\mathbf{G}_W]\{\mathbf{V}_W\} + [\mathbf{G}_S]\{\mathbf{V}_S\} + [\mathbf{G}_E]\{\mathbf{V}_E\} + [\mathbf{G}_N]\{\mathbf{V}_N\} + [\mathbf{G}_j]\{\mathbf{V}_j\} = \begin{Bmatrix} 2 \\ 0 \\ 0 \end{Bmatrix}, \tag{5-43}$$

where

$$[\mathbf{G}_W] = [\mathbf{G}_E] = \begin{bmatrix} 0 & 0 & 0 \\ -1 & \frac{1}{12} & 0 \\ 0 & 0 & 0 \end{bmatrix}, \quad [\mathbf{G}_S] = [\mathbf{G}_N] = \begin{bmatrix} 0 & 0 & 0 \\ 0 & 0 & 0 \\ -1 & 0 & \frac{1}{12} \end{bmatrix},$$

$$[\mathbf{G}_j] = \begin{bmatrix} 0 & 1 & 1 \\ 2 & \frac{5}{6} & 0 \\ 2 & 0 & \frac{5}{6} \end{bmatrix}, \quad \text{and } \{\mathbf{V}\} = \begin{Bmatrix} \phi \\ M^x \\ M^y \end{Bmatrix}.$$

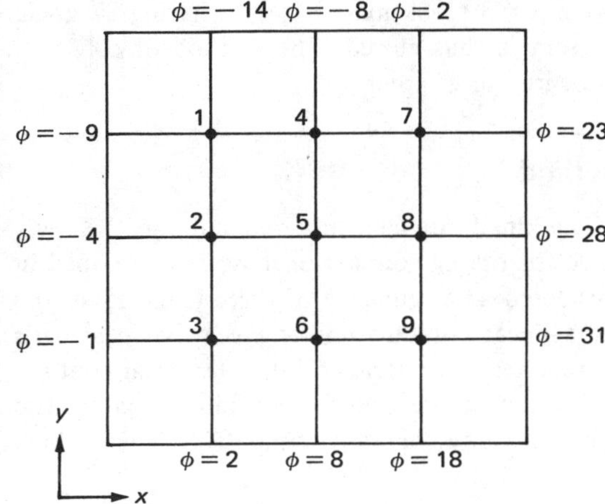

Figure 5-2 Mesh system with 3 × 3 interior grid points.

Equation (5-43), however, is not applicable at the other eight grid points because the second derivatives M_p, $p = W, S, E, N$, are not specified on the boundary.† For this reason, we are forced to replace Eqs. (5-41*a*) and (5-41*b*) with the following two equations of lower-order accuracy:

$$M_j^x = \frac{1}{h^2}(\phi_W - 2\phi_j + \phi_E) + O(h^2) \tag{5-44a}$$

and

$$M_j^y = \frac{1}{h^2}(\phi_S - 2\phi_j + \phi_N) + O(h^2). \tag{5-44b}$$

Along with the governing equation (5-42), these equations can also be written in matrix form as

$$[\mathbf{H}_W]\{\mathbf{V}_W\} + [\mathbf{H}_S]\{\mathbf{V}_S\} + [\mathbf{H}_E]\{\mathbf{V}_E\} + [\mathbf{H}_N]\{\mathbf{V}_N\} + [\mathbf{H}_j]\{\mathbf{V}_j\} = \begin{Bmatrix} 2 \\ 0 \\ 0 \end{Bmatrix}, \tag{5-45}$$

where

$$[\mathbf{H}_W] = [\mathbf{H}_E] = \begin{bmatrix} 0 & 0 & 0 \\ -1 & 0 & 0 \\ 0 & 0 & 0 \end{bmatrix}, \quad [\mathbf{H}_S] = [\mathbf{H}_N] = \begin{bmatrix} 0 & 0 & 0 \\ 0 & 0 & 0 \\ -1 & 0 & 0 \end{bmatrix},$$

$$[\mathbf{H}_j] = \begin{bmatrix} 0 & 1 & 1 \\ 2 & 1 & 0 \\ 2 & 0 & 1 \end{bmatrix}, \quad \text{and } \{\mathbf{V}\} = \begin{Bmatrix} \phi \\ M^x \\ M^y \end{Bmatrix}.$$

†In practice, it is acceptable to apply Eq. (5-43) to the grid points adjacent to the boundaries provided M^x and M^y are somehow approximated at the boundary points.

Equations (5-43) and (5-44) constitute a 3×3 block matrix system having 27 nodal unknowns and 27 equations. The system is thus closed. The method of solving a closed block matrix system will be presented in Chapter 7.

5-2*c* Galerkin Finite-Element Method

The logic and procedure of the Galerkin finite-element method for the two-dimensional system are similar to those for the one-dimensional system described in Section 1-2. This method can be considered as a sequence of three tasks. First, the function $\phi(x, y)$ is approximated by a linear combination of piecewise local basis functions. Then a weak equation that requires less restricted inter-elemental continuity is derived. Finally, the physical coordinates are transformed into isoparametric coordinates so that the algebraic formulation is greatly simplified. These three strategies are explained in further detail as follows.

First task: approximation of the function $\phi(x, y)$ For easy illustration, we will refer to Fig. 5-3, which shows the domain of interest Ω and a few square finite elements surrounding a typical interior point j. In element e_{NW}, for example, the approximate function $\tilde{\phi}(x, y)$ can be written as

$$\tilde{\phi}(x, y) = \phi_j N_j(x, y) + \phi_N N_N(x, y) + \phi_{NW} N_{NW}(x, y) + \phi_W N_W(x, y), \quad \text{if } x, y \in \text{element } e_{NW}, \tag{5-46a}$$

and, in element e_{SW},

$$\tilde{\phi}(x, y) = \phi_j N_j(x, y) + \phi_W N_W(x, y) + \phi_{SW} N_{SW}(x, y) + \phi_S N_S(x, y), \quad \text{if } x, y \in \text{element } e_{SW}, \tag{5-46b}$$

and so forth. The basis function $N_k(x, y)$ is constructed such that it is unity at $x = x_k$ and $y = y_k$ and vanishes at the other three vertices of the same square element. In general, if a standard element is considered, as shown in Fig. 5-4, these four basis functions can be derived (see Section 8-5*b*) as

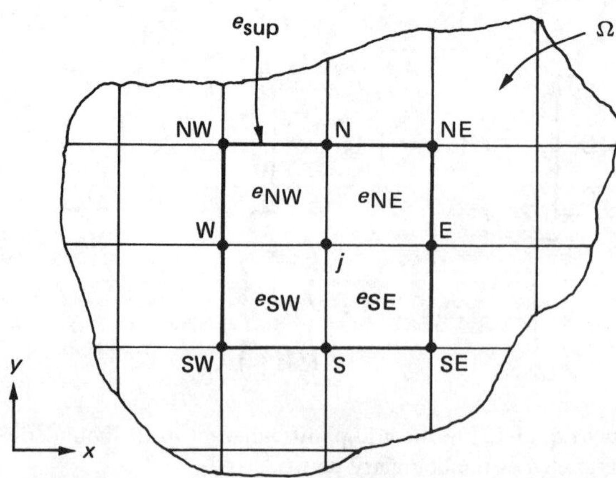

Figure 5-3 Four square elements surrounding a typical interior point j.

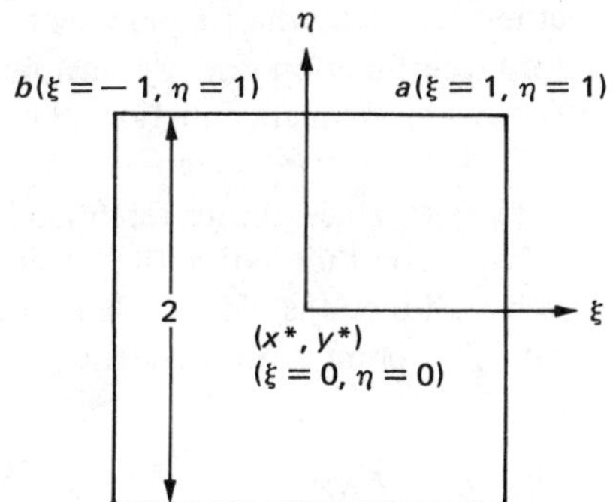

Figure 5-4 The square isoparametric element.

$$N_a(\xi, \eta) = \tfrac{1}{4}(1 + \xi)(1 + \eta), \tag{5-47a}$$

$$N_b(\xi, \eta) = \tfrac{1}{4}(1 - \xi)(1 + \eta), \tag{5-47b}$$

$$N_c(\xi, \eta) = \tfrac{1}{4}(1 - \xi)(1 - \eta), \tag{5-47c}$$

and

$$N_d(\xi, \eta) = \tfrac{1}{4}(1 + \xi)(1 - \eta), \tag{5-47d}$$

where

$$\xi = \frac{2(x - x^*)}{h}, \quad \eta = \frac{2(y - y^*)}{h}, \quad \text{and } |\xi|, |\eta| \leqslant 1.$$

Furthermore, these basis functions have local support only in the region $|\xi|, |\eta| \leqslant 1$. As will be seen in later chapters, they are very useful in the finite-element formulation.

Second task: derivation of the weak equation Since $\tilde{\phi}(x, y)$ given in Eqs. (5-46*a*) and so on is only an approximation to the exact solution $\phi(x, y)$, its substitution into Eq. (5-36*a*) naturally will produce a residual, namely

$$-\frac{\partial^2 \tilde{\phi}}{\partial x^2} - \frac{\partial^2 \tilde{\phi}}{\partial y^2} + 2 = R(x, y). \tag{5-48}$$

One way to reduce the residual $R(x, y)$ is to make $R(x, y)$ orthogonal to all the basis functions $N_j(x, y)$, i.e.,

$$(R, N_j) = (-\nabla^2 \tilde{\phi} + 2, N_j) = \iint_\Omega (-\nabla^2 \tilde{\phi} + 2) N_j(x, y)\, dx\, dy = 0. \tag{5-49}$$

Using integration by parts (or Green's identities) and assuming the Dirichlet boundary conditions leads to

$$(-\nabla^2 \tilde{\phi} + 2, N_j) = \iint_\Omega \left(\frac{\partial \tilde{\phi}}{\partial x} \frac{\partial N_j}{\partial x} + \frac{\partial \tilde{\phi}}{\partial y} \frac{\partial N_j}{\partial y} + 2N_j \right) dx\, dy = 0. \tag{5-50}$$

One remarkable feature of the Galerkin method is that the requirement of inter-elemental continuity is lowered. In Eq. (5-36*a*), the function $\phi(x, y)$ must be at least twice differentiable, whereas in Eq. (5-50) the approximate function $\tilde{\phi}(x, y)$ only needs to be continuous in itself (or its first derivative integrable in Ω). Equation (5-50) is called the weak form of Eq. (5-36*a*), since the differentiability of $\tilde{\phi}(x, y)$ is weaker than that of $\phi(x, y)$. Another attractive feature of the finite-element method is that the piecewise basis functions $N_j(x, y)$ have local support only in the elements centered around the grid point j. Consequently, Eq. (5-50) can be reduced to a sum of only four integrals:

$$(-\nabla^2\tilde{\phi} + 2, N_j) = K_{NW} + K_{SW} + K_{SE} + K_{NE} = 0, \tag{5-51}$$

where

$$K_p = \iint_{e_p} \left(\frac{\partial\tilde{\phi}}{\partial x}\frac{\partial N_j}{\partial x} + \frac{\partial\tilde{\phi}}{\partial y}\frac{\partial N_j}{\partial y} + 2N_j \right) dx\, dy, \qquad p = NW, SW, SE, NE.$$

The last task is to compute the integrals given in Eq. (5-51).

Third task: isoparametric transformation To simplify the algebra required to compute Eq. (5-51), we transform the physical coordinates into isoparametric coordinates and compare all the finite elements with the single isoparametric element shown in Fig. 5-4 (see Section 1-2*a* for the definition of isoparametric element). Take element e_{NW}, for example. If we define

$$\xi = \frac{2x - (x_N + x_{NW})}{h}$$

and

$$\eta = \frac{2y - (y_W + y_{NW})}{h},$$

then we can readily change

$$K_{NW} = \int_{x_W}^{x_j} \int_{y_W}^{y_{NW}} \left(\frac{\partial\tilde{\phi}}{\partial x}\frac{\partial N_j}{\partial x} + \frac{\partial\tilde{\phi}}{\partial y}\frac{\partial N_j}{\partial y} + 2N_j \right) dx\, dy \tag{5-52a}$$

into

$$K_{NW} = \int_{-1}^{1} \int_{-1}^{1} \left(\frac{\partial\tilde{\phi}}{\partial \xi}\frac{\partial N_j}{\partial \xi} + \frac{\partial\tilde{\phi}}{\partial \eta}\frac{\partial N_j}{\partial \eta} + \frac{h^2}{2} N_j \right) d\xi\, d\eta. \tag{5-52b}$$

A comparison of element e_{NW} with the isoparametric element reveals the correspondence $j \to d$, $N \to a$, $NW \to b$, and $W \to c$. Therefore, Eq. (5-52*b*) may be reduced to an algebraic equation

$$K_{NW} = A_{dd}\phi_j + A_{ad}\phi_N + A_{bd}\phi_{NW} + A_{cd}\phi_W + \frac{h^2}{2} B_d, \tag{5-53}$$

where

$$A_{pq} = \int_{-1}^{1} \int_{-1}^{1} \left(\frac{\partial N_p}{\partial \xi} \frac{\partial N_q}{\partial \xi} + \frac{\partial N_p}{\partial \eta} \frac{\partial N_q}{\partial \eta} \right) d\xi \, d\eta$$

and

$$B_d = \int_{-1}^{1} \int_{-1}^{1} N_d \, d\xi \, d\eta.$$

These integrals have been evaluated and their numerical values are listed in Table 5-3.

In some finite-element formulations the exact integration could be troublesome, and a numerical integration such as Gauss-Legendre quadrature [26-28] is preferred. Likewise, for the other three elements we obtain

Table 5-3 Numerical values of some integrals appearing in the square finite-element formulation

Integral expression	Numerical value
$A_{pq} = \int_{-1}^{1} \int_{-1}^{1} \left(\frac{\partial N_p}{\partial \xi} \frac{\partial N_q}{\partial \xi} + \frac{\partial N_p}{\partial \eta} \frac{\partial N_q}{\partial \eta} \right) d\xi \, d\eta,$ p-q are the pairs on the same edge a-d, a-b, b-c, c-d	$-\frac{1}{6}$
A_{pq}, p-q are diagonally opposite pairs a-c, b-d	$-\frac{1}{3}$
A_{pp}, $p = a, b, c, d$	$\frac{2}{3}$
$C_{pq} = \int_{-1}^{1} \int_{-1}^{1} N_p N_q \, d\xi \, d\eta,$ p-q are pairs on the same edge a-d, a-b, b-c, c-d	$\frac{2}{9}$
C_{pq}, p-q are diagonally opposite pairs a-c, b-d	$\frac{1}{9}$
C_{pp}, $p = a, b, c, d$	$\frac{4}{9}$
$B_r = \int_{-1}^{1} \int_{-1}^{1} N_r \, d\xi \, d\eta,$ $r = a, b, c, d$	1

$$K_{SW} = A_{aa}\phi_j + A_{ba}\phi_W + A_{ca}\phi_{SW} + A_{da}\phi_S + \frac{h^2}{2} B_a,$$

$$K_{SE} = A_{bb}\phi_j + A_{cb}\phi_S + A_{db}\phi_{SE} + A_{ab}\phi_E + \frac{h^2}{2} B_b,$$

and

$$K_{NE} = A_{cc}\phi_j + A_{dc}\phi_E + A_{ac}\phi_{NE} + A_{bc}\phi_N + \frac{h^2}{2} B_c.$$

With the aid of Table 5-3, Eq. (5-51) is finally simplified into the following algebraic equation:

$$8\phi_j - \sum_{p=W}^{NW} \phi_p + 6h^2 = 0, \tag{5-54}$$

which is the nine-point relation at grid point j. It is noted that the summation here is over the eight positions $W, SW, S, \ldots, NW$, whereas the summation in Eq. (5-38) is only over the four positions W, S, E, and N. For illustration, Eq. (5-54) is applied to the mesh system shown in Fig. 5-1, resulting in two simultaneous equations

$$8\phi_1 - \phi_2 = -17 \tag{5-55a}$$

$$-\phi_1 + 8\phi_2 = 10 \tag{5-55b}$$

which are solved to yield $\phi_1 = -2$ and $\phi_2 = 1$, a result identical to the exact solution.

In addition to square elements, triangular elements are commonly used. In particular, the local basis functions of C^0 triangular elements are linear and the associated mathematics in the discretization is thus quite simple. Figure 5-5 shows two types of isoparametric triangular elements. The approximate solution $\tilde{\phi}(x, y)$ can be interpolated as

$$\tilde{\phi}(x, y) = \phi_a N_a(\xi, \eta) + \phi_b N_b(\xi, \eta) + \phi_c N_c(\xi, \eta), \quad \text{if } x, y \in e_{abc}, \tag{5-56a}$$

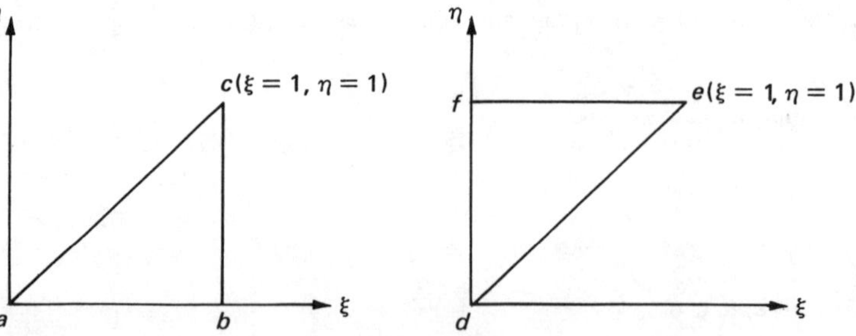

Figure 5-5 Two isoparametric triangular elements.

and $\quad \tilde{\phi}(x, y) = \phi_d N_d(\xi, \eta) + \phi_e N_e(\xi, \eta) + \phi_f N_f(\xi, \eta), \quad$ if $x, y \in e_{def}$,

(5-56b)

where the basis functions $N(\xi, \eta)$ can be derived as

$$N_a(\xi, \eta) = 1 - \xi, \quad N_b(\xi, \eta) = \xi - \eta, \quad N_c(\xi, \eta) = \eta \tag{5-57a}$$

and

$$N_d(\xi, \eta) = 1 - \eta, \quad N_e(\xi, \eta) = \xi, \quad N_f(\xi, \eta) = \eta - \xi. \tag{5-57b}$$

Now the inner product corresponding to Eq. (5-51) can be written as

$$(-\nabla^2\tilde{\phi} + 2, N_j) = K_A + K_B + K_C + K_D + K_E + K_F, \tag{5-58}$$

where

$$K_p = \iint_{e_p} \left(\frac{\partial \tilde{\phi}}{\partial x} \frac{\partial N_j}{\partial x} + \frac{\partial \tilde{\phi}}{\partial y} \frac{\partial N_j}{\partial y} + 2N_j \right) dx\, dy, \quad p = A, B, C, D, E, F. \tag{5-59}$$

The elements $A, \ldots, F$ are shown in Fig. 5-6 to surround the grid point j. We will take only element e_A to illustrate the discretization procedure. Using the isoparametric transformation and noting the correspondence $j \to b$, $N \to c$, and $W \to a$, we derive

$$K_A = D_{bb}\phi_j + D_{cb}\phi_N + D_{ab}\phi_W + 2h^2 E_b, \tag{5-60}$$

where

$$D_{pq} = \int_0^1 \int_0^1 \left(\frac{\partial N_p}{\partial \xi} \frac{\partial N_q}{\partial \xi} + \frac{\partial N_p}{\partial \eta} \frac{\partial N_q}{\partial \eta} \right) d\xi\, d\eta \tag{5-61a}$$

and

$$E_b = \int_0^1 \int_0^1 N_b\, d\xi\, d\eta. \tag{5-61b}$$

These integrals were evaluated and their numerical values are listed in Table 5-4. The evaluation of the other five integrals $K_B, \ldots, K_F$ is left to the reader as an exercise. The final discretized algebraic equation is found to be identical to Eq. (5-38).

5-2*d* Ritz (Variational) Method with Global Basis Functions

As mentioned in Section 1-2, the Ritz method leads to the same discretized equation as the Galerkin method when the differential operator is linear, self-adjoint, and positive-definite. Since the Galerkin finite-element method was described in the preceding section, the Ritz method, used in conjunction with the

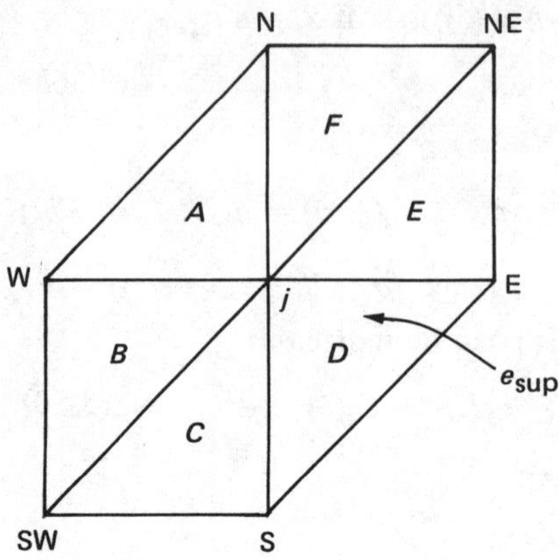

Figure 5-6 Six triangular elements $e_{\rm sup}$ surrounding grid point j.

finite-element formulation, will not be repeated here. Rather, we will present this method with the use of global basis functions.

Upon construction of the global basis functions, it is generally preferred that a boundary-value problem subject to inhomogeneous Dirichlet boundary conditions be transformed into one with homogeneous Dirichlet boundary conditions. Consider, for example,

$$-\nabla^2\phi = f(x, y), \qquad x, y \in R \tag{5-62a}$$

subject to

$$\phi = g(x, y), \qquad x, y \in \partial R, \tag{5-62b}$$

where ∂R denotes the boundary of the domain R. It is possible to introduce a new function

Table 5-4 Numerical values of some integrals appearing in the triangular finite-element formulation

Integral expression	Numerical value
$D_{pq} = \int_0^1 \int_0^1 \left(\frac{\partial N_p}{\partial \xi} \frac{\partial N_q}{\partial \xi} + \frac{\partial N_p}{\partial \eta} \frac{\partial N_q}{\partial \eta} \right) d\xi \, d\eta$ a-b, b-c, d-f, e-f	-1
D_{pq}, a-c, d-e	0
D_{pq}, $p = a, c, d, e$	1
D_{pq}, $p = b, f$	2
$E_r = \int_0^1 \int_0^1 N_r \, d\xi \, d\eta$, $r = a, c, d, e$	$\frac{1}{2}$
$r = b, f$	0

$$\psi(x, y) = \phi(x, y) - g(x, y), \tag{5-63}$$

provided $g(x, y) \epsilon C^2(R)$. This new function satisfies

$$-\nabla^2 \psi = F(x, y), \quad x, y \in R \tag{5-64a}$$

now subject to

$$\psi = 0, \quad x, y \in \partial R, \tag{5-64b}$$

where

$$F(x, y) = f(x, y) + \frac{\partial^2 g}{\partial x^2} + \frac{\partial^2 g}{\partial y^2}. \tag{5-65}$$

With the homogeneous boundary condition (5-64*b*), the task of constructing the global basis functions for domains of simple geometry, such as rectangles, becomes easy. For example, one set of global basis functions for Fig. 5-1 can be

$$\{st(1-s)(1-t), s^2t^2(1-s)(1-t), \ldots, s^nt^n(1-s)(1-t)\}, \tag{5-66}$$

where $s = x/2$ and $t = y/3$. Now we seek the variational functional corresponding to Eqs. (5-64*a*) and (5-64*b*). The inner product produced by the variation of $\psi(x, y)$ is written as

$$(-\nabla^2\psi - F, \delta\psi) = \iint_\Omega (-\nabla^2\psi - F)\,\delta\psi\, dx\, dy = 0.$$

Using integration by parts (or Green's identities) enables us to identify the variational principle as

$$I = \iint_\Omega \left[\frac{1}{2}\left(\frac{\partial\psi}{\partial x}\right)^2 + \frac{1}{2}\left(\frac{\partial\psi}{\partial y}\right)^2 - F\psi\right] dx\, dy, \tag{5-67}$$

which is made stationary when $\psi(x, y)$ satisfies Eqs. (5-64*a*) and (5-64*b*). If $\psi(x, y)$ is approximated by

$$\tilde{\psi}(x, y) = e_1 v_1(x, y) + e_2 v_2(x, y) + \cdots + e_n v_n(x, y), \tag{5-68}$$

where $v_n(x, y)$ are the basis functions that vanish on the boundary ∂R, then $I(e_1, e_2, \ldots, e_n)$ becomes a function of the undetermined constants and these constants should be chosen such that

$$\frac{\partial I}{\partial e_k} = 0, \quad k = 1, 2, \ldots n \tag{5-69}$$

or

$$\iint_\Omega \left[\frac{\partial\tilde{\psi}}{\partial x}\frac{\partial v_k}{\partial x} + \frac{\partial\tilde{\psi}}{\partial y}\frac{\partial v_k}{\partial y} - F v_k\right] dx\, dy = 0. \tag{5-70}$$

Substitution of Eq. (5-68) into Eq. (5-70) yields

$$\sum_{j=1}^{n} e_j \iint_{\Omega} \left(\frac{\partial v_j}{\partial x} \frac{\partial v_k}{\partial x} + \frac{\partial v_j}{\partial y} \frac{\partial v_k}{\partial y} \right) dx\, dy$$

$$= \iint_{\Omega} F v_k \, dx\, dy, \qquad k = 1, 2, \ldots, n. \tag{5-71}$$

After evaluation of the double integrals (see Table 5-5), this set of algebraic equations can be solved simultaneously to yield the undetermined coefficients e_j.

When the geometry of the domain is not as simple as a rectangle, it is inconvenient to find suitable global basis functions that satisfy the boundary conditions. Besides, even if the basis functions are available, the evaluation of the integrals is cumbersome. Therefore, in general, we prefer the finite-element formulation to the global-basis-function formulation for two-dimensional problems, although the latter does have certain advantages [29] such as higher convergence rates for problems of simple geometry.

Another method for dealing with the inhomogeneous Dirichlet boundary condition is to introduce the so-called penalty functional [30-32] into the variational principle, namely

Table 5-5 Convenient values of the double integrals in Eq. (5-71); $v_1 = st(1-s)(1-t)$ and $v_2 = s^2t^2(1-s)(1-t)$

Integral, $\int_0^1 \int_0^1 (\,)\, ds\, dt$	Value
$\left(\frac{\partial v_1}{\partial s}\right)^2$	$\frac{1}{90}$
$\left(\frac{\partial v_1}{\partial t}\right)^2$	$\frac{1}{90}$
$\left(\frac{\partial v_2}{\partial s}\right)^2$	$\frac{2}{1575}$
$\left(\frac{\partial v_2}{\partial t}\right)^2$	$\frac{2}{1575}$
$\frac{\partial v_1}{\partial s} \frac{\partial v_2}{\partial s}$	$\frac{1}{360}$
$\frac{\partial v_1}{\partial t} \frac{\partial v_2}{\partial t}$	$\frac{1}{360}$

$$I = \iint_{\Omega} \left[\frac{1}{2} \left(\frac{\partial \phi}{\partial x} \right)^2 + \frac{1}{2} \left(\frac{\partial \phi}{\partial y} \right)^2 - f\phi \right] dx\, dy + \lambda \int_{\partial \Omega} (\phi - g)\, ds, \tag{5-72}$$

where ϕ satisfies Eqs. (5-62) and λ is the penalty constant, whose value is chosen by numerical experiments. With this penalty functional included, neither the global basis functions nor the local basis functions need to satisfy the boundary conditions.

5-2*e* Iterative Matrix-Solving Schemes

When the domain is subdivided into fine mesh systems and the matrix system containing the nodal unknowns becomes large, it is essential to adopt an efficient scheme for solving the matrix system. Matrix inversion requires a great deal of memory storage and computation time, as demonstrated in Example 4-21, and therefore has been gradually abandoned. In Section 4-3 we presented an iterative scheme—the Gauss-Seidel method for a 3 X 3 matrix system. In the present section this technique will be used again to solve a "large" matrix system consisting of the finite-difference equation

$$-\sum_{p=W}^{N} \phi_p + 4\phi_j + 2 = 0, \qquad j = 1, 2, \ldots, 9. \tag{5-38}$$

The corresponding mesh system with boundary conditions is shown in Fig. 5-2, and the solution procedure is illustrated in the following example.

Example 5-2 Use the Gauss-Seidel method to solve Eq. (5-38).

Solution: Let us take initial guesses to be

$$\phi_1^{[0]} = \phi_2^{[0]} = \cdots = \phi_9^{[0]} = 0. \tag{a}$$

Beginning with $j = 1$, we rewrite Eq. (5-38) as

$$\begin{aligned} \phi_1^{[1]} &= -\frac{1}{2} + \frac{1}{4}(\phi_W^{[0]} + \phi_S^{[0]} + \phi_E^{[0]} + \phi_N^{[0]}) \\ &= -\frac{1}{2} + \frac{1}{4}(-9 + 0 + 0 - 14) = -6.25. \end{aligned} \tag{b}$$

For $j = 2, 5$, the updated values are used, i.e.,

$$\phi_2^{[1]} = -\frac{1}{2} + \frac{1}{4}(-4 + 0 + 0 - 6.25) = -3.0625 \tag{c}$$

and $\quad \phi_5^{[1]} = -\dfrac{1}{2} + \dfrac{1}{4}(\phi_2^{[1]} + \phi_6^{[0]} + \phi_8^{[0]} + \phi_4^{[1]}) = -2.2813.$

The result of the first three iterations is listed in Table 5-6*a*. This solution converges to the exact solution at the 18th iteration. We may first show that the matrix

$$\begin{bmatrix} 4 & -1 & 0 & -1 & 0 & 0 & 0 & 0 & 0 \\ -1 & 4 & -1 & 0 & -1 & 0 & 0 & 0 & 0 \\ 0 & -1 & 4 & 0 & 0 & -1 & 0 & 0 & 0 \\ -1 & 0 & 0 & 4 & -1 & 0 & -1 & 0 & 0 \\ 0 & -1 & 0 & -1 & 4 & -1 & 0 & -1 & 0 \\ 0 & 0 & -1 & 0 & -1 & 4 & 0 & 0 & -1 \\ 0 & 0 & 0 & -1 & 0 & 0 & 4 & -1 & 0 \\ 0 & 0 & 0 & 0 & -1 & 0 & -1 & 4 & -1 \\ 0 & 0 & 0 & 0 & 0 & -1 & 0 & -1 & 4 \end{bmatrix}$$

is symmetric and positive-definite, and then, by means of the extremum principle, prove that the Gauss-Seidel method converges when applied to Eq. (5-38). The details can be found in [33].

The Gauss-Seidel method, if convergent, generally does not converge sufficiently fast. Fortunately, it is possible to accelerate the convergence by successive relaxation [34-36], which is shown mathematically by

$$\phi_j^{[k]} = \phi_j^{[k-1]} + \rho(\phi_{j,GS}^{[k]} - \phi_j^{[k-1]}). \tag{5-73}$$

Here the subscript *GS* denotes Gauss-Seidel and ρ is the relaxation parameter, with the ranges $1 \leqslant \rho \leqslant 2$ for over-relaxation and $0 < \rho < 1$ for under-relaxation. When $\rho = 1$, Eq. (5-73) reduces to the Gauss-Seidel method. A proof that Eq. (5-73) converges faster than the Gauss-Seidel method would be quite lengthy. Readers who are interested in this proof may refer to [33, pp. 119-125]. We also note that the choice of ρ determines the speed of convergence, but the task of choosing an optimal ρ is often difficult. In the following example, Eq. (5-73) will be used to solve Eq. (5-38). The corresponding mesh system is shown in Fig. 5-2.

Example 5-3 Construct a table similar to Table 5-6*a*, but using Eq. (5-73) with $\rho = 1.2$. Does the solution converge faster than that shown in Table 5-6*a*?

Solution: Again, the initial guesses are taken to be

$$\phi_1^{[0]} = \phi_2^{[0]} = \cdots = \phi_9^{[0]} = 0. \tag{a}$$

For $j = 1$, Eq. (5-38) yields (see Example 5-2) $\phi_{1,GS}^{[1]} = -6.25$. Hence, Eq. (5-73) becomes

$$\phi_1^{[1]} = 0 + 1.2(-6.25 - 0) = -7.5 \tag{b}$$

Table 5-6 Solution of Eq. (5-38) and Fig. 5-2
(*a*) by the Gauss-Seidel method

k	$\phi_1^{(k)}$	$\phi_2^{(k)}$	$\phi_3^{(k)}$	$\phi_4^{(k)}$	$\phi_5^{(k)}$	$\phi_6^{(k)}$	$\phi_7^{(k)}$	$\phi_8^{(k)}$	$\phi_9^{(k)}$
0	0	0	0	0	0	0	0	0	0
1	−6.25	−3.0625	−1.0156	−4.0625	−2.2813	0.6758	4.7344	7.1133	13.6973
2	−8.0313	−4.3320	−1.1641	−3.8946	−0.6094	4.4810	6.5547	11.4106	15.7229
3	−8.3067	−4.020	−0.1348	−3.0904	1.6953	5.8209	7.8300	12.8121	16.4083
...	...	...	...	...	...	...	...	...	...
18	−7	−2	1	−1	4	7	9	14	17

(*b*) by the successive over-relaxation method with $\rho = 1.2$

k	$\phi_1^{(k)}$	$\phi_2^{(k)}$	$\phi_3^{(k)}$	$\phi_4^{(k)}$	$\phi_5^{(k)}$	$\phi_6^{(k)}$	$\phi_7^{(k)}$	$\phi_8^{(k)}$	$\phi_9^{(k)}$
0	0	0	0	0	0	0	0	0	0
1	−7.50	−4.050	−1.5150	−5.250	−3.390	0.3285	5.3250	8.3805	16.7127
2	−8.790	−5.0985	−1.4280	−4.0065	−0.0408	6.3075	7.1472	13.2696	16.6306
3	−8.4735	−3.7630	0.7489	−2.6088	3.3697	6.7633	8.6688	13.7468	16.9269
...	...	...	...	...	...	...	...	...	...
9	−7	−2	1	−1	4	7	9	14	17

Using the same procedure, we construct Table 5-6*b*. Occurance of convergence at the 9th iteration suggests that, for this problem, the successive over-relaxation method converges faster than the Gauss-Seidel method.

Other well-known iterative matrix solvers [37] are (1) successive over-relaxation by lines [38], in which the nodal variables in one line are solved simultaneously while the west and east nodal values are the current ones and the previously iterated ones, respectively, (2) the alternating direction implicit (ADI) scheme [10], in which the direction of lines alternates after each sweep, and (3) the conjugate gradients method [39–41], in which the residuals generated by substituting the trial solutions into the algebraic equations are squared. The basic procedures of the ADI scheme applied to Fig. 5-2 are listed in Table 5-7.

5-3 TWO-DIMENSIONAL TRANSIENT SYSTEMS

For two-dimensional transient systems with constant thermal conductivity and without heat sources, Eq. (5-1) reduces to

$$L\phi = \frac{\partial \phi}{\partial t} - \alpha\left(\frac{\partial^2 \phi}{\partial x^2} + \frac{\partial^2 \phi}{\partial y^2}\right) = 0. \tag{5-74}$$

First, we will use the finite-difference scheme to discretize Eq. (5-74).

5-3*a* Finite-Difference Methods and Stability

With uniform mesh size $\Delta x = \Delta y = h$, Eq. (5-74) can be discretized explicitly to

$$\phi_j^{(n+1)} = r \sum_{p=W}^{N} \phi_p^{(n)} + (1 - 4r)\phi_j^{(n)} \tag{5-75}$$

and implicitly to

$$(1 + 4r)\phi_j^{(n+1)} + r \sum_{p=W}^{N} \phi_p^{(n+1)} = \phi_j^{(n)}, \tag{5-76}$$

where $r = \alpha\Delta t/h^2$. In the explicit equation (5-75), since the information on the right-hand side is available, the solution procedure is as straightforward as that for the one-dimensional transient system.

Example 5-4 Consider the mesh system shown in Fig. 5-7 with boundary conditions as indicated. Initially, $\phi_1^{(0)} = \phi_2^{(0)} = 1$. Find the solution $\phi_1^{(n)}$ and $\phi_2^{(n)}$ at $n = 1, 2, 3, 4$. Take (*a*) $r = \frac{1}{4}$ and (*b*) $r = \frac{1}{2}$, respectively.

Table 5-7 Basic procedures of the ADI method applied to Fig. 5-2[a]

Procedure	Equation computed
Initial-guess setup	$\phi_j = 0,\ 1 \leqslant j \leqslant 9$
First sweep in x direction	$-\phi_N^{[1]} + 4\phi_j^{[1]} - \phi_S^{[1]} = \phi_W^{[1]} + \phi_E^{[0]} + 2$
First sweep in y direction	$-\phi_E^{[1.5]} + 4\phi_j^{[1.5]} - \phi_W^{[1.5]} = \phi_S^{[1.5]} + \phi_N^{[1]} + 2$
Second sweep in x direction	$-\phi_N^{[2]} + 4\phi_j^{[2]} - \phi_S^{[2]} = \phi_W^{[2]} + \phi_E^{[1.5]} + 2$
Second sweep in y direction	$-\phi_E^{[2.5]} + 4\phi_j^{[2.5]} - \phi_W^{[2.5]} = \phi_S^{[2.5]} + \phi_N^{[2]} + 2$
Convergence check	$\dfrac{(\phi_j^{[k+1]} - \phi_j^{[k]})}{(\phi_j^{[k]} + 10^{-7})} \leqslant 10^{-4}$

[a] With necessary modifications, the ADI method listed here can be applied to transient two-dimensional problems [10].

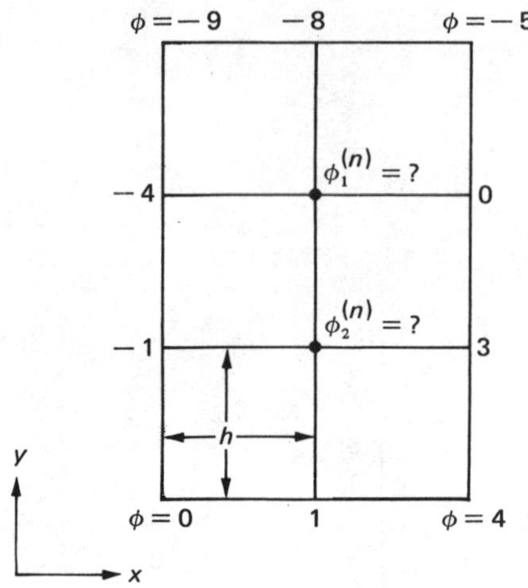

Figure 5-7 Network of a two-dimensional transient system, $\Delta x = \Delta y = h = 1$.

Solution Both solutions are listed in Table 5-8. For $r = \frac{1}{4}$, the transient solution behaves well and gradually approaches the exact steady-state solution $\phi(x, y) = x^2 - y^2$. For $r = \frac{1}{2}$, the values of $\phi_1^{(n)}$ and $\phi_2^{(n)}$ oscillate and will become unpredictably unbounded as $t \to \infty$. This example suggests that there exists an r value governing the stability of the scheme (5-75).

The stability criterion of Eq. (5-75) can, in principle, be examined by either the von Neumann method or the matrix method (see Section 4-1). For two-dimensional initial-value problems, however, these two methods appear to be inconvenient because it is difficult to find the amplification factor in the von Neumann method and the matrix eigenvalues in the matrix method. For this reason, we will investigate the stability criterion of Eq. (5-75) by means of the extremum principle [42]. A rigorous proof will not be attempted here. Instead, we will prove that as $n \to \infty$ in Eq. (5-75), nodal values are bounded if $q = 1 - 4r \geqslant 0$. Under the condition $q \geqslant 0$, we have

Table 5-8 Solutions of various difference schemes

	Explicit, $r = \frac{1}{4}$		Explicit, $r = \frac{1}{2}$		Implicit, $r = \frac{1}{2}$	
$t(n\Delta t)$	$\phi_1^{(n)}$	$\phi_2^{(n)}$	$\phi_1^{(n)}$	$\phi_2^{(n)}$	$\phi_1^{(n)}$	$\phi_2^{(n)}$
0	1	1	1	1	1	1
Δt	−2.750	0.0630	−6.5	1	−1.5520	0.5710
$2\Delta t$	−2.984	0.0039	1	−2.75	−2.5005	0.2785
$3\Delta t$	−2.999	0.0002	−8.375	4.75	−2.8605	0.1240
$4\Delta t$	−3.000	0.0000	4.75	−7.4375	−2.9996	0.0425
∞	−3	0	?	?	−3	0

$$\phi_j^{(n+1)} \leqslant r \sum_{p=W}^{N} |\phi_p^{(n)}| + q|\phi_j^{(n)}|^{\dagger} \leqslant (4r+q)M^{(n)} = M^{(n)}, \qquad (5\text{-}77)$$

where $$M^{(n)} = \max\,\{|\phi_W^{(n)}|, |\phi_S^{(n)}|, |\phi_E^{(n)}|, |\phi_N^{(n)}|, |\phi_j^{(n)}|\}.$$

The possibility that any position among W, S, E, and N is on the boundary should also be included. Since j can represent any nodal point, Eq. (5-77) remains valid even if $\phi_j^{(n+1)} = M^{(n+1)}$. Therefore,

$$M^{(n+1)} \leqslant M^{(n)} \qquad (5\text{-}78a)$$

and likewise,

$$M^{(n)} \leqslant M^{(n-1)}. \qquad (5\text{-}78b)$$

Successive applications of Eq. (5-78b) at $n = 1, 2, \ldots$ yield

$$|\phi_j^{(n+1)}| \leqslant M^{(n)} \leqslant M^{(n-1)} \leqslant \cdots \leqslant M^{(0)}, \qquad (5\text{-}79)$$

which indicates that any nodal solution cannot exceed the initial and boundary values. Such a condition is sometimes considered a stability criterion of the difference scheme, since the departure caused by a line of errors is itself a solution of the difference equations.

Table 5-8 also lists the difference solution obtained by the implicit scheme (5-76) with $r = \frac{1}{2}$. As in the transient one-dimensional case, the implicit scheme is always stable for any value of r.

5-3*b* Galerkin Finite-Element Method

Once we use the finite-element formulation to assume

$$\tilde{\phi}(x, y, t) = \sum_{j=1}^{J} N_j(x, y)\phi_j(t), \qquad j = 1, 2, \ldots, J \qquad (5\text{-}80)$$

where J is the total number of nodal points, the weak Galerkin form of Eq. (5-74) can be written, after integration by parts, as

$$(L\tilde{\phi}, N_j) = T_1 + \alpha T_2 = 0, \qquad (5\text{-}81)$$

where $$T_1 = \frac{\partial}{\partial t}\iint_{e_{\text{sup}}} \tilde{\phi} N_j \, dx\, dy \qquad (5\text{-}82a)$$

and $$T_2 = \iint_{e_{\text{sup}}} \left(\frac{\partial \tilde{\phi}}{\partial x}\frac{\partial N_j}{\partial x} + \frac{\partial \tilde{\phi}}{\partial y}\frac{\partial N_j}{\partial y}\right) dx\, dy. \qquad (5\text{-}82b)$$

†Instead of the absolute values, other types of norms such as the Sobolev norm may also be taken.

The element e_{sup} denotes the region within which the basis function $N_j(x, y)$ has local support. For example, if the square elements are considered, then in reference to Fig. 5-3 the element e_{sup} consists of e_{NW}, e_{SW}, e_{SE}, and e_{NE}. After the isoparametric transformation, namely $d\xi = 2dx/h$ and $d\eta = 2dy/h$ with the correspondence $j \to d$, $N \to a$, $NW \to b$, and $W \to c$ for the element e_{NW} and so forth, Eqs. (5-82*a*) and (5-82*b*) can be written as

$$4\frac{T_1}{h^2} = (C_{aa} + C_{bb} + C_{cc} + C_{dd})\frac{d\phi_j}{dt} + (C_{cd} + C_{ba})\frac{d\phi_W}{dt} + (C_{da} + C_{cb})\frac{d\phi_S}{dt}$$
$$+ (C_{dc} + C_{ab})\frac{d\phi_E}{dt} + (C_{ad} + C_{bc})\frac{d\phi_N}{dt} + C_{bd}\frac{d\phi_{NW}}{dt}$$
$$+ C_{ca}\frac{d\phi_{SW}}{dt} + C_{db}\frac{d\phi_{SE}}{dt} + C_{ac}\frac{d\phi_{NE}}{dt} \tag{5-83a}$$

and

$$T_2 = (A_{aa} + A_{bb} + A_{cc} + A_{dd})\phi_j + (A_{cd} + A_{ba})\phi_W + (A_{da} + A_{cb})\phi_S$$
$$+ (A_{dc} + A_{ab})\phi_E + (A_{ad} + A_{bc})\phi_N + A_{bd}\phi_{NW} + A_{ca}\phi_{SW}$$
$$+ A_{db}\phi_{SE} + A_{ac}\phi_{NE}. \tag{5-83b}$$

With the aid of Table 5-3, we can reduce Eq. (5-81) to

$$\frac{h^2}{4}\left(\frac{16}{9}\frac{d\phi_j}{dt} + \frac{4}{9}\sum_{p=W}^{N}\frac{d\phi_p}{dt} + \frac{1}{9}\sum_{q=NW}^{SW}\frac{d\phi_q}{dt}\right) + \alpha\left(\frac{8}{3}\phi_j - \frac{1}{3}\sum_{r=W}^{NW}\phi_r\right) = 0, \tag{5-84}$$

where the indices p, q, and r are counted for 4, 4, and 8 positions, respectively. In order to derive Eq. (5-84), we have carried out a considerable amount of algebra. Obviously, it would be desirable to use a quick method to check this equation. One method is to assume $\phi(x, y, t) = t$. Then the exact value of the inner product $(L\phi, N_j)$ over a unit square element is

$$(L\tilde{\phi}, N_j) = \int_0^1\int_0^1 \left(\frac{\partial\tilde{\phi}}{\partial t} - \alpha\nabla^2\tilde{\phi}\right)N_j\,dx\,dy$$
$$= \int_0^1\int_0^1 (1 - \alpha \times 0)N_j\,dx\,dy = \frac{1}{4}. \tag{5-85a}$$

On the other hand, Eq. (5-84), which is discretized with $h = \frac{1}{2}$ from such an inner product after lengthy algebra, becomes

$$\left(\frac{1}{2}\right)^2\frac{1}{4}\left(\frac{16}{9} + \frac{4}{9} \times 4 + \frac{4}{9}\right) + \alpha\left(\frac{8}{3}t - \frac{8}{3}t\right) = \frac{1}{4}. \tag{5-85b}$$

The fact that the results in Eqs. (5-85*a*) and (5-85*b*) are identical suggests that Eq. (5-84) is probably correct.

In the investigation of transient problems, it is also possible to use the space-time element in the finite-element formulation [43–45]. In this case the time is considered, for example, as the fourth dimension for three-dimensional systems.

5-3*c* Eigenvalue Method

We mentioned in Section 5-1*a* that there exist (at least) explicit, implicit, and hybrid schemes for discretizing the temporal derivative. However, the explicit schemes are restricted by the stability criteria and the implicit schemes generally must be followed by some iterative procedures to solve the matrix systems. When the transient time is long and the matrix system becomes large, both explicit and implicit schemes may require considerable computation time.

In this section, a different approach that utilizes the eigenvalues of the conductance matrix is described. This method (hereafter called the eigenvalue method) has been used by researchers in structure mechanics for some time, but for some reason has been relatively unpopular in the heat transfer community. Even though it has been rarely applied to heat transfer problems, the eigenvalue method has an attractive feature. When the number of nodal unknowns is large, it is found that only a few dominant eigenvalues need to be computed and retained to yield reasonably good approximations. Consequently, a great deal of computation time is saved.

Without loss of generality, we will describe the eigenvalue method in terms of its application to a transient two-dimensional conduction problem governed by Eq. (5-74).

Discretization of Eq. (5-74) in a uniform space grid by a finite-difference scheme of second-order accuracy yields

$$\rho c_v \frac{\partial T_j}{\partial t} = \frac{k}{h^2}(T_W + T_S + T_E + T_N - 4T_j), \quad j = 1, 2, \ldots, n$$

or, in matrix form,

$$[\mathbf{C}]\{\dot{\mathbf{T}}\} = [\mathbf{K}]\{\mathbf{T}\} + \{\mathbf{F}\}, \tag{5-86}$$

where $[\mathbf{C}]$ is an $n \times n$ identity matrix multiplied by $\rho h c_v$, $[\mathbf{K}]$ is an $n \times n$ sparse conductance matrix whose diagonal elements are $-4k/h$, and $\{\mathbf{F}\}$ is an n-component vector associated with boundary conditions.

We first multiply both sides of Eq. (5-86) by $[\mathbf{C}]^{-1}$ to yield

$$\{\dot{\mathbf{T}}\} = [\mathbf{D}]\{\mathbf{T}\} + \{\mathbf{G}\}, \tag{5-87}$$

where

$$[\mathbf{D}] = [\mathbf{C}]^{-1}[\mathbf{K}] \tag{5-88a}$$

and

$$\{\mathbf{G}\} = [\mathbf{C}]^{-1}\{\mathbf{F}\}. \tag{5-88b}$$

Next, let $\{\mathbf{T}\}$ be split into

$$\{\mathbf{T}\} = \{\phi(t)\} + \{\mathbf{T}_f\}, \tag{5-89}$$

where $\{\mathbf{T}_f\}$ is a vector independent of time; the subscript f means "fixed." In many cases, $\{\mathbf{T}_f\}$ is actually the steady-state solution. Substituting Eq. (5-89) and its time derivative into Eq. (5-87) leads to

$$\{\dot{\phi}\} = [\mathbf{D}]\,\{\phi\}, \tag{5-90}$$

provided $\{\mathbf{T}_f\}$ satisfies

$$\{\mathbf{T}_f\} = -\,[\mathbf{D}]^{-1}\,\{\mathbf{G}\}.$$

Equation (5-90) is a set of first-order, linear, ordinary differential equations. For the sake of clear description, we will apply its solution procedure to a 2×2 matrix system, realizing that this procedure can be extended to larger matrices. Thus, we rewrite Eq. (5-90) as

$$\begin{Bmatrix} \dot{\phi}_1 \\ \dot{\phi}_2 \end{Bmatrix} = \begin{bmatrix} d_{11} & d_{12} \\ d_{21} & d_{22} \end{bmatrix} \begin{Bmatrix} \phi_1 \\ \phi_2 \end{Bmatrix}. \tag{5-91}$$

Note that it is unnecessary to restrict ourselves to symmetric matrices; i.e., d_{12} may not equal d_{21}. Now, assume that the solution to Eq. (5-91) can be expressed as

$$\begin{Bmatrix} \phi_1 \\ \phi_2 \end{Bmatrix} = e^{\lambda_1 t} \begin{Bmatrix} x_1 \\ x_2 \end{Bmatrix} + e^{\lambda_2 t} \begin{Bmatrix} y_1 \\ y_2 \end{Bmatrix}, \tag{5-92}$$

where λ_1, λ_2, x_1, x_2, y_1, and y_2 are to be determined. Substituting Eq. (5-92) and its time derivatives into Eq. (5-91) yields, after simplification,

$$e^{\lambda_1 t} \begin{bmatrix} d_{11} - \lambda_1 & d_{12} \\ d_{21} & d_{22} - \lambda_1 \end{bmatrix} \begin{Bmatrix} x_1 \\ x_2 \end{Bmatrix} + e^{\lambda_2 t} \begin{bmatrix} d_{11} - \lambda_2 & d_{12} \\ d_{21} & d_{22} - \lambda_2 \end{bmatrix} \begin{Bmatrix} y_1 \\ y_2 \end{Bmatrix} = 0. \tag{5-93}$$

Equations (5-93) holds if λ_1 and λ_2 are the two eigenvalues of $[\mathbf{D}]$ and

$$\begin{Bmatrix} x_1 \\ x_2 \end{Bmatrix} \quad \text{and} \quad \begin{Bmatrix} y_1 \\ y_2 \end{Bmatrix}$$

are the two corresponding eigenvectors. Therefore, once the eigenvalues and the eigenvectors are somehow computed, Eq. (5-91) is equivalently solved. The proportionalities of the two eigenvectors can be fixed, by imposing the initial condition and utilizing Eqs. (5-89) and (5-92), to be

$$\begin{Bmatrix} \phi_{1,i} \\ \phi_{2,i} \end{Bmatrix} = \begin{Bmatrix} T_{1,i} \\ T_{2,i} \end{Bmatrix} - \begin{Bmatrix} T_{1,f} \\ T_{2,f} \end{Bmatrix} = \begin{Bmatrix} x_1 + y_1 \\ x_2 + y_2 \end{Bmatrix}. \tag{5-94}$$

More details of the eigenvalue method can be found in [46, 47].

SYMBOLS

A_{pq}	integrals [Eq. (5-53)]
D_{pq}	integrals [Eq. (5-60)]
M^x	$\partial^2 \phi/\partial x^2$
M^y	$\partial^2 \phi/\partial y^2$
$N(\xi, \eta)$	bilinear basis functions [Eqs. (5-47*a*)-(5-47*d*)]
r	mesh-size ratio $(\alpha\Delta t/h^2)$
[S]	standard matrix [Eq. (5-14*b*)]
η	isoparametric coordinate $[= 2(y - y^*)/h]$
ξ	isoparametric coordinate $[= 2(x - x^*)/h]$ or amplification factor
ρ	relaxation parameter ($1 \leqslant \rho \leqslant 2$ for over-relaxation and $0 \leqslant \rho \leqslant 1$ for under-relaxation)
τ	truncated higher-order terms

Subscripts

a, b, c, d	vertices of square isoparametric element (Fig. 5-4)
CN	Crank-Nicolson
ex	explicit
GS	Gauss-Seidel
im	implicit
sup	support

REFERENCES

1. E. Schmidt, *Über die Anwendung der Differenzenrechnung auf technische Anheiz-und Abkühlungsprobleme, A. Föppl Festschrift,* pp. 179–189, Springer-Verlag, Berlin, 1924.
2. S. H. Crandall, *Engineering Analysis,* pp. 376–380, McGraw-Hill, New York, 1956.
3. G. G. O'Brien, M. A. Hyman, and S. J. Kaplan, A Study of the Numerical Solution of Partial Differential Equations, *J. Math. and Phys.*, vol. 29, pp. 223–251, 1951.
4. J. Crank and P. Nicolson, A Practical Method for Numerical Evaluation of Solutions of Partial Differential Equations of the Heat-Conduction Type, *Proc. Camb. Philos. Soc.*, vol. 43, pp. 50–67, 1947.
5. R. D. Richtmyer and K. W. Morton, *Difference Methods for Initial-Value Problems,* pp. 189–191, Wiley-Interscience, New York, 1967.
6. K. Farnia and J. V. Beck, Numerical Solution of Transient Heat Conduction Equation for Heat-treatable Alloys Whose Thermal Properties Change with Times and Temperature, *ASME J. Heat Transfer,* vol. 99, pp. 471–478, 1977.
7. W. L. Wood, Control of Crank-Nicolson Noise in the Numerical Solution of the Heat Conduction Equation, *Int. J. Numer. Methods Eng.*, vol. 11, pp. 1059–1065, 1977.
8. P. Egerton, J. A. Howarth, and G. Poots, A Theoretical Investigation of Heat Transfer in a Ladle of Molten Steel during Pouring, *Int. J. Heat Mass Transfer,* vol. 22, pp. 1525–1532, 1979.
9. S. Orivuori, Efficient Method for Solution of Nonlinear Heat Conduction Problems, *Int. J. Numer. Methods Eng.*, vol. 14, pp. 1461–1476, 1979.
10. D. W. Peaceman and H. H. Rachford, Jr., The Numerical Solution of Parabolic and Elliptic Differential Equations, *SIAM (Soc. Ind. Appl. Math.) J.*, vol. 3, pp. 28–41, 1955.

11. L. V. Kantorovich, A Direct Method of Solving the Problem of the Minimum of a Double Integral, *Bull. Acad. Sci. USSR,* vol. 5, pp. 646–652, 1933.
12. L. V. Kantorovich and V. I. Krylov, *Approximate Methods of Higher Analysis,* Wiley-Interscience, New York, 1958.
13. I. S. Berezin and N. P. Zhidkov, *Computing Methods II,* Pergamon, Oxford, 1965.
14. P. J. Roache, *Computational Fluid Dynamics,* p. 312, Hermosa, Albuquerque, N.M., 1976.
15. E. A. Rukos, Continuous Elements in the Finite Element Method, *Int. J. Numer. Methods Eng.,* vol. 12, pp. 11–33, 1978.
16. L. T. Fan and K. B. Wang, Simulation of Transient, Multiphase Axial Dispersion Model for Mass Transfer Operations, *Numer. Heat Transfer,* vol. 1, pp. 299–314, 1978.
17. G. R. Otey and H. A. Dwyer, Numerical Study of the Interaction of Fast Chemistry and Diffusion, *AIAA J.,* vol. 17, pp. 606–613, 1979.
18. T. R. Hopkins and R. Wait, A Comparison of Galerkin, Collocation and the Method of Lines for Partial Differential Equations, *Int. J. Numer. Methods Eng.,* vol. 12, pp. 1081–1107, 1978.
19. L. A. Kurtz, R. E. Smith, C. L. Parks, and L. R. Boney, A Comparison of the Method of Lines to Finite Difference Techniques in Solving Time-Dependent Partial Differential Equations, *Comput. Fluids,* vol. 6, pp. 49–70, 1978.
20. M. Suzuki, S. Matsumoto, and S. Maeda, New Analytical Method for a Non-linear Diffusion Problem, *Int. J. Heat Mass Transfer,* vol. 20, pp. 883–889, 1977.
21. E. A. Thornton and A. R. Wieting, A Finite Element Thermal Analysis Procedure for Several Temperature-dependent Parameters, *ASME J. Heat Transfer,* vol. 100, pp. 551–553, 1978.
22. M. Suzuki and S. Maeda, Nonlinear Diffusion Problems with Variable Diffusivity and Time-dependent Flux Boundary Conditions, *Int. J. Heat Mass Transfer,* vol. 21, pp. 653–654, 1978.
23. Meric, R. A., Finite Element Analysis of Optimal Heating of a Slab with Temperature Dependent Thermal Conductivity, *Int. J. Heat Mass Transfer,* vol. 22, pp. 1347–1353, 1979.
24. V. Seeniraj and M. Arunachalam, Heat Conduction in a Semi-infinite Solid with Variable Thermophysical Properties, *Int. J. Heat Mass Transfer,* vol. 22, pp. 1455–1456, 1979.
25. S. H. Lin, Transient Heat Conduction in a Composite Slab with Variable Thermal Conductivity, *Int. J. Numer. Methods Eng.,* vol. 14, pp. 1726–1731, 1979.
26. J. M. Leone, Jr., P. M. Gresho, S. T. Chan, and R. L. Lee, A Note on the Accuracy of Gauss-Legendre Quadrature in the Finite Element Method, *Int. J. Numer. Methods Eng.,* vol. 14, pp. 769–784, 1979.
27. R. W. Thatcher, The Order of Numerical Quadrature in Two Dimensions, *Int. J. Numer. Methods Eng.,* vol. 14, pp. 1085–1089, 1979.
28. R. K. Ahluwalia and K. H. Im, Combined Conduction, Convection, Gas Radiation and Particle Radiation in MHD Diffusers, *Int. J. Heat Mass Transfer,* vol. 24, pp. 1421–1430, 1981.
29. B. Davies and J. A. Hendry, The Indirect Design of Fluid Channels Using a Global Variational Method with Trial Functions Not Satisfying the Prescribed Boundary Conditions, *Int. J. Numer. Methods Eng.,* vol. 11, pp. 579–591, 1977.
30. J. H. Bramble and A. H. Schatz, Rayleigh-Ritz-Galerkin Methods for Dirichlet's Problem Using Subspaces without Boundary Conditions, *Commun. Pure Appl. Math.,* vol. 23, pp. 653–675, 1970.
31. J. Nitsche, Uber ein Variationsprinzip zur Lösung von Dirichlet-Problemen bei Verwendung von Teilräumen, die keinen Randbedingungen unterworfen sind, *Ham. Univ. Math. Sem. Abh,* vol. 36, pp. 9–15, 1971.
32. I. Babuska, The Finite Element Method with Lagrangian Multipliers, *Numer. Math.,* vol. 20, pp. 179–192, 1973.
33. W. F. Ames, *Numerical Methods for Partial Differential Equations,* p. 113, Academic, New York, 1977.
34. S. P. Frankel, Convergence Rates of Iterative Treatments of Partial Differential Equations, *Math. Tables. Other Aids Comput.,* vol. 4, pp. 65–75, 1950.
35. A. Ostrowski, On Over and Under Relaxation in the Theory of the Cyclic Single Step Iteration, *Math. Tables Other Aids Comput.,* vol. 7, pp. 152–159, 1953.
36. D. M. Young, Iterative Method for Solving Differential Equations of Elliptic Type, *Trans. Am. Math. Soc.,* vol. 276, pp. 92–111, 1954.

37. R. S. Varga, *Matrix Iterative Analysis*, Prentice-Hall, Englewood Cliffs, N.J., 1962.
38. Y. H. Pao and R. J. Daugherty, Time-Dependent Viscous Incompressible Flow Past a Finite Flat Plate, *Boeing Sci. Res. Lab. Rept. D1-82-0822*, Jan. 1969.
39. M. R. Hestenes and E. Stiefel, Methods of Conjugate Gradients for Solving Linear Systems, *J. Res. Natl. Bur. Stand.*, vol. 49, pp. 409–436, 1952.
40. A. L. Yettram and M. J. S. Hirst, The Solution of Structural Equilibrium Equations by the Conjugate Gradient Method with Particular Reference to Plane Stress Analysis, *Int. J. Numer. Methods Eng.*, vol. 3, pp. 349–360, 1971.
41. I. Fried and J. Metzler, SOR vs. Conjugate Gradients in a Finite Element Discretization, *Int. J. Numer. Methods Eng.*, vol. 12, pp. 1329–1342, 1978.
42. L. Collatz, *The Numerical Treatment of Differential Equations*, pp. 396–405, Springer-Verlag, Berlin, 1960.
43. L. C. Wellford, Jr., and R. M. Ayer, A Finite Element Free Boundary Formulation for the Problem of Multiphase Heat Conduction, *Int. J. Numer. Methods Eng.*, vol. 11, pp. 933–943, 1977.
44. E. Varoglu and W. D. Liam Finn, Finite Elements Incorporating Characteristic for One-dimensional Diffusion-Convection Equation, *J. Comput. Phys.*, vol. 34, pp. 371–389, 1980.
45. K. S. Chung, The Fourth-Dimension Concept in the Finite Element Analysis of Transient Heat Transfer Problems, *Int. J. Numer. Methods Eng.*, vol. 17, pp. 315–325, 1981.
46. K. A. Rathjen, CAVE: A Computer Code for Two-dimensional Transient Heating Analysis of Conceptual Thermal Protection Systems for Hypersonic Vehicles, *NASA Contract. Rept. 2897*, 1977.
47. J. V. Palmieri and K. A. Rathjen, CAVE 3—A General Transient Heat Transfer Computer Code Utilizing Eigenvectors and Eigenvalues, *NASA Contract. Rept. 145290*, 1978.

PROBLEMS

5-1 (*a*) Compute $\phi(0.2L,\ 0.02L^2/\alpha)$, $\phi(0.4L,\ 0.02L^2/\alpha)$, $\phi(0.2L,\ 0.03L^2/\alpha)$, and $\phi(0.4L,\ 0.03L^2/\alpha)$ from Eq. (5-3) with the leading three terms ($m = 1, 3, 5$).

(*b*) Find the errors of the nodal solution $|\phi - \phi_j^{(n)}|$, where $\phi_1^{(2)} = \frac{3}{4}$, $\phi_2^{(2)} = 1$, $\phi_1^{(3)} = \frac{5}{8}$, and $\phi_2^{(3)} = \frac{15}{16}$.

(*c*) With $r = \frac{1}{4}$ and Eq. (5-4*c*), calculate $\tau_{\text{ex}}/\Delta t$ at the four locations given in (*a*).

From these results, do we observe that large nodal-solution errors correspond to large truncated errors?

5-2 Use the matrix method to prove that Eq. (5-7) is stable.

5-3 Use the von Neumann analysis to examine the stability of Eq. (5-10). What value of r is expected to yield the most accurate result?

5-4 Use the Runge-Kutta method to solve Eq. (5-24). Let $\phi_1(0) = \phi_2(0) = \phi_3(0) = 1$, $\phi_0(t) = 0$, and $\phi_2(t) = \phi_3(t)$, and integrate the system of equations from $t = 0$ to $0.75h^2/\alpha$. Compare $\phi_2(0.75h^2/\alpha)$ with the exact value $\phi(0.4L,\ 0.03L^2/\alpha)$.

5-5 Use the von Neumann analysis to prove that Eq. (5-31) is stable.

5-6 Prove that, by use of the piecewise linear weighting function $N_j(x)$, Eq. (5-33) can also be derived by the Galerkin method, i.e.,

$$-\frac{\displaystyle\int_{x_{j-1}}^{x_{j+1}} N_j \frac{d}{dx}\left(\phi \frac{d\phi}{dx}\right) dx}{\displaystyle\int_{x_{j-1}}^{x_{j+1}} N_j \, dx} = \frac{1}{2h^2}(-\phi_{j-1}^2 + 2\phi_j^2 - \phi_{j+1}^2).$$

Note that the algebra will be simplified if integration by parts is used.

5-7 Derive Eq. (5-34) by linearizing Eq. (5-33) and then discretizing the result.

5-8 Prove that the truncation error of Eq. (5-33) is $O(h^2)$.

5-9 Solve Eqs. (5-43) and (5-45) with the following boundary conditions:

$$\phi(0, y) = -y^2, \quad \phi(4, y) = 32 - y^2, \quad \phi(x, 0) = 2x^2$$

and

$$\phi(x, 4) = 2x^2 - 16.$$

The scheme that may be used for such a problem is the Gauss-Seidel method modified for block iterations. If no computer is accessible, carry out only the first iteration cycle.

5-10 With the nodal values $\phi_j = -2$, $\phi_W = -4$, $\phi_{SW} = -1$, and $\phi_S = 1$, estimate $\tilde{\phi}(\frac{1}{2}, \frac{3}{2})$ using Eqs. (5-46*b*) and (5-47*a*–*d*). Compare the result with the exact solution $\phi(\frac{1}{2}, \frac{3}{2}) = -\frac{7}{4}$.

5-11 Use isoparametric transformation to derive a discretized equation analogous to Eq. (5-54) for rectangular elements ($\Delta x \neq \Delta y$). Does the derived equation reduce to Eq. (5-54) if $\Delta x = \Delta y = h$?

5-12 Discretize the Laplace equation for the grid point j surrounded by five *irregular* triangular elements. The discretized result is considered satisfactory if the coefficients of the nodal unknowns are functions of grid coordinates only.

5-13 Repeat Example 5-3 for $\rho = 0.8$. How is the convergence rate in comparison with that for $\rho = 1$?

5-14 Illustrate Eqs. (5-77)–(5-79) with numerical values from Table 5-8 for $r = \frac{1}{4}$.

5-15 It is desirable to use a quick method to check the lengthy algebra leading to Eq. (5-84). Thus let

$$\tilde{\phi}(x, y, t) = (x^2 + y^2)t. \qquad \text{(P5-15}a\text{)}$$

(*a*) Substitute Eq. (P5-15*a*) into Eq. (5-81) and then set $x = y = 1$.
(*b*) Set $x_j = y_j = 1$ and then substitute $\phi_j = 2t$, $\phi W = t$, etc. into Eq. (5-84).
(*c*) Compare the results obtained in (*a*) and (*b*).

5-16 In addition to the classical iterative methods such as Gauss-Seidel, SOR, and ADI, a new method was proposed to solve an $n \times n$ matrix system efficiently. In this method, a matrix system of much smaller dimension ($m \times m$, $m < n$) is solved. Now consider, for example,

$$\begin{aligned} x_1 + x_2 + x_3 &= 6 \\ x_1 + x_2 - 2x_3 &= -3 \qquad \text{of } [\mathbf{A}]\{\mathbf{X}\} = \{\mathbf{B}\} \\ x_1 - x_2 + x_3 &= 2 \end{aligned}$$

and attempt to find x_1, x_2, and x_3 by following this proposed method step by step:

(*a*) Choose two ($m = 2$) trial vectors

$$\mathbf{U}_1^{[0]} = \begin{Bmatrix} 0 \\ 1 \\ 1 \end{Bmatrix} \quad \text{and } \mathbf{U}_2^{[0]} = \begin{Bmatrix} 1 \\ 0 \\ 1 \end{Bmatrix}.$$

(*b*) Let

$$\begin{Bmatrix} x_1^{[0]} \\ x_2^{[0]} \\ x_3^{[0]} \end{Bmatrix} = \alpha_1 \mathbf{U}_1^{[0]} + \alpha_2 \mathbf{U}_2^{[0]},$$

where α_1 and α_2 are constants to be determined.

(*c*) Find α_1 and α_2 by solving $\partial V_0/\partial \alpha_1 = 0$ and $\partial V_0/\partial \alpha_2 = 0$, where

$$V_0 = (x_1^{[0]} + x_2^{[0]} + x_3^{[0]} - 6)^2 + (x_1^{[0]} + x_2^{[0]} - 2x_3^{[0]} + 3)^2 + (x_1^{[0]} - x_2^{[0]} + x_3^{[0]} - 2)^2.$$

(*d*) Choose the next two trial vectors $\mathbf{U}_1^{[1]} = \mathbf{X}^{[0]}$ and $\mathbf{U}_2^{[1]} = \mathbf{U}_2^{[0]}$.
(*e*) Repeat (*b*) and (*c*) for the first iteration cycle [1]. What are $x_1^{[1]}$, $x_2^{[1]}$, and $x_3^{[1]}$?

CHAPTER

SIX

LAMINAR FORCED CONVECTION: HYDRODYNAMIC BOUNDARY LAYER (I)

In addition to its physical and mathematical simplicity, laminar forced convection is a common phenomenon in which other more complicated mechanisms such as buoyancy force, gaseous radiation, turbulence, and combustion coexist. Topics in laminar forced convection have been comprehensively investigated and reported by, among others, Rohsenow and Choi [1], Kays [2], Schlichting [3], Gebhart [4], Eckert and Drake [5], Kreith [6], and Holman [7]. In their work, the emphasis was mainly on the description of physical phenomena, use of *analytical* methods, interpretation of *analytical* solutions, and experimental results. Unfortunately, as the area of heat transfer is further explored, there are fewer and fewer physical problems that can be solved analytically. Therefore, we are frequently forced to rely on numerical analysis to obtain an approximate solution. For heat transfer researchers, it is desirable to be able to select a suitable numerical method for a given physical problem, use this method to solve the problem, and finally analyze the numerical results in terms of properties such as stability, consistency, convergence, accuracy, and efficiency. In this chapter, we use some numerical methods to solve the continuity and momentum equations describing laminar boundary-layer flow.

In Section 6-1, the phenomenon of laminar forced convection is classified into three types of flows, which are named after their physical characteristics and respective governing partial differential equations. These are: (1) boundary-layer flows (parabolic type), (2) fully developed internal flows (analytically integrable), and (3) streamwise diffusion flows (elliptic type).† The generalized partial differen-

†Many authors prefer to call this type of flow convection-diffusion flow, but both convection and diffusion also coexist in boundary-layer flow.

tial equations governing the streamwise and transverse momentum balances of laminar forced-convection flow are introduced. They are further reduced to equations governing boundary-layer flow.

In Section 6-2, the famous Blasius similarity equation, which constitutes a nonlinear two-point boundary-value problem, is solved by an iterative shooting method and a parametric expansion method.

The finite-difference method with the x-ω^n transformation, where ω is the normalized stream function, is presented in Section 6-3. The widely used Patankar-Spalding method and the version of this method revised by Denny and Landis correspond to $n = 1$ and $n = 2$, respectively.

Section 6-4 concerns the numerical stability, consistency, and accuracy of the finite-difference scheme used to discretize the momentum differential equation.

6-1 BRIEF DESCRIPTION OF THE PHYSICAL PHENOMENA

Laminar forced convection is a phenomenon in which a fluid driven by some external force flows over or inside a solid surface such that the streamlines of the flow appear smooth and parallel. From the mathematical viewpoint, this phenomenon can be approximately categorized into three regimes as follows.

6-1*a* Boundary-Layer Flows (Parabolic Type)

When a viscous fluid moves rapidly over a solid body, the friction between the fluid and the solid surface causes the fluid to be retarded within a thin region immediately adjacent to the solid surface. This thin region, in which large flow velocity gradients exist, is called the boundary layer. Figures 6-1 and 6-2*a* show, respectively, boundary layers formed externally over a solid surface and internally inside a channel. Several important features of laminar forced convection are summarized as follows:

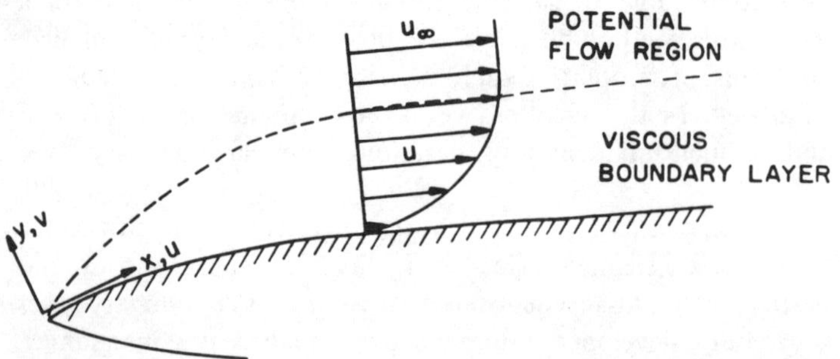

Figure 6-1 Hydrodynamic boundary layer externally formed.

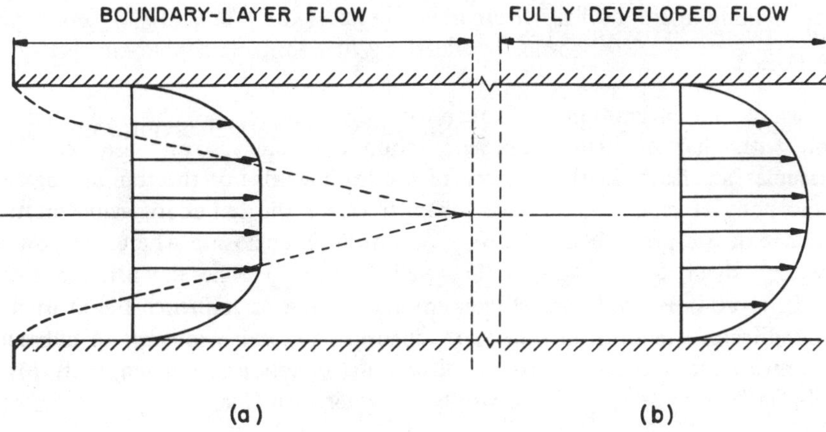

Figure 6-2 (*a*) Hydrodynamic boundary layer internally formed; (*b*) fully developed flow in a channel.

1. The concept of the hydrodynamic boundary layer, introduced by Prandtl in 1904 [8], can be extended to studies of thermal and species boundary layers.
2. Although the thickness of the boundary layer cannot be precisely defined, it is conventionally said that at the edge of the boundary layer u/u_∞ approaches unity and $\partial u/\partial y$ vanishes, where u is the local streamwise flow velocity, u_∞ is the velocity of the ambient potential flow, and y is the transverse direction.
3. For laminar flows, the thickness of the hydrodynamic boundary layer can be determined uniquely by the Reynolds number, $u_\infty x/\nu$.
4. Inside the boundary layer, except near the leading edge, the assumptions $u \gg v$, $\partial u/\partial y \gg \partial u/\partial x \approx \partial v/\partial y \gg \partial v/\partial x$ are valid.
5. Because of feature 4, the elliptic governing transport equations are reduced to parabolic equations, for which the solution procedures are relatively simple.
6. The results of primary interest are the streamwise velocity profile inside the boundary layer, the thickness of the boundary layer, and the shear stress distribution at the solid surface.

6-1*b* Fully Developed Channel Flows (Analytically Integrable)

Far downstream in a channel ($x \cong D_h\text{Re}/20$), the transverse velocity vanishes and the axial velocity becomes independent of x. Under such circumstances, the pressure difference between two streamwise locations is completely balanced by the shear stress. The flow in this regime is said to be hydrodynamically fully developed and is shown in Fig. 6-2*b*. The analysis of hydrodynamically fully developed flow is relatively elementary and, in most cases, can yield closed-form analytical solutions [1-7]. It is therefore of minor interest to us and will be omitted here.

6-1*c* Streamwise Diffusion Flows (Elliptic Type)

Flows in which the magnitude of streamwise diffusion cannot be neglected in comparison with that of mainstream convection are called streamwise diffusion flows. Examples are flows in the vicinity of the leading edge or the trailing edge of a streamline body (Fig. 6-3), in a short channel where the boundary values at the entrance and exit are prescribed (Fig. 6-4), or within an enclosure where the flow is driven by the shear force of a sliding wall (Fig. 6-5). The downstream flow conditions in these cases will "travel" against the stream to influence the upstream through the diffusion process. If transverse diffusion also exists and is as significant as the streamwise counterpart, the partial differential equations governing such flows become elliptic. This type of problem will be considered in Chapter 8.

6-1*d* Governing Partial Differential Equations

The Navier-Stokes equations governing steady-state, two-dimensional, forced-convection laminar flow can be written in Cartesian ($n = 0$, $r = y$) or cylindrical ($n = 1$) coordinates as follows.

Continuity:

$$\frac{\partial}{\partial x}(\rho u) + \frac{1}{r^n}\frac{\partial}{\partial r}(r^n \rho v) = 0, \tag{6-1}$$

x-direction momentum:

$$\rho u \frac{\partial u}{\partial x} + \rho v \frac{\partial u}{\partial r} = \frac{1}{r^n}\frac{\partial}{\partial r}\left(r^n \mu \frac{\partial u}{\partial r}\right) + \frac{\partial}{\partial x}\left(\mu \frac{\partial u}{\partial x}\right) - \frac{\partial p}{\partial x}, \tag{6-2}$$

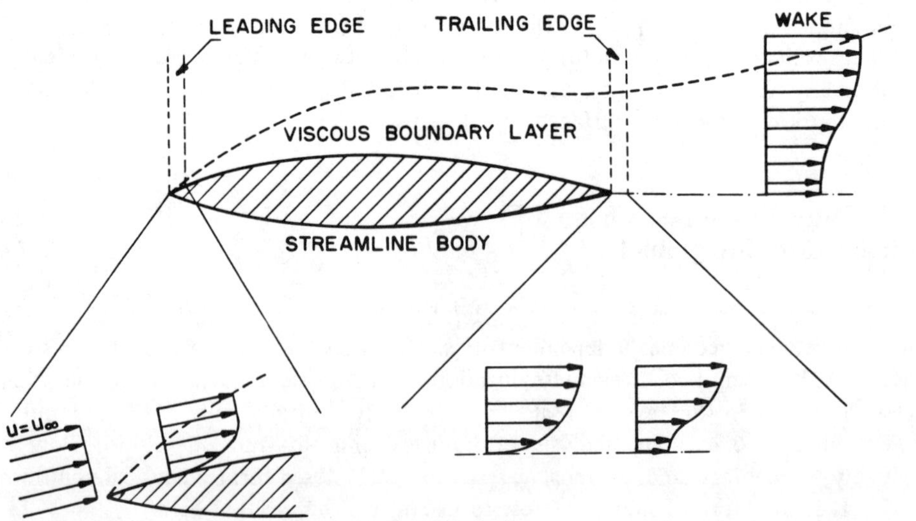

Figure 6-3 External streamwise diffusion flows.

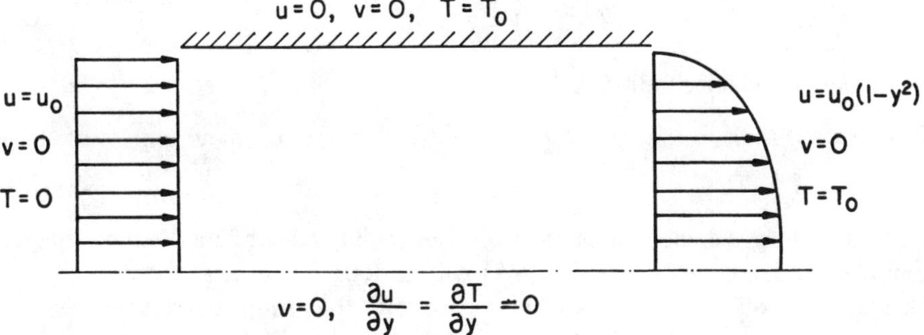

Figure 6-4 Internal streamwise diffusion flows.

and r-direction momentum:

$$\rho u \frac{\partial v}{\partial x} + \rho v \frac{\partial v}{\partial r} = \frac{1}{r^n} \frac{\partial}{\partial r}\left(r^n \mu \frac{\partial v}{\partial r}\right) - n\mu \frac{v}{r^2} + \frac{\partial}{\partial x}\left(\mu \frac{\partial v}{\partial x}\right) - \frac{\partial p}{\partial r}. \tag{6-3}$$

Several simpler cases can be deduced from Eqs. (6-1)–(6-3). We will discuss them in the following sections and the next two chapters. The first simple case is that of laminar boundary-layer flow.

Consider an incompressible fluid of constant viscosity μ flowing over an impermeable flat plate with a constant free-stream velocity u_∞. Following the boundary-layer assumptions, namely $u \gg v$ and $\partial/\partial y \gg \partial/\partial x$, we can simplify Eq. (6-3) into $\partial p/\partial y \approx 0$, which implies $\partial p/\partial x \approx dp_\infty/dx$. In the potential flow outside the boundary layer, however, dp_∞/dx is equal to $-\rho_\infty u_\infty du_\infty/dx$ according to Bernoulli's equation. Since u_∞ is constant, $\partial p/\partial x$ vanishes. Consequently, in Cartesian coordinates, Eqs. (6-1) and (6-2) reduce to

$$\frac{\partial u}{\partial x} + \frac{\partial v}{\partial y} = 0 \tag{6-4}$$

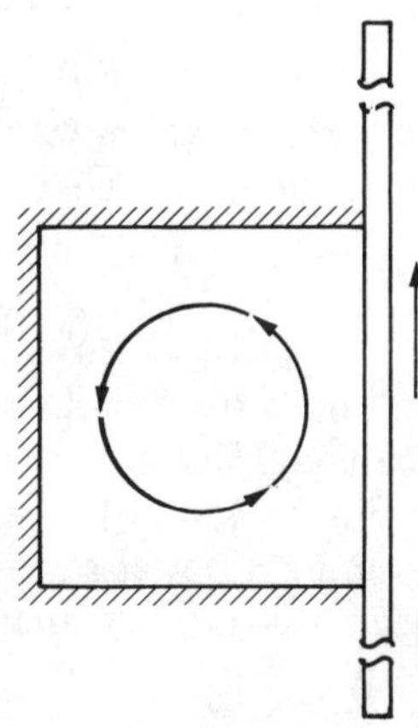

Figure 6-5 Elliptic flow in an enclosure in which the circulation is driven by the sliding motion of the plate.

and
$$u \frac{\partial u}{\partial x} + v \frac{\partial u}{\partial y} = \nu \frac{\partial^2 u}{\partial y^2} \tag{6-5}$$

subject to the boundary conditions

$$u(x, 0) = v(x, 0) = 0, \quad u(x, \infty) = u_\infty, \quad u(x_0, y) = s_1(y), \quad \text{and } v(x_0, y) = s_2(y), \tag{6-6}$$

where x_0 is a short distance measured from the leading edge of the boundary layer. It is improper to let $x_0 = 0$ since Eq. (6-5) will no longer hold near there.

The system of Eqs. (6-4)–(6-6) is one of the simplest versions of the Navier-Stokes equations. Yet its solution procedures are not without difficulties. First, Eq. (6-5) is nonlinear; whenever a nonlinearity arises, an iterative scheme is usually needed. Second, the edge of the boundary layer, which is treated as the boundary of the domain, is not prescribed a priori; in using any numerical method, some specification must be artificially made. Third, the boundary conditions at $x = x_0$ (sometimes called the initial conditions) must be given; mathematically this prescription is legitimate, but physically the initial velocity distributions $u(x_0, y)$ and $v(x_0, y)$ are unique and not known. For these three reasons, the set of Eqs. (6-4)–(6-6) is a fundamental and nontrivial numerical problem. It may be worthwhile to devote extensive effort to the numerical analysis of this problem.

6-2 BLASIUS SIMILARITY EQUATION

Using the similarity transformation

$$\left(\frac{\partial}{\partial x}\right)_y = \left(\frac{\partial}{\partial x}\right)_\eta - \frac{\eta}{2x}\frac{\partial}{\partial \eta} \quad \text{and} \quad \frac{\partial}{\partial y} = \left(\frac{u_\infty}{x\nu}\right)^{1/2} \frac{\partial}{\partial \eta}, \tag{6-7a, b}$$

Blasius in 1908 [9] reduced Eqs. (6-4)–(6-6) to the nonlinear ordinary differential equation

$$f''' + \tfrac{1}{2} f f'' = 0, \tag{6-8}$$

where $f(\eta) \equiv \psi(x, y)/(u_\infty \nu x)^{1/2}$; $\psi(x, y)$ is the stream function related to the flow velocities as $u = \partial\psi/\partial y$ and $v = -\partial\psi/\partial x$. The primes denote differentiation with respect to the similarity variable $\eta \equiv \mathrm{Re}_x^{1/2} y/x$. The corresponding boundary conditions in terms of the similarity variable become

$$f(0) = f'(0) = 0 \quad \text{and } f'(\eta_\infty) = 1. \tag{6-9}$$

Some useful relations between the primitive functions and the similarity functions are listed in Table 6-1. With the aid of Eqs. (6-7*a*) and (6-7*b*) and Table 6-1, the reader will readily be able to derive Eqs. (6-8) and (6-9) from Eqs. (6-4)–(6-6). The first solution of Eqs. (6-8) and (6-9) was obtained by Blasius, using a power series expansion [9]. Since then, this set of equations has continued to attract research interest [10–14].

Table 6-1 Useful relations between primitive functions and Blasius similarity functions

Primitive function	Similarity function
$u(x, y)$	$u_\infty f'(\eta)$
$v(x, y)$	$u_\infty \mathrm{Re}_x^{-1/2} \dfrac{\eta f'(\eta) - f(\eta)}{2}$
$\dfrac{\partial u(x, y)}{\partial x}$	$-\dfrac{u_\infty \eta}{2x} f''(\eta)$
$\dfrac{\partial u(x, y)}{\partial y}$	$\left(\dfrac{u_\infty^3}{\nu x}\right)^{1/2} f''(\eta)$
$\dfrac{\partial^2 u(x, y)}{\partial y^2}$	$\dfrac{u_\infty^2}{\nu x} f'''(\eta)$
$\psi(x, y)$	$(\nu x u_\infty)^{1/2} f(\eta)$

Before proceeding to describe various numerical methods, it may be of interest to assess the validity of the boundary-layer assumptions.

Example 6-1 At $x = 5$ cm and $y = 0.1$ cm with $\mathrm{Re}_x = 10^4$, are the boundary-layer assumptions $u \gg v$ and $\partial^2 u/\partial y^2 \gg \partial^2 u/\partial x^2$ valid? The similarity solution listed in Appendix A or obtained in [10] can be used to estimate the values of these terms.

Solution: Let the point of interest be indexed by (i, j) and take $\Delta x = \Delta y = 0.01$ cm. Since $\mathrm{Re}_x = 10^4$ at $x = 5$ cm, we obtain $u_\infty/\nu = 2000$ cm^{-1}. From the definition

$$\eta \equiv \frac{y}{x^{1/2}} \left(\frac{u_\infty}{\nu}\right)^{1/2}, \tag{a}$$

we calculate

$$\eta_{i,j} = \frac{0.1}{\sqrt{5}} (2000)^{1/2} = 2. \tag{b}$$

Similarly,

$$\eta_{i-1,j} = \frac{0.1}{(4.99)^{1/2}} (2000)^{1/2} = 2.0020, \tag{c}$$

$$\eta_{i+1,j} = \frac{0.1}{(5.01)^{1/2}} (2000)^{1/2} = 1.9980, \tag{d}$$

$$\eta_{i,j-1} = \frac{0.09}{\sqrt{5}} (2000)^{1/2} = 1.8000, \tag{e}$$

and

$$\eta_{i,j+1} = \frac{0.11}{\sqrt{5}}(2000)^{1/2} = 2.2000. \tag{f}$$

With these computed η's, we are able to construct Fig. 6-6 based on the similarity results. Note that values at grid points $(i-1, j)$ and $(i+1, j)$ are interpolated. Now, let us define $U = u/u_\infty$ and $V = v/u_\infty$. At grid point (i, j), the following quantities can be computed with the aid of Table 6-1:

$$U\frac{\partial U}{\partial x} = f'_{i,j} \times \frac{f'_{i+1,j} - f'_{i-1,j}}{2\Delta x} = (0.62977)\frac{0.62922 - 0.63029}{2(0.01)} = -0.03369 \tag{g}$$

and

$$V = \frac{\eta_{i,j}f'_{i,j} - f_{i,j}}{2\,\mathrm{Re}_x^{1/2}} = \frac{2 \times 0.62977 - 0.65003}{2(100)} = 0.00305. \tag{h}$$

Therefore,

$$\frac{U}{V} = 206.65. \tag{i}$$

Indeed, it is found that $U \gg V$. We continue to compute

$$V\frac{\partial U}{\partial y} = (0.00305)\frac{0.68132 - 0.57477}{2(0.01)} = 0.01625, \tag{j}$$

$$\frac{\nu}{u_\infty}\frac{\partial^2 U}{\partial y^2} = \left(\frac{1}{2000}\right)\frac{0.68132 - 2(0.62977) + 0.57477}{(0.01)^2} = -0.01725, \tag{k}$$

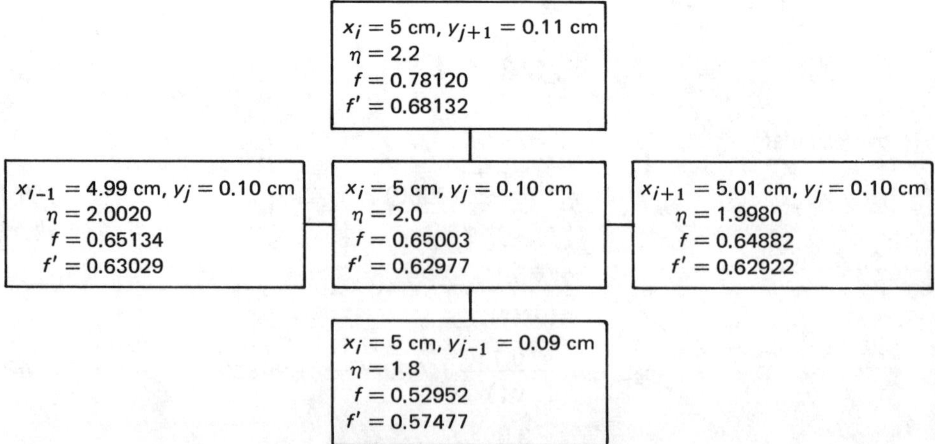

Figure 6-6 Values of the similarity functions at five points in the x-y plane.

and

$$\frac{\nu}{u_\infty}\frac{\partial^2 U}{\partial x^2}=\left(\frac{1}{2000}\right)\frac{0.63029-2(0.62977)+0.62922}{(0.01)^2}=-0.00015. \qquad (l)$$

It is obvious that

$$34.5=\left|\frac{\partial^2 U}{\partial y^2}\right| \gg \left|\frac{\partial^2 U}{\partial x^2}\right|=0.3. \qquad (m)$$

Thus, the validity of the boundary-layer assumption is well demonstrated by Eqs. (*i*) and (*m*). For clarity, the streamwise momentum equation and the corresponding magnitude of each term are rewritten below:

$$U\frac{\partial U}{\partial x}+V\frac{\partial U}{\partial y}=\frac{\nu}{u_\infty}\frac{\partial^2 U}{\partial y^2}+\frac{\nu}{u_\infty}\frac{\partial^2 U}{\partial x^2}, \qquad (n)$$

corresponding to

$$-0.03369+0.01625\approx-0.01725-0.00015. \qquad (o)$$

Since $f''(0)$, which is related to the wall shear stress, is not known a priori, Eqs. (6-8) and (6-9) constitute a two-point boundary-value problem. In addition to the Blasius equation, we will soon realize that numerous heat transfer problems dealing with boundary-layer flow can often, but not always, be transformed into two-point boundary-value problems by using a similarity transformation. It is therefore important to become familiar with a certain numerical technique that solves problems of this type. We will first introduce a classical iterative method.

6-2*a* Shooting Method

The shooting method is an iterative numerical method in which the missing initial conditions are systematically guessed with the hope of shooting the terminal targets after many direct integrations. The procedure of the shooting method is well illustrated by its application to Eqs. (6-8) and (6-9).

Let an arbitrarily guessed value of $f''(0)$ be denoted by κ and the correct value of $f''(0)$ be denoted by κ^*. Using an integration scheme–the Runge-Kutta method, for example–with the three initial conditions $f(0)=f'(0)=0$ and $f''(0)=\kappa$, we can integrate Eq. (6-8) from $\eta=0$ to $\eta=\eta_\infty$. When a certain κ is tried, a corresponding incorrect value of $f'(\eta_\infty)$ will be generated. Thus, let us postulate that there exists a single-valued function $G(\kappa)$ defined as

$$f'(\eta_\infty)=G(\kappa). \qquad (6\text{-}10)$$

Consequently,

$$1=G(\kappa^*). \qquad (6\text{-}11)$$

We may expand $G(\kappa)$ in a Taylor's series about κ^* to yield

$$G(\kappa)=1+\left(\frac{dG}{d\kappa}\right)_{\kappa^*}(\kappa-\kappa^*)+\text{higher-order terms.} \qquad (6\text{-}12)$$

If the higher-order terms are dropped, then the equality in Eq. (6-12) no longer holds. Instead, we introduce κ_{imp} such that

$$G(\kappa) = 1 + \left(\frac{dG}{d\kappa}\right)_{\kappa_{\text{imp}}} (\kappa - \kappa_{\text{imp}}), \tag{6-13}$$

where κ_{imp} is an improved missing initial condition. Using two arbitrary initial missing conditions κ_1 and κ_2 to obtain the corresponding terminal values $G(\kappa_1)$ and $G(\kappa_2)$, and then eliminating $(dG/d\kappa)_{\kappa_{\text{imp}}}$ with the aid of Eq. (6-13), we obtain

$$\kappa_{\text{imp}} = \frac{\kappa_2\,[G(\kappa_1) - 1] - \kappa_1\,[G(\kappa_2) - 1]}{G(\kappa_1) - G(\kappa_2)}, \tag{6-14}$$

in which $G(\kappa_1) \neq G(\kappa_2)$ provided $\kappa_1 \neq \kappa_2$. This improved value κ_{imp} and the generated $G(\kappa_{\text{imp}})$ are then used to replace a less accurate pair in Eq. (6-14). The procedure is repeated until the iterated solution converges. It will be interesting to determine whether $G(\kappa)$ is indeed a single-valued function of κ and, if it is, to find an explicit expression for the function $G(\kappa)$ for a very simple scheme and a very crude step size.

Example 6-2 Find a polynomial expression $G(\kappa)$ for a step size $h = \frac{1}{5}\,\eta_\infty$.

Solution: For simplicity, the forward difference (known as the Euler method) with uniform mesh size h will be used here; i.e.,

$$\phi_{j+1} = \phi_j + \phi_j' h, \tag{a}$$

where the subscript j denotes the location $\eta = jh$ and the function ϕ represents f, f', and f''. Starting with the initial values $f(0) = f'(0) = 0$ and $f''(0) = \kappa$, we can integrate Eq. (6-8) with respect to η, using Eq. (*a*). Results from $\eta = 0$ to $\eta = 5h$ are listed in Table 6-2. Therefore, from Table 6-2, we derive

$$G(\kappa) = f'(\eta_\infty) = \left(\frac{3}{4}\,h^7\right)\kappa^3 - \left(\frac{5}{2}\,h^4\right)\kappa^2 + (5h)\kappa. \tag{b}$$

Figure 6-7 shows $G(\kappa)$ versus κ for various values of $J = \eta_\infty/h$. Here η_∞ is taken as 7. The intersection of $G(\kappa^*) = 1$ and the curve gives the correction missing initial conditions. The procedure of the shooting method is also illustrated graphically in Fig. 6-8. We start by choosing two arbitrary missing initial conditions κ_1 and κ_2 to obtain $G(\kappa_1)$ at A and $G(\kappa_2)$ at B. Points A and B are connected with a straight line, which is then extended to intersect the line $G(\kappa) = 1$ at C. The abscissa of C is κ_{imp}, which can be used to obtain $G(\kappa_{\text{imp}})$ at D. Again, a straight line is drawn passing points B and D (the guessed solution at A, being worse, is dropped) and intersecting the line $G(\kappa) = 1$ at E. This procedure can be repeated until κ^* is obtained.

Generally, for problems with only one missing initial condition, the shooting

Table 6-2 Sample derivation of the polynomial function $G(h, \kappa)$

η	f	f'	f''	$f'''\left(=-\frac{ff''}{2}\right)$
0	0	0	κ	0
h	0	$h\kappa$	κ	0
$2h$	$h^2\kappa$	$2h\kappa$	κ	$-\frac{h^2\kappa^2}{2}$
$3h$	$3h^2\kappa$	$3h\kappa$	$\kappa-\frac{h^3\kappa^2}{2}$	$\frac{3}{4}h^5\kappa^3-\frac{3}{2}h^2\kappa^2$
$4h$	$6h^2\kappa$	$4h\kappa-\frac{h^4\kappa^2}{2}$	$\frac{3}{4}h^6\kappa^3-2h^3\kappa^2+\kappa$	$-\frac{9}{4}h^8\kappa^4+6h^5\kappa^3-3h^2\kappa^2$
$5h$	$10h^2\kappa-\frac{h^4\kappa^2}{2}$	$5h\kappa-\frac{5}{2}h^4h^2+\frac{3}{4}h^7\kappa^3$	–	–

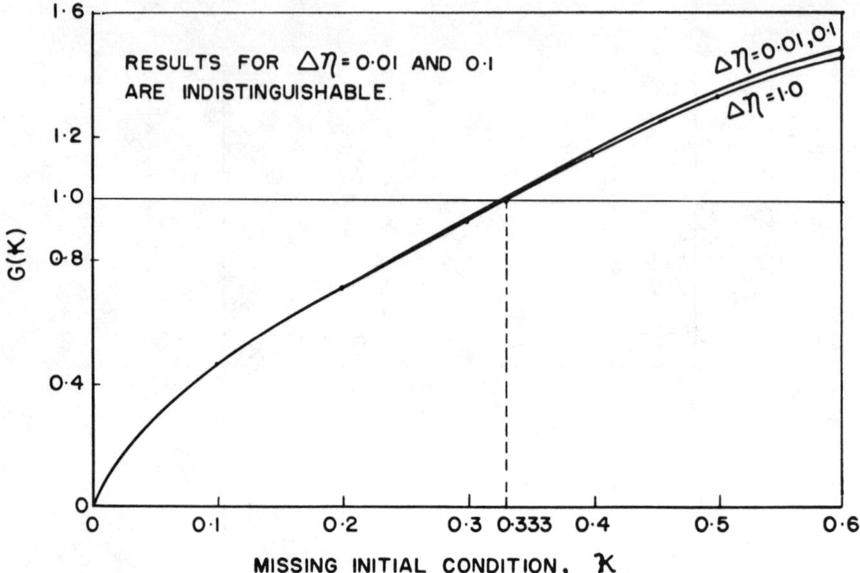

Figure 6-7 Terminal value $f'(\infty)$ versus $\kappa = f''(0)$.

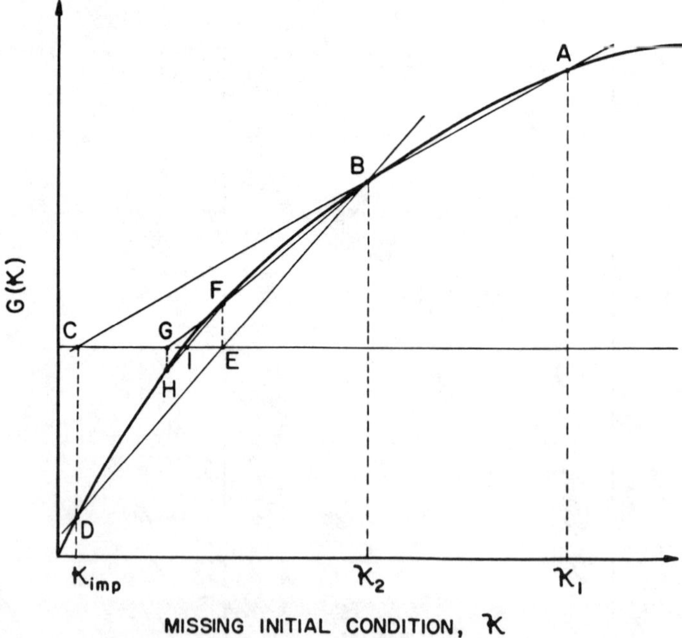

Figure 6-8 Graphic representation of the shooting method.

method is simple and the computed solution may converge. We will introduce an example to illustrate once again the procedure of the shooting method.

Example 6-3 The similarity transformation of the partial differential equation governing the momentum balance in laminar stagnation flow yields $\phi''' + \phi\phi'' - \phi'^2 + 1 = 0$ with $\phi(0) = \phi'(0) = 0$ and $\phi'(5) = 1$ [15]. Use the shooting method to find the missing initial condition $\phi''(0)$.

Solution: Let us start with two guessed missing initial conditions $\phi_1''(0) = 1$ and $\phi_2''(0) = 2$. Using an integration scheme, we can obtain $\phi_1'(5)$ and $\phi_2'(5)$, which most likely are not equal to 1. Substituting $\phi_1''(0)$, $\phi_2''(0)$, $\phi_1'(5)$, and $\phi_2'(5)$ into Eq. (6-14) leads to $\phi_{\text{imp}}''(0)$. This improved value, along with the generated $\phi_{\text{imp}}'(5)$, are then used to replace a less accurate pair in Eq. (6-14). This procedure is repeated, and all the iterated values are listed in Table 6-3. The final converged answer is $\phi''(0) = 1.23259$.

For more complicated problems such as free-convection boundary-layer flows over a vertical isothermal plate in which both the shear stress and the heat flux at the wall are unknown, it is frequently possible to obtain a divergent solution with the shooting method. In those cases, some other methods are more satisfactory. Among them are the finite-difference, multiple shooting, quasi-linearization, adjoint [16, 17], and parametric differentiation [18-23] methods. Although the quasi-linearization and adjoint methods enable the computed solution to clamp at both initial and terminal points, these methods have the difficulty that the computed solution may converge to one that does not satisfy the differential equations. In other words, the numerical results obtained by the quasi-linearization and adjoint methods will satisfy the boundary conditions, but may not satisfy the governing equations. The parametric differentiation method is noniterative (in a certain sense) and is very economical when solutions for many parametric values are desired. There are, however, two major shortcomings of this method that have not been

Table 6-3 Iterated missing initial conditions of the stagnation flow equation obtained by the shooting method

κ_1	κ_2	κ_{imp}
1.0	2.0	1.05328
2.0	1.05328	1.10158
1.05328	1.10158	1.37165
1.10158	1.37165	1.19693
1.37165	1.19693	1.22446
1.19693	1.22446	1.23321
1.22446	1.23321	1.23258
1.23321	1.23258	1.23259

reported in the literature: (1) a starting solution must always be given; that is, the method is not capable of generating a starting solution by itself and therefore can be used only when a solution at a certain parametric value is available; and (2) as the parameter is varied further from its starting value, the truncation error resulting from the first-order finite-difference approximation is greatly amplified. These two deficiencies of the parametric differentiation method are eliminated in the parametric expansion method.

6-2*b* Parametric Expansion Method

In the parametric expansion method [24], solutions at a new parametric value $\zeta + \Delta\zeta$ are expressed in terms of those at ζ by using a Taylor's series expansion, where ζ stands for either an inherent parameter of physical significance, such as the Reynolds number or the Prandtl number, or an artificially embedded parameter without physical meaning. The first-order and second-order terms are computed and the remainder term is estimated. Inclusion of the second-order term and the remainder term cannot be overlooked, since this is the main reason why the parametric expansion method yields accurate results. In contrast with the first shortcoming of the parametric differentiation method, the parametric expansion method is able to generate a set of starting solutions if an artificial parameter is embedded as a coefficient into the nonlinear ordinary differential equation. In other words, a starting solution can be obtained by analytically integrating the embedded differential equation with this parameter set to zero. (This embedding procedure is similar to the continuation method [17, pp. 15–19].) Then by varying the artificial parameter from zero to unity with the aid of the parametric expansion method, one recovers the original differential equation and its solutions. More details of the method are given in [24]. In this section, we will present the procedure by solving the Blasius similarity equation (6-8) subject to Eq. (6-9).

In the first step, an artificial parameter ζ is embedded into Eq. (6-8) as a coefficient of the nonlinear term,

$$f''' + \frac{\zeta}{2} ff'' = 0. \tag{6-15}$$

It is noteworthy that, unless ζ is equal to unity, Eq. (6-15) may not have any physical significance. However, we are not concerned about the lack of physical significance of Eq. (6-15). The purpose of introducing ζ is to seek an analytically integrable solution at a particular value of ζ, normally $\zeta = 0$. Indeed, the reduced equation $f'''(\eta) = 0$ with the boundary conditions $f(0) = f'(0) = 0$ and $f'(\eta_\infty) = 1$ can be easily integrated to yield

$$f(\eta) = \frac{\eta^2}{2\eta_\infty}. \tag{6-16}$$

The next step is to find the solutions at $\zeta = \Delta\zeta$. The function $f(\eta, \zeta + \Delta\zeta)$ can be expanded in a Taylor's series about $f(\eta, \zeta)$ as

$$f(\eta, \zeta + \Delta\zeta) = f(\eta, \zeta) + \left(\frac{\partial f}{\partial \zeta}\right)_\zeta (\Delta\zeta) + \left(\frac{\partial^2 f}{\partial \zeta^2}\right)_\zeta \frac{(\Delta\zeta)^2}{2} + Q(\eta, \zeta), \tag{6-17}$$

where $Q(\eta, \zeta)$ is the remainder containing all the higher-order terms and will be estimated. Denoting the first and second derivatives as $g(\eta, \zeta)$ and $h(\eta, \zeta)$, respectively, leads to

$$f(\eta, \zeta + \Delta\zeta) = f(\eta, \zeta) + g(\eta, \zeta)(\Delta\zeta) + h(\eta, \zeta)\frac{(\Delta\zeta)^2}{2} + Q(\eta, \zeta). \tag{6-18}$$

The other two functions $f'(\eta, \zeta + \Delta\zeta)$ and $f''(\eta, \zeta + \Delta\zeta)$ can be expressed in terms of f', g', h', f'', g'', and h'' at ζ in a similar fashion. Here primes denote differentiation with respect to η. Thus, if g, h, and Q at ζ are somehow obtained, $f(\eta, \zeta + \Delta\zeta)$ at the new location can be evaluated from Eq. (6-18). The procedure for obtaining g, h, and Q at ζ is described below.

We differentiate Eq. (6-15) with respect to ζ to yield

$$Lg + \tfrac{1}{2} ff'' = 0 \tag{6-19}$$

subject to

$$g(0) = g'(0) = 0 \quad \text{and } g'(\eta_\infty) = 0, \tag{6-20}$$

where L is a linear operator (since f, f', and f'' are known at ζ) defined as

$$L = \frac{d^3}{d\eta^3} + \frac{\zeta}{2}\left(f\frac{d^2}{d\eta^2} + f''\right). \tag{6-21}$$

Differentiating Eq. (6-19) with respect to ζ again yields

$$Lh + \zeta gg'' + fg'' + gf'' = 0 \tag{6-22}$$

subject to

$$h(0) = h'(0) = 0 \quad \text{and } h'(\eta_\infty) = 0. \tag{6-23}$$

Equations (6-19) and (6-22) are linear in g and h, respectively. This linearity permits the superposition of two auxiliary functions to describe g and h:

$$g(\eta, \zeta) = \alpha g_1(\eta, \zeta) + g_2(\eta, \zeta) \tag{6-24a}$$

and

$$h(\eta, \zeta) = \beta h_1(\eta, \zeta) + h_2(\eta, \zeta), \tag{6-24b}$$

where α and β are constants. The purpose of introducing g_1, g_2, h_1, and h_2 is to transform the two-point boundary-value problems Eqs. (6-19)–(6-23) into initial-value problems (see also Section 3-4*a*). Substituting Eqs. (6-24*a*) and (6-24*b*) into Eqs. (6-19) and (6-22) and grouping terms with and without α and β, we obtain

$$Lg_1 = 0, \tag{6-25a}$$

$$Lg_2 + \tfrac{1}{2} ff'' = 0, \tag{6-25b}$$

$$Lh_1 = 0, \tag{6-26a}$$

and

$$Lh_2 + \zeta gg'' + fg'' + gf'' = 0. \tag{6-26b}$$

Before proceeding further, let us pause for a moment and introduce an example to show the derivation of Eqs. (6-26*a*) and (6-26*b*).

Example 6-4 Derive Eqs. (6-26*a*) and (6-26*b*) from Eqs. (6-22) and (6-24*b*).

Solution: Omitting the independent variables in parentheses for clarity, we have, from Eq. (6-24*b*),

$$h = \beta h_1 + h_2, \qquad h' = \beta h_1' + h_2', \qquad h'' = \beta h_1'' + h_2'', \qquad h''' = \beta h_1''' + h_2'''. \tag{a–d}$$

Here we restate that the primes denote differentiation with respect to η. Substituting Eqs. (*a*)-(*d*) into Eq. (6-22), we obtain

$$\beta h_1''' + h_2''' + \frac{\zeta}{2}\,[f(\beta h_1'' + h_2'') + f''(\beta h_1 + h_2) + 2gg''] + fg'' + gf'' = 0$$

or

$$\beta\left[h_1''' + \frac{\zeta}{2}(fh_1'' + f''h_1)\right] + h_2''' + \frac{\zeta}{2}(fh_2'' + f''h_2) + \zeta gg'' + fg'' + gf'' = 0. \tag{e}$$

Since β is an arbitrary constant, Eq. (*e*) is a sufficient condition for Eqs. (6-26*a*) and (6-26*b*).

If α and β are chosen to be $g''(0, \zeta)$ and $h''(0, \zeta)$, respectively, we can derive, from Eqs. (6-24*a*) and (6-24*b*), the following initial conditions for the auxiliary Eqs. (6-25) and (6-26):

$$g_1(0, \zeta) = g_1'(0, \zeta) = 0, \qquad g_1''(0, \zeta) = 1, \tag{6-27a}$$

$$g_2(0, \zeta) = g_2'(0, \zeta) = g_2''(0, \zeta) = 0, \tag{6-27b}$$

$$h_1(0, \zeta) = h_1'(0, \zeta) = 0, \qquad h_1''(0, \zeta) = 1, \tag{6-27c}$$

and

$$h_2(0, \zeta) = h_2'(0, \zeta) = h_2''(0, \zeta) = 0. \tag{6-27d}$$

Equations (6-25), (6-26), and (6-27*a–d*) constitute a linear initial-value problem and can be readily integrated by a documented integration scheme, e.g., the Runge-Kutta method [25]. When $g_1(\eta_\infty, \zeta)$, $g_2(\eta_\infty, \zeta)$, $h_1(\eta_\infty, \zeta)$, and $h_2(\eta_\infty, \zeta)$ are computed, the constants α and β can be determined from

$$\alpha = \frac{g'(\eta_\infty, \zeta) - g_2'(\eta_\infty, \zeta)}{g_1'(\eta_\infty, \zeta)} \tag{6-28a}$$

and

$$\beta = \frac{h'(\eta_\infty, \zeta) - h_2'(\eta_\infty, \zeta)}{h_1'(\eta_\infty, \zeta)}. \tag{6-28b}$$

Since, at the terminal point η_∞, the local velocity u approaches u_∞, we have

$$f'(\eta_\infty) = 1 \neq \text{function of } \zeta.$$

Consequently, $g'(\eta_\infty) = \partial f'(\eta_\infty)/\partial\zeta$ and $h'(\eta_\infty) = \partial g'(\eta_\infty)/\partial\zeta$ vanish. Equations (6-28*a*) and (6-28*b*) therefore reduce to, respectively,

$$\alpha = -\frac{g_2'(\eta_\infty, \zeta)}{g_1'(\eta_\infty, \zeta)} \quad \text{and } \beta = -\frac{h_2'(\eta_\infty, \zeta)}{h_1'(\eta_\infty, \zeta)}. \tag{6-29}$$

Once α and β are determined, the functions $g(\eta, \zeta)$ and $h(\eta, \zeta)$ can be obtained from Eqs. (6-24*a*) and (6-24*b*). Referring to Eq. (6-18), we are pleased to see that, in order to find $f(\eta, \zeta + \Delta\zeta)$, the only task left is to estimate the remainder term $Q(\eta, \zeta)$. Differentiating Eq. (6-18) twice with respect to η and then setting η to zero yields

$$f''(0, \zeta + \Delta\zeta) = f_s''(0, \zeta + \Delta\zeta) + Q''(0, \zeta), \tag{6-30}$$

where

$$f_s''(0, \zeta + \Delta\zeta) = f''(0, \zeta) + \alpha\Delta\zeta + \frac{\beta(\Delta\zeta)^2}{2}.$$

The subscript s means "second-order term included." To estimate $Q''(0, \zeta)$, we can assume, as in Eq. (6-10), that the terminal value $f'(\eta_\infty, \zeta + \Delta\zeta)$ is a single-valued function of the missing initial value $f''(0, \zeta + \Delta\zeta)$, which will be denoted as κ temporarily.

Thus,

$$f'(\eta_\infty, \zeta + \Delta\zeta) = G(\kappa), \tag{6-31}$$

where G is generally a single-valued function; its specific form is not essential here. On the basis of Eq. (6-31), we can expand $f_s'(\eta_\infty, \zeta + \Delta\zeta)$ about the correct terminal condition, namely 1, in a Taylor's series as

$$f_s'(\eta_\infty, \zeta + \Delta\zeta) \cong 1 + \frac{\partial G}{\partial\kappa}\,[f_s''(0, \zeta + \Delta\zeta) - f''(0, \zeta + \Delta\zeta)] = 1 - Q''(0, \zeta)\,\frac{\partial G}{\partial\kappa}. \tag{6-32}$$

The second-order missing initial condition $f_s''(0, \zeta + \Delta\zeta)$ is generally so close to the true missing initial condition $f''(0, \zeta + \Delta\zeta)$ that the higher-order terms in Eq. (6-32) are negligible and the partial derivative $\partial G/\partial\kappa$ can be well approximated by

$$\frac{\partial G}{\partial\kappa} \cong \frac{G[f_s''(0, \zeta + \Delta\zeta) + \epsilon] - G[f_s''(0, \zeta + \Delta\zeta)]}{\epsilon}, \tag{6-33}$$

where ϵ is a small positive number, $\mathbf{O}(10^{-3})$.† For example, if $f_s''(0, 0.1)$ has been found to be 0.2914803, then another trial missing initial condition can be taken as 0.2924803. Therefore, the remainder $Q''(0, \zeta)$ is estimated from Eqs. (6-32) and

†This process may be regarded as the Richardson extrapolation and may be used safely if ϵ is small relative to $f_s''(0, \zeta + \Delta\zeta)$.

(6-33) and $f''(0, \zeta + \Delta\zeta)$ is computed from Eq. (6-30). With $f''(0, \zeta + \Delta\zeta)$ available, all the information at $\zeta + \Delta\zeta$ can be acquired without difficulty.

This procedure can be applied at any value of ζ and certainly at $\zeta = 0$, where $f(\eta, 0) = \eta^2/2\eta_\infty$, $f'(\eta, 0) = \eta/\eta_\infty$, and $f''(\eta, 0) = 1/\eta_\infty$ as given by Eq. (6-16). We first obtain all the profiles at $\zeta = \Delta\zeta$ and repeat the procedure from $\zeta = 2\Delta\zeta$ to $\zeta = 1$. At $\zeta = 1$, the original Blasius equation is recovered as indicated by Eq. (6-15) with the missing initial condition $f''(0, 1)$ obtained from

$$f''(0, 1) = f''(0, 1 - \Delta\zeta) + g''(0, 1 - \Delta\zeta)(\Delta\zeta) + h''(0, 1 - \Delta\zeta)\frac{(\Delta\zeta)^2}{2} + Q''(0, 1 - \Delta\zeta). \tag{6-34}$$

The procedure for the parametric expansion method is summarized in Table 6-4.

The intermediate values of $f''(0, \zeta)$ and the true nondimensionalized shear stress of $f''(0, 1)$ are listed in Table 6-5. Here η_∞ is taken as 7.0 and $\Delta\zeta$ as 0.1. Note that, once $\Delta\zeta$ is fixed, the number of marching steps is known a priori, whereas for the iterative method, the number of iterations cannot be predicted. The parametric expansion method is therefore noniterative in a certain sense. Let us introduce one more example before turning to other topics.

Example 6-5 Repeat Example 6-3, using the parametric expansion method. Take $\Delta\zeta = 0.1$.

Solution: We will follow the instructions in Table 6-4.

Table 6-4 Procedures of the parametric expansion method

Step	Procedure	Corresponding equation
1	Insert an artificial parameter into the original differential equation(s)	(6-15)
2	Analytically find $f(\eta, 0)$	(6-16)
3	Derive the linearized equations for g and h and their boundary conditions	(6-19)–(6-23)
4	Introduce the auxiliary functions g_1, g_2, h_1, and h_2	(6-24*a*) and (6-24*b*)
5	Derive the auxiliary equations and corresponding initial conditions	(6-25)–(6-27)
6	Find $g_k(\eta_\infty, 0)$ and $h_k(\eta_\infty, 0)$, where $k = 1$ and 2	An efficient integration scheme
7	Find α and β	(6-29)
8	Find $Q''(0, 0)$	(6-32) and (6-33)
9	Find $f''(0, \Delta\zeta)$	(6-30)
10	Repeat steps 6–9 at $\zeta = 2\Delta\zeta, 3\Delta\zeta, \ldots, 1.0$	

Table 6-5 Missing initial conditions of the parameterized Blasius equation for various ζ[a]

ζ	$f''(0)$
0	0.14286
0.1	0.15855
0.2	0.17642
0.3	0.19611
0.4	0.21690
0.5	0.23797
0.6	0.25860
0.7	0.27840
0.8	0.29724
0.9	0.31511
1.0	0.33209

[a]Only that at $\zeta = 1.0$ has physical significance.

Step 1:

$$\phi''' + \zeta(\phi\phi'' - \phi'^2) + 1 = 0.$$

Step 2:

At $\zeta = 0$, $\quad \phi'''(\eta, 0) = -1, \quad \phi''(\eta, 0) = -\eta + \frac{27}{10}, \quad \phi'(\eta, 0) = -\frac{\eta^2}{2} + \frac{27\eta}{10},$

and

$$\phi(\eta, 0) = -\frac{\eta^3}{6} + \frac{27\eta^2}{20}.$$

Step 3:

$$g''' + \zeta(\phi g'' - 2\phi' g' + \phi'' g) + \phi\phi'' - \phi'^2 = 0$$

and $h''' + \zeta(\phi h'' - 2\phi' h' + \phi'' h + 2gg'' - 2g'^2) + \phi g'' - 2\phi' g' + \phi'' g = 0.$

Step 4: Same as Eqs. (6-24*a*) and (6-24*b*).

Step 5:

$$g_1''' + \zeta(\phi g_1'' - 2\phi' g_1' + \phi'' g_1) = 0,$$

$$g_2''' + \zeta(\phi g_2'' - 2\phi' g_2' + \phi'' g_2) + \phi\phi'' - \phi'^2 = 0,$$

$$h_1''' + \zeta(\phi h_1'' - 2\phi' h_1' + \phi'' h_1) = 0,$$

and

$$h_2''' + \zeta(\phi h_2'' - 2\phi' h_2' + \phi'' h_2 + 2gg'' - 2g'g') + \phi g'' - 2\phi' g' + \phi'' g = 0.$$

Table 6-6 Missing initial conditions of the parameterized stagnation flow equation for various ζ[a]

ζ	$\phi''(0)$
0	2.70000
0.1	1.98062
0.2	1.74439
0.3	1.60729
0.4	1.51252
0.5	1.44101
0.6	1.38408
0.7	1.33709
0.8	1.29726
0.9	1.26282
1.0	1.23259

[a]Only that at $\zeta = 1.0$ has physical significance.

The initial conditions are the same as in Eqs. (6-27*a*)-(6-27*d*).

Step 6: Compute $g_k(5, 0)$ and $h_k(5, 0)$, where $k = 1$ and 2.

Steps 7-10: As shown in Table 6-4.

The numerical results for the missing initial condition at intermediate ζ values and the true nondimensionalized wall shear stress at $\zeta = 1$ are listed in Table 6-6.

The drawback of the parametric expansion method is that the mathematical formulation is somewhat tedious. Therefore, if there is only one missing initial condition associated with the boundary-value problem, we will generally prefer to use the simple and straightforward shooting method described in Section 6-2*a*. However, the parametric expansion method is more advantageous than other numerical methods when (1) there are at least two missing initial conditions, e.g., the wall shear stress and the wall heat flux in free-convection boundary-layer flow over a vertical isothermal plate; (2) the solution is highly sensitive to the initial conditions, e.g., in the Shvab-Zeldovich combustion system [26]; and (3) solutions for a wide range of parametric values are desired.

Unfortunately, the parametric expansion method cannot be successfully applied to partial differential equations. The reason is that setting the artificial parameter ζ to zero may completely change the characteristics of the partial differential equations. For example, if ζ is embedded in Eq. (6-5) as

$$\zeta\left(u\frac{\partial u}{\partial x} + v\frac{\partial u}{\partial y}\right) = \nu\frac{\partial^2 u}{\partial y^2}, \tag{6-35}$$

then Eq. (6-35) becomes, at $\zeta = 0$,

$$\nu \frac{\partial^2 u}{\partial y^2} = 0,$$

where the x dependence vanishes. On the other hand, if ζ is embedded as

$$u \frac{\partial u}{\partial x} + \zeta v \frac{\partial u}{\partial y} = \nu \frac{\partial^2 u}{\partial y^2}, \tag{6-36}$$

Eq. (6-36) reduces to, at $\zeta = 0$,

$$u \frac{\partial u}{\partial x} = \nu \frac{\partial^2 u}{\partial y^2},$$

which remains analytically unsolvable.

Because of this deficiency, an improved version of the parametric expansion method that can be applied to both ordinary and partial differential equations was developed. We will present this improved version in Sections 7-5 and 11-2*a*. Now we will direct our attention to the finite-difference method.

6-3 FINITE–DIFFERENCE METHOD WITH x-ω^n TRANSFORMATION

Boundary-layer flow has a similarity character, but the existence of a similarity equation is not automatically guaranteed for other physical problems. Therefore, whenever a partial differential equation cannot be reduced to an ordinary differential equation, some other numerical techniques are called for. Among them are the two most powerful techniques: the finite-difference method and the finite-element method. It is undoubtedly helpful to familiarize ourselves with these two techniques so that we are well prepared when nonsimilar problems are to be solved. In the present section we will first introduce the finite-difference method incorporating the von Mises transformation [27].

6-3*a* x-ω Transformation

Since the domain of the boundary layer is not of regular geometry and the transverse flow velocity is generally of minor interest, it is inconvenient to directly discretize the original differential equation. Instead, the x-y physical coordinates are transformed into x-ω coordinates, where ω is the normalized stream function defined by

$$\omega = \frac{\psi - \psi_I}{\psi_E - \psi_I} = \frac{\int_0^y u\,dy}{\int_0^\delta u\,dy}. \tag{6-37}$$

Here the subscripts I and E denote, respectively, the inner and external edges of the boundary layer. With the definition (6-37), the von Mises transformation can be represented by

$$\left(\frac{\partial}{\partial x}\right)_y = \left(\frac{\partial \omega}{\partial x}\right)_y \frac{\partial}{\partial \omega} + \left(\frac{\partial}{\partial x}\right)_\omega \quad \text{and} \quad \left(\frac{\partial}{\partial y}\right)_x = \left(\frac{\partial \omega}{\partial y}\right)_x \frac{\partial}{\partial \omega}. \tag{6-38a, b}$$

From Eqs. (6-37) and (6-38), we can transform all the terms in the x-y plane into the x-ω plane. For convenience, we list several relations in Table 6-7. Utilizing Table 6-7, we can transform Eq. (6-5) into

$$\frac{\partial u}{\partial x} + (a + b\omega)\frac{\partial u}{\partial \omega} = \frac{\partial}{\partial \omega}\left(c\,\frac{\partial u}{\partial \omega}\right), \tag{6-39}$$

where

$$a = -\frac{(d\psi_I/dx)}{\psi_{EI}}, \quad b = -\frac{(d\psi_{EI}/dx)}{\psi_{EI}},$$

$$c = \frac{\nu u}{\psi_{EI}^2}, \quad \text{and } \psi_{EI} = \psi_E - \psi_I$$

according to the notation in [28, 29]. Here x continues to be the streamwise coordinate; ω is the new transverse coordinate, ranging from 0 to 1. The new boundary conditions are

$$u(x, 0) = 0, \quad u(x, 1) = u_\infty,$$

and

$$u(x_0, \omega) = s(\omega). \tag{6-40}$$

Before discretizing Eq. (6-39), we wish to make three remarks about the von Mises transformation and the transformed equation. First, we observe that the transverse velocity v does not appear in Eq. (6-39). This is a significant advantage of

Table 6-7 Convenient relations between terms in the x-y plane and the x-ω plane

x-y plane	x-ω plane
$\dfrac{\partial \omega}{\partial x}$	$-\dfrac{(v + d\psi_I/dx + \omega\, d\psi_{EI}/dx)}{\psi_{EI}}$
$\dfrac{\partial \omega}{\partial y}$	$\dfrac{u}{\psi_{EI}}$
$u\,\dfrac{\partial u}{\partial x}$	$u\left(\dfrac{\partial u}{\partial x}\right)_\omega - \left[\dfrac{uv}{\psi_{EI}} + \dfrac{u}{\psi_{EI}}\dfrac{d\psi_I}{dx} + \dfrac{u\omega}{\psi_{EI}}\dfrac{d\psi_{EI}}{dx}\right]\dfrac{\partial u}{\partial \omega}$
$v\,\dfrac{\partial u}{\partial y}$	$\dfrac{uv}{\psi_{EI}}\dfrac{\partial u}{\partial \omega}$
$\nu\,\dfrac{\partial^2 u}{\partial y^2}$	$u\,\dfrac{\partial}{\partial \omega}\left[\dfrac{\nu u}{\psi_{EI}^2}\dfrac{\partial u}{\partial \omega}\right]$

Eq. (6-39) over Eq. (6-5) since the computation of one unknown is clearly more efficient than that of two. Second, it is noteworthy that $-d\psi_E/dx$ and v_∞ are different in their values and physical meanings. This will be explained in detail in the following example. Third, if the initial condition $u(x_0, \omega)$ is chosen to be the similarity solution, the x derivative $\partial u/\partial x$ in Eq. (6-39) will vanish for the case of impermeable walls. The explanation for this will be postponed until the discretization of Eq. (6-39) is presented.

Example 6-6 A little puzzle is presented in this example. From the definition of the stream function, we know that v is equal to $-\partial\psi/\partial x$. Therefore, at $y=\delta$, v_∞ is equal to $-d\psi_E/dx$. Let us designate this as statement A. However, our physical intuition tells us that (B) v_∞ is positive and (C) ψ_E increases along x, i.e., $d\psi_E/dx > 0$. Statements A, B, and C do not seem to be consistent. Which of them is false?

Solution: The continuity equation can be written as

$$\frac{\partial u}{\partial x}+\frac{\partial v}{\partial y}=0 \tag{a}$$

or

$$v_\infty=-\int_0^\delta \frac{\partial u}{\partial x}\,dy. \tag{b}$$

Here an impermeable wall has been assumed. By using Leibnitz's rule, Eq. (*b*) can be changed to

$$v_\infty=u_\infty\frac{d\delta}{dx}-\frac{d}{dx}\int_0^\delta u\,dy=u_\infty\frac{d\delta}{dx}-\frac{d\psi_E}{dx}. \tag{c}$$

Let us further assume that ρ is constant and that the normalized velocity profile is $u/u_\infty = 1.5(y/\delta)-0.5(y/\delta)^3$. Then Eq. (*c*) can be reduced to

$$\frac{v_\infty}{u_\infty}=\frac{d\delta}{dx}-0.625\frac{d\delta}{dx}. \tag{d}$$

Since δ is approximately equal to $5x\mathrm{Re}_x^{-1/2}$ [1-7], the solution of the puzzle is about to be revealed from Eq. (*d*). Let us start with statement C. From Eqs. (*c*) and (*d*), we find

$$\frac{d\psi_E}{dx}=0.625\,u_\infty\frac{d\delta}{dx}>0. \tag{e}$$

Therefore, statement C is true. Again, from Eq. (*d*) we deduce

$$\frac{v_\infty}{u_\infty}=0.375\frac{d\delta}{dx}>0. \tag{f}$$

Statement B is also proved to be true. Finally, from Eq. (*c*) we find that, instead of being equal to $-d\psi_E/dx$ only, v_∞ is equal to the sum of $-d\psi_E/dx$ and $u_\infty d\delta/dx$. The latter is the mass flux into the boundary layer from the free stream as shown schematically in Fig. 6-9. It can be seen from Fig. 6-9 that $v_\infty \Delta x$ is equal to $\psi_E(x) - \int_0^{y_C} u\, dy$, not $\psi_E(x) - \int_0^{y_B} u\, dy$. Expressing this in an equation, we write

$$-v_\infty = \left(\frac{\partial \psi}{\partial x}\right)_{y=\delta} \neq \frac{d\psi_E}{dx}. \tag{g}$$

The equality in Eq. (*g*) arises from the definition of the stream function. The inequality is the key solution to the puzzle; whenever we evaluate a partial derivative, attention must be paid to what is being kept constant. Therefore, from Eq. (*g*), we know that statement A is false.

There are several schemes that can discretize Eq. (6-39). To discretize ω derivatives, for example, we may adopt the central-difference, control volume, or Galerkin scheme. In the control volume or the Galerkin scheme, ϕ can be assumed linear in $\omega \epsilon [\omega_{j-1}, \omega_{j+1}]$. The corresponding discretized expressions for uniform mesh size are listed in Table 6-8 for central-difference and control volume schemes. Similarly, to discretize first-order x derivatives, we may use the central difference or the upwind difference. The latter can be represented by

$$\frac{\partial \phi}{\partial x} = \frac{1}{\Delta x}(\phi_j - \phi_W) \tag{6-41}$$

where the subscript W indicates west to the j grid point.

With the aid of Table 6-8, the final three-point algebraic equation can be derived as

$$Au_{j-1} + Bu_j + Cu_{j+1} = D, \quad j = 2, 3, \ldots, J-1 \tag{6-42}$$

where the coefficients A, B, C, and D are listed in Table 6-9 for the control volume and Galerkin schemes. A sample derivation of an expression in Table 6-8 is presented in the following example.

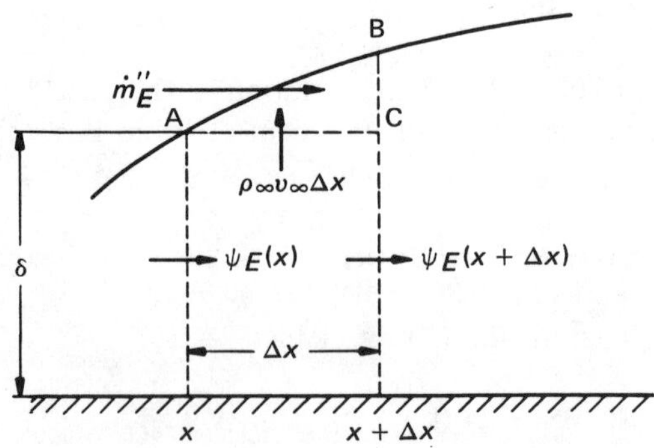

Figure 6-9 Mass balance over a control volume inside the boundary layer.

Table 6-8 Discretized expressions for central-difference and control volume schemes

Differential	Central difference	Control volume, $\frac{1}{\Delta\omega}\int_{\omega_{j-1/2}}^{\omega_{j+1/2}}(\cdot)d\omega$
ϕ	ϕ_j	$\frac{1}{8}\phi_{j-1}+\frac{3}{4}\phi_j+\frac{1}{8}\phi_{j+1}$
$\frac{\partial\phi}{\partial\omega}$	$\frac{1}{2\Delta\omega}(\phi_{j+1}-\phi_{j-1})$	$\frac{1}{2\Delta\omega}(\phi_{j+1}-\phi_{j-1})$
$\omega\frac{\partial\phi}{\partial\omega}$	$\frac{\omega_j}{2\Delta\omega}(\phi_{j+1}-\phi_{j-1})$	$\left(\frac{1}{8}+\frac{\omega_j}{2\Delta\omega}\right)\phi_{j+1}-\frac{1}{4}\phi_j-\left(\frac{1}{8}+\frac{\omega_{j-1}+\omega_j}{4\Delta\omega}\right)\phi_{j-1}$
$\frac{\partial}{\partial\omega}\left(c\frac{\partial\phi}{\partial\omega}\right)$	$\frac{1}{2(\Delta\omega)^2}[(c_j+c_{j+1})\phi_{j+1}-(c_{j+1}+2c_j+c_{j-1})\phi_j+(c_j+c_{j-1})\phi_{j-1}]$	

Example 6-7 Use the linear approximation to derive a discretized expression for

$$I\equiv\frac{1}{\Delta\omega}\int_{\omega_{j-1/2}}^{\omega_{j+1/2}}\phi\,d\omega.$$

by the control volume method.

Solution:

$$I\equiv\frac{1}{\Delta\omega}\int_{\omega_{j-1/2}}^{\omega_{j+1/2}}\phi\,d\omega=\frac{1}{\Delta\omega}\int_{\omega_{j-1/2}}^{\omega_j}\phi\,d\omega+\frac{1}{\Delta\omega}\int_{\omega_j}^{\omega_{j+1/2}}\phi\,d\omega. \tag{a}$$

But $$\tilde{\phi}=N_{j-1}\phi_{j-1}+N_j\phi_j \quad \text{if } \omega\in[\omega_{j-1},\omega_j] \tag{b}$$

and $$\tilde{\phi}=N_j\phi_j+N_{j+1}\phi_{j+1} \quad \text{if } \omega\in[\omega_j,\omega_{j+1}], \tag{c}$$

where N_k, $k=j-1,\ j,\ j+1$, are pyramid functions defined in Eq. (1-16). Therefore,

$$I=\frac{\phi_{j-1}}{\Delta\omega}\left(\int_{\omega_{j-1/2}}^{\omega_j}N_{j-1}\,d\omega\right)+\frac{\phi_j}{\Delta\omega}\left(\int_{\omega_{j-1/2}}^{\omega_{j+1/2}}N_j\,d\omega\right)$$
$$+\frac{\phi_{j+1}}{\Delta\omega}\left(\int_{\omega_j}^{\omega_{j+1/2}}N_{j+1}\,d\omega\right). \tag{d}$$

The integrals in Eq. (d) can be evaluated by measuring the trapezoidal areas beneath the lines as shown in Fig. 6-10. Thus,

Table 6-9 Coefficients in Eqs. (6-42) and (6-49) for the Galerkin and control volume schemes[a]

Coefficient	Galerkin	Control volume
A	$\frac{h_j}{6\Delta x} - \frac{b(2\omega_j + \omega_{j-1})}{6} - \frac{c^*(\bar{u}_{j-1} + \bar{u}_j)}{2h_j}$	$\frac{h_j}{8\Delta x} - \frac{b(3\omega_j + \omega_{j-1})}{8} - \frac{c^*(\bar{u}_{j-1} + \bar{u}_j)}{2h_j}$
B	$\frac{h_{j+1} + h_j}{3\Delta x} - \frac{b(h_{j+1} + h_j)}{6} + \frac{c^*}{2}\left[\frac{\bar{u}_{j-1} + \bar{u}_j}{h_j} + \frac{\bar{u}_j + \bar{u}_{j+1}}{h_{j+1}}\right]$	$\frac{3(h_{j+1} + h_j)}{8\Delta x} - \frac{b(h_{j+1} + h_j)}{8} + \frac{c^*}{2}\left[\frac{(\bar{u}_{j-1} + \bar{u}_j)}{h_j} + \frac{\bar{u}_j + \bar{u}_{j+1}}{h_{j+1}}\right]$
C	$\frac{h_{j+1}}{6\Delta x} + \frac{b(2\omega_j + \omega_{j+1})}{6} - \frac{c^*(\bar{u}_j + \bar{u}_{j+1})}{2h_{j+1}}$	$\frac{h_{j+1}}{8\Delta x} + \frac{b(3\omega_j + \omega_{j+1})}{8} - \frac{c^*(\bar{u}_j + \bar{u}_{j+1})}{2h_{j+1}}$
D	$\frac{h_j\bar{u}_{j-1}}{6\Delta x} + \frac{(h_{j+1} + h_j)\bar{u}_j}{3\Delta x} + \frac{h_{j+1}\bar{u}_{j+1}}{6\Delta x}$	$\frac{h_j\bar{u}_{j-1}}{8\Delta x} + \frac{3(h_{j+1} + h_j)\bar{u}_j}{8\Delta x} + \frac{h_{j+1}\bar{u}_{j+1}}{8\Delta x}$
A_1	$\frac{\omega_1}{6\Delta x} - \frac{b\omega_1}{4} - \frac{c^*\bar{u}_1}{2\omega_1}$	$\frac{\omega_1}{3\Delta x} - \frac{b\omega_1}{3} - \frac{c^*\bar{u}_1}{2\omega_1}$
B_1	$\frac{\omega_1 + 2\omega_2}{6\Delta x} - \frac{b(\omega_1 + 2\omega_2)}{12} + \frac{c^*}{2}\left(\frac{\bar{u}_1}{\omega_1} + \frac{\bar{u}_1 + \bar{u}_2}{h_2}\right)$	$\frac{7\omega_1 + 9\omega_2}{24\Delta x} - \frac{b(\omega_1 + 3\omega_2)}{24} + \frac{c^*}{2}\left(\frac{\bar{u}_1}{\omega_1} + \frac{\bar{u}_1 + \bar{u}_2}{h_2}\right)$
C_1	Same as C (with $j = 1$)	Same as C (with $j = 1$)
D_1	$\frac{\omega_1\bar{\phi}_I}{6\Delta x} + \frac{(\omega_1 + 2\omega_2)\bar{\phi}_1}{6\Delta x} + \frac{h_2\bar{\phi}_2}{6\Delta x}$	$\frac{\omega_1\bar{\phi}_I}{3\Delta x} + \frac{(7\omega_1 + 9\omega_2)\bar{\phi}_1}{24\Delta x} + \frac{h_2\bar{\phi}_2}{8\Delta x}$

[a]The terms b and c^* should be specified carefully. See [28, 29]

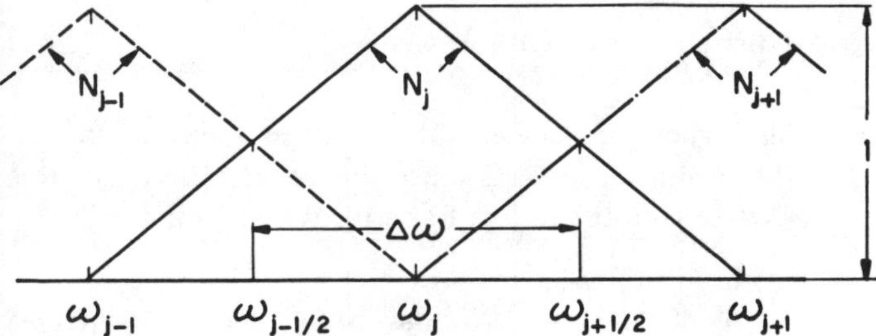

Figure 6-10 Linear basis functions.

$$\int_{\omega_{j-1/2}}^{\omega_j} N_{j-1}\, d\omega = \left(\frac{1}{2}\right)\left(\frac{1}{2}\right)(\omega_j - \omega_{j-1/2}) = \frac{\Delta\omega}{8}. \tag{e}$$

Similarly,

$$\int_{\omega_{j-1/2}}^{\omega_{j+1/2}} N_j\, d\omega = \frac{3}{4}\,\Delta\omega \quad \text{and} \quad \int_{\omega_j}^{\omega_{j+1/2}} N_{j+1}\, d\omega = \frac{\Delta\omega}{8}.$$

Finally, we obtain

$$I = \frac{1}{8}\,\phi_{j-1} + \frac{3}{4}\,\phi_j + \frac{1}{8}\,\phi_{j+1}. \tag{f}$$

In the special case $\phi_{j-1} = \phi_j = \phi_{j+1}$, I reduces to ϕ_j—the expression that would have been obtained by the finite-difference method.

Equation (6-42) is not valid for the grid node at $j = 1$ because, in the region immediately adjacent to the wall, it has been found that $\partial u/\partial x \approx 0$ and $v \approx 0$ for impermeable walls. Consequently, from Eq. (6-5), we obtain

$$\frac{\partial^2 u}{\partial y^2} \approx 0 \qquad \text{as } y \to 0. \tag{6-43}$$

Equation (6-43) suggests that u is linear in y near the wall. This linearity was also confirmed by experimental results (e.g., [30]). Thus,

$$d\omega \propto u\, dy \propto u\, du. \tag{6-44}$$

Instead of being linear in ω for $\omega \epsilon [\omega_1, 1]$, the velocity distribution, at least on the interval $[0, \omega_1]$, should be approximated by

$$u = u_1 \left(\frac{\omega}{\omega_1}\right)^{1/2} \qquad \text{if } \omega \epsilon [0, \omega_1]. \tag{6-45}$$

Similarly, for other dependent variables ϕ,

$$\phi = \left(\frac{\omega}{\omega_1}\right)^{1/2} (\phi_1 - \phi_I) + \phi_I. \tag{6-46}$$

Equation (6-45) not only yields higher accuracy in the numerical solution but also removes the singularity at the wall. If u is assumed linear on $[0, \omega_1]$, this singularity arises when the transverse coordinate y is to be recovered from ω, i.e.,

$$y_1 \propto \int_0^{\omega_1} \frac{d\omega}{u} \propto \int_0^{\omega_1} \frac{d\omega}{\omega} \to \infty. \tag{6-47a}$$

Instead, if u is assumed to be given by Eq. (6-45), we will have

$$y_1 \propto \omega_1^{1/2}. \tag{6-47b}$$

With Eq. (6-45), we can discretize Eq. (6-39) over the interval $[0, \omega_{1.5}]$ in the control volume method and over $[0, \omega_1]$ in the Galerkin method. In the central-difference formulation, we approximate, according to Eq. (6-45),

$$\left(\frac{\partial u}{\partial \omega}\right)_1 \approx \frac{1}{2}\frac{u_1}{\omega_1} \tag{6-48a}$$

and

$$\left[\frac{\partial}{\partial \omega}\left(c\,\frac{\partial u}{\partial \omega}\right)\right]_1 \approx \frac{\nu u_1^2}{4\psi_{EI}^2 \omega_1^2}. \tag{6-48b}$$

The final algebraic equation at $j = 1$ can be written as

$$B_1 u_1 + C_1 u_2 = D_1, \tag{6-49}$$

where the coefficients B, C, and D for two schemes are also listed in Table 6-9. Equations (6-42) and (6-49) form a set of simultaneous algebraic equations with a tridiagonal coefficient matrix. This set of equations can be efficiently solved for u_1, $u_2, \ldots, u_{j-1}$ by the Gaussian elimination scheme [31] or its modified version called the Thomas algorithm (see Section 5-1b), provided the initial condition of $u(x_0, \omega)$ is given. A typical initial distribution of $u(x_0, \omega)$ takes the form

$$\frac{u(x_0, \omega)}{u_\infty} = 1.5\omega - 0.5\omega^3. \tag{6-50}$$

Many other profiles that are commonly assumed in the integral method [32] are equally acceptable. Upon recalling that the similarity solution is available, some readers may be curious about the outcome of using this solution as the initial condition for flows over a nonpermeable wall. This is an interesting point, which we will discuss in the following example.

Example 6-9 If the similarity solution $u_\infty f'(\eta)$ is used as the initial profile $u(x_0, \omega)$ for the flow velocity $u(x, \omega)$, show that Eq. (6-39) is not needed for the calculation of downstream velocity distributions for the case of impermeable walls.

Solution: From Table 6-1, we have

$$\psi(x, y) = f(\eta)(xu_\infty \nu)^{1/2}. \tag{a}$$

Therefore, by the definition of ω with $\psi_I = 0$, we obtain

$$\omega = \frac{\psi}{\psi_E} = \frac{f(\eta)(xu_\infty \nu)^{1/2}}{f(\eta_\infty)(xu_\infty \nu)^{1/2}} = \frac{f(\eta)}{f(\eta_\infty)}. \tag{b}$$

Since $f(\eta_\infty)$ is simply a constant, Eq. (b) implies that a given ω corresponds to a fixed value of $f(\eta)$ and thereby a fixed value of u/u_∞. In other words, u/u_∞ is not a function of x. Thus, Eq. (6-39) is reduced to

$$b\omega \frac{du}{d\omega} = \frac{d}{d\omega}\left(c \frac{du}{d\omega}\right). \tag{c}$$

That is, a partial differential equation of the parabolic type is reduced to an ordinary differential equation which the initial profile $u(x_0, \omega)$ should automatically satisfy. Therefore, Eq. (c) is not needed. It provides no new information other than the initial profile.

At first glance, Eq. (c) in Example 6-9 appears different from the Blasius similarity equation (6-8). Actually, however, Eq. (c) can be shown to be identical to Eq. (6-8). The derivation is presented as follows. The differentiation $d/d\omega$ can be transformed into

$$\frac{d}{d\omega} = \frac{d}{d\eta}\frac{d\eta}{df}\frac{df}{d\omega} = \frac{f(\eta_\infty)}{f'}\frac{d}{d\eta}, \tag{6-51}$$

where the prime denotes differentiation with respect to η. With Eq. (6-51) and the definitions

$$b = \frac{-1}{\psi_E}\frac{d\psi_E}{dx} = -\frac{1}{2x} \tag{6-52a}$$

and

$$c = \frac{\nu u}{\psi_E^2} = \frac{f'}{x[f(\eta_\infty)]^2}, \tag{6-52b}$$

we can derive

$$\frac{du}{d\omega} = \frac{f(\eta_\infty)u_\infty f''}{f'} \tag{6-53a}$$

and

$$\frac{d}{d\omega}\left(c\frac{du}{d\omega}\right) = c\frac{d^2u}{d\omega^2} + \frac{dc}{d\omega}\frac{du}{d\omega} = \frac{u_\infty}{x}\frac{d}{d\eta}\left(\frac{f''}{f'}\right) + \left(\frac{f''}{f'}\right)^2. \tag{6-53b}$$

Substituting Eqs. (6-52) and (6-53) into Eq. (*c*) in Example 6-9 leads to

$$f''' + \frac{1}{2} ff'' = 0. \qquad (6\text{-}8)$$

Now we will present a simple example to illustrate the calculation of nodal streamwise velocities.

Example 6-10 Assign the similarity solution (see Appendix A) to the upstream flow velocities $\bar{u}_j$, $j = 1, 2, 3, 4$, and to u_4. The mesh system is shown in Fig. 6-11. Take $\Delta\omega = 0.02$, $\mathrm{Re}_x = 10^4$, $x = 5$ cm, and $u_\infty = 3$ m/s. Use Eqs. (6-42) and (6-49) to find u_1, u_2, and u_3 for $\Delta x = 0.1$, 0.5, and 2 cm, respectively.

Solution: At $x = 5$ cm the velocity distribution based on the similarity solution is obtained and is listed in Table 6-10. To calculate the downstream velocities, we need to know the values of *a, b,* and *c.* For impermeable walls, the term *a* vanishes. Terms *b* and *c* can be calculated from Eqs. (6-52*a*) and (6-52*b*). The tridiagonal matrix system for u_1, u_2, and u_3 thus can be written as

$$B_1 u_1 + C_1 u_2 = D_1, \qquad (6\text{-}54a)$$

$$A_2 u_1 + B_2 u_2 + C_2 u_3 = D_2, \qquad (6\text{-}54b)$$

$$A_3 u_2 + B_3 u_3 = D_3 - C_3 u_4. \qquad (6\text{-}54c)$$

The final results for u_1, u_2, and u_3 are listed in Table 6-11 for various step sizes. It is seen that these results are insensitive to the choice of Δx. This insensitivity suggests that $\partial u/\partial x \approx 0$. The fact that $\partial u/\partial x$ is not exactly equal to zero is partially due to the substitution of upstream values u_W in the term *c.*

After the distributions $u(x, \omega)$ are computed, the next task is to recover the physical coordinate y from ω at the same x location. Recalling the definition of ω, we obtain

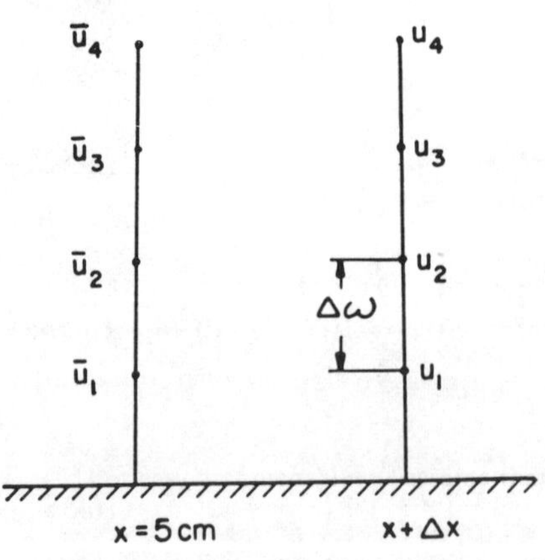

Figure 6-11 Mesh system for the nodal velocities in the x-ω plane.

Table 6-10 Upstream velocity distribution obtained by utilizing the similarity solution

Grid node	ω	$f(\eta)$	$f'(\eta)$	$\bar{u}$
1	0.02	0.1056	0.2640	0.7919
2	0.04	0.2111	0.3700	1.1103
3	0.06	0.3168	0.4517	1.3551
4	0.08	0.4223	0.5178	1.5535

$$dy = \frac{\psi_{EI} d\omega}{u}. \tag{6-55}$$

Integration of Eq. (6-55) by using the trapezoidal rule leads to y. Of particular interest is the transverse distance at the grid point 1, y_1. From Eq. (6-55), we write

$$y_1 = \frac{\omega_1 \psi_{EI}}{u_1} \int_0^1 \frac{d(\omega/\omega_1)}{(\omega/\omega_1)^{1/2}} = \frac{2\omega_1 \psi_{EI}}{u_1}. \tag{6-56}$$

If y_1 is chosen to be sufficiently small, then the shear stress at the wall can be approximated by

$$\tau_w = \mu \left(\frac{\partial u}{\partial y}\right)_{y=0} \approx \frac{\mu u_1}{y_1}. \tag{6-57}$$

An alternative way to obtain the shear stress or the heat flux at the wall is to follow the so-called Couette flow analysis. Interested readers can consult [28, 29] for details.

6-3b x-ω_D^2 Transformation

Mainly because of the singularity existing at the wall after the x-ω transformation, another finite-difference scheme with an improved x-ω_D^2 transformation was proposed by Denny and co-workers [33-35], where

Table 6-11 Nodal values of downstream velocity

u_∞	3.0 (m/s)	3.0	3.0
u_4	1.55350	1.55350	1.55350
u_3	1.35484	1.35496	1.35496
u_2	1.11257	1.11271	1.11267
u_1	0.79064	0.79011	0.78989
u_0	0	0	0
x	5.1 cm	5.5 cm	7.0 cm

$$\omega_D^2 = \frac{\int_0^y u\,dy}{\int_0^{y_E} u\,dy.}. \tag{6-58}$$

(The subscript D stands for Denny.) Using Eq. (6-58), we can transform Eq. (6-5) into

$$\frac{\partial u}{\partial x} + \frac{a + b\omega_D}{2\omega_D}\frac{\partial u}{\partial \omega_D} = \frac{1}{4\omega_D}\frac{\partial}{\partial \omega_D}\left(\frac{c}{\omega_D}\frac{\partial u}{\partial \omega_D}\right). \tag{6-59}$$

Assuming that ϕ is linear in $\omega_D \in [\omega_{D,j-1}, \omega_{D,j+1}]$ and employing either the control volume or the central-difference scheme as well as the upwind scheme, we can further discretize Eq. (6-59) into a linearized algebraic equation involving unknowns u_{j-1}, u_j, and u_{j+1}. It is noteworthy that Eq. (6-58) implies

$$\omega_D = \omega^{1/2}. \tag{6-60}$$

Since we assume the linearity $\phi \sim \omega_D$ in all subintervals within $[0, 1]$, we actually recover the approximation $\phi \sim \omega^{1/2}$, as discussed in the previous section. This is why the singularity at the wall can be eliminated by the x-ω_D^2 transformation. Furthermore, the x-ω_D^2 transformation, according to [33, 34] generally yields improved accuracy and its numerical results are relatively insensitive to the choice of difference scheme.

Before proceeding to the next section, we will make a few remarks here.

1. In both the x-ω and x-ω_D^2 transformations, the net mass flux entering the boundary-layer edge, i.e., $d\psi_E/dx$, must be prescribed a priori.
2. Keller and Cebeci [36] proposed a finite-difference scheme in which the normalized stream function and its first and second derivatives with respect to a similarity variable are solved simultaneously. The streamwise derivatives are also discretized based on the upwind difference. In the investigation of boundary-layer flows with nonsimilar profiles, their scheme has been widely used [37–45].
3. Although convenient, it is not always necessary to transform the x-y plane into x-η, x-ω, or x-ω_D^2 planes before the discretization of differential equations. In [3, pp. 181–184, and 46], for example, the boundary-layer momentum equation was directly discretized in the x-y plane, and the primitive variables u and v were sought (see also Section 7-5).

6-4 NUMERICAL STABILITY, CONSISTENCY, AND ACCURACY

In Chapter 4 we described the numerical properties of various discretization schemes. In this section we will utilize this information to focus our attention on the application of a finite-difference scheme to the boundary-layer equation (6-39).

6-4a Numerical Stability

For convenience, Eq. (6-39) may be cast in the standard form

$$\frac{\partial \phi}{\partial x} = c \frac{\partial^2 \phi}{\partial \omega^2} + e \frac{\partial \phi}{\partial \omega}, \tag{6-61}$$

where

$$e = \frac{\partial c}{\partial \omega} - a - b\omega.$$

Approximating the ω derivatives and $\partial\phi/\partial x$ by the central difference and the upwind difference, respectively, we obtain

$$\frac{\phi_j^{(n+1)} - \phi_j^{(n)}}{\Delta x} = \frac{c_j^{(n+1)}}{(\Delta\omega)^2} (\phi_{j-1}^{(n+1)} - 2\phi_j^{(n+1)} + \phi_{j+1}^{(n+1)}) + \frac{e_j^{(n+1)}}{2\Delta\omega} (\phi_{j+1}^{(n+1)} - \phi_{j-1}^{(n+1)}), \tag{6-62}$$

where $n = x/\Delta x$ and $j = \omega/\Delta\omega$. Rearranging Eq. (6-62), we obtain

$$(q_j^{(n+1)} - s_j^{(n+1)})\phi_{j-1}^{(n+1)} + (2s_j^{(n+1)} + 1)\phi_j^{(n+1)} - (s_j^{(n+1)} + q_j^{(n+1)})\phi_{j+1}^{(n+1)} = \phi_j^{(n)}, \tag{6-63}$$

where

$$s_j^{(n+1)} = \frac{c_j^{(n+1)}\Delta x}{(\Delta\omega)^2} \quad \text{and } q_j^{(n+1)} = \frac{e_j^{(n+1)}\,\Delta x}{2\Delta x}.$$

At the expense of some generality, it is relatively simple to establish the stability criterion for Eq. (6-63) if the coefficients $q_j^{(n+1)}$ and $s_j^{(n+1)}$ are assumed to be constants. In this section the von Neumann stability analysis will be used to test whether Eq. (6-63) is stable.

As usual, we assume $\phi_j^{(n)} = A_m \xi^n \exp(imj\pi\Delta\omega)$, where A_m is the coefficient that is determined by the initial condition ($x = 0$ in this case), ξ is the amplification factor, and i is the imaginary number. Substituting this assumed expression into Eq. (6-63), we obtain

$$\xi = [1 + 2s(1 - \cos\beta) + 2iq \sin\beta]^{-1}, \tag{6-64}$$

where s and q are constants as previously assumed and $\beta = m\pi\Delta\omega$. From Eq. (6-64), we derive

$$|\xi| = \{[1 + 2s(1 - \cos\beta)]^2 + 4q^2 \sin^2\beta\}^{-1/2}. \tag{6-65}$$

In the special case $q = 0$, Eq. (6-65) reduces to

$$|\xi| = \frac{1}{1 + 2s(1 - \cos\beta)}, \tag{6-66}$$

which implies that, for any positive value of s, we always have

$$|\xi| \leqslant 1. \tag{6-67}$$

This is the same result derived in Section 4-1*c* for the implicit scheme applied to the one-dimensional transient heat conduction equation. If q is nonzero, then

$$|\xi| = \frac{1}{\{[1 + 2s(1 - \cos\beta)]^2 + 4q^2 \sin^2\beta\}^{1/2}} < \frac{1}{1 + 2s(1 - \cos\beta)} \leqslant 1 \tag{6-68}$$

for any positive value of s. Therefore, we may conclude that, based on Eqs. (6-67) and (6-68) the implicit finite-difference equation (6-63) is unconditionally stable and the influence of a nonzero q on the stability is minor.

6-4*b* Numerical Consistency

After establishing the stability of Eq. (6-62), we proceed to examine its consistency. With the aid of Table 4-2, in which the finite-difference approximation for various schemes and corresponding truncated leading terms are listed, we can discretize Eq. (6-61) into the following difference equation, which also includes the leading higher-order terms:

$$\frac{1}{\Delta x}[\phi(x, \omega) - \phi(x^-, \omega)] + \frac{\partial^2\phi}{\partial x^2}\frac{\Delta x}{2} = \frac{c(x, \omega)}{(\Delta\omega)^2}[\phi(x, \omega^+) - 2\phi(x, \omega) + \phi(x, \omega^-)]$$
$$- \frac{c(x, \omega)}{12}\frac{\partial^4\phi}{\partial\omega^4}(\Delta\omega)^2 + \frac{e(x, \omega)}{2\Delta\omega}[\phi(x, \omega^+) - \phi(x, \omega^-)] - \frac{e(x, \omega)}{6}\frac{\partial^3\phi}{\partial\omega^3}(\Delta\omega)^2. \tag{6-69}$$

If we let Δx and $\Delta\omega$ approach zero, Eq. (6-69) reduces to Eq. (6-62). This indicates that Eq. (6-62) is consistent with the original partial differential equation (6-61).

6-4*c* Numerical Accuracy

If a finite-difference scheme is considered accurate, the truncation errors associated with the discretization must be small. To examine the accuracy of the scheme in which the central difference in ω and the upwind difference in x were adopted, we rearrange Eq. (6-69) to

$$A^-\phi(x, \omega^-) + A\phi(x, \omega) + A^+\phi(x, \omega^+) = \phi(x^-, \omega) + R, \tag{6-70}$$

where

$$A^- = \frac{e\gamma\Delta\omega}{2}, \quad A = 1 + 2c\gamma, \quad A^+ = \frac{-e\gamma\Delta\omega}{2} - c\gamma,$$

and

$$R = -\frac{\gamma^2(\Delta\omega)^4}{2}\frac{\partial^2\phi}{\partial x^2} - \frac{e\gamma(\Delta\omega)^4}{6}\frac{\partial^3\phi}{\partial\omega^3} - \frac{c\gamma(\Delta\omega)^4}{12}\frac{\partial^4\phi}{\partial\omega^4} \tag{6-71}$$

or, in operator notation,

$$R = -\gamma^2(\Delta\omega)^4\left(\frac{\gamma}{2}D_x^2 + \frac{e}{6}D_\omega^3 + \frac{c}{12}D_\omega^4\right)\phi, \tag{6-72}$$

where

$$\gamma = \frac{\Delta x}{(\Delta\omega)^2}, \quad D_x = \frac{\partial}{\partial x}, \quad \text{and } D_\omega = \frac{\partial}{\partial\omega}.$$

It is noteworthy that, for example, $D_\omega^4\phi$ is equal to $\partial^4\phi/\partial\omega^4$ but not $(\partial\phi/\partial\omega)^4$, and that the dependence of c and e on x and ω has not been explicitly written for clarity. In order to achieve high accuracy, we wish to minimize $|R|$. The operator for the original partial differential equation (6-61) can be written

$$D_x - eD_\omega - cD_\omega^2 = 0. \tag{6-73}$$

Eliminating D_x from Eqs. (6-72) and (6-73) yields

$$|R| = \gamma(\Delta\omega)^4|F\phi|, \tag{6-74}$$

where

$$F = \frac{\gamma e^2 D_\omega^2}{2} + \left(\gamma ec + \frac{e}{6}\right)D_\omega^3 + \left(\frac{\gamma c^2}{2} + \frac{c}{12}\right)D_\omega^4. \tag{6-75}$$

The possible minimum value of $|F\phi|$ is zero, which enables us to express γ as

$$\gamma = -\left[\frac{cD_\omega^2 + 2eD_\omega}{6(c^2D_\omega^2 + 2ceD_\omega + e^2)}\right]\phi. \tag{6-76}$$

Therefore, in principle, if the local mesh-size ratio γ is chosen according to Eq. (6-76), the truncation error R may be reduced to zero, and high accuracy of the difference scheme can be achieved.

Table 4-2 is provided for the readers' convenience when examination of the accuracy of a certain finite-difference scheme is desired.

SYMBOLS

a	term in x-ω momentum equation, m^{-1} $[= -(d\psi_I/dx)/\psi_{EI}]$
A_m	amplitude used in von Neumann stability analysis
b	term in x-ω momentum equation, m^{-1} $[= -(d\psi_{EI}/dx)/\psi_{EI}]$
c	term in x-ω momentum equation, m^{-1} $(= \nu u/\psi_{EI}^2)$
c^*	convenient term, s/m^2 $(= \nu/\psi_{EI}^2)$
D_x, D_ω	differential operators ($= \partial/\partial x$ and $\partial/\partial\omega$, respectively)
e	coefficient in Eq. (6-61), $m^{-1}(= \partial c/\partial\omega - a - b\omega)$
f	similarity function $[= \psi(x, y)(\nu x u_\infty)^{-1/2}]$
g	parametrically differentiated function $(= \partial f/\partial\zeta)$
G	terminal-value function [Eq. (6-10)]

h	parametrically differentiated function $(=\partial^2 f/\partial\zeta^2)$ or uniform mesh size (e.g., $\Delta\omega$, Δx, and $\Delta\eta$)
J	largest integer value of index j
L	linear operator [Eq. (6-21)]
$Q(\eta, \zeta)$	remainder term in Taylor's series [Eq. (6-17)]
R	sum of truncated higher-order terms [Eqs. (6-71) and (6-72)]
s_j^{n+1}	convenient term $[= c_j^{n+1}\ \Delta x/(\Delta\omega)^2]$
U	normalized streamwise velocity $(= u/u_\infty)$
V	normalized transverse velocity $(= v/u_\infty)$
x_0	streamwise distance from leading edge of boundary layer, m
y_1	transverse coordinate at grid point 1, m
α	superposition constant [Eq. (6-24*a*)]
β	superposition constant [Eq. (6-24*b*)] or convenient expression $(= m\pi\Delta\omega)$ [Eq. (6-64)]
γ	mesh-size ratio, m $[= \Delta x/(\Delta\omega)^2]$
$\delta(x)$	boundary-layer thickness, m
ζ	artificial parameter inserted into nonlinear differential equation $(0 \leqslant \zeta \leqslant 1)$
η	similarity variable $(= y\mathrm{Re}_x^{1/2}/x)$
κ	missing initial condition $[= f''(0)]$
κ^*	correct missing initial condition (e.g., 0.332058 for the Blasius equation)
ξ	amplification factor in von Neumann stability analysis
τ_w	shear stress at wall, $\mathrm{NT/m^2}$
ϕ	field variables such as velocity, temperature, and species mass fraction
ψ	stream function, $\mathrm{m^2/s}$ $(\partial\psi/\partial y = u$ and $\partial\psi/\partial x = -v)$
ω	normalized stream function $[= (\psi - \psi_I)/(\psi_E - \psi_I)]$

Subscripts

D	Denny and Landis
E	outer edge of boundary layer $(y = \delta)$
i	x-direction grid-node index
I	inner edge of boundary layer $(y = 0)$
imp	improved
j	y-direction grid-node index (when a one-dimensional problem is considered, j is preferable to i because the latter is used to denote the imaginary number in von Neumann stability analysis)
s	second-order term included
∞	ambient free stream

Superscripts

$\bar{}$	overbar denoting upstream
$+$	plus an increment
$-$	minus an increment

REFERENCES

1. W. M. Rohsenow and H. Y. Choi, *Heat, Mass and Momentum Transfer,* Prentice-Hall, Englewood Cliffs, N.J., 1961.
2. W. M. Kays, *Convective Heat and Mass Transfer,* McGraw-Hill, New York, 1966.
3. H. Schlichting, *Boundary-Layer Theory,* McGraw-Hill, New York, 1968.
4. B. Gebhart, *Heat Transfer,* McGraw-Hill, New York, 1971.
5. E. R. G. Eckert and R. M. Drake, Jr., *Heat and Mass Transfer,* McGraw-Hill, New York, 1972.
6. F. Kreith, *Principles of Heat Transfer,* Intext, New York, 1976.
7. J. P. Holman, *Heat Transfer,* McGraw-Hill, New York, 1976.
8. L. Prandtl, Über Flüssigkeitsbewegung bei sehr kleiner Reibung, *Proc. 3d Int. Math. Congr.,* pp. 484–491, Heidelberg, 1904.
9. H. Blasius, Grenzschichten in Flüssigkeiten mit kleiner Reibung, *Z. Angew. Math. Phys.,* vol. 56, pp. 1–37, 1908.
10. L. Howarth, On the Solution of the Laminar Boundary Layer Equations, *Proc. R. Soc. London Ser. A,* vol. 164, pp. 547–579, 1938.
11. G. V. Rao, K. K. Raju, N. Muthiyalu, and J. Venkataramana, Least Square and Galerkin Finite Element Solution of Flow Past a Flat Plate, *Int. J. Numer. Methods Eng.,* vol. 11, pp. 185–190, 1977.
12. R. N. Tadros and J. Kirkhope, Galerkin-Finite Element Formulation in Viscous Flow, *Comput. Fluids,* vol. 6, pp. 293–298, 1978.
13. S. Y. Tsay and Y. P. Shih, Blasius Series Solution for Heat Transfer with Nonisothermal Wall Temperatures, *ASME J. Heat Transfer,* vol. 102, pp. 366–368, 1980.
14. R. Hunt and G. Wilks, Continuous Transformation Computation of Boundary Layer Equations between Similarity Regions, *J. Comput. Phys.,* vol. 40, pp. 478–490, 1981.
15. L. Howarth, On the Calculation of the Steady Flow in the Boundary Layer Near the Surface of a Cylinder in a Stream, ARC RM, 1935, p. 1632.
16. S. M. Roberts and J. S. Shipman, *Two-Point Boundary Value Problems: Shooting Methods,* American Elsevier, New York, 1972.
17. H. B. Keller, *Numerical Solution of Two-Point Boundary Value Problems,* Society for Industrial and Applied Mathematics, Philadelphia, 1976.
18. P. E. Rubbert and M. T. Landahl, Solution of Nonlinear Flow Problems, through Parametric Differentiation, *Phys. Fluids,* vol. 10, pp. 831–835, 1967.
19. C. W. Tan and R. DiBano, A Parametric Study of Falkner-Skan Problem with Mass Transfer, *AIAA J.,* vol. 10, pp. 923–925, 1972.
20. T. Y. Na and I. S. Habib, Solution of the Natural Convection Problem by Parametric Differentiation, *Int. J. Heat Mass Transfer,* vol. 17, pp. 457–459, 1974.
21. G. Nath and M. Muthanna, Laminar Hypersonic Boundary Layer Flow at a Three-dimensional Stagnation Point with Slip and Mass Transfer, *Int. J. Heat Mass Transfer,* vol. 20, pp. 177–180, 1977.
22. W. Schneider, A Similarity Solution for Combined Forced and Free Convection Flow over a Horizontal Plate, *Int. J. Heat Mass Transfer,* vol. 22, pp. 1401–1406, 1979.
23. N. T. Sivaneri and W. L. Harris, Numerical Solution of Transonic Flows by Parametric Differentiation and Integral Equation Techniques, *AIAA J.,* vol. 18, pp. 1534–1536, 1980.
24. T. M. Shih, A Method to Solve Two-Point Boundary-Value Problems in Boundary-Layer Flows or Flames, *Numer. Heat Transfer,* vol. 2, pp. 177–191, 1979.
25. C. Runge and W. Kutta, Beitrag zur näherungsweisen Integration totaler Differentialgleichungen *Z. Angew. Math. Phys.,* vol. 46, pp. 435–453, 1901.
26. T. M. Shih and P. J. Pagni, Laminar Mixed-Mode, Forced and Free, Diffusion Flames, *J. Heat Transfer,* vol. 100, pp. 253–259, 1978.
27. W. F. Ames, *Nonlinear Partial Differential Equations in Engineering,* p. 78, Academic, New York, 1965.

28. S. V. Patankar and D. B. Spalding, *Heat and Mass Transfer in Boundary Layers,* 2d ed., Intertext, London, 1970.
29. D. B. Spalding, *A General Computer Program for Two-dimensional Parabolic Phenomena,* vol. 1, Pergamon, Elmsford, N.Y., 1977.
30. J. Nikuradse, Monograph, Laminare reibungsschichten an der längsangetrömten platte, *Zentral. Wiss. Berichtswes.,* Berlin, 1942.
31. E. Isaacson and H. B. Keller, *Analysis of Numerical Methods,* p. 29, Wiley, New York, 1966.
32. R. von Mises, Bemerkungen zur Hydrodynamik, *Z. Angew. Math. Mech.,* vol. 7, pp. 425–431, 1927.
33. V. E. Denny, A. F. Mills, and V. J. Jusionis, Laminar Film Condensation from a Stream-Air Mixture Undergoing Forced Flow down a Vertical Surface, *ASME J. Heat Transfer,* vol. 93, pp. 297–304, 1971.
34. V. E. Denny and R. B. Landis, An Improved Transformation of the Patankar-Spalding Type for Numerical Solution of Two-dimensional Boundary Layer Flows, *Int. J. Heat Mass Transfer,* vol. 14, pp. 1859–1862, 1971.
35. A. T. Wassel and I. Catton, Diffusion from a Line Source in a Neutral or Stably Stratified Atmospheric Surface Layer, *Int. J. Heat Mass Transfer,* vol. 20, pp. 383–391, 1977.
36. H. B. Keller and T. Cebeci, Accurate Numerical Methods for Boundary-Layer Flows, II. Two-dimensional Turbulent Flows, *AIAA J.,* vol. 10, pp. 1193–1199, 1972.
37. A. Mucoglu and T. S. Chen, Mixed Convection across a Horizontal Cylinder with Uniform Surface Heat Flux, *ASME J. Heat Transfer,* vol. 99, pp. 679–682, 1977.
38. H. B. Keller, Numerical Methods in Boundary Layer Theory, *Annu. Rev. Fluid Mech.,* vol. 10, pp. 417–433, 1978.
39. T. Cebeci, Heat Transfer from a Circular Cylinder Impulsively Started from Rest, *Numer. Heat Transfer,* vol. 1, pp. 557–567, 1978.
40. B. J. Venkatachala and G. Nath, Finite Difference Solution of Free Convection Problem with Non-uniform Gravity, *ASME J. Heat Transfer,* vol. 101, pp. 745–747, 1979.
41. T. Cebeci, F. Thiele, P. G. Williams, and K. Stewartson, On the Calculation of Symmetric Wakes, 1. Two-dimensional Flows, *Numer. Heat Transfer,* vol. 2, pp. 35–60, 1979.
42. T. Cebeci, The Laminar Boundary Layer on a Circular Cylinder Started Impulsively from Rest, *J. Comput. Phys.,* vol. 31, pp. 153–172, 1979.
43. S. G. Rubin and P. K. Khosla, An Integral Spline Method for Boundary Layer Equations, *Comput. Fluids,* vol. 7, pp. 75–78, 1979.
44. B. J. Venkatachala and G. Nath, Nonsimilar Laminar Natural Convection in a Thermally Stratified Fluid, *Int. J. Heat Mass Transfer,* vol. 24, pp. 1848–1850, 1981.
45. R. Hunt and G. Wilks, Low Prandtl Number Magnetohydrodynamic Natural Convection in a Strong Cross Field, *Numer. Heat Transfer,* vol. 4, pp. 305–316, 1981.
46. T. M. Shih and H. J. Huang, A Method for Solving Nonlinear Differential Equations for Boundary-Layer Flows, *Numer. Heat Transfer,* vol. 4, pp. 159–178, 1981.

PROBLEMS

6-1 Utilize the similarity solution given in Appendix A to assess the validity of the boundary-layer assumptions $u \gg v$ and $\partial^2 u/\partial y^2 \gg \partial^2 u/\partial x^2$ with $\mathrm{Re}_x = 5 \times 10^4$ at the location $x = 10$ cm and $y = 0.1$ cm. In the vicinity of the leading edge, e.g., $x = 0.01$ cm and $y = 0.001$ cm, are the assumptions still valid? Why?

6-2 Consider the two-point boundary-value equation $f'' + ff' = 0$ subject to $f(0) = 0$ and $f(3) = 2$. Use the shooting method and forward-difference integration with $\Delta\eta = 0.2$ to find $f'(0)$.

6-3 It is postulated that there also exists a function $G^{-1}[f'(\infty)]$ such that $\kappa = G^{-1}[f'(\infty)]$ and $0.33206 = G^{-1}(1)$. Derive an equation similar to Eq. (6-14).

6-4 Repeat Problem 6-2 by the parametric expansion method. Take $\Delta\zeta = 0.5$.

6-5 Derive the expressions listed in Table 6-7.

6-6 Derive Eq. (*d*) of Example 6-6, using an assumed velocity profile $u/u_\infty = 2(y/\delta) - 2(y/\delta)^3 + (y/\delta)^4$, where $\delta = 5.84 \times \mathrm{Re}_x^{-1/2}$. Then compare your result for v_∞ with the similarity solution $0.8604\ u_\infty \mathrm{Re}_x^{-1/2}$.

6-7 Derive the expressions listed in Table 6-9.

6-8 A wind of velocity u_∞ blows over the surface of a pond, causing the water to drift with an unknown velocity. Assuming that the surface shear stress is τ_0, what is the modified form of Eq. (6-49)?

6-9 Using Table 6-9 and the data given in Example 6-10 with $\Delta x = 0.1$ cm, compute the distribution $u(6\text{cm}, y)$ and the wall shear stress. The latter should be normalized for comparison with the Blasius solution $f''(0) = 0.332058$.

CHAPTER

SEVEN

LAMINAR FORCED CONVECTION: HYDRODYNAMIC BOUNDARY LAYER (II)

In the present chapter we will describe a few more important numerical methods that can be applied to the discretization of the boundary-layer equation.

In Section 7-1, splines are introduced. They can be used as local basis functions to approximate unknown solutions, and also as convenient tools to produce three-point expressions among the nodal functions and their derivatives.

In Section 7-2, the Galerkin finite-element method and the variational finite-element method are described. Discretization of the transformed momentum equation in the x-ω plane is presented. The difficulty of obtaining a variational form for the dissipative first derivative is removed by using the so-called adjoint variational principle.

Section 7-3 describes the application of a few methods to problems in which a nonzero streamwise pressure gradient exists.

Section 7-4 presents various numerical methods applied to problems with nonzero mass transfer across the solid surface.

Finally, a method that does not require any coordinate transformation is described in Section 7-5. The merits and shortcomings of this method are discussed.

7-1 SPLINES—APPROXIMATIONS AND HIGHER-ORDER RELATIONS

Hermite interpolation (see Section 3-1*b*) increases the smoothness of approximate functions, but the final expressions involve the functions' nodal derivatives, whose existence enlarges the system of algebraic equations. It is thus natural to seek to construct piecewise polynomial approximations by merely requiring that the approximate functions have certain continuity properties at the grid points such that the derivatives will not appear in the final expressions. In this section we will consider such polynomials.

A *piecewise* polynomial of degree k is said to be a spline of degree k if it has continuous derivatives up to order $k-1$. Thus a linear spline (or a spline of degree one) is a linear function consisting of piecewise straight-line segments, as shown in Fig. 7-1*a;* a quadratic spline is a polynomial that consists of piecewise parabola segments, and is continuous in both the function itself and the first derivative, as shown in Fig. 7-1*b.* Of particular interest is the cubic spline. However, we will first focus on quadratic splines since they are simple and can be used for comparison with cubic splines later.

7-1*a* Quadratic-Spline Formulation

Let the interval $[a, b]$ be uniformly divided into J subintervals $[x_{j-1}, x_j]$, $j=1, 2, \ldots, J$. A quadratic spline that has the minimum support $[x_{j-1}, x_{j+1}]$ can be determined by assuming two polynomials

$$P_I(x) = a_0 + a_1 x + a_2 x^2, \qquad x \epsilon [x_{j-1}, x_j] \tag{7-1a}$$

and

$$P_{II}(x) = b_0 + b_1 x + b_2 x^2, \qquad x \epsilon [x_j, x_{j+1}], \tag{7-1b}$$

and then imposing six constraints

$$P_I(x_j) = P_{II}(x_j), \quad P_I'(x_j) = P_{II}'(x_j), \tag{7-2a, b}$$

$$P_I(x_{j-1}) = P_{II}(x_{j+1}) = 0, \tag{7-2c, d}$$

$$P_I(x_j) = 1, \quad \text{and } P_I'(x_j) = 0. \tag{7-2e, f}$$

The subscripts I and II denote the intervals $[x_{j-1}, x_j]$ and $[x_j, x_{j+1}]$, respectively; the constraints in Eqs. (7-2*a*) and (7-2*b*) follow the definition for a quadratic spline; the constraints in Eqs. (7-2*c*)-(7-2*f*) are arbitrary but convenient. After algebra, we derive a quadratic spline at node j as

$$S_j(x) = P_I(x) = 1 - \left(\frac{x - x_j}{h}\right)^2, \qquad x \epsilon [x_{j-1}, x_{j+1}]$$

or

$$S_j(t) = 1 - t^2, \quad t = \frac{x - x_j}{h}, \qquad x \epsilon [x_{j-1}, x_{j+1}]. \tag{7-3}$$

We will find later that the quadratic spline given in Eq. (7-3) is not very useful because its support is too narrow. A better quadratic spline has the support $[x_{j-2}, x_{j+2}]$. To seek the expression for this spline, we notice that there now exist 3×4 undetermined coefficients and only 3×2 mandatory constraints, i.e.,

$$P_I(x_{j-1}) = P_{II}(x_{j-1}), \quad P_I'(x_{j-1}) = P_{II}'(x_{j-1}), \tag{7-4a, b}$$

$$P_{II}(x_j) = P_{III}(x_j), \quad P_{II}'(x_j) = P_{III}'(x_j), \tag{7-4c, d}$$

$$P_{III}(x_{j+1}) = P_{IV}(x_{j+1}), \quad \text{and } P_{III}'(x_{j+1}) = P_{IV}'(x_{j+1}). \tag{7-4e, f}$$

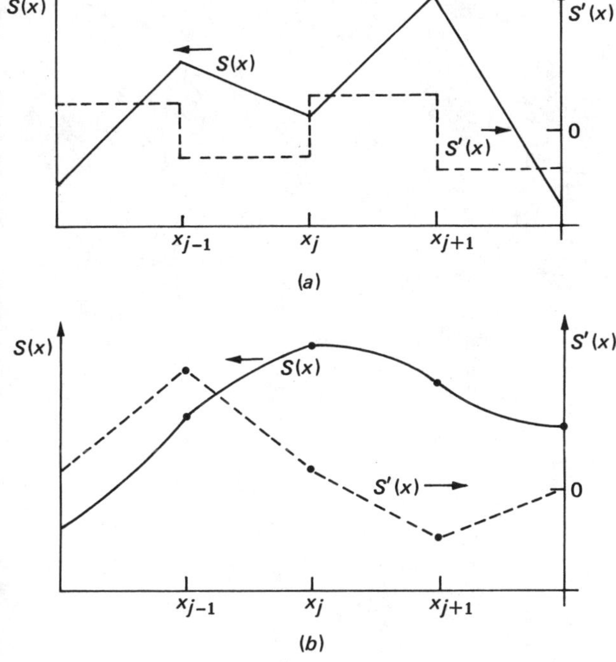

Figure 7-1 Typical linear (*a*) and quadratic (*b*) splines.

The additional six constraints can be specified such that the final spline expression becomes simple. We thus choose

$$\begin{aligned} P_I(x_{j-2}) &= P_I'(x_{j-2}) = 0, \\ P_{IV}(x_{j+2}) &= P_{IV}'(x_{j+2}) = 0, \end{aligned} \tag{7-5a-f}$$

along with

$$P_I(x_{j-1}) = P_{III}(x_{j+1}) = 1.$$

After lengthy but straightforward algebra, we finally derive

$$S_j(t) = \begin{cases} P_I(t) = t^2, & t = \dfrac{x - x_{j-2}}{h}, \quad x \epsilon I, \\ P_{II}(t) = 1 + 2t - t^2, & t = \dfrac{x - x_{j-1}}{h} \quad x \epsilon II, \\ P_{III}(t) = 2 - t^2, & t = \dfrac{x - x_j}{h}, \quad x \epsilon III, \\ P_{IV}(t) = (1 - t)^2, & t = \dfrac{x - x_{j+1}}{h}, \quad x \epsilon IV, \end{cases} \tag{7-6}$$

which is plotted in Fig. 7-2*a*.

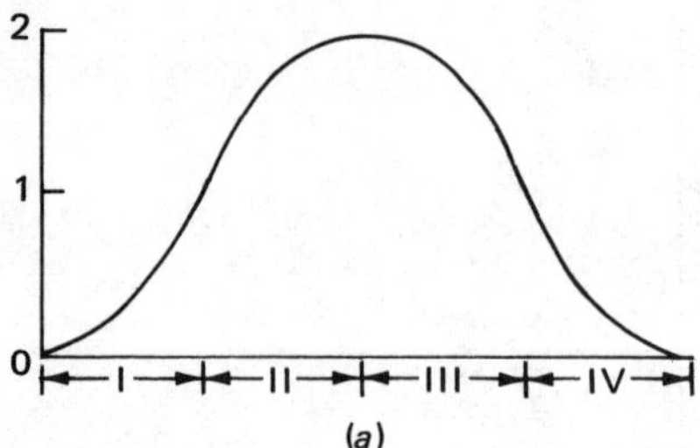

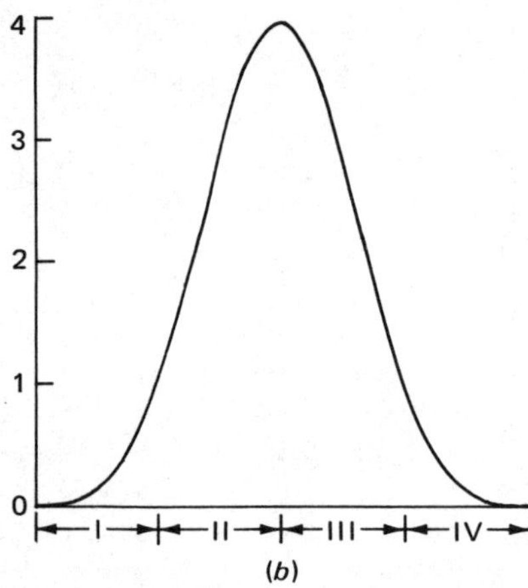

Figure 7-2 Quadratic (*a*) and cubic (*b*) splines.

One important application of splines treats them as the basis functions and uses a linear combination of these basis functions to approximate unknown solutions. This application can be related to the spline collocation method (to be described in Section 9-2). A preview of this treatment is provided when we use splines to approximate a given function.

Example 7-1 Approximate the function

$$f(x) = \sin \frac{\pi x}{2}, \qquad x \epsilon [0, 2] \tag{a}$$

by a linear combination of five quadratic splines given in Eq. (7-6).

Solution: Assume

$$\tilde{f}_q(x) = d_0 S_0(x) + d_{1/2} S_{1/2}(x) + d_1 S_1(x) + d_{3/2} S_{3/2}(x) + d_2 S_2(x), \tag{b}$$

where the subscript q denotes quadratic. Since there are five undetermined coefficients, we need to construct five algebraic equations. At $x = 0$, we obtain

$$\tilde{f}_q(0) = d_0 S_0(0) + d_{1/2} S_{1/2}(0) + d_1 S_1(0) + d_{3/2} S_{3/2}(0) + d_2 S_2(0). \tag{c}$$

Using Table 7-1 with $h = \frac{1}{2}$ leads to

$$0 = 2d_0 + d_{1/2}. \qquad (d)$$

Similarly, we can also derive, at $x = \frac{1}{2}$, 1,

$$0.7071 = d_0 + 2d_{1/2} + d_1 \qquad (e)$$

and

$$1 = 2d_{1/2} + 2d_1, \qquad (f)$$

in which the symmetry of the sine function with respect to $x = 1$ (or $d_{1/2} = d_{3/2}$) has been used. Solving Eqs. (*d*)-(*f*) simultaneously yields

$$d_0 = -0.2071, \quad d_{1/2} = 0.4142, \text{ and} \quad d_1 = 0.0858.$$

Therefore, Eq. (*b*) can be rewritten as

$$\tilde{f}_q(x) = -0.2071S_0(x) + 0.4142S_{1/2}(x) + 0.0858S_1(x) + 0.4142S_{3/2}(x) - 0.2071S_2(x).$$

Another important application of splines is in deriving higher-order finite-difference relations (see Section 1-1b for a different approach). A quadratic spline leads to the second-order three-point relation for the second derivative.

Table 7-1 Nodal values and their derivatives of both the quadratic splines $S_k(x)$ and the cubic splines $B_k(x)$

Splines	x_{j-3}	x_{j-2}	x_{j-1}	x_j	x_{j+1}	x_{j+2}	x_{j+3}
$S_j(x)$	0	0	1	2	1	0	0
$B_j(x)$	0	0	1	4	1	0	0
$S'_j(x)$	0	0	$\frac{2}{h}$	0	$-\frac{2}{h}$	0	0
$B'_j(x)$	0	0	$\frac{3}{h}$	0	$-\frac{3}{h}$	0	0
$B''_j(x)$	0	0	$\frac{6}{h^2}$	$-\frac{12}{h^2}$	$\frac{6}{h^2}$	0	0
$S_{j-1}(x)$	0	1	2	1	0	0	0
$B_{j-1}(x)$	0	1	4	1	0	0	0
$S'_{j-1}(x)$	0	$\frac{2}{h}$	0	$-\frac{2}{h}$	0	0	0
$B'_{j-1}(x)$	0	$\frac{3}{h}$	0	$-\frac{3}{h}$	0	0	0
$B''_{j-1}(x)$	0	$\frac{6}{h^2}$	$-\frac{12}{h^2}$	$\frac{6}{h^2}$	0	0	0
$S_{j+1}(x)$	0	0	0	1	2	1	0
$B_{j+1}(x)$	0	0	0	1	4	1	0
$S'_{j+1}(x)$	0	0	0	$\frac{2}{h}$	0	$-\frac{2}{h}$	0
$B'_{j+1}(x)$	0	0	0	$\frac{3}{h}$	0	$-\frac{3}{h}$	0
$B''_{j+1}(x)$	0	0	0	$\frac{6}{h^2}$	$-\frac{12}{h^2}$	$\frac{6}{h^2}$	0

Example 7-2 Derive the second-order three-point relation for the second derivative $\partial^2\phi/\partial x^2$, using a quadratic spline.

Solution: On the interval $[x_{j-1}, x_j]$ we wish to derive a quadratic spline $S_j(x)$ subject to

$$\text{(i) } S_j(x_{j-1}) = \phi_{j-1}, \quad \text{(ii) } S_j(x_j) = \phi_j, \quad \text{(iii) } \left(\frac{d^2S_j}{dx^2}\right)_{x_j} = M_j.$$

After straightforward algebra, we obtain

$$S_j(t) = \phi_{j-1} + \left(\phi_j - \phi_{j-1} - \frac{1}{2}M_jh_j^2\right)t + \frac{1}{2}M_jh_j^2t^2, \tag{a}$$

where

$$h_j = x_j - x_{j-1}, \quad t = \frac{x - x_{j-1}}{h_j} \quad \text{and } x\epsilon[x_{j-1}, x_j].$$

Similarly, on the interval $[x_j, x_{j+1}]$ the quadratic spline $S_{j+1}(x)$ subject to

$$\text{(i) } S_{j+1}(x_j) = \phi_j, \quad \text{(ii) } S_{j+1}(x_{j+1}) = \phi_{j+1}, \quad \text{(iii) } \left(\frac{d^2S_{j+1}}{dx^2}\right)_{x_j} = M_j$$

is derived as

$$S_{j+1}(t) = \phi_j + \left(\phi_{j+1} - \phi_j - \frac{1}{2}M_jh_{j+1}^2\right)t + \frac{1}{2}M_jh_{j+1}^2t^2, \tag{b}$$

where

$$h_{j+1} = x_{j+1} - x_j, \quad t = \frac{x - x_j}{h_{j+1}}, \quad \text{and } x\epsilon[x_j, x_{j+1}].$$

The continuity of the first derivative at $x = x_j$ dictates that

$$\frac{1}{h_j}\left(\phi_j - \phi_{j-1} - \frac{1}{2}M_jh_j^2\right) + M_jh_j = \frac{1}{h_{j+1}}\left(\phi_{j+1} - \phi_j - \frac{1}{2}M_jh_{j+1}^2\right) \tag{c}$$

or

$$M_j = \frac{\sigma\phi_{j-1} - (1 + \sigma)\phi_j + \phi_{j+1}}{h_jh_{j+1}}, \tag{d}$$

where $\sigma_j = h_{j+1}/h_j$.

The previous examples provide a preview of the spline formulation. We may now understand how splines can be determined and how they can be used to approximate a given function as well as to derive a three-point relation. Quadratic splines, however, cannot be used for problems where C^2 continuity is required. Therefore, higher-order splines must be sought.

7-1*b* Cubic-Spline Formulation

A cubic spline is a piecewise polynomial of third degree that possesses continuous derivatives up to second order. Because of its practical importance, it has been extensively used and studied [1-12]. In this section, we will first show how a cubic spline can be determined. With this cubic spline, we will then show how a given function can be approximated and how the higher-order relations can be derived.

A typical cubic spline on $[x_{j-2}, x_{j+2}]$ can be determined as follows. First, we write

$$P_I(t) = a_0 + a_1 t + a_2 t^2 + a_3 t^3, \qquad t = \frac{x - x_{j-2}}{h}, \; x \epsilon I \tag{7-7a}$$

$$\cdots \qquad\qquad \cdots$$

$$P_{IV}(t) = d_0 + d_1 t + d_2 t^2 + d_3 t^3, \qquad t = \frac{x - x_{j+1}}{h}, \; x \epsilon IV. \tag{7-7d}$$

There are 4×4 undetermined coefficients subject to 3×3 mandatory constraints, i.e.,

$$P_k(1) = P_{k+1}(0), \quad P_k'(1) = P_{k+1}'(0), \quad \text{and } P_k''(1) = P_{k+1}''(0), \tag{7-8a–i}$$

where $k = I, II, \text{ and } III.$

For convenience in algebraic manipulation, we further set

$$P_I(0) = P_I'(0) = P_I''(0) = P_{IV}(1) = P_{IV}'(1) = P_{IV}''(1) = 0. \tag{7-8j–o}$$

The last degree of freedom is constrained arbitrarily (but conveniently) by

$$P_I(1) = 1. \tag{7-8p}$$

Omitting the lengthy algebra, we present the final expression for the cubic spline, designated by the special symbol $B_j(x)$, as

$$B_j(x) = \begin{cases} P_I(t) = t^3, & t = \dfrac{x - x_{j-2}}{h}, \quad x \epsilon I \\ P_{II}(t) = 1 + 3t + 3t^2 - 3t^3, & t = \dfrac{x - x_{j-1}}{h}, \quad x \epsilon II \\ P_{III}(t) = 4 - 6t^2 + 3t^3, & t = \dfrac{x - x_j}{h}, \quad x \epsilon III \\ P_{IV}(t) = (1 - t)^3, & t = \dfrac{x - x_{j+1}}{h}, \quad x \epsilon IV \end{cases} \tag{7-9}$$

which is plotted in Fig. 7-2*b*.† When this cubic spline is used, it is often convenient to construct a table, such as Table 7-1, that lists the nodal values and their first and second derivatives of the splines.

The procedure for using cubic splines to approximate a function is similar to that described in Example 7-1.

Example 7-3 Approximate the given function

$$f(x) = \sin\frac{\pi x}{2}, \quad x \in [0, 2] \tag{a}$$

by a linear combination of five cubic splines given in Eq. (7-9).

Solution: The approximate function $\tilde{f}_c(x)$ can be written as

$$\tilde{f}_c(x) = e_{-1}B_{-1}(x) + e_0B_0(x) + e_1B_1(x) + e_2B_2(x) + e_3B_3(x), \tag{b}$$

where the subscript c denotes cubic. The five undetermined coefficients can be calculated by solving the following five equations:

At $x = 0$:

$$\frac{\pi}{2} = -3e_{-1} + 0 + 3e_1, \tag{c}$$

$$0 = e_{-1} + 4e_0 + e_1, \tag{d}$$

At $x = 1$:

$$1 = e_0 + 4e_1 + e_2, \tag{e}$$

At $x = 2$:

$$0 = e_1 + 4e_2 + e_3, \tag{f}$$

and

$$-\frac{\pi}{2} = -3e_1 + 0 + 3e_3. \tag{g}$$

The matrix system

$$\begin{bmatrix} -3 & 0 & 3 & 0 & 0 \\ 1 & 4 & 1 & 0 & 0 \\ 0 & 1 & 4 & 1 & 0 \\ 0 & 0 & 1 & 4 & 1 \\ 0 & 0 & -3 & 0 & 3 \end{bmatrix} \begin{Bmatrix} e_{-1} \\ e_0 \\ e_1 \\ e_2 \\ e_3 \end{Bmatrix} = \begin{Bmatrix} \frac{\pi}{2} \\ 0 \\ 1 \\ 0 \\ -\frac{\pi}{2} \end{Bmatrix} \tag{h}$$

†Since it looks like a bell, the cubic spline defined in Eq. (7-9) is also called B spline.

can be solved to yield

$$e_{-1} = -0.2775, \quad e_0 = 0.0079, \quad e_1 = 0.2461, \quad e_2 = 0.0079,$$

and

$$e_3 = -0.2775.$$

The approximate function, along with the contribution of five superimposed cubic splines, is plotted in Fig. 7-3 in comparison with the given sine function.

Having obtained the approximations of quadratic splines and cubic splines in Examples 7-1 and 7-3, it may be interesting to compare their accuracies. If we compare the error of the function values at, for example, $x = 0.75$,

$$|\tilde{f}_q(0.75) - f(0.75)| = |0.9269 - 0.9239| = 0.0030 \tag{7-10a}$$

and

$$|\tilde{f}_c(0.75) - f(0.75)| = |0.9176 - 0.9239| = 0.0063, \tag{7-10b}$$

the quadratic spline appears to be more accurate. If, however, we compare the error of the derivative values at, for example, $x = 0$,

$$|\tilde{f}_q'(0) - f'(0)| = |1.6568 - 1.5708| = 0.086 \tag{7-10c}$$

and

$$|\tilde{f}'_c(0) - f'(0)| = |1.5708 - 1.5708| = 0, \tag{7-10d}$$

the cubic spline appears to be more accurate. Therefore, making such comparisons cannot yield a conclusive answer. Figure 7-4 shows a fictitious exact function (solid line) and two approximations (dashed and dotted lines). The dashed-line approximation may have less errors in the function values, while the dotted-line solution may be more accurate in the derivative values. It is thus difficult to judge which

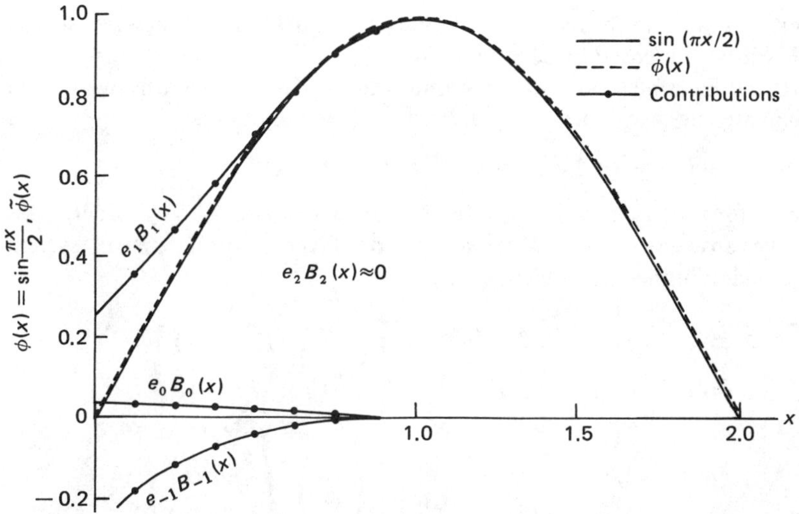

Figure 7-3 The sine function approximated by the cubic splines. Superposition is shown for $x\epsilon[0,1]$ only.

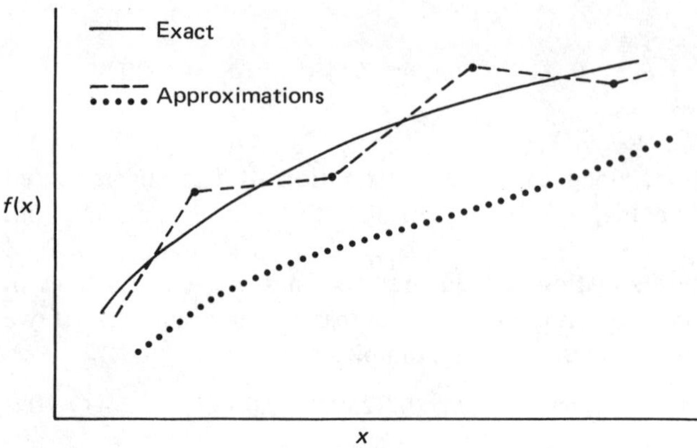

Figure 7-4 Comparison of accuracies of two approximate functions.

approximation is better. A proper way to estimate the error of an approximate solution is to use the concept of various norms such as

$$\|\tilde{f}(x) - f(x)\| = \left\{ \int_0^2 [\tilde{f}(x) - f(x)]^2 \, dx \right\}^{1/2},$$

called the $L_2\,[0, 2]$ norm, or

$$\|\tilde{f}(x) - f(x)\| = \left\{ \int_0^2 \{[\tilde{f}(x) - f(x)]^2 + [\tilde{f}'(x) - f'(x)]^2\} \, dx \right\}^{1/2},$$

called the Sobolev norm. The concept and usage of these norms deserve special attention and will be presented in Section 14-1.

After studying Examples 7-1 and 7-3, some readers may wonder why we do not simply approximate the sine function by a fourth-degree polynomial

$$\tilde{f}(x) = b_0 + b_1 x + b_2 x^2 + b_3 x^3 + b_4 x^4,$$

which is simple to construct and is continuous in derivatives of any order. This question can be answered by a close inspection of the resulting matrix system containing the undetermined coefficients, i.e.,

$$\begin{bmatrix} 1 & x_{j-2} & x_{j-2}^2 & x_{j-2}^3 & x_{j-2}^4 \\ 1 & x_{j-1} & x_{j-1}^2 & x_{j-1}^3 & x_{j-1}^4 \\ 1 & x_j & x_j^2 & x_j^3 & x_j^4 \\ 1 & x_{j+1} & x_{j+1}^2 & x_{j+1}^3 & x_{j+1}^4 \\ 1 & x_{j+2} & x_{j+2}^2 & x_{j+2}^3 & x_{j+2}^4 \end{bmatrix} \begin{Bmatrix} b_0 \\ b_1 \\ b_2 \\ b_3 \\ b_4 \end{Bmatrix} = \begin{Bmatrix} \tilde{f}(x_{j-2}) \\ \tilde{f}(x_{j-1}) \\ \tilde{f}(x_j) \\ \tilde{f}(x_{j+1}) \\ \tilde{f}(x_{j+2}), \end{Bmatrix}. \qquad (7\text{-}11)$$

The matrix in Eq. (7-11) is a *full* matrix, which is undesirable from the computational viewpoint. If the matrix system is enlarged, the computation of a large full-matrix system becomes quite cumbersome. On the other hand, the spline formulation leads to tridiagonal matrix systems such as Eq. (h) in Example 7-3, which can be handled by the Thomas algorithm (see Section 5-1b) after the first and last rows are adjusted to contain two nonzero elements.

In addition to approximating given or unknown functions, cubic splines can be used to derive three-point expressions that interrelate the nodal functions and nodal derivatives. We will present the derivation of these expressions as follows. Let the sought solution $\phi(x)$ be approximated by

$$\tilde{\phi}(x) = \cdots + e_{j-2}B_{j-2}(x) + e_{j-1}B_{j-1}(x) + e_jB_j(x) + e_{j+1}B_{j+1}(x) + e_{j+2}B_{j+2}(x) + \cdots. \tag{7-12}$$

Differentiating Eq. (7-12) with respect to x once and twice yields

$$m(x) = \cdots + e_{j-2}B'_{j-2}(x) + e_{j-1}B'_{j-1}(x) + e_jB'_j(x) + e_{j+1}B'_{j+1}(x) + e_{j+2}B'_{j+2}(x) + \cdots \tag{7-13}$$

and

$$M(x) = \cdots + e_{j-2}B''_{j-2}(x) + e_{j-1}B''_{j-1}(x) + e_jB''_j(x) + e_{j+1}B''_{j+1}(x) + e_{j+2}B''_{j+2}(x) + \cdots, \tag{7-14}$$

where

$$m = \frac{d\phi}{dx} \quad \text{and } M = \frac{d^2\phi}{dx^2}.$$

If the relation among ϕ_k and m_k, $k = j-1, j, j+1$, is desired, Eqs. (7-12) and (7-13) can be used (with the aid of Table 7-1) to generate the following six equations:

$$\phi_k = e_{k-1} + 4e_k + e_{k+1} \qquad \text{(7-15}a\text{–}c\text{)}$$

and

$$m_k = -3e_{k-1} + 3e_{k+1}. \qquad \text{(7-16}a\text{–}c\text{)}$$

Eliminating the five undetermined constants e_{j-2}, e_{j-1}, e_j, e_{j+1}, and e_{j+2} from Eqs. (7-15) and (7-16) finally yields

$$m_{j-1} + 4m_j + m_{j+1} = 3(-\phi_{j-1} + \phi_{j+1}), \tag{7-17}$$

which is identical to Eq. (1-10a) with $h = 1$. Similarly, if the relation among ϕ_k and M_k, $k = j-1, j, j+1$, is sought, Eq. (7-14) can be used to generate the following three equations:

$$M_k = 6e_{k-1} - 12e_k + 6e_{k+1}. \qquad \text{(7-18}a\text{–}c\text{)}$$

Eliminating the five constants from Eqs. (7-15) and (7-18) yields

$$M_{j-1} + 4M_j + M_{j+1} = 6(\phi_{j-1} - 2\phi_j + \phi_{j+1}). \tag{7-19}$$

It is always advisable to conduct some simple tests to partially check if Eqs. (7-17) and (7-19) are derived correctly.

Example 7-4 Let $\phi(x) = x^3$ and $x_j = 2$ with unit interval size. Evaluate the nodal values ϕ_k, m_k, and M_k to check whether they satisfy Eqs. (7-17) and (7-19).

Solution: We tabulate

Grid point	ϕ	m	M
$j-1$	1	3	6
j	8	12	12
$j+1$	27	27	18

Substituting these values into Eqs. (7-17) and (7-19), we obtain

$$3 + 4 \times 12 + 27 = 3(-1 + 27)$$

and

$$6 + 4 \times 12 + 18 = 6(1 - 2 \times 8 + 27).$$

In many practical problems regarding boundary-layer flow, it is common that certain boundary conditions of ϕ, m, or M are not prescribed. Therefore, some two-point relations are needed to ensure closure of the equation system. Suppose, for example, that M_J is not prescribed. Then a two-point relation that expresses M_J in terms of ϕ_J, ϕ_{J-1}, m_J, and m_{J-1} is needed. We first write

$$\phi_J = e_{J-1} + 4e_J + e_{J+1}, \tag{7-20a}$$

$$\phi_{J-1} = e_{J-2} + 4e_{J-1} + e_J, \tag{7-20b}$$

$$m_J = -3e_{J-1} + 3e_{J+1}, \tag{7-21a}$$

$$m_{J-1} = -3e_{J-2} + 3e_J, \tag{7-21b}$$

and

$$M_J = 6e_{J-1} - 12e_J + 6e_{J+1}. \tag{7-22}$$

Then eliminating the four undetermined constants yields

$$M_J = 2(m_{J-1} + 2m_J) + 6(\phi_{J-1} - \phi_J). \tag{7-23}$$

Following a similar procedure, we can derive

$$M_0 = -2(m_1 + 2m_0) + 6(\phi_1 - \phi_0), \tag{7-24}$$

$$m_J = \phi_J - \phi_{J-1} + \frac{1}{3}\left(\frac{1}{2}M_{J-1} + M_J\right), \tag{7-25}$$

and

$$m_0 = \phi_1 - \phi_0 - \frac{1}{3}\left(\frac{1}{2}M_1 + M_0\right). \tag{7-26}$$

Once Eqs. (7-17), (7-19), and (7-23)-(7-26) are provided, we are able to solve a two-point boundary-value problem. It must be noted that these equations are valid for uniform and unit mesh intervals only.

7-1*c* Blasius Similarity Equation

The application of Eqs. (7-17), (7-19), and (7-23)–(7-26) to a two-point boundary-value problem can be illustrated by solving the Blasius similarity equations (6-8) and (6-9). We will describe the procedure in detail as follows. Consider

$$m'' + \tfrac{1}{2}\,\phi M = 0 \tag{7-27}$$

subject to

$$\phi_0 = m_0 = 0, \qquad m_J = 1, \qquad \text{and } M_J = 0. \tag{7-28}$$

It is noted that M_0 and ϕ_J are two missing boundary conditions and, consequently, Eq. (7-24) and either Eq. (7-23) or Eq. (7-25) in the rearranged form must be used at the boundaries.

The nonlinear term $\frac{1}{2}\,\phi M$ in Eq. (7-27) can be linearized as

$$\phi M \approx -\,\bar{\phi}\bar{M} + \bar{M}\phi + \bar{\phi}M, \tag{7-29}$$

which is then substituted into Eq. (7-27) with the 3-point difference for m'' to yield

$$2m_{j-1} + \bar{M}_j\phi_j - 4m_j + \bar{\phi}_j M_j + 2m_{j+1} = \bar{\phi}_j\bar{M}_j. \tag{7-30}$$

Here the overbar denotes "previously iterated"; the mesh size $\Delta\eta$ is taken to be unity. This governing equation (7-30) along with the two three-point relations (7-17) and (7-19) can be written in matrix form as

$$[\mathbf{A}_j]\,\{\mathbf{V}_{j-1}\} + [\mathbf{B}_j]\,\{\mathbf{V}_j\} + [\mathbf{C}_j]\,\{\mathbf{V}_{j+1}\} = \{\mathbf{D}_j\}, \qquad j = 2, 3, \ldots, J-2^{\dagger} \tag{7-31}$$

where

$$[\mathbf{A}_j] = \begin{bmatrix} 0 & 2 & 0 \\ -3 & -1 & 0 \\ 6 & 0 & -1 \end{bmatrix}, \qquad [\mathbf{B}_j] = \begin{bmatrix} \bar{M}_j & -4 & \bar{\phi}_j \\ 0 & -4 & 0 \\ -12 & 0 & -4 \end{bmatrix},$$

$$[\mathbf{C}_j] = \begin{bmatrix} 0 & 2 & 0 \\ 3 & -1 & 0 \\ 6 & 0 & -1 \end{bmatrix}, \qquad \{\mathbf{D}_j\} = \begin{Bmatrix} \bar{\phi}_j\bar{M}_j \\ 0 \\ 0 \end{Bmatrix},$$

and
$$\{\mathbf{V}_k\}^T = \{\phi_k \;\; m_k \;\; M_k\}, \qquad k = j-1, j, j+1.$$

Equation (7-31) is valid for all the interior points $j = 2, 3, \ldots, J-2$. When it is used for the grid points $j = 1, J-1$, the two boundary conditions M_0 and ϕ_J that

†For programming convenience, an alternative is to apply Eq. (7-31) also at grid points $j = 1$ and $J-1$ and treat $[\mathbf{A}_1]\{\mathbf{V}_0\}$ and $[\mathbf{C}_{J-1}]\{\mathbf{V}_J\}$ as knowns.

are not prescribed will be involved. Therefore, its form must be changed such that M_0 and ϕ_J will disappear. To achieve this, we may substitute Eq. (7-24) into Eq. (7-31) for $j = 1$ and substitute a rearranged form of Eq. (7-23)

$$\phi_J = \phi_{J-1} + \frac{1}{3} m_{J-1} + \frac{2}{3} m_J - \frac{1}{6} M_J$$

into Eq. (7-31) for $j = J - 1$ to obtain

$$[\mathbf{B}_1]\{\mathbf{V}_1\} + [\mathbf{C}_1]\{\mathbf{V}_2\} = \{\mathbf{D}_1\} \tag{7-32}$$

and

$$[\mathbf{A}_{J-1}]\{\mathbf{V}_{J-2}\} + [\mathbf{B}_{J-1}]\{\mathbf{V}_{J-1}\} = \{\mathbf{D}_{J-1}\}, \tag{7-33}$$

where

$$[\mathbf{B}_1] = \begin{bmatrix} \bar{M}_1 & -4 & \bar{\phi}_1 \\ 0 & -4 & 0 \\ -18 & 2 & -4 \end{bmatrix}, \quad [\mathbf{C}_1] = \begin{bmatrix} 0 & 2 & 0 \\ 3 & -1 & 0 \\ 6 & 0 & -1 \end{bmatrix},$$

$$\{\mathbf{D}_1\} = \begin{Bmatrix} \bar{\phi}_1 \bar{M}_1 - 2m_0 \\ 3\phi_0 + m_0 \\ -12\phi_0 - 4m_0 \end{Bmatrix}, \quad [\mathbf{A}_{J-1}] = \begin{bmatrix} 0 & 2 & 0 \\ -3 & -1 & 0 \\ 6 & 0 & -1 \end{bmatrix},$$

$$[\mathbf{B}_{J-1}] = \begin{bmatrix} \bar{M}_{J-1} & -4 & \bar{\phi}_{J-1} \\ 3 & -3 & 0 \\ -6 & 2 & -4 \end{bmatrix}, \quad \text{and } \{\mathbf{D}_{J-1}\} = \begin{Bmatrix} \bar{\phi}_{J-1}\bar{M}_{J-1} - 2m_J \\ -m_J + \frac{1}{2}M_J \\ -4m_J + 2M_J \end{Bmatrix}.$$

There are $3(J-1)$ unknowns, namely ϕ_j, m_j, and M_j, $j = 1, 2, \ldots, J-1$, which must satisfy $3(J-1)$ equations, namely Eqs. (7-31)-(7-33). The system is finally closed. When these equations are put together, a 3×3 block tridiagonal system

$$\begin{bmatrix} B_1 & C_1 & & & & & \\ A_2 & B_2 & C_2 & & & & \\ & A_3 & B_3 & C_3 & & & \\ & & \cdot & \cdot & \cdot & & \\ & & & A_j & B_j & C_j & \\ & & & & \cdot & \cdot & \cdot \\ & & & & A_{J-2} & B_{J-2} & C_{J-2} \\ & & & & & A_{J-1} & B_{J-1} \end{bmatrix} \begin{Bmatrix} V_1 \\ V_2 \\ V_3 \\ \cdot \\ V_j \\ \cdot \\ V_{J-2} \\ V_{J-1} \end{Bmatrix} = \begin{Bmatrix} D_1 \\ D_2 \\ D_3 \\ \cdot \\ D_j \\ \cdot \\ D_{J-2} \\ D_{J-1} \end{Bmatrix} \tag{7-34}$$

is formed. Several algorithms for solving the block tridiagonal system are available [13]. They are generally analogous to the algorithm for solving the element tridiagonal system. We will use the modified Thomas algorithm to solve Eq. (7-34) in the following example.

Example 7-5 Modify the Thomas algorithm for a 3 X 3 block tridiagonal system (7-34). Note that the matrix multiplication is noncommutative.

Solution: The procedure is described step by step as follows.

Step 1: Multiply both sides of Eq. (7-32) by B_1^{-1}, i.e.,

$$V_1 + B_1^{-1} C_1 V_2 = B_1^{-1} D_1. \qquad (a)$$

Step 2: Evaluate $B_1^{-1} C_1$ and $B_1^{-1} D_1$.

Step 3: Multiply both sides of Eq. (7 31) by A_2^{-1} with $j = 2$, i.c.,

$$V_1 + A_2^{-1} B_2 V_2 + A_2^{-1} C_2 V_3 = A_2^{-1} D_2. \qquad (b)$$

Step 4: Evaluate $A_2^{-1} B_2$, $A_2^{-1} C_2$, and $A_2^{-1} D_2$.

Step 5: Subtract Eq. (*b*) from Eq. (*a*), i.e.,

$$(B_1^{-1} C_1 - A_2^{-1} B_2) V_2 - A_2^{-1} C_2 V_3 = B_1^{-1} D_1 - A_2^{-1} D_2$$

or

$$E_2 V_2 + F_2 V_3 = G_2, \qquad (c)$$

where

$$E_2 = B_1^{-1} C_1 - A_2^{-1} B_2, \quad F_2 = - A_2^{-1} C_2, \quad \text{and } G_2 = B_1^{-1} D_1 - A_2^{-1} D_2. \qquad (d)$$

Step 6: Repeat increasing the indices in Eqs. (*c*) and (*d*), i.e.,

$$E_j V_j + F_j V_{j+1} = G_j, \quad j = 3, 4, \ldots, J-2 \qquad (e)$$

where

$$E_j = E_{j-1}^{-1} F_{j-1} - A_j^{-1} B_j, \quad F_j = - A_j^{-1} C_j, \quad \text{and } G_j = E_{j-1}^{-1} G_{j-1} - A_j^{-1} D_j. \qquad (f)$$

Step 7: Evaluate V_{J-1}, i.e.,

$$V_{J-1} = E_{J-1}^{-1} G_{J-1}, \qquad (g)$$

where

$$G_{J-1} = E_{J-2}^{-1} G_{J-2} - A_{J-1}^{-1} D_{J-1}. \qquad (h)$$

Step 8: Obtain V_{J-2} through Eq. (*e*), i.e.,

$$V_{J-2} = E_{J-2}^{-1} G_{J-2} - E_{J-2}^{-1} F_{J-2} V_{J-1}. \qquad (i)$$

Step 9: Repeat the evaluation of V_j, $j = J-3, J-4, \ldots, 2$, i.e.,

$$V_j = E_j^{-1}G_j - E_j^{-1}F_jV_{j+1}. \qquad (j)$$

Step 10: Determine V_1, using Eq. (*a*).

Using the procedures described in Example 7-5, we can obtain ϕ_j, m_j, and M_j for $j = 0, 1, \ldots, J$. To evaluate A_j^{-1}, B_j^{-1}, C_j, D_j, and thereby E_j, F_j, and G_j, however, we must start with some initially guessed solutions for $\phi_j^{[0]}$, $m_j^{[0]}$, and $M_j^{[0]}$ that are to satisfy the boundary conditions. For example, we may assume

$$\phi^{[0]}(\eta) = \frac{\eta^2}{2\eta_\infty}, \qquad m^{[0]}(\eta) = \frac{\eta}{\eta_\infty}, \qquad \text{and } M^{[0]}(\eta) = \frac{1}{\eta_\infty}.$$

The procedures can then be repeated with over-relaxation until the solution converges, e.g., $|(M_0^{[k+1]} - M_0^{[k]})/M_0^{[k]}| \leqslant 10^{-4}$. Table 7-2*a* lists the solution ϕ_j, m_j, and M_j computed by following example 7-5 with $\Delta\eta = 1$ and $\eta_\infty = 7$. The solution converges at the 160th iteration. If Eq. (7-19) is replaced with Eq. (1-10*b*), the solution listed in Table 7-2*b* converges faster and is more accurate than that listed in Table 7-2*a*.

7-1*d* Orthogonal Collocation

It was mentioned in Section 2-4*c* that it is not necessary to choose the collocation points to be evenly spaced grid nodes. In fact, we showed that better accuracy was achieved when the collocation points were coincident with the Gaussian quadrature knots. In this section we will confirm this by using cubic splines as the basis functions.

Example 7-6 Approximate the given function

$$f(x) = \sin\frac{\pi x}{2}, \qquad x \in [0, 1] \qquad (a)$$

by a linear combination of three cubic splines and by arranging the collocation points at (*a*) $x = 0, \frac{1}{2}, 1$ and (*b*) $x = 0.112702, \frac{1}{2}, 0.887298$.

Table 7-2 Blasius solution obtained by the cubic spline formulation

	(*a*) with Eq. (7-19)			(*b*) with Eq. (1-10*b*)		
$j(\eta)$	ϕ_j	m_j	M_j	ϕ_j	m_j	M_j
0	0	0	0.3271	0	0	0.3310
1	0.1640	0.3286	0.3301	0.1655	0.3311	0.3264
2	0.6482	0.6301	0.2729	0.6532	0.6352	0.2704
3	1.3948	0.8432	0.1532	1.4064	0.8510	0.1571
4	2.2989	0.9494	0.0593	2.3183	0.9562	0.0610
5	3.2709	0.9875	0.0170	3.2953	0.9908	0.0160
6	4.2647	0.9979	0.0039	4.2911	0.9991	0.0034
7	5.2640	1.0	0.0	5.2910	1.0	0.0

Solution: Let us first write

$$\tilde{f}(x) = aB_0(x) + bB_{1/2}(x) + cB_1(x). \qquad (b)$$

(*a*) Collocation at $x = 0, \frac{1}{2}, 1$ yields

$$4a + b = 0,$$

$$a + 4b + c = 0.70711,$$

$$b + 4c = 1,$$

of which the solution is

$$a = -0.03265, \quad b = 0.13060, \text{ and} \quad c = 0.21735.$$

(*b*) Collocation at $x = 0.112702, \frac{1}{2}, 0.887298$ yields

$$3.7295a + 1.79428b + 0.01145c = 0.17611,$$

$$a + 4b + c = 0.70711,$$

$$0.01145a + 1.79428b + 3.72950c = 0.98437,$$

of which the solution is

$$a = -0.01617, \quad b = 0.13052, \text{ and} \quad c = 0.20120.$$

We may examine the accuracies of these two approximations by, for example, comparing $|\int_0^1 [f(x) - \tilde{f}(x)]\, dx|$. Thus,

$$\left| \int_0^1 [f(x) - \tilde{f}_a(x)]\, dx \right| = |0.63662 - 0.63620| = 4.2 \times 10^{-4}$$

and

$$\left| \int_0^1 [f(x) - \tilde{f}_b(x)]\, dx \right| = |0.63662 - 0.63648| = 1.4 \times 10^{-4}.$$

It is seen that the approximation of orthogonal collocation has a smaller error.

7-2 FINITE-ELEMENT METHOD

So far, we have directed considerable attention to methods for solving the Blasius similarity equation, which is a typical nonlinear two-point boundary-value problem. We also applied the control volume method to discretize the transformed partial differential equation (6-39) in the ω coordinate. In this section we will continue to introduce some finite-element schemes that can be used to discretize Eq. (6-39). Since the Galerkin method (see Section 1-2*a*) has been widely used and can be applied to a wide class of differential equations, it will be described first.

7-2a Galerkin Method

In the Galerkin technique, which falls in the category of weighted-residuals methods, the weighting function is chosen to be the basis function. We will use this method to discretize Eq. (6-39) in ω coordinate. First, let $\phi(\omega)$ be approximated by

$$\tilde{\phi}(\omega) = \sum_{j=0}^{J} N_j(\omega)\phi_j, \tag{7-35}$$

where ϕ_j is the nodal streamwise velocity and $N_j(\omega)$ is the pyramid basis function defined as

$$N_j(\omega) = \begin{cases} \dfrac{\omega - \omega_{j-1}}{\omega_j - \omega_{j-1}} & \text{if } \omega \in [\omega_{j-1}, \omega_j] \\ \dfrac{\omega_{j+1} - \omega}{\omega_{j+1} - \omega_j} & \text{if } \omega \in [\omega_j, \omega_{j+1}] \\ 0 & \text{elsewhere.} \end{cases} \tag{7-36a}$$

The index j of the basis function represents 2, 3, . . . , J. Based on the argument described in Section 6-3*a*, the basis functions at $j = 0$, 1 are defined differently as

$$N_1(\omega) = \begin{cases} \dfrac{\omega_2 - \omega}{\omega_2 - \omega_1} & \text{if } \omega \in [\omega_1, \omega_2] \\ \left(\dfrac{\omega}{\omega_1}\right)^{1/2} & \text{if } \omega \in [0, \omega_1] \\ 0 & \text{elsewhere} \end{cases} \tag{7-36b}$$

and

$$N_0(\omega) = \begin{cases} 1 - \left(\dfrac{\omega}{\omega_1}\right)^{1/2} & \text{if } \omega \in [0, \omega_1] \\ 0 & \text{elsewhere.} \end{cases} \tag{7-36c}$$

We then rewrite the transformed boundary-layer momentum equation (6-39) as

$$A\phi = \frac{\partial \phi}{\partial x} + (a + b\omega)\frac{\partial \phi}{\partial \omega} - \frac{\partial}{\partial \omega}\left(c\,\frac{\partial \phi}{\partial \omega}\right) = 0, \tag{7-37}$$

where A is a differential operator in the domain $x > 0$ and $0 \leqslant \omega \leqslant 1$. Further, let us define an inner product as

$$(p, q) = \int_0^1 pq\, d\omega. \tag{7-38}$$

Other types of inner products will be introduced in Chapter 14. Following Eq. (7-38), we obtain

$$(N_j, A\bar{\phi}) = \int_0^1 N_j \frac{\partial \bar{\phi}}{\partial x} d\omega + \int_0^1 N_j(a + b\omega) \frac{\partial \bar{\phi}}{\partial \omega} d\omega$$

$$- \int_0^1 N_j \frac{\partial}{\partial \omega}\left(c \frac{\partial \bar{\phi}}{\partial \omega}\right) d\omega. \tag{7-39}$$

The first term on the right-hand side of Eq. (7-39) can be discretized to

$$\int_0^1 N_j \frac{\partial \bar{\phi}}{\partial x} d\omega = \int_0^{\omega_{j-1}} 0 \times \frac{\partial \bar{\phi}}{\partial x} d\omega + \int_{\omega_{j-1}}^{\omega_j} N_j \frac{\partial \bar{\phi}}{\partial x} d\omega$$

$$+ \int_{\omega_j}^{\omega_{j+1}} N_j \frac{\partial \bar{\phi}}{\partial x} d\omega + \int_{\omega_{j+1}}^1 0 \times \frac{\partial \bar{\phi}}{\partial x} d\omega$$

$$= \Lambda_{j-1,j} \frac{d\phi_{j-1}}{dx} + \Lambda_{j,j} \frac{d\phi_j}{dx} + \Lambda_{j,j+1} \frac{d\phi_{j+1}}{dx}, \tag{7-40}$$

where the coefficients are defined, evaluated, and listed in Table 1-4.

Similarly, the second and third terms on the right-hand side of Eq. (7-39) can be derived as

$$\int_0^1 N_j(a + b\omega) \frac{\partial \bar{\phi}}{\partial \omega} d\omega = D_{j-1,j}\phi_{j-1} + E_j\phi_j + D_{j,j+1}\phi_{j+1} \tag{7-41}$$

and

$$\int_0^1 N_j \frac{\partial}{\partial \omega}\left(c \frac{\partial \bar{\phi}}{\partial \omega}\right) d\omega = - F_{j-1,j}\phi_{j-1} - G_j\phi_j - F_{j,j+1}\phi_{j+1}, \tag{7-42}$$

where all the coefficients are listed in Table 7-3 for convenience. Next, we discretize the streamwise derivative by using the upwind difference, i.e.,

$$\frac{d\phi_j}{dx} = \frac{1}{\Delta x}(\phi_j - \phi_W), \tag{7-43}$$

where the subscript W stands for west (i.e., upstream). Because of the adoption of the finite-difference approximation in the x direction, this technique is also called the semidiscrete Galerkin method [14, 15]. With Eqs. (7-40)-(7-43), we are now able to rewrite Eq. (7-39) in a linear algebraic form as

Table 7-3 Convenient expressions for several integrals; $j = 2, 3, \ldots, J-1$

Symbol	Integral	Final expression
$D_{j-1,j}$	$\int_{\omega_{j-1}}^{\omega_j} (a + b\omega)\frac{dN_{j-1}}{d\omega} N_j \, d\omega$	$-\left(\frac{a}{2} + \frac{b\omega_{j-1}}{2} + \frac{bh_j}{3}\right)$
$D_{j,\,j+1}$	$\int_{\omega_j}^{\omega_{j+1}} (a + b\omega)\frac{dN_{j+1}}{d\omega} N_j \, d\omega$	$\frac{a + b\omega_{j+1}}{2} + \frac{bh_{j+1}}{6}$
E_j	$\int_{\omega_{j-1}}^{\omega_{j+1}} (a + b\omega)N_j \frac{dN_j}{d\omega} d\omega$	$-\frac{bh_j}{2} + \frac{b}{6}(2h_j - h_{j+1})$
$F_{j-1,j}$	$\int_{\omega_{j-1}}^{\omega_j} c\left(\frac{dN_{j-1}}{d\omega}\right)\left(\frac{dN_j}{d\omega}\right) d\omega$	$-\frac{c^*}{2h_j}(\bar{u}_{j-1} + \bar{u}_j)$
$F_{j,\,j+1}$	$\int_{\omega_j}^{\omega_{j+1}} c\left(\frac{dN_j}{d\omega}\right)\left(\frac{dN_{j+1}}{d\omega}\right) d\omega$	$-\frac{c^*}{2h_{j+1}}(\bar{u}_j + \bar{u}_{j+1})$
G_j	$\int_{\omega_{j-1}}^{\omega_{j+1}} c\left(\frac{dN_j}{d\omega}\right)^2 d\omega$	$\frac{c^*}{2h_j}\bar{u}_{j-1} + \frac{c^*\bar{u}_j}{2}\left(\frac{1}{h_j} + \frac{1}{h_{j+1}}\right) + \frac{c^*}{2h_{j+1}}\bar{u}_{j+1}$

$$A\phi_{j-1} + B\phi_j + C\phi_{j+1} = D, \quad j = 2, 3, \ldots, J-1 \tag{7-44}$$

where

$$A = \frac{\Lambda_{j-1,j}}{\Delta x} + D_{j-1,j} + F_{j-1,j} \quad \text{(not to be confused with the operator } A\text{)},$$

$$B = \frac{\Lambda_{j,j}}{\Delta x} + G_j + E_j,$$

$$C = \frac{\Lambda_{j,j+1}}{\Delta x} + D_{j,j+1} + F_{j,j+1},$$

and

$$D = \frac{\phi_{SW}\Lambda_{j-1,j}}{\Delta x} + \frac{\phi_W\Lambda_{j,j}}{\Delta x} + \frac{\phi_{NW}\Lambda_{j,j+1}}{\Delta x}.$$

These coefficients are listed in Table 6-9 and can be compared with their counterparts obtained by the control volume method. At grid point $j = 1$, we also obtain

$$A_1\phi_1 + B_1\phi_1 + C_1\phi_2 = D_1. \tag{7-45}$$

Here the coefficients A_1, B_1, C_1, and D_1 are similar to those at the interior points and can be evaluated by using Eqs. (7-36*b*) and (7-36*c*). The final expressions are also listed in Table 6-9. Equations (7-44) and (7-45) thus constitute a tridiagonal matrix system, which can be efficiently solved by, for example, the Thomas algorithm (see Section 5-1*b*).

Instead of discretizing $d\phi/dx$ by upwind difference, as shown in Eq. (7-43), we may also directly integrate a set of first-order ordinary differential equations with respect to x [16, 17]. This alternative is sometimes referred to as the Galerkin-Kantorovich method or the method of lines.

Before closing this section, we wish to demonstrate an algebraic derivation of an integral given in Table 7-3.

Example 7-7 Derive the final expression for the integral $F_{j-1,j}$ listed in Table 7-3.

Solution: For the purpose of linearizing Eq. (7-37), let

$$c = \frac{\nu\bar{u}(\omega)}{\psi_{EI}^2} = c^*\bar{u}(\omega), \tag{a}$$

where $c^* = \nu/\psi_{EI}^2$ and the overbar denotes the upstream velocity distribution. We then approximate $\bar{u}(\omega)$ as

$$\bar{u}(\omega) = N_{j-1}(\omega)u_{SW} + N_j(\omega)u_W, \qquad \omega \in [\omega_{j-1}, \omega_j]. \tag{b}$$

Therefore,

$$F_{j-1,j} = \int_{\omega_{j-1}}^{\omega_j} c\,\frac{dN_{j-1}}{d\omega}\frac{dN_j}{d\omega}\,d\omega = c^*u_{SW}\int_{\omega_{j-1}}^{\omega_j} N_{j-1}\,\frac{dN_{j-1}}{d\omega}\frac{dN_j}{d\omega}\,d\omega$$

$$+\, c^*u_W\int_{\omega_{j-1}}^{\omega_j} N_j\,\frac{dN_{j-1}}{d\omega}\frac{dN_j}{d\omega}\,d\omega. \tag{c}$$

In reference to Fig. 6-10, it is easy to find that

$$\frac{dN_{j-1}}{d\omega} = -\frac{1}{h_j}, \qquad \frac{dN_j}{d\omega} = \frac{1}{h_j},$$

and

$$\int_{\omega_{j-1}}^{\omega_j} N_{j-1}\,d\omega = \int_{\omega_{j-1}}^{\omega_j} N_j\,d\omega = \frac{h_j}{2}.$$

Thus, Eq. (*c*) becomes

$$F_{j-1,j} = -\frac{c^*}{2h_j}(u_{SW} + u_W), \tag{d}$$

which is the final expression for $F_{j-1,j}$.

7-2b Adjoint Variational Principle

In Chapters 1 and 5, the variational finite-element method was applied to the heat conduction equations, which rarely contain a first derivative with respect to the space coordinate. Under those circumstances, the variational principle corresponding to the linear differential equation normally exists. For example, the variational form of a second derivative can be derived as

$$-\int_0^1 \frac{d^2\phi}{d\omega^2}\,\delta\phi\; d\omega = \frac{\delta}{2}\int_0^1 \left(\frac{d\phi}{d\omega}\right)^2 d\omega, \tag{7-46a}$$

where Dirichlet boundary conditions have been assumed. Similarly, for ϕ itself and a constant k_1, we can also obtain

$$\int_0^1 \phi\;\delta\phi\; d\omega = \frac{\delta}{2}\int_0^1 \phi^2\; d\omega \tag{7-46b}$$

and

$$\int_0^1 k_1\;\delta\phi\; d\omega = k_1\delta\int_0^1 \phi\; d\omega. \tag{7-46c}$$

If a differential equation is a linear combination of $d^2\phi/d\omega^2$, ϕ, and k_1, e.g.,

$$L_1\phi = -a\frac{d^2\phi}{d\omega^2} + b\phi + k_1, \tag{7-47}$$

then the variational principle of $L_1\phi$ can be shown, according to Eqs. (7-46*a*)-(7-46*c*), to be

$$I(\phi) = \frac{a}{2}\int_0^1 \left(\frac{d\phi}{d\omega}\right)^2 d\omega + \frac{b}{2}\int_0^1 \phi^2\; d\omega + k_1\int_0^1 \phi\; d\omega. \tag{7-48}$$

In the present chapter we are dealing with the first derivatives $\partial\phi/\partial x$ and $\partial\phi/\partial\omega$, which originate from convective momentum transport. Unfortunately, the variational principle corresponding to a differential equation containing $\partial\phi/\partial x$ or $\partial\phi/\partial\omega$ does not exist (see Sections 1-2*c* and 2-1*c*). Therefore, a modified version of the variational principle must be used if we wish to adopt the variational method. This modified version is sometimes called the adjoint variational principle or the pseudovariational principle method [14, pp. 33-34] and will be presented next.

Referring to Eq. (7-37), we write the inner product

$$(A\phi, \delta\phi^*) = \int_0^L\int_0^1 \frac{\partial\phi}{\partial x}\,\delta\phi^*\; d\omega\; dx + \int_0^L\int_0^1 (a + b\omega)\frac{\partial\phi}{\partial\omega}\,\delta\phi^*\; d\omega\; dx \tag{7-49}$$

$$-\int_0^L \int_0^1 \frac{\partial}{\partial \omega}\left(c \frac{\partial \phi}{\partial \omega}\right) \delta\phi^* \, d\omega \, dx = 0, \tag{7-49 (cont.)}$$

where the domain of the x direction is assumed to be $[0, L]$; $\delta\phi^*$ is the variation of the adjoint function ϕ^*, which satisfies the adjoint differential equation

$$A^*\phi^* = -\frac{\partial \phi^*}{\partial x} - \frac{\partial}{\partial \omega}\left[(a + b\omega)\phi^*\right] - \frac{\partial}{\partial \omega}\left(c \frac{\partial \phi^*}{\partial \omega}\right) = 0. \tag{7-50}$$

We note that the signs of the first derivatives are changed and that the variable coefficient $(a + b\omega)$ is moved inside the differentiation. Equation (7-49) can be further integrated by parts to become

$$\int_0^L \int_0^1 \frac{\partial \phi}{\partial x} \delta\phi^* \, d\omega \, dx + \int_0^L \int_0^1 (a + b\omega) \frac{\partial \phi}{\partial \omega} \delta\phi^* \, d\omega \, dx$$

$$+ \int_0^L \int_0^1 c \frac{\partial \phi}{\partial \omega} \delta\left(\frac{\partial \phi^*}{\partial \omega}\right) d\omega \, dx = 0, \tag{7-51}$$

in which Dirichlet boundary conditions have been assumed at $\omega = 0, 1$. Similarly, from Eq. (7-50), we obtain the inner product

$$(A^*\phi^*, \delta\phi) = -\int_0^L \int_0^1 \frac{\partial \phi^*}{\partial x} \delta\phi \, d\omega \, dx - \int_0^L \int_0^1 \frac{\partial}{\partial \omega}\left[(a + b\omega)\phi^*\right] \delta\phi \, d\omega \, dx$$

$$-\int_0^L \int_0^1 \frac{\partial}{\partial \omega}\left(c \frac{\partial \phi^*}{\partial \omega}\right) \delta\phi \, d\omega \, dx = 0 \tag{7-52}$$

or, by means of integration by parts,

$$\int_0^L \int_0^1 \phi^* \, \delta\left(\frac{\partial \phi}{\partial x}\right) d\omega \, dx + \int_0^L \int_0^1 (a + b\omega)\phi^* \, \delta\left(\frac{\partial \phi}{\partial \omega}\right) d\omega \, dx$$

$$+ \int_0^L \int_0^1 c \frac{\partial \phi^*}{\partial \omega} \delta\left(\frac{\partial \phi}{\partial \omega}\right) d\omega \, dx - \left[\int_0^1 \phi^* \, \delta\phi \, d\omega\right]_{x=0}^{L} = 0. \tag{7-53a}$$

It is desirable that the last boundary-condition term be eliminated. At $x = 0$, if ϕ is prescribed, then $\delta\phi$ vanishes. At $x = L$, however, ϕ is not known. One way to make the boundary-condition term vanish there is to set $\phi^* = 0$ at $x = L$. This suggests that the integration of the adjoint equation with respect to x must be performed in the opposite direction to the integration of the primary equation. As a result, Eq. (7-53*a*) becomes

$$\int_0^L \int_0^1 \phi^* \,\delta\left(\frac{\partial \phi}{\partial x}\right) d\omega\, dx + \int_0^L \int_0^1 (a + b\omega)\phi^* \,\delta\left(\frac{\partial \phi}{\partial \omega}\right) d\omega\, dx$$

$$+ \int_0^L \int_0^1 c \frac{\partial \phi^*}{\partial \omega} \,\delta\left(\frac{\partial \phi}{\partial \omega}\right) d\omega\, dx = 0. \tag{7-53b}$$

Finally, Eqs. (7-51) and (7-53*b*) can be added to yield

$$\delta I(\phi, \phi^*) = 0, \tag{7-54}$$

where $I(\phi, \phi^*)$, which is a functional of ϕ and ϕ^*, takes the form

$$I(\phi, \phi^*) = \int_0^L \int_0^1 \frac{\partial \phi}{\partial x} \phi^* \, d\omega\, dx + \int_0^L \int_0^1 (a + b\omega) \frac{\partial \phi}{\partial \omega} \phi^* \, d\omega\, dx$$

$$+ \int_0^L \int_0^1 c \frac{\partial \phi}{\partial \omega} \frac{\partial \phi^*}{\partial \omega} \, d\omega\, dx. \tag{7-55}$$

Instead of solving the original differential equation (7-37), we now seek the functions ϕ and ϕ^* such that Eq. (7-55) is made stationary. Analytically, it may be difficult to find the exact solution that makes $I(\phi, \phi^*)$ stationary and has integrable first derivatives. Numerically, however, it is relatively easy to do so. The procedure involves allowing this approximate solution to belong to a certain finite-element space. We approximate $\phi(x, \omega)$ and $\phi^*(x, \omega)$ by

$$\tilde{\phi} = \sum_{j=0}^{J} N_j(\omega)\phi_j(x) \tag{7-56}$$

and

$$\tilde{\phi}^* = \sum_{j=0}^{J} N_j(\omega)\phi_j^*(x), \tag{7-57}$$

in which the basis function $N_j(\omega)$ was defined in Eqs. (7-36*a*)-(7-36*c*). With these approximations, Eq. (7-54) can be rewritten to stress the dependence of I on the $2(J-1)$ nodal functions as

$$\delta I(\phi_1, \phi_2, \ldots, \phi_{J-1}, \phi_1^*, \phi_2^*, \ldots, \phi_{J-1}^*) = 0 \tag{7-58}$$

or

$$\frac{\partial I}{\partial \phi_1}\,\delta\phi_1 + \frac{\partial I}{\partial \phi_2}\,\delta\phi_2 + \cdots + \frac{\partial I}{\partial \phi_{J-1}}\,\delta\phi_{J-1} + \frac{\partial I}{\partial \phi_1^*}\,\delta\phi_1^* + \frac{\partial I}{\partial \phi_2^*}\,\delta\phi_2^* + \cdots + \frac{\partial I}{\partial \phi_{J-1}^*}\,\delta\phi_{J-1}^* = 0. \quad (7\text{-}59)$$

Since all the variations of the functions are arbitrary, Eq. (7-59) implies that

$$\frac{\partial I}{\partial \phi_1} = \frac{\partial I}{\partial \phi_2} = \cdots = \frac{\partial I}{\partial \phi_{J-1}} = \frac{\partial I}{\partial \phi_1^*} = \frac{\partial I}{\partial \phi_2^*} = \cdots = \frac{\partial I}{\partial \phi_{J-1}^*} = 0. \quad (7\text{-}60)$$

Consequently, a system of $2(J-1)$ algebraic equations with $2(J-1)$ nodal unknowns is generated. The first set of $J-1$ equations leads to the solution of the adjoint equation (7-50), whereas the second set leads to the solution of the primary equation (7-37). Fortunately, these two sets of equations are decoupled. We will demonstrate the derivation of the jth algebraic equation in the following example.

Example 7-8 Consider the functional $I(\phi, \phi^*)$ defined in Eq. (7-55). Start with the extremal condition $\partial I/\partial \phi_j^* = 0$ and use the approximations (7-56) and (7-57) to derive the jth three-point algebraic equation.

Solution: It is convenient to subdivide the integration range of ω into

$$I(\phi, \phi^*) = \int_0^L \int_{\omega_{j-1}}^{\omega_{j+1}} \frac{\partial \phi}{\partial x}\,\phi^*\,d\omega\,dx + \int_0^L \int_{\omega_{j-1}}^{\omega_{j+1}} (a + b\omega)\frac{\partial \phi}{\partial \omega}\,\phi^*\,d\omega\,dx + \int_0^L \int_{\omega_{j-1}}^{\omega_{j+1}} c\,\frac{\partial \phi}{\partial \omega}\frac{\partial \phi^*}{\partial \omega}\,d\omega\,dx + R, \quad (a)$$

where R represents the integrals that do not contain ϕ_j^*. Therefore,

$$\frac{\partial I}{\partial \phi_j^*} = \int_0^L \int_{\omega_{j-1}}^{\omega_{j+1}} \frac{\partial \phi}{\partial x}\frac{\partial \phi^*}{\partial \phi_j^*}\,d\omega\,dx + \int_0^L \int_{\omega_{j-1}}^{\omega_{j+1}} (a + b\omega)\frac{\partial \phi}{\partial \omega}\frac{\partial \phi^*}{\partial \phi_j^*}\,d\omega\,dx + \int_0^L \int_{\omega_{j-1}}^{\omega_{j+1}} c\,\frac{\partial \phi}{\partial \omega}\frac{\partial}{\partial \omega}\left(\frac{\partial \phi^*}{\partial \phi_j^*}\right)d\omega\,dx. \quad (b)$$

But Eq. (7-57) indicates that

$$\frac{\partial \phi^*}{\partial \phi_j^*} = N_j(\omega). \quad (c)$$

Substitution of Eq. (7-35) and Eq. (c) into Eq. (b) leads to

$$\int_0^L \left[\left(\Lambda_{j-1,j} \frac{d\phi_{j-1}}{dx} + \Lambda_{j,j} \frac{d\phi_j}{dx} + \Lambda_{j,j+1} \frac{d\phi_{j+1}}{dx} \right) \right.$$

$$+ (D_{j-1,j}\phi_{j-1} + E_j\phi_j + D_{j,j+1}\phi_{j+1})$$

$$\left. + (F_{j-1,j}\phi_{j-1} + G_j\phi_j + F_{j,j+1}\phi_{j+1}) \right] dx = 0. \tag{d}$$

If the upper and lower bounds of the integral are replaced by x_j and x_{j-1}, respectively, Eq. (*d*) can be simplified to

$$A\phi_{j-1} + B\phi_j + C\phi_{j+1} = D,$$

which is identical to Eq. (7-44) derived by the Galerkin method.

7-3 STREAMWISE PRESSURE GRADIENT

In Chapter 6 and in the present chapter, we have considered the momentum transport of boundary-layer flow over a flat impermeable plate. Although this problem is classical and deserves attention, it is slightly idealized. In reality, objects immersed in a fluid are often nonflat and, in many cases, transverse mass transfer such as evaporation and condensation takes place at the solid surfaces. It is therefore our objective in this section to describe numerical methods that can be efficiently applied to a system with nonzero streamwise pressure gradient. The topic of nonzero mass transfer across the solid surface will be considered in the next section.

When fluids flow over a nonflat object, the stream function $\psi(x, y)$ can be obtained from potential flow theory [18] and thereby the flow velocity in the free stream can be deduced. Generally, this flow velocity becomes a function of the space coordinates and, from Bernoulli's equation, the streamwise pressure gradient dp/dx becomes nonzero.

The partial differential equations governing steady-state, two-dimensional, incompressible, forced-convection, laminar, boundary-layer flow with constant properties now take the form

$$\frac{\partial u}{\partial x} + \frac{\partial v}{\partial y} = 0 \tag{7-61}$$

and

$$u \frac{\partial u}{\partial x} + v \frac{\partial u}{\partial y} = \nu \frac{\partial^2 u}{\partial y^2} - \frac{1}{\rho} \frac{dp}{dx}, \tag{7-62}$$

where

$$\frac{1}{\rho} \frac{dp}{dx} = - u_\infty \frac{du_\infty}{dx}.$$

Equations (7-61) and (7-62) can be numerically solved by a number of methods [19–24], among which three will be described subsequently.

7-3*a* Parametric Expansion Method

Under the assumption that the free-stream velocity u_∞ can be prescribed as a certain power of x, i.e.,

$$u_\infty = Ux^m, \tag{7-63}$$

where U is a constant and m is a real number ranging from -0.091 to 1 [25], it is possible to transform Eqs. (7-61) and (7-62) into an ordinary differential equation [25] by using the similarity variable η

$$\eta = \frac{y}{x}\,\mathrm{Re}_x^{1/2} \qquad \text{and } \mathrm{Re}_x = \frac{u_\infty x}{\nu}, \tag{7-64}$$

with the similarity function $f(\eta)$ defined as

$$f(\eta) = \psi(x, y)(\nu x u_\infty)^{-1/2} \tag{7-65}$$

and the transformation differentiation defined as

$$\left(\frac{\partial}{\partial x}\right)_y = \left(\frac{\partial}{\partial x}\right)_\eta + \frac{(m-1)\eta}{2x}\frac{\partial}{\partial \eta} \quad \text{and} \quad \frac{\partial}{\partial y} = \left(\frac{u_\infty}{x\nu}\right)^{1/2}\frac{\partial}{\partial \eta}. \tag{7-66a, b}$$

The ordinary differential equation, called the Falkner-Skan equation after its two originators, is rearranged to

$$f''' + \tfrac{1}{2}ff'' + m(1 - f'^2 + \tfrac{1}{2}ff'') = 0 \tag{7-67}$$

subject, for impermeable surfaces, to the boundary conditions

$$f(0) = 0, \qquad f'(0) = 0, \qquad \text{and } f'(\eta_\infty) = 1. \tag{7-68}$$

Some useful relations between the primitive functions and the similarity functions are listed in Table 7-4, which is constructed according to Eqs. (7-64)-(7-66). Referring to this table, readers can derive Eq. (7-67) from Eqs. (7-61) and (7-62). When solutions for a wide range of the parametric value m are sought, it is convenient to adopt the parametric expansion method described in Section 6-2*b*. The well-documented solution at $m = 0$–i.e., $u_\infty = U$–can naturally be used as the starting solution.

7-3*b* Perturbation Method

When the object immersed in the fluid is nearly flat–i.e., $m \approx 0$–the perturbation method can be used. To proceed, we assume that the function $f(\eta)$ can be expanded in terms of an asymptotic series as

$$f(\eta) = f_0(\eta) + mf_1(\eta) + m^2 f_2(\eta) + o(m^2), \tag{7-69}$$

Table 7-4 Useful relations between the primitive functions and the Falkner-Skan similarity functions

Primitive function	Similarity function
$u(x, y)$	$u_\infty f'(\eta)$
$v(x, y)$	$u_\infty \mathrm{Re}_x^{-1/2} \dfrac{(1-m)\eta f'(\eta)-(1+m)f(\eta)}{2}$
$\dfrac{\partial u(x, y)}{\partial x}$	$\dfrac{mf'(\eta)-(1-m)\eta f''(\eta)}{2}\dfrac{u_\infty}{x}$
$\dfrac{\partial u(x, y)}{\partial y}$	$\left(\dfrac{u_\infty^3}{\nu x}\right)^{1/2} f''(u)$
$\dfrac{\partial^2 u(x, y)}{\partial y^2}$	$\left(\dfrac{u_\infty^2}{\nu x}\right)^{1/2} f'''(\eta)$
$\psi(x, y)$	$(\nu x u_\infty)^{1/2} f(\eta)$

where $\mathbf{o}(t)$ stands for a value much smaller than t. Substituting Eq. (7-69) into Eq. (7-67) and collecting terms with the same power of m, we derive

$$f_0''' + \tfrac{1}{2} f_0 f_0'' = 0, \tag{7-70a}$$

$$f_1''' + \tfrac{1}{2} f_0 f_1'' + \tfrac{1}{2} f_0'' f_1 + 1 - f_0'^2 + \tfrac{1}{2} f_0 f_0'' = 0, \tag{7-70b}$$

and

$$f_2''' + \tfrac{1}{2} f_0 f_2'' + \tfrac{1}{2} f_0'' f_2 + \tfrac{1}{2} f_1 f_1'' + \tfrac{1}{2} f_0 f_1'' - 2f_0' f_1' + \tfrac{1}{2} f_0'' f_1 = 0. \tag{7-70c}$$

The corresponding boundary conditions for $f_0(\eta)$, $f_1(\eta)$, and $f_2(\eta)$ can also be obtained from Eqs. (7-68) and (7-69) as

$$f_0(0) = 0, \quad f_0'(0) = 0, \quad f_0'(\eta_\infty) = 1, \tag{7-71a}$$

$$f_1(0) = 0, \quad f_1'(0) = 0, \quad f_1'(\eta_\infty) = 0, \tag{7-71b}$$

and

$$f_2(0) = 0, \quad f_2'(0) = 0, \quad f_2'(\eta_\infty) = 0. \tag{7-71c}$$

Although the number of governing differential equations is tripled, we do gain the following advantages: (1) Eq. (7-70*a*) is exactly the Blasius similarity equation, which we fortunately do not need to solve; (2) Eq. (7-70*b*) is a linear equation in $f_1(\eta)$ since $f_0(\eta)$ and its derivatives are known; and (3) Eq. (7-70*c*) becomes a linear equation in $f_2(\eta)$ once $f_1(\eta)$ and its derivatives are obtained. From Section 3-4*a*, we know that it is relatively easy to solve linear two-point boundary-value problems.

Example 7-9 Assume that $f_0(\eta)$ is given. Transform the two-point boundary-value problem of Eqs. (7-70*b*) and (7-71*b*) into an initial-value problem.

Solution: Since Eq. (7-70*b*) is linear, the function $f_1(\eta)$ can be expressed in terms of two auxiliary functions as

$$f_1(\eta) = \alpha p(\eta) + q(\eta), \tag{a}$$

where α is an arbitrary constant to be determined. Substituting Eq. (a) into Eq. (7-70b) and grouping terms with and without α, we obtain

$$p''' + \tfrac{1}{2} f_0 p'' + \tfrac{1}{2} f_0'' p = 0 \tag{b}$$

and
$$q''' + \tfrac{1}{2} f_0 q'' + \tfrac{1}{2} f_0'' q + 1 - f_0'^2 + \tfrac{1}{2} f_0 f_0'' = 0. \tag{c}$$

Furthermore, if α is chosen as $f_1''(0)$, then, according to Eq. (7-71b) and Eq. (a), we should have the following initial conditions for p and q:

$$p(0) = 0, \quad p'(0) = 0, \quad p''(0) = 1, \tag{d}$$

and
$$q(0) = 0, \quad q'(0) = 0, \quad q''(0) = 0. \tag{e}$$

Equations (b) and (c) with the known initial conditions (d) and (e) can thus be integrated forward to yield solutions at the terminal point $\eta = \eta_\infty$. From these solutions and Eq. (a), the arbitrary constant α can be determined from

$$\alpha = \frac{1 - q'(\eta_\infty)}{p'(\eta_\infty)}. \tag{f}$$

Once α is evaluated, Eq. (7-70b) becomes an initial-value problem with $f_1(0) = f_1'(0) = 0$ and $f_1''(0) = \alpha$ as the prescribed initial conditions.

Finally with regard to the perturbation method, we wish to make the following remarks (see also Section 3-3):

1. When the parameter embedded in the differential equation is $\approx \mathbf{O}(1)$, the perturbation method may not be used since an asymptotic series such as Eq. (7-69) no longer exists.
2. When the dependent variable $f(\eta)$ can be expressed in terms of a single asymptotic series throughout the field of interest, the method is called the regular perturbation method. Generally, this case arises when the perturbed parameter is contained in lower-order derivatives [26]. The regular perturbation method is relatively easy to apply, as previously demonstrated.
3. When the dependent variable $f(\eta)$ must be represented in terms of both an inner-expansion and an outer-expansion asymptotic series, the method is called the singular perturbation method. Such problems often arise when the perturbed parameter is embedded in the highest-order derivative. Two methods, called the method of matched asymptotic expansions and the method of strained coordinates, have been developed to solve singular perturbation problems. Interested readers may refer to [26, 27] for details.
4. Some researchers modified Eqs. (7-61)–(7-63) to account for the curvature effect of the cambered surfaces. These equations are referred to as higher-order boundary-layer equations and are generally solved by using the perturbation method [28–30].

7-3c Finite-Difference Method

The finite-difference method with the x-ω transformation [31] can also be used to solve Eqs. (7-61) and (7-62), which, with the aid of Table 6-7, can be transformed into

$$\frac{\partial u}{\partial x} + (a + b\omega)\frac{\partial u}{\partial \omega} = \frac{\partial}{\partial \omega}\left(c\,\frac{\partial u}{\partial \omega}\right) + S(u), \tag{7-72}$$

where

$$S(u) = -\frac{1}{\rho u}\frac{dp}{dx}.$$

Upon discretization of Eq. (7-72), we may also wish to linearize $S(u)$. One linearization method is to expand it about u_W (upstream flow velocity) in a Taylor's series as

$$S(u) = S(u_W) + \left(\frac{dS}{du}\right)_W (u - u_W) = \frac{u_\infty}{u_W}\frac{du_\infty}{dx}\left(2 - \frac{u}{u_W}\right). \tag{7-73}$$

With Eq. (7-73), the final finite-difference approximation of Eq. (7-72) can be obtained by slightly modifying Eq. (6-42) to yield

$$Au_{j-1} + B^*u_j + Cu_{j+1} = D^*, \quad j = 2, 3, \ldots, J-1, \tag{7-74}$$

where

$$B^* = B + \frac{u_\infty}{u_W^2}\frac{du_\infty}{dx} \quad \text{and } D^* = D + \frac{2u_\infty}{u_W}\frac{du_\infty}{dx}.$$

Special attention must be paid again to the discretization at $j = 1$. In Section 6-3, since u is proportional to y when $y \approx 0$, the basis function $N_1(\omega)$ was derived as $(\omega/\omega_1)^{1/2}$ if $0 \leqslant \omega \leqslant \omega_1$. When a nonzero streamwise pressure gradient exists in the boundary layer, the velocity u near the wall is not necessarily proportional to y. We will discuss this problem in the following section.

7-4 MASS TRANSFER ACROSS A SOLID SURFACE

When the surface to which the boundary layer is attached is porous, the fluid stored within the solid or flowing inside the boundary layer is permitted to seep through the surface. In this situation, the transverse flow velocity v becomes nonzero at the surface. If the injected fluid happens to be of the same kind as that in the boundary layer, then only the boundary condition of the momentum differential equation should be changed. In the present section we will focus on this case. If the injected fluid is different from the ambient fluid so that the boundary-layer flow becomes a multispecies system, additional differential equations governing the conservation of species must be considered simultaneously. We will postpone consideration of multispecies systems until Chapter 9, where species diffusion is coupled with energy transport. In the following section, a few numerical methods are presented to solve Eqs. (6-4) and (6-5) subject to the boundary condition of nonzero transverse flow velocity.

7-4*a* Parametric Expansion Method

As mentioned at the beginning of Section 6-3, the elegant similarity equations exist only for a limited number of physical problems. If the mass transfer parameter λ defined as $v_w \mathrm{Re}_x^{1/2}/u_\infty$ (note that the subscript w denotes wall, not west), is free of x dependence, then we are able to reduce Eqs. (6-4) and (6-5) to

$$f''' + \tfrac{1}{2} ff'' = 0 \tag{7-75}$$

subject to

$$f(0) = -2\lambda, \quad f'(0) = 0, \quad \text{and } f'(\infty) = 1, \tag{7-76}$$

where the prime denotes differentiation with respect to η defined as $y\,\mathrm{Re}_x^{1/2}/x$. If λ is a function of x, then the similarity function f is not a function of η alone and derivatives in x such as $\partial f'/\partial x$ will appear in Eq. (7-75). The elegance of the similarity is thus destroyed. For convenience here, we will follow the classical assumption that v_w is proportional to $x^{-1/2}$. For the treatment of nonsimilar flows, see Section 11-2*b*.

Equations (7-75) and (7-76) constitute a two-point boundary-value problem, which can be solved by many numerical techniques [32]. When solutions for a wide range of λ values are desired, however, the parametric expansion method proves most convenient. With regard to Eqs. (7-75) and (7-76), since the Blasius solution at $\lambda = 0$ has already been well documented, we can start expanding the unknown solution at $\lambda = \Delta\lambda$ in Taylor's series about $\lambda = 0$ and generate solutions at other λ values. The procedures for the parametric expression method are listed in Table 6-4. Taking advantage of the availability of the Blasius solution, we can skip steps 1 and 2. Only step 5, which involves finding the initial conditions for g and h, deserves additional attention. According to Eq. (7-76), we deduce

$$g(0, \lambda) = \frac{\partial f(0, \lambda)}{\partial \lambda} = -2 \tag{7-77a}$$

and

$$h(0, \lambda) = \frac{\partial g(0, \lambda)}{\partial \lambda} = 0. \tag{7-77b}$$

Consequently, the corresponding initial conditions for the two auxiliary functions are

$$g_1(0, \lambda) = 0, \quad g_1'(0, \lambda) = 0, \quad g_1''(0, \lambda) = 1 \tag{7-78a}$$

and

$$g_2(0, \lambda) = -2, \quad g_2'(0, \lambda) = 0, \quad g_2''(0, \lambda) = 0. \tag{7-78b}$$

The missing initial conditions for the parametric values within $-2.5 \leqslant \lambda \leqslant 0.619$ (separation) are given in Table 7-5. The results reported in [33] are included for comparison.

7-4*b* Integral Method

The integral method initiated by von Karman and Pohlhausen [34] is a well-known approximation technique in which the boundary-layer thickness generally becomes the dependent variable in the ordinary differential equation reduced from the original partial differential equation [35–38]. This technique is well described in most heat transfer textbooks and therefore will be presented here briefly with emphasis on accounting for the mass transfer effect. The governing integral momentum equation with mass transfer through the solid surface can be derived as

$$\frac{d}{dx}\int_0^\delta (u_\infty - u)\,u\,dy = \nu\left(\frac{\partial u}{\partial y}\right)_0 + v_w u_\infty, \tag{7-79}$$

where δ is the boundary-layer thickness. If it is assumed that the normalized velocity u/u_∞ can be expressed as a function of y/δ, Eq. (7-79) can be rewritten as

$$d_1\frac{d\delta}{dx} = \frac{d_2\nu}{u_\infty\delta} + \frac{\lambda}{\mathrm{Re}_x^{1/2}}, \tag{7-80}$$

where

$$d_1 = \int_0^1 [1 - F(\xi)]F(\xi)\,d\xi, \quad d_2 = \left(\frac{dF}{d\xi}\right)_0, \quad F(\xi) = \frac{u}{u_\infty}, \quad \text{and } \xi = \frac{y}{\delta}.$$

Table 7-5 Missing initial conditions obtained by using the parametric expansion method and a classical shooting method for various parametric values λ

λ	$f''(0)$ by parametric expansion	$f''(0)$ in [33]
0.619	0	0
0.50	0.043607	0.0356
0.375	0.09369	0.0937
0.250	0.164952	0.165
0	0.33206	0.332
−0.250	0.52303	0.523
−0.750	0.94511	0.945
−2.50	2.59054	2.59

In the case of impermeable surfaces, a typical velocity profile is assumed to be $F(\xi) = 1.5\xi - 0.5\xi^3$, which is determined by $F(0) = F''(0) = F'(1) = 0$ and $F(1) = 1$. For problems with nonzero mass transfer across the surface, the condition $F''(0) = 0$ no longer holds. It is essential to determine how this condition should be modified to account for the mass transfer effect. Consider the Couette flow ($\partial/\partial x \approx 0$) near the wall. The momentum equation is reduced to

$$v_w \frac{\partial u}{\partial y} \approx \nu \frac{\partial^2 u}{\partial y^2} \quad \text{at } y \approx 0 \tag{7-81}$$

or

$$v_w \left(\frac{dF}{d\xi}\right)_0 \approx \frac{\nu}{\delta}\left(\frac{d^2F}{d\xi^2}\right)_0 . \tag{7-82}$$

Since we are attempting to determine a well-assumed profile, it may be acceptable to assume $\delta \approx 5x\,\mathrm{Re}_x^{1/2}$ temporarily, which is derived from the case of an impermeable surface. A formal solution for δ is obtained by solving Eq. (7-80). Hence, Eq. (7-82) reduces to

$$5\lambda\left(\frac{dF}{d\xi}\right)_0 = \left(\frac{d^2F}{d\xi^2}\right)_0 \tag{7-83}$$

and the assumed velocity profile should satisfy the following conditions:

$$F(0) = 0, \qquad \text{Eq. (7-83)}, \qquad F(1) = 1, \qquad \text{and } F'(1) = 0. \tag{7-84}$$

After straightforward algebra, a cubic polynomial can be determined by using Eq. (7-84):

$$F(\xi) = \left(\frac{6}{4+5\lambda}\right)\xi + \left(\frac{15\lambda}{4+5\lambda}\right)\xi^2 - \left(\frac{2+10\lambda}{4+5\lambda}\right)\xi^3 . \tag{7-85}$$

At $\lambda = 0$, Eq. (7-85) reduces to

$$F(\xi) = 1.5\xi - 0.5\xi^3$$

as expected. Once the assumed velocity profile is obtained, Eq. (7-80), which is a first-order ordinary differential equation, can be integrated to yield the solution $\delta(x)$. Through the relation

$$\tau_w = \mu\left(\frac{\partial u}{\partial y}\right)_0 = \frac{\mu u_\infty}{\delta}\left(\frac{dF}{d\xi}\right)_0 , \tag{7-86}$$

the shear stress at the wall can thus be obtained.

7-4*c* Finite-Difference Method

In Section 6-3*a* we mentioned that the discretized three-point algebraic equation at grid point $j = 1$ takes a different expression from those at $j = 2, 3, \ldots, J-1$. The reason is that $u \propto y \propto \omega^{1/2}$ near an impermeable wall. When the fluid is allowed to cross the wall, Eq. (6-45) may be invalid. Therefore, in the present section we will seek a reasonable approximation $u(\omega)$ on the interval $[0, \omega_1]$.

Let us first assume that the transverse flow velocity v is fairly constant along y if $y \approx 0$. The justification for this assumption is presented in Example 7-11. Then Eq. (7-81) can be integrated analytically to yield

$$\frac{u}{u_1} = \frac{1 - e^{R_1 y/y_1}}{1 - e^{R_1}}, \tag{7-87}$$

where

$$R_1 = \frac{v_w y_1}{\nu}.$$

If $R_1 \ll 1$, i.e., $y_1 \ll x/\lambda \, \mathrm{Re}_x^{1/2}$, then Eq. (7-87) can be simplified to

$$\frac{u}{u_1} \approx \frac{2}{2 + R_1}\left(\frac{y}{y_1}\right) + \frac{R_1}{2 + R_1}\left(\frac{y}{y_1}\right)^2, \tag{7-88}$$

where the approximation $e^x \approx 1 + x + x^2/2 + \cdots$ has been used. Next, using the definition of ω, we can obtain a relation between ω and y as

$$\frac{\omega}{\omega_1} = (2 + R_1)\int_0^{y/y_1} \frac{1 - e^{R_1 z}}{1 - e^{R_1}}\, dz$$

or

$$\frac{\omega}{\omega_1} \approx \left(\frac{y}{y_1}\right)^2, \tag{7-89}$$

where

$$\omega_1 = \frac{u_1 y_1}{\psi_{EI}(2 + R_1)}.$$

A by-product of Eq. (7-89) is

$$y_1 = \frac{\omega_1 \psi_{EI}(2 + R_1)}{u_1}, \tag{7-90}$$

which becomes identical to Eq. (6-56) if $R_1 = 0$. Eliminating y/y_1 from Eqs. (7-88) and (7-89) leads to

$$\frac{u}{u_1} = \frac{2}{2 + R_1}\left(\frac{\omega}{\omega_1}\right)^{1/2} + \frac{R_1}{2 + R_1}\left(\frac{\omega}{\omega_1}\right), \tag{7-91}$$

which is reduced to Eq. (6-45) if $R_1 = 0$. Equation (7-91) represents an appropriate expression for the velocity distribution on the interval $[0, \omega_1]$. With this equation we can integrate Eq. (6-39) with respect to ω over the interval $[0, (\omega_1 + \omega_2)/2]$ to yield a two-point algebraic equation containing u_1 and u_2. The system of algebraic equations for nodal unknowns $u_1, u_2, \ldots, u_{J-1}$ is thus closed.

Example 7-10 Consider a boundary-layer flow with coexisting mass transfer at the wall and streamwise pressure gradient. What is a good assumption for the velocity distribution near the wall?

Solution: The momentum equation governing Couette flow near the wall is

$$v\,\frac{\partial u}{\partial y} \approx \nu\,\frac{\partial^2 u}{\partial y^2} - \frac{1}{\rho}\frac{dp}{dx}. \tag{a}$$

If v is assumed to be constant, Eq. (a) is a second-order ordinary differential equation, which can be analytically integrated to yield

$$\frac{u}{u_1} = \frac{1 - e^{R_1 y/y_1}}{1 - e^{R_1}} - \frac{y_1^2(dp/dx)}{\mu u_1 R_1}\left(\frac{y}{y_1}\right) \tag{b}$$

or, if $R_1 \ll 1$,

$$\frac{u}{u_1} = \left[\frac{2}{2+R_1} - \frac{y_1^2(dp/dx)}{\mu u_1 R_1}\right]\left(\frac{y}{y_1}\right) + \frac{R_1}{2+R_1}\left(\frac{y}{y_1}\right)^2, \tag{c}$$

which can be reduced to Eq. (7-88) if $dp/dx = 0$.

Example 7-11 Using the similarity solution listed in Table 7-6, assess the validity of the assumption that v remains nearly constant at $y \approx 0$. Let λ be 0.25.

Solution: Referring to Table 6-1, we obtain

$$\frac{v(x, y)}{v_w(x)} = \frac{u_\infty \mathrm{Re}_x^{-1/2}[\eta f'(\eta) - f(\eta)]/2}{u_\infty \mathrm{Re}_x^{-1/2}[-f(0)]/2} = \frac{\eta f'(\eta) - f(\eta)}{-f(0)}. \tag{a}$$

Based on Eq. (a) and the similarity solutions $f(\eta)$ and $f'(\eta)$ from Table 7-5, we are able to construct the rightmost column in Table 7-6. Judging the distribution in this column, we find that the variation of $v(\eta)$ is within 1% of $v(0)$ if $\eta \leqslant 0.22$. The assumption that the transverse flow velocity is approximately constant near the wall should be considered valid.

Example 7-12 In comparison with the velocity distribution listed in Table 7-6, is the approximation (7-88) valid?

Solution: Based on the definitions of R_1, η_1, and λ, we can derive

$$R_1 = \eta_1 \lambda. \tag{a}$$

For $\eta_1 = 0.22$ and $\lambda = 0.25$, we find $R_1 = 0.055 \ll 1$. Equation (7-88) can thus be rewritten as

Table 7-6 Similarity solution ($\eta \leqslant 0.22$) of the equation $f''' + \frac{1}{2} ff'' = 0$ with the mass transfer parameter $\lambda = 0.25$

η	$f(\eta)$	$f'(\eta)$	$f''(\eta)$	$f'''(\eta)$	$\frac{v(\eta)}{v(0)}$
0	−0.5	0	0.1650	0.04125	1
0.01	−0.5	0.00165	0.16541	0.04135	1.00003
0.02	−0.49998	0.00330	0.16583	0.04146	1.00010
0.04	−0.49992	0.00662	0.16666	0.04166	1.00036
0.06	−0.49979	0.00995	0.16749	0.04185	1.00076
0.1	−0.49932	0.01667	0.16917	0.04223	1.00197
0.14	−0.49865	0.02344	0.17086	0.04260	1.00387
0.18	−0.49772	0.03027	0.17256	0.04294	1.00633
0.22	−0.49650	0.03717	0.17428	0.04327	1.00937

$$\frac{u}{u_1} \approx 0.9732 \left(\frac{\eta}{\eta_1}\right) + 0.0268 \left(\frac{\eta}{\eta_1}\right)^2. \tag{b}$$

The velocity distribution for $0 \leqslant \eta \leqslant 0.22$ is computed from Eq. (*b*) and compared with the similarity solution in Table 7-7. We find that Eq. (7-88) is indeed a reasonable approximation.

7-5 A METHOD WITHOUT COORDINATE TRANSFORMATIONS

In Chapter 6 and in the preceding sections of the present chapter, the numerical methods introduced have been associated with either the similarity transformation or the von Mises transformation. Other types of transformations, of course, are also available [17, 39]. Generally, the purposes of a transformation in a boundary-layer problem are to suppress the role played by the transverse velocity v, to reduce (but not eliminate) the uncertainty in the boundary-layer thickness, and to construct a new domain of regular geometry. If the flow is of boundary-layer type, these transformations indeed simplify the mathematical formulation or the computational

Table 7-7 Comparison of the assumed velocity profile and the similarity solution, $\eta_1 = 0.22$

η	0	0.02	0.04	0.06	0.08	0.1	0.14	0.18	0.22
$\frac{u}{u_1}$ (similarity)	0	0.0888	0.1781	0.2677	0.3578	0.4485	0.6306	0.8144	1
$\frac{u}{u_1}$ [Eq. (*b*) of Example 7-12]	0	0.0887	0.1778	0.2674	0.3574	0.4479	0.6302	0.8142	1

process. When elliptic behavior exists in the governing equations, however, these transformations may no longer be incorporated. Therefore, it is useful to know of the availability of a numerical method that does not require coordinate transformation.

We will consider the following classical equations:

$$\frac{\partial u}{\partial x} + \frac{\partial v}{\partial y} = 0 \tag{6-4}$$

and

$$u\,\frac{\partial u}{\partial x} + v\,\frac{\partial u}{\partial y} - \nu\,\frac{\partial^2 u}{\partial y^2} = 0 \tag{6-5}$$

subject to the boundary conditions

$$u(x, 0) = v(x, 0) = 0, \qquad u(x, \infty) = u_\infty, \qquad u(x_0, y) = s_1(y)$$

and

$$v(x_0, y) = s_2(y), \tag{6-6}$$

where x_0 is a short distance measured from the leading edge of the boundary layer. Equations (6-4) and (6-5) can be either discretized first and linearized later, or linearized first and discretized later. We will choose the former sequence since it is algebraically simpler (see Section 3-4*c*). Further, we will adopt the control volume method and the upwind finite-difference scheme for the discretizations in y and x coordinates, respectively. In the control volume method (see Section 2-2), the governing equations are integrated over the intervals $[y_{j-1/2}, y_{j+1/2}]$ and the integrals are then evaluated by using the piecewise linear approximations for u and v. In the upwind scheme, we approximate

$$\frac{\partial u}{\partial x} \approx \frac{u - u_W}{\Delta x}$$

to ensure stability (see Section 4-1*b*). The resulting discretized nonlinear equations at nodal point j can be written as

$$\frac{\gamma}{8}\,u_{j-1} - \frac{1}{2}\,v_{j-1} + \frac{3\gamma}{4}\,u_j + \frac{\gamma}{8}\,u_{j+1} + \frac{1}{2}\,v_{j+1} = \gamma\left(\frac{1}{8}\,u_{SW} + \frac{3}{4}\,u_W + \frac{1}{8}\,u_{NW}\right), \tag{7-92}$$

where $\gamma = \Delta y/\Delta x$, and

$$H(u_k, v_k) - a_j u_{j-1} - b_j u_j - c_j u_{j+1} = 0, \qquad k = j-1, j, j+1, \tag{7-93}$$

where $H(u_k, v_k)$, which is a nonlinear expression, and the coefficients a_j, b_j, and c_j are listed in Table 7-8. Although these algebraic results are lengthy, we can partially check the algebra by assuming $u_{NW} = u_W = u_{SW} = u_{j-1} = u_j = u_{j+1}$ and then examining whether Eqs. (7-92) and (7-93) indeed vanish. For $j = 1$, Eqs. (7-92) and (7-93) are integrated over $[0, y_{3/2}]$. No difficulty regarding the singularity at the wall due to the von Mises transformation arises in this case.

The initial condition is customarily prescribed as

Table 7-8 Expressions for the coefficients given in Eqs. (7-92) and (7-93), $\mathrm{Re} = u_\infty L/\nu$

$H(u_k, v_k)$	$\frac{\gamma}{6}\left(\frac{1}{4}u_{j-1}^2 + \frac{7}{2}u_j^2 + \frac{1}{4}u_{j+1}^2 + u_{j-1}u_j + u_ju_{j+1}\right) + \frac{1}{8}[v_{j-1}(u_j - u_{j-1}) + 3v_j(u_{j+1} - u_{j-1}) + v_{j+1}(u_{j+1} - u_j)]$
a_j	$\frac{\gamma}{12}\left(\frac{1}{2}U_{SW} + U_W\right) + \frac{1}{\mathrm{Re}\,\Delta y}$
b_j	$\frac{\gamma}{12}(U_{SW} + 7U_W + U_{NW}) - \frac{2}{\mathrm{Re}\,\Delta y}$
c_j	$\frac{\gamma}{12}\left(U_j + \frac{1}{2}U_{j+1}\right) + \frac{1}{\mathrm{Re}\,\Delta y}$

$$\frac{u(x_0, y)}{u_\infty} = 1.5\left[\frac{y}{\delta(x_0)}\right] - 0.5\left[\frac{y}{\delta(x_0)}\right]^3, \tag{7-94}$$

where

$$\delta(x_0) = 7\left(\frac{\nu x_0}{u_\infty}\right)^{1/2}.$$

The next step is to seek the solution at $x = x_0 + \Delta x$. We write an approximate solution that satisfies the boundary conditions of Eq. (6-6), but not the discretized governing equations (7-92) and (7-93):

$$\frac{\tilde{u}_j}{u_\infty} = 1.5\left[\frac{y_j}{\delta(x_0 + \Delta x)}\right] - 0.5\left[\frac{y_j}{\delta(x_0 + \Delta x)}\right]^3, \tag{7-95}$$

where $y_j = j\Delta y$. Substituting Eq. (7-95) into Eq. (7-92) yields an approximate solution for $\tilde{v}_j$. These $\tilde{u}_j$ and $\tilde{v}_j$ are then substituted into Eq. (7-93) and, according to the parameterized-residuals method [40], an artificial parameter $1 - \lambda$ is embedded to yield

$$H(\tilde{u}_k, \tilde{v}_k) - a_j\tilde{u}_{j-1} - b_j\tilde{u}_j - c_j\tilde{u}_{j+1} = (1 - \lambda)R_j, \quad k = j-1, j, j+1. \tag{7-96}$$

Some selected nodal residuals R_j are listed in Table 7-9. Obviously, Eq. (7-95) is the solution of Eqs. (6-6) and (7-96) at $\lambda = 0$ and the true solution is that of the same equations at $\lambda = 1$. To relate the solution at $\lambda = 0$ to that at $\lambda = 1$, we expand $\tilde{u}_k$ and $\tilde{v}_k$ at $\lambda + \Delta\lambda$ about those at λ in Taylor's series as

$$\tilde{u}_k(\lambda + \Delta\lambda) = \tilde{u}_k(\lambda) + g_k(\lambda)\Delta\lambda + h_k(\lambda)\frac{(\Delta\lambda)^2}{2} \tag{7-97a}$$

and $$\tilde{v}_k(\lambda + \Delta\lambda) = \tilde{v}_k(\lambda) + p_k(\lambda)\Delta\lambda + q_k(\lambda)\frac{(\Delta\lambda)^2}{2}, \tag{7-97b}$$

where

$$g_k = \frac{d\tilde{u}}{d\lambda}, \quad h_k = \frac{d^2\tilde{u}}{d\lambda^2}, \quad p_k = \frac{d\tilde{v}_k}{d\lambda}, \quad \text{and } q_k = \frac{d^2\tilde{v}_k}{d\lambda^2}.$$

We may wonder whether the deletion of remainders in Eqs. (7-97*a*) and (7-97*b*) will cause noticeable errors. Table 7-9 lists such errors e_j, defined as the computed value of the right-hand side of Eq. (7-96) when the nodal solutions u_j and v_j are substituted into Eq. (7-96) at $\lambda = 1$. The fact that $|e_j|_{max}$ is only $\mathbf{O}(10^{-8})$ suggests that the deletion of remainders is relatively safe.

We will now attempt to compute u_k and v_k at $\lambda = \Delta\lambda$. Equations (7-92) and (7-96) can be linearized by applying parametric differentiation (with respect to λ) once to yield

$$[\mathbf{A}_j]\begin{Bmatrix} g_{j-1} \\ p_{j-1} \end{Bmatrix} + [\mathbf{B}_j]\begin{Bmatrix} g_j \\ p_j \end{Bmatrix} + [\mathbf{C}_j]\begin{Bmatrix} g_{j+1} \\ p_{j+1} \end{Bmatrix} = \begin{Bmatrix} 0 \\ -R_j \end{Bmatrix}. \tag{7-98}$$

The lengthy expressions for $[\mathbf{A}_j]$, $[\mathbf{B}_j]$, and $[\mathbf{C}_j]$ are listed in Table 7-10. They are 2×2 matrices which are functions of Δx, Δy, $\tilde{u}_k$, and $\tilde{v}_k$, with $k = j-1, j, j+1$, and not of g_k and p_k. Equation (7-98) is therefore a set of linear algebraic equations, whose coefficient matrix is tridiagonally banded and consists of

Table 7-9 Nodal residuals R_j at $\lambda = 0$ and nodal errors e_j at $\lambda = 1$

Grid point j	$R_j \times 10^4$	$\lvert e_j\rvert \times 10^8$
2	0.02537	0.04657
10	0.05612	0.09313
20	−0.25136	0.74506
30	−1.1983	1.1176
40	−2.7502	2.2352
50	−4.4771	1.1176
60	−5.6956	0.74506
80	−3.6047	2.2352
100	3.7386	0.74506

where

$$-r_j = \frac{da_{21}}{d\lambda}g_{j-1} + \frac{da_{22}}{d\lambda}p_{j-1} + \frac{db_{21}}{d\lambda}g_j + \frac{db_{22}}{d\lambda}p_j + \frac{dc_{21}}{d\lambda}g_{j+1} + \frac{dc_{22}}{d\lambda}p_{j+1}$$

Note that a_{21}, a_{22}, etc. are listed in Table 7-10.

Table 7-10 Expressions for A_j, B_j, and C_j given in Eq. (7-98)

$$\mathbf{A}_j = \begin{bmatrix} a_{11} & a_{12} \\ a_{21} & a_{22} \end{bmatrix}, \quad \mathbf{B}_j = \begin{bmatrix} b_{11} & b_{12} \\ b_{21} & b_{22} \end{bmatrix}, \quad \mathbf{C}_j = \begin{bmatrix} c_{11} & c_{12} \\ c_{21} & c_{22} \end{bmatrix},$$

where

$$a_{11} = \frac{\gamma}{8}, \quad a_{12} = -\frac{1}{2},$$

$$a_{21} = \frac{\gamma}{12} u_{j-1} + \frac{\gamma}{6} u_j - \frac{v_{j-1}}{8} - \frac{3}{8} v_j - a_j$$

$$a_{22} = \frac{u_j}{8} - \frac{u_{j-1}}{8},$$

$$b_{11} = \frac{3\gamma}{4}, \quad b_{12} = 0,$$

$$b_{21} = \frac{7}{6} \gamma u_j + \frac{\gamma}{6} (u_{j-1} + u_{j+1}) + \frac{v_{j-1}}{8} - \frac{v_{j+1}}{8} - b_j$$

$$b_{22} = \frac{3}{8} u_{j+1} - \frac{3}{8} u_{j-1},$$

$$c_{11} = \frac{\gamma}{8}, \quad c_{12} = \frac{1}{2},$$

$$c_{21} = \frac{\gamma}{12} u_{j+1} + \frac{\gamma}{6} u_j + \frac{v_{j+1}}{8} + \frac{3}{8} v_j - c_j$$

$$c_{22} = \frac{u_{j+1}}{8} - \frac{u_j}{8}.$$

a_j, b_j, and c_j are given in Table 7-8.

2×2 – submatrix elements. Such a system can be efficiently solved by using the Thomas algorithm modified for block systems (see Section 7-1*b*). Next, Eq. (7-98) can be differentiated with respect to λ once again to yield

$$[\mathbf{A}_j] \begin{Bmatrix} h_{j-1} \\ q_{j-1} \end{Bmatrix} + [\mathbf{B}_j] \begin{Bmatrix} h_j \\ q_j \end{Bmatrix} + [\mathbf{C}_j] \begin{Bmatrix} h_{j+1} \\ q_{j+1} \end{Bmatrix} = \begin{Bmatrix} 0 \\ r_j \end{Bmatrix}, \tag{7-99}$$

where r_j can also be found in Table 7-9. The modified Thomas algorithm is again used to calculate h_k and q_k. From Eqs. (7-97*a*) and (7-97*b*) solutions $\tilde{u}(x_0 + \Delta x, y, \Delta\lambda)$ and $\tilde{v}(x_0 + \Delta x, y, \Delta\lambda)$ can be obtained. These solutions will satisfy the boundary condition of Eq. (6-6) and the governing equation (7-96) with $\lambda = \Delta\lambda$ and, most likely, do not have any physical significance. Repeating the procedure from $2\Delta\lambda$ to 1, we finally obtain the solutions $u(x_0 + \Delta x, y, 1)$ and $v(x_0 + \Delta x, y, 1)$. These profiles then become the initial conditions for the downstream solutions at $x_0 + 2\Delta x$, which are subsequently sought. The step size Δx is chosen such that

$$\Delta x = \frac{\Delta y}{d\delta/dx}, \tag{7-100}$$

which ensures that the outermost grid point will lie on the edge of the boundary layer. Consequently, the number of grid points will increase by one per streamwise step. The numerical result thus computed converges to the Blasius similarity solution at $x \cong x_0 + \Sigma_{i=1}^{10} \Delta x_i$.

When a lengthy expression is derived, it is advisable to employ a quick method to check the algebra (the shortcoming of the parameterized-residuals method is the lengthy algebraic formulation). In the present situation, we may first assume u_k and v_g to be simple functions of λ, and then substitute these functions into Eqs. (7-92) and (7-93). If our algebraic manipulation is correct, the differentiated (with respect to λ) result should be the same as Eq. (7-98) after substitution of the corresponding $g_k(\lambda)$ and $p_k(\lambda)$.

Example 7-13 Set

$$u_{j-1} = v_{j-1} = \lambda, \quad u_j = v_j = \lambda^2, \quad \text{and } u_{j+1} = v_{j+1} = \lambda^3. \tag{a}$$

Substitute these functions into Eqs. (7-92) and (7-93), differentiate the expression with respect to λ, and examine whether the differentiated result is the same as Eq. (7-98) after substitution of

$$g_{j-1} = p_{j-1} = 1, \quad g_j = p_j = 2\lambda, \quad \text{and } g_{j+1} = p_{j+1} = 3\lambda^2. \tag{b}$$

Note that the derivatives of a_j, b_j and c_j equal zero.

Solution: We rearrange Eqs. (7-92) and (7-93) into

$$A_1 = \gamma u_{j-1} - 4v_{j-1} + 6\gamma u_j + \gamma u_{j+1} + 4v_{j+1} - \gamma(u_{SW} + 6u_W + u_{NW}) \tag{c}$$

and

$$A_2 = H(u_k, v_k) - a_j u_{j-1} - b_j u_j - c_j u_{j+1}. \tag{d}$$

Substituting Eq. (*a*) into Eqs. (*c*) and (*d*) and differentiating the result, we obtain

$$\frac{dA_1}{d\lambda} = \gamma(1 + 12\lambda + 3\lambda^2) + 12\lambda^2 - 4 \tag{e}$$

and

$$\frac{dA_2}{d\lambda} = \frac{\gamma}{6}\left(\frac{\lambda}{2} + 3\lambda^2 + 14\lambda^3 + 5\lambda^4 + \frac{3}{2}\lambda^5\right) + \frac{1}{8}(-2\lambda - 6\lambda^2 + 10\lambda^4 + 6\lambda^5)$$
$$- a_j - 2\lambda b_j - 3\lambda^2 c_j. \tag{f}$$

On the other hand, substitution of Eqs. (*a*) and (*b*) into Eq. (7-98) yields the identical result as given in Eqs. (*e*) and (*f*). Note that this consistency is only a necessary condition and not a sufficient condition for a correct algebraic manipulation.

Despite the lengthy mathematical formulation, the parameterized-residuals method has several merits. First, it is applicable to both ordinary and partial differential equations, while the parametric expansion method is only available for ordinary ones. Second, for nonlinear problems that are sensitive to missing initial conditions, it can yield accurate results due to inclusion of the second-order term in the Taylor's series expansion. Such an accuracy does not seem to exist in the parametric differentiation method. Third, the discretization that is part of the parameterized-residuals procedure can be performed by using finite-difference, finite-element, splines, or other discretization schemes. Therefore, the method is considered to be of broader scope than the discretization schemes, and not parallel to them. Finally, once a computer program is written, the subroutines could be repeatedly used for most problems with minor modifications.

SYMBOLS

a	term defined in x-ω momentum equation $[= -(d\psi_I/dx)/\psi_{EI}]$, m^{-1}
$B_j(x)$	cubic spline (Fig. 7-2*b*)
c	term defined in x-ω momentum equation $(= \nu u/\psi_{EI}^2)$, m^{-1}
c^*	convenient term $(= \nu/\psi_{EI}^2)$, s/m
d_1	constant in the integral method $\{= \int_0^1 [1 - F(\xi)]F(\xi)\,d\xi\}$
d_2	constant in the integral method $[= (dF/d\xi)_0]$
f_0, f_1, f_2	perturbation functions [Eq. (7-69)]
$F(\xi)$	normalized velocity profile $(= u/u_\infty)$ used in the integral method
h_j	grid interval $(= x_j - x_{j-1})$
L	differential operator or length of the plate, m
L^*	adjoint differential operator of L
m	first-order derivative of ϕ, or exponent of x for variable free-stream velocity in Eq. (7-63)
M	second-order derivative of ϕ
p, q	superposition functions [Eq. (*a*) of Example 7-9]
$P(x)$	polynomial
R_1	mass transfer Reynolds number $(= v_w y_1/\nu)$
$S(u)$	source term in x-direction momentum equation $[= -(dp/dx)/\rho u]$, s^{-1}
$S_j(x)$	quadratic spline (Fig. 7-2*a*)
V	vector whose components are nodal unknowns, their first-order derivatives, and second-order derivatives
δ	operator denoting variation of a function, or boundary-layer thickness, m
λ	mass transfer parameter $(= v_w \mathrm{Re}_x^{1/2}/u_\infty)$

$\Lambda_{k,j}$ integrals, $k = j - 1, j, j + 1$ (Table 1-4)
ξ normalized transverse distance used in integral method $(= y/\delta)$
ψ stream function ($u = \partial\psi/\partial y$ and $v = -\partial\psi/\partial x$), m^2/s
σ_j mesh size ratio of two adjacent mesh intervals $(= h_{j+1}/h_j)$
τ_w shear stress at wall, NT/m^2
ω normalized stream function $[= (\psi - \psi_I)/(\psi_E - \psi_I)]$

Subscripts

c cubic spline
E outer edge of boundary layer ($y = \delta$)
I inner edge of boundary layer ($y = 0$)
q quadratic spline

Superscripts

$*$ adjoint
$-$ overbar denoting upstream or previously iterated values

REFERENCES

1. J. H. Ahlberg, E. N. Nilson, and J. L. Walsh, *The Theory of Splines and Their Applications,* Academic, New York, 1967.
2. K. E. Atkinson, On the Order of Convergence of Natural Cubic Spline Interpolation, *SIAM J. Numer. Anal.,* vol. 5, pp. 89–101, 1968.
3. A. Meir and A. Sharma, On Uniform Approximation by Cubic Splines, *J. Approx. Theory,* vol. 2, pp. 270–274, 1969.
4. R. S. Varga, Error Bounds for Spline Interpolation, in I. J. Schoenberg (ed.), *Approximations with Special Emphasis on Spline Functions,* pp. 367–388, Academic, New York, 1969.
5. I. J. Schoenberg, *Spline Analysis,* Prentice-Hall, Englewood Cliffs, N.J., 1973.
6. S. G. Rubin and R. A. Graves, Jr., Viscous Flow Solutions using Cubic Splines Approximation, *Comput. Fluids,* vol. 3, pp. 1–36, 1975.
7. S. G. Rubin and P. K. Khosla, Polynomial Interpolation Methods for Viscous Flow Calculations, *J. Comput. Phys.,* vol. 24, pp. 217–244, 1977.
8. P. M. Prenter, *Splines and Variational Methods,* Wiley-Interscience, New York, 1975.
9. C. deBoor, Package for Calculating with B Splines, *SIAM J. Numer. Anal.,* vol. 14, pp. 441–472, 1977.
10. T. C. Chawla and S. H. Chan, Solution of Radiation-Conduction Problems with Collocation Method using B-Splines as Approximating Functions, *Int. J. Heat Mass Transfer,* vol. 22, pp. 1657–1667, 1979.
11. G. K. Leaf, T. C. Chawla, and W. J. Minkowycz, Numerical Method for Hydraulic Transients, *Numer. Heat Transfer,* vol. 2, pp. 1–34, 1979.
12. U. Ascher, Solving Boundary-Value Problems with a Spline-Collocation Code, *J. Comput. Phys.,* vol. 34, pp. 401–413, 1980.
13. E. Isaacson and H. B. Keller, *Analysis of Numerical Methods,* pp. 58–61, Wiley, New York, 1966.
14. A. R. Mitchell and R. Wait, *The Finite Element Method in Partial Differential Equations,* pp. 53–55, Wiley, New York, 1977.

15. M. N. Bismarck-Nasr, A Finite Difference-Galerkin Finite Element Solution of Compressible Laminar Boundary-Layer Flows, *Int. J. Numer. Methods Eng.*, vol. 12, pp. 711-719, 1978.
16. L. V. Kantorovich and V. I. Krylov, *Approximate Methods of Higher Analysis* (translated), Wiley-Interscience, New York, 1958.
17. D. A. MacDonald, Solution of the Incompressible Boundary Layer Equations via the Galerkin-Kantorovich Technique, *J. Inst. Math. Appl.*, vol. 6, pp. 115-130, 1970.
18. S. W. Yuan, *Foundations of Fluid Mechanics,* pp. 202-206, Prentice-Hall, Englewood Cliffs, N.J., 1967.
19. J. C. Gottifredi and O. D. Quiroga, A Simple Analysis of Unsteady Heat Transfer in Impulsive Falkner-Skan Flows, *Int. J. Heat Mass Transfer,* vol. 21, pp. 662-664, 1978.
20. T. Cebeci, F. Thiele, P. G. Williams, and K. Stewartson, On the Calculation of Symmetric Wakes. 1. Two-dimensional Flows, *Numer. Heat Transfer,* vol. 2, pp. 35-60, 1979.
21. F. Thiele, Accurate Numerical Solutions of Boundary Layer Flows by the Finite Difference Method of Hermitian Type, *J. Comput. Phys.*, vol. 27, pp. 138-159, 1978.
22. T. M. El-Mistikawy and M. J. Werle, Numerical Method for Boundary Layers with Blowing–the Exponential Box Scheme, *AIAA J.*, vol. 16, pp. 749-751, 1978.
23. T. Suno, Unsteady Stagnation Point Heat Transfer with Blowing or Suction, *ASME J. Heat Transfer,* vol. 103, pp. 448-452, 1981.
24. A. Aziz and T. Y. Na, New Approach to the Solution of Falkner-Skan Equation, *AIAA J.*, vol. 19, pp. 1242-1244, 1981.
25. V. M. Falkner and S. W. Skan, Some Approximate Solutions of the Boundary-Layer Equations, *Philos. Mag.*, vol. 12, pp. 865-896, 1931.
26. M. Van Dyke, *Perturbation Methods in Fluid Mechanics,* chapters 5 and 6, Academic, New York, 1964.
27. A. H. Nayfeh, *Perturbation Methods,* chapters 3 and 4, Wiley, New York, 1973.
28. F. Schultz-Grunow and W. Breuer, Laminar Boundary Layers on Cambered Walls, Basic Developments in Fluid Dynamics, vol. 1, p. 377, Academic, New York, 1965.
29. K. Gersten, H. D. Papenfuss, and J. F. Gross, Influence of the Prandtl Number on Second-Order Heat Transfer due to Surface Curvature at a Three-Dimensional Stagnation Point, *Int. J. Heat Mass Transfer,* vol. 21, pp. 275-284, 1978.
30. T. Cebeci, R. S. Hirsh, and J. H. Whitelaw, On the Calculation of Laminar and Turbulent Boundary Layers on Longitudinally Curved Surfaces, *AIAA J.*, vol. 17, pp. 434-436, 1979.
31. D. B. Spalding, *GENMIX–A Generalized Computer Program for Two-dimensional Parabolic Phenomena,* Pergamon, New York, 1977.
32. S. M. Roberts and J. S. Shipman, *Two-Point Boundary Value Problems: Shooting Methods,* American Elsevier, New York, 1972.
33. W. M. Kays, *Convective Heat and Mass Transfer,* p. 88, McGraw-Hill, New York, 1966.
34. H. Schlichting, *Boundary-Layer Theory,* pp. 192-201, McGraw-Hill, New York, 1968.
35. M. Miyamoto, Influence of Variable Properties upon Transient and Steady-State Free Convection, *Int. J. Heat Mass Transfer,* vol. 20, pp. 1258-1261, 1977.
36. M. Alamgir, Over-all Heat Transfer from Vertical Cones in Laminar Free Convection: An Approximate Method, *ASME J. Heat Transfer,* vol. 101, pp. 174-176, 1979.
37. J. Sucec, Extension of a Modified Integral Method to Boundary Conditions of Prescribed Surface Heat Flux, *Int. J. Heat Mass Transfer,* vol. 22, pp. 771-774, 1979.
38. R. S. Rath and G. Jena, Combined Free and Forced Convection near a Vertical Wall due to Oscillations in the Wall Velocity and Temperature, *Int. J. Heat Mass Transfer,* vol. 23, pp. 1598-1601, 1980.
39. H. B. Keller and T. Cebeci, Accurate Numerical Methods for Boundary Layer Flows. II. Two-dimensional Turbulent Flows, *AIAA J.*, vol. 10, pp. 1193-1199, 1972.
40. T. M. Shih and H. J. Huang, A Method of Solving Nonlinear Differential Equations for Boundary-Layer Flows, *Numer. Heat Transfer,* vol. 4, pp. 159-178, 1981.

PROBLEMS

7-1 Consider a transformed Eq. (1-1) given as

$$\phi'' + 4\phi' - 32\phi = 0, \qquad \phi(0) = 0, \qquad \phi(1) = 1. \tag{P7-1a}$$

Approximate the solution by

$$\tilde{\phi}(x) = d_{-1/2}B_{-1/2}(x) + d_0 B_0(x) + d_{1/2}B_{1/2}(x) + d_1 B_1(x) + d_{3/2}B_{3/2}(x)$$

which it is hoped will satisfy the two boundary conditions. Use the collocation method to find $\tilde{\phi}(x)$ with collocation points at (*a*) $x = 0, \frac{1}{2}, 1$ and (*b*) $x = 0.112702, \frac{1}{2}, 0.887298$. Note that case (*b*) is related to orthogonal collocation and Table 7-1 is valid only for unit interval sizes.

7-2 We wish to approximate

$$f(x) = e^x, \qquad x \in [0, 1]$$

by a certain linear combination of cubic splines and examine the error ξ as a function of h, where

$$\xi = \int_0^1 [e^x - \tilde{f}(x)]\, dx.$$

(*a*) Starting with $h = 1$ and

$$f(x) = d_0 B_0(x) + d_1 B_1(x),$$

calculate the cases $h = \frac{1}{2}$ and $\frac{1}{3}$ and finally $h = \frac{1}{4}$ with

$$\tilde{f}(x) = d_0 B_0(x) + d_{1/4}B_{1/4}(x) + d_{1/2}B_{1/2}(x) + d_{3/4}B_{3/4}(x) + d_1 B_1(x).$$

(*b*) Plot ξ versus h. If we assume $\xi = Mh^\nu$, what are the values of M and ν?

7-3 Follow the procedure described in Example 7-5 with $\Delta\eta = 1$ to solve the similarity equation governing laminar stagnation flow

$$\phi''' + \phi\phi'' - \phi'^2 + 1 = 0, \qquad \phi(0) = \phi'(0) = 0, \qquad \text{and } \phi'(5) = 1.$$

7-4 Follow the procedure described in Example 7-5 with $\Delta\eta = 1$ and $\eta_\infty = 7$ to solve Eq. (7-67) and (7-68) with $m = 0.2$.

7-5 Consider the nonlinear differential equation

$$\phi'' + m\phi\phi' = 0, \tag{P7-5a}$$

where m is a parameter with a small positive value. Linearize Eq. (P7-5*a*) by Taylor's series expansion, parametric differentiation, and the perturbation method. Comment on the similarities and differences among these linearized results.

7-6 A fourth-order differential equation takes the form

$$\phi'''' + f(\phi''', \phi'', \phi', \phi, x) = 0 \tag{P7-6a}$$

subject to

$$\phi(0) = \phi''(0) = 0 \quad \text{and } \phi'(1) = \phi''(1) = 1. \tag{P7-6b}$$

Transform Eqs. (P7-6*a*) and (P7-6*b*) into three initial-value problems. Explain how the original two-point boundary-value problem can be solved after the transformation. Note that f is a linear function.

7-7 Air flows over a nonflat body such that $u_\infty = Ux^{0.2}$. Use the data leading to Table 6-11 to find u_1, u_2, and u_3 at $x = 5.5$ cm. Does the wall shear increase along x as expected?

7-8 Derive the three algebraic equations for ϕ_1, ϕ_2, and ϕ_3, which are the nodal velocities in the boundary-layer flow over a permeable wall with nonzero mass transfer parameter $\lambda \neq 0$. Refer to Fig. 7-5.

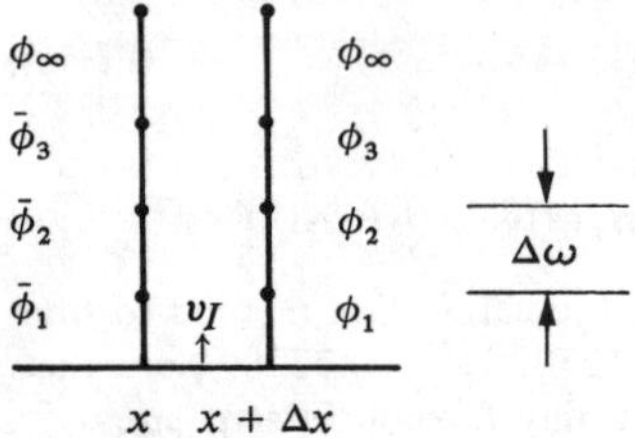

Figure 7-5 Grid system.

7-9 Integrate Eq. (7-80) analytically to obtain a closed form for $\delta(x)$. How does your result for the shear stress at the wall compare with that in Table 7-5?

7-10 Use the method of parameterized residuals to solve the following set of two nonlinear equations:

$$f(x, y) = x^2 - x + 2y^2 - 3y - 11 = 0$$

and

$$g(x, y) = 2x^2 - 3x + y^2 - 4y + 1 = 0.$$

7-11 Replace Eq. (7-27) with

$$M' + \tfrac{1}{2}\phi M = 0 \tag{P7-11a}$$

and follow example 7-5 to solve Eq. (7-34). Does any difficulty arise? Explain why.

CHAPTER

EIGHT

LAMINAR STREAMWISE DIFFUSION FLOWS

In the preceding two chapters we considered the application of some numerical techniques to problems of the boundary-layer type, in which the magnitude of streamwise diffusion is negligible in comparison with that of transverse diffusion. As described in Section 6-1*c*, several flow systems do not fall into this category; in these systems streamwise diffusion contributes to the momentum or energy transport at least as significantly as streamwise convection or transverse diffusion. From both the physical and computational viewpoints, these flows are quite different from boundary-layer flows. We will devote this chapter exclusively to flows of this type.

In Section 8-1, the governing Navier-Stokes equations are written. Difficulty caused by the existence of pressure-gradient terms is briefly described.

Section 8-2 describes several numerical schemes that are capable of solving the elliptic-type streamwise diffusion equation. Iteration procedures are applied to a classical problem of shear-driven recirculating flows. Listings of computer programs are provided in Appendix B.

In Section 8-2 it is taken for granted that the numerical scheme described in that section is stable. Actually, if the Reynolds number becomes too large, there is a risk that the scheme may become unstable. In Section 8-3, we attempt to explain why the numerical instability arises and show how to remove it. One-dimensional problems are considered first. The upwind finite-element method with asymmetric weighting function appears to be a very attractive scheme for handling the instability problem.

In Section 8-4 the variational formulations applied to a simple problem subject to a convective boundary condition.

Finally, the numerical oscillation appearing in the solution of two-dimensional problems is examined in Section 8-5. The algebra involved in the finite-element formulation is seen to be lengthier than that involved in the finite-difference formulation.

8-1 GOVERNING PARTIAL DIFFERENTIAL EQUATIONS

The Navier-Stokes equations governing steady-state, two-dimensional, forced-convection, laminar, incompressible flow can be written in nonconservative form as

$$A(u, v) = \frac{\partial u}{\partial x} + \frac{\partial v}{\partial y} = 0\,, \tag{8-1}$$

$$B(u, v, p) = u\,\frac{\partial u}{\partial x} + v\,\frac{\partial u}{\partial y} - \left(\nu\,\frac{\partial^2 u}{\partial x^2} + \nu\,\frac{\partial^2 u}{\partial y^2}\right) + \frac{1}{\rho}\,\frac{\partial p}{\partial x} = 0\,, \tag{8-2}$$

and

$$C(u, v, p) = u\,\frac{\partial v}{\partial x} + v\,\frac{\partial v}{\partial y} - \left(\nu\,\frac{\partial^2 v}{\partial x^2} + \nu\,\frac{\partial^2 v}{\partial y^2}\right) + \frac{1}{\rho}\,\frac{\partial p}{\partial y} = 0\,. \tag{8-3}$$

There are three primitive variables u, v, and p that must satisfy three equations (8-1)–(8-3). This nonlinear system is thus closed, provided that proper boundary conditions are specified. A similar system subject to gravitation will be considered in Chapter 11.

Before introducing a few numerical methods capable of solving Eqs. (8-1)–(8-3), we wish to outline the differences between this set of equations and the set governing boundary-layer flow. Table 8-1 lists several differences, among which the presence of the pressure term deserves special attention.

We note that in Eqs. (8-2) and (8-3) the pressure gradients appear, but not the pressure itself. If we adopt the central-difference scheme to approximate $\partial p/\partial x$ and $\partial p/\partial y$ at grid point j by

$$\frac{\partial p}{\partial x} \approx \frac{p_E - p_W}{2h} \tag{8-4a}$$

and

$$\frac{\partial p}{\partial y} \approx \frac{p_N - p_S}{2h}\,, \tag{8-4b}$$

Table 8-1 Differences between Streamwise Diffusion Equations and Boundary-Layer Equations

Aspect	Streamwise diffusion equations	Boundary-layer equations
Equation type	Elliptic; downstream conditions influence upstream conditions	Parabolic; downstream conditions do not influence upstream conditions
Pressure gradients	Existent	Nonexistent for flat plates without body force
Magnitude of y-direction flow velocity	As significant as that of x-direction velocity	Negligible in comparison with that of x-direction velocity
Similarity transformation	Cannot be applied	Can generally be applied

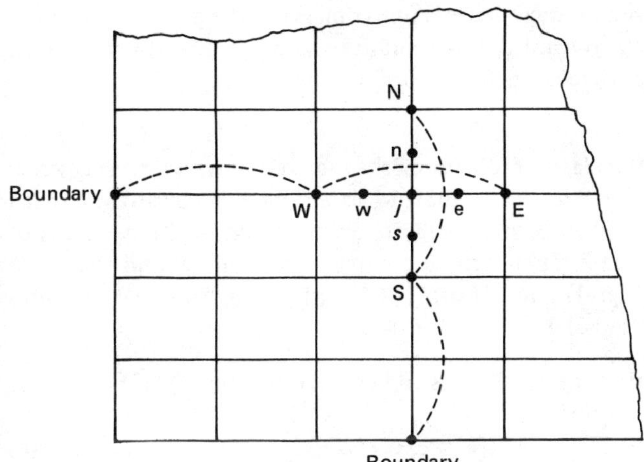

Figure 8-1 Ill behavior caused by the connection between only the alternate nodal pressures.

it can be seen from Fig. 8-1 that only the nodal pressures at the *alternate* grid points, and not those at the *adjacent* grid points, are related to one another. The undesirable consequence is that the value p_j is hardly influenced by the boundary conditions. In other words, if the pressure at the boundary changes say, from 1 to 10 atm, the nodal pressure p_j may only change from 0.9 to 0.91 atm. This insensitivity, which is physically unmeaningful, is clearly caused by the inadequacy of the numerical scheme. There are several ways to remove this difficulty or to avoid directly solving for the pressure field. We will discuss them in the following sections, considering the shear-driven recirculating flow [1] as our example problem.

8-2 NUMERICAL SCHEMES EXAMINING RECIRCULATING FLOWS

The subject of recirculating flows has attracted intense interest from numerical analysts and engineers [2]. It is noted that, at the moment of writing, the numerical schemes analyzing this subject continue to be developed and improved.

8-2*a* Primitive-Variables Finite-Difference Method with Staggered Grid

Since the approximations (8-4*a*) and (8-4*b*) only connect the alternate points, it was proposed [3-5] that they be replaced by

$$\frac{\partial p}{\partial x} \approx \frac{p_E - p_j}{h} \tag{8-5a}$$

and

$$\frac{\partial p}{\partial y} \approx \frac{p_N - p_j}{h} \tag{8-5b}$$

as also shown in Fig. 8-1. Based on these better approximations, the grids for the nodal velocities "stagger" halfway (u toward the east and v toward the north) between the grids for the nodal pressures. Details of this method will be presented in the following example.

Example 8-1 Consider the classic problem of shear-driven flows whose staggered-grid domain is shown in Fig. 8-2*a*. The right-side plate is moving upward at a speed v_∞. In reference to the shaded control volume, discretize Eq. (8-2) by means of the central-difference scheme. Then write a computer program to find the nodal velocities u (on 10×11 grid) and v (on 11×10 grid) and the nodal pressures (on 11×11 grid). Take $\mathrm{Re} = 10$ and $h = \bar{h}/L = 0.1$.

Solution: Let us first write a dimensionless and conservative version of Eq. (8-2) as

$$\frac{\partial}{\partial X} U^2 + \frac{\partial}{\partial Y} UV = \frac{1}{\mathrm{Re}} \nabla^2 U - \frac{\partial P}{\partial X} . \tag{a}$$

Each term in Eq. (*a*) can be discretized, for example, by

$$\frac{\partial}{\partial X} U^2 \approx \frac{1}{h}(U_c^2 - U_a^2) = \frac{1}{h}(U_c - U_a)(U_c + U_a) \approx \frac{2}{h}(U_c - U_a)\bar{U}_j , \tag{b}$$

$$\frac{\partial}{\partial Y} UV \approx \frac{1}{h}(U_d V_d - U_b V_b) \approx \frac{1}{h}(\bar{V}_d U_d - \bar{V}_b U_b) , \tag{c}$$

$$\nabla^2 U \approx \frac{1}{h^2}(U_w + U_s + U_e + U_n - 4U_j) , \tag{d}$$

and

$$\frac{\partial P}{\partial X} \approx \frac{1}{h}(P_c - P_a) , \tag{e}$$

where the nodal unknowns that are located off the grid points can be assumed to be the averages of the two adjacent nodal values. For example,

$$\bar{U}_a \approx \tfrac{1}{2}(\bar{U}_w + \bar{U}_j) , \quad U_b \approx \tfrac{1}{2}(U_s + U_j) , \quad \text{and} \quad \bar{V}_d \approx \tfrac{1}{2}(\bar{V}_1 + \bar{V}_4) .$$

Substituting Eqs. (*b*)-(*e*) into Eq. (*a*) and rearranging the result, we obtain

$$A_j U_j = A_w U_w + A_s U_s + A_e U_e + A_n U_n + P_a - P_c , \tag{f}$$

where

$$A_j = \frac{4}{h\,\mathrm{Re}} + \frac{1}{2}(\bar{V}_d - \bar{V}_b) ,$$

$$A_w = \frac{1}{h\,\mathrm{Re}} + \bar{U}_j , \quad A_s = \frac{1}{h\,\mathrm{Re}} + \frac{\bar{V}_b}{2} , \quad A_e = \frac{1}{h\,\mathrm{Re}} - \bar{U}_j ,$$

and

$$A_n = \frac{1}{h\,\mathrm{Re}} - \frac{\bar{V}_d}{2} .$$

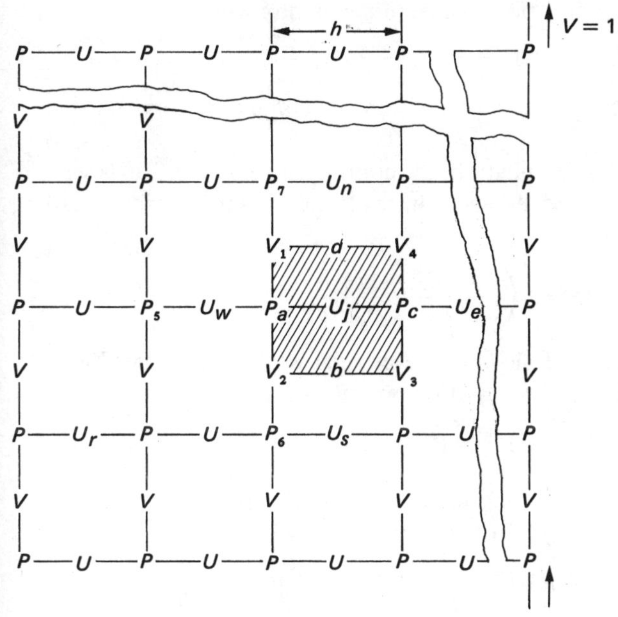

(*a*) staggered grid

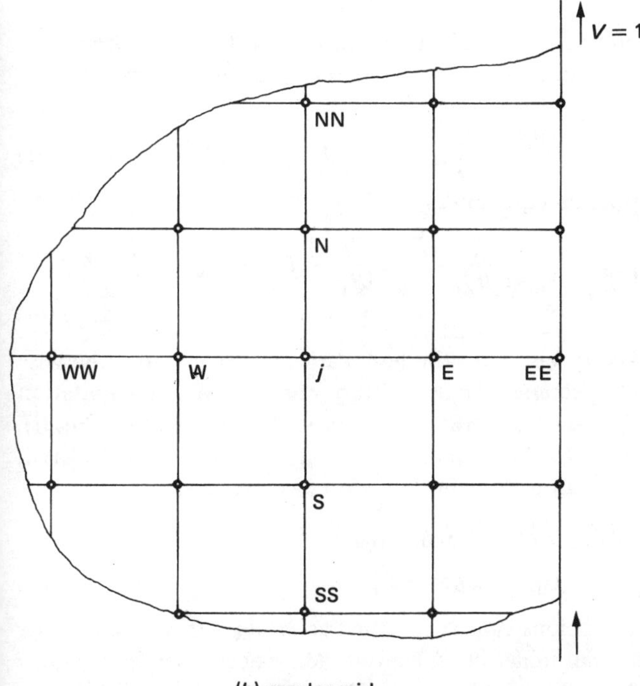

(*b*) regular grid

Figure 8-2 (a, b) Grid systems for the shear-driven recirculating flow.

Similarly, discretization of y-direction momentum equation will yield

$$B_j V_j = B_w V_w + B_s V_s + B_e V_e + B_n V_n + P_c - P_7 \,, \qquad (g)$$

in which the coefficients can be found readily in the computer program listed in Appendix B-1.

In the primitive-variables formulation, the continuity equation (8-1) is seldom directly used as mentioned in Section 8-1. Instead, the pressure Poisson equation (also see Problem 8-3)

$$-\nabla^2 P = 2\left(\frac{\partial V}{\partial X}\frac{\partial U}{\partial Y} - \frac{\partial U}{\partial X}\frac{\partial V}{\partial Y}\right) \qquad (h)$$

can be adopted. Alternatively, following the original work of the SIMPLE code [3], we can also use Eq. (8-1) indirectly. First we introduce

$$U = U^* + U' \,, \qquad V = V^* + V' \,, \qquad \text{and} \qquad P = P^* + P' \,, \qquad (i)$$

where * denotes "guessed" or "being iterated" and the prime denotes "correction." During the iteration, Eq. (f) should be replaced by

$$A_j U_j^* = A_w U_w^* + A_s U_s^* + A_e U_e^* + A_n U_n^* + P_a^* - P_c^* \,. \qquad (j)$$

Subtracting Eq. (j) from Eq. (f) yields

$$A_j U_j' = P_a' - P_c' \,, \qquad (k)$$

where the neighboring terms $\Sigma_{k=w}^{n} A_k U_k'$ are intentionally dropped to facilitate the iteration. We then substitute $U_j = U_j^* + (P_a' - P_c')/A_j$ and others into the continuity equation

$$U_j - U_w + V_1 - V_2 = 0 \qquad (l)$$

to derive the pressure-correction equation as

$$\left(\frac{1}{A_j} + \frac{1}{A_w} + \frac{1}{B_1} + \frac{1}{B_2}\right) P_a' = \frac{P_5'}{A_w} + \frac{P_6'}{B_2} + \frac{P_c'}{A_j} + \frac{P_7'}{B_1} - (U_j^* - U_w^* + V_1^* - V_2^*) \,. \qquad (m)$$

It is recommended that the P' values on the boundaries be set to zero otherwise the P value will ultimately become a large arbitrary number. After computation of P', we can compute U_j' from Eq. (k) and the improved quantities from Eq. (i). When the grid point adjacent to the boundaries, e.g. grid point r, are considered, the west point now lies beyond the domain. Therefore, we may assume

$$U_{\text{wall}} = \tfrac{1}{2}(U_r + U_{\text{fictitious}}) = 0$$

or

$$U_{\text{fictitious}} = -U_r \,. \qquad (n)$$

The pressure boundary conditions can be specified as $\partial P/\partial n = 0$, where n is the coordinate normal to the wall under consideration. More accurate specifications can be achieved by applying Eqs. (8-2) and (8-3) to the boundaries. Finally, as remarked in [5], it is essential to under-relax Eqs. (f), (g), and (m) otherwise the solution may diverge. For details of the computational procedure, see the computer program listed in Appendix B-1.

8-2*b* Primitive-Variables Finite-Element Method

It is natural to suspect that one possible way to link both the alternate and adjacent grid points is to adopt a discretization scheme that leads to a nine-point relation for the nodal pressures. The Galerkin finite-element method with square elements appears to be a good choice.

For convenience, let us consider only Eq. (8-2). Following the Galerkin formulation, we choose the weighting function to be the bilinear piecewise basis function $N_j(\xi, \eta)$ (or interpolation functions for velocities) defined as

$$N_j(\xi,\eta) = \begin{cases} N_d = \frac{1}{4}(1+\xi)(1-\eta), x, y \in e_{NW}, \\ N_a = \frac{1}{4}(1+\xi)(1+\eta), x, y \in e_{SW}, \\ N_b = \frac{1}{4}(1-\xi)(1+\eta), x, y \in e_{SE}, \\ N_c = \frac{1}{4}(1-\xi)(1-\eta), x, y \in e_{NE}, \\ 0, \text{ elsewhere}, \end{cases}$$

where

$$\xi = \frac{2(x-x^*)}{h} \quad \text{and} \quad \eta = \frac{2(y-y^*)}{h}.$$

The four elements are shown in Fig. 8-3. Routinely, we can generate the following weak form by using integration by parts:

$$(A(u, v, p), N_j) = \iint_{e_{\text{sup}}} \left(\cdots - \frac{2}{h} \tilde{p} \frac{\partial N_j}{\partial \xi} \right) d\xi \, d\eta \,,$$

in which only the pressure term was written since we will temporarily focus on this term. The subscript "sup" stands for the support possessed by the basis function $N_j(\xi, \eta)$. In this case,

$$e_{\text{sup}} = e_{NW} + e_{SW} + e_{SE} + e_{NE} \,.$$

Also, Dirichlet boundary conditions have been assumed for the pressure. For the element e_{NW}, the integral can be evaluated as

$$\int_{-1}^{1}\int_{-1}^{1} \tilde{p} \frac{\partial N_d}{\partial \xi} d\xi \, d\eta = \int_{-1}^{1}\int_{-1}^{1} (p_N N_a + p_{NW} N_b + p_W N_c + p_j N_d) \frac{\partial N_d}{\partial \xi} d\xi \, d\eta$$

$$= \frac{1}{6} p_N + \frac{1}{6} p_{NW} + \frac{1}{3} p_W + \frac{1}{3} p_j \,.$$

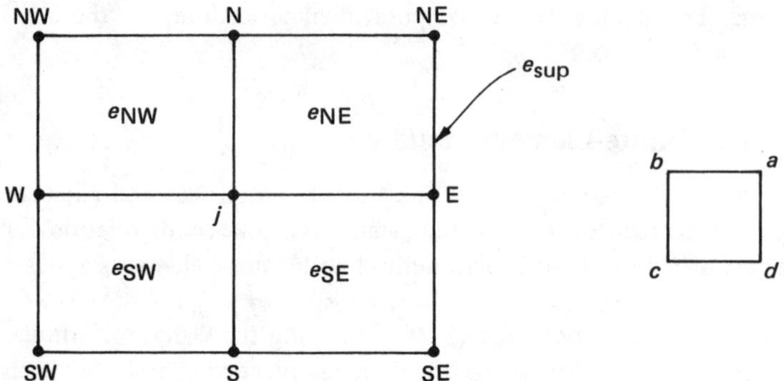

Figure 8-3 Four square elements surrounding grid point j.

The isoparametric mapping $N \to a$, $NW \to b$, $W \to c$, and $j \to d$ has been performed. In a similar way we can evaluate the integrals for the other three elements e_{SW}, e_{SE}, and e_{NE}. The final result is

$$\iint_{e_{sup}} \tilde{p}\,\frac{\partial N_j}{\partial \xi}\,d\xi\,d\eta = \frac{2}{3}\,(p_W - p_E) + \frac{1}{6}\,(p_{NW} + p_{SW} - p_{SE} - p_{NE})\,. \tag{8-6}$$

From Eq. (8-6), it is seen that not all nine nodal pressures are involved. In fact, the alternating behavior similar to that exhibited by the central-difference scheme remains.

In constructing the weak form of Eq. (8-1), it is a common practice [6-12] to choose the interpolation function for the pressure to be the weighting function. This weighting procedure can be interpreted as one requiring that no net work be done on the fluid by the mean normal stress within the fluid. For illustration, let us refer to Fig. 8-4, which is a crude version of Fig. 8-2b with only four interior grid points 1, 2, 3, and 4. The boundary of the system at $X = 1$ is moving at a constant speed v_∞, and this movement drives the fluid to circulate counterclockwise in the enclosure. Now suppose we wish to solve for u_j, v_j, and p_k, $j = 1, 2, 3, 4$, $k = 1, 2, \ldots, 8$.† Following the Galerkin method and applying a certain linearization scheme to the convective terms, we write

$$(A(\tilde{u}, \tilde{v}), M_k) = 0\,, \tag{8-7a}$$

$$(B(\tilde{u}, \tilde{p}), N_j) = 0\,, \tag{8-7b}$$

and

$$(C(\tilde{v}, \tilde{p}), N_j) = 0\,, \tag{8-7c}$$

where $M_k(x, y)$ and $N_j(x, y)$ are, respectively, the interpolation functions for the pressure and the velocities. After integration over the entire domain consisting of five squares and four triangles, Eqs. (8-7a)–(8-7c) can be derived as

†For easy argument, let us assume that pressures at grid points 9, 10, 11, and 12 are given.

$$\begin{bmatrix} 8\times 4 \\ \text{matrix} \end{bmatrix} \begin{Bmatrix} u_1 \\ u_2 \\ u_3 \\ u_4 \end{Bmatrix} + \begin{bmatrix} 8\times 4 \\ \text{matrix} \end{bmatrix} \begin{Bmatrix} v_1 \\ v_2 \\ v_3 \\ v_4 \end{Bmatrix} = 0\,, \tag{8-8a}$$

$$\begin{bmatrix} 4\times 4 \\ \text{matrix} \end{bmatrix} \begin{Bmatrix} u_1 \\ u_2 \\ u_3 \\ u_4 \end{Bmatrix} + \begin{bmatrix} 4\times 8 \\ \text{matrix} \end{bmatrix} \begin{Bmatrix} p_1 \\ p_2 \\ \vdots \\ p_8 \end{Bmatrix} = \begin{Bmatrix} f_1 \\ f_2 \\ f_3 \\ f_4 \end{Bmatrix}, \tag{8-8b}$$

and

$$\begin{bmatrix} 4\times 4 \\ \text{matrix} \end{bmatrix} \begin{Bmatrix} v_1 \\ v_2 \\ v_3 \\ v_4 \end{Bmatrix} + \begin{bmatrix} 4\times 8 \\ \text{matrix} \end{bmatrix} \begin{Bmatrix} p_1 \\ p_2 \\ \vdots \\ p_8 \end{Bmatrix} = \begin{Bmatrix} g_1 \\ g_2 \\ g_3 \\ g_4 \end{Bmatrix}, \tag{8-8c}$$

where f_j and g_j, $j = 1, 2, 3, 4$, are known expressions generated from linearization of the convective terms as well as from incorporation of the boundary conditions.

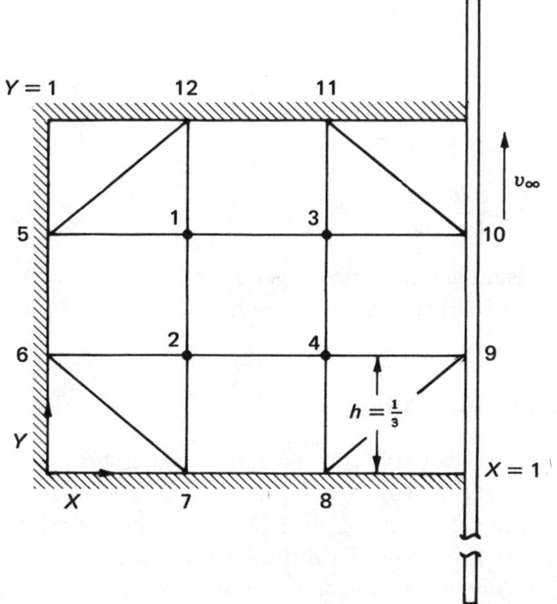

Figure 8-4 A coarse mesh system of the circulating flow within a square enclosure.

If the number of the unknown nodal pressures is more than twice that of the unknown velocities, i.e., $k > 2j$, it can be shown that the coefficient matrix assembled from matrices in Eqs. (8-8*a*)–(8-8*c*) becomes singular. It is due to this singularity that the conventional finite-element formulation dictates that the order of interpolation for pressure be one lower than that for velocities [10, 12].

Since some trouble may be caused by imposing the incompressibility constraint, it is possible to impose this constraint only globally, eliminate the pressure as a dependent variable, and retain the velocities as the fundamental dependent variables. This approach is called the "penalty function" finite-element method, in which a penalty term $F(\tilde{u}, \tilde{v})$ is added to Eq. (8-7*b*) to give

$$(B(\tilde{u}), N_j) + \lambda F(\tilde{u}, \tilde{v}) = 0 \,. \tag{8-9}$$

It is noted that the pressure no longer appears in Eq. (8-9). For details of this method, see [13–18].

8-2*c* Streamfunction-Vorticity Method

It is also possible to eliminate the pressure term by cross-differentiation of Eqs. (8-2) and (8-3). Introducing the vorticity and the stream function, defined as

$$\omega = -\frac{\partial u}{\partial y} + \frac{\partial v}{\partial x}$$

and

$$u = \frac{\partial \psi}{\partial y}\,, \qquad v = -\frac{\partial \psi}{\partial x}\,,$$

we can derive

$$\nabla^2 \psi = -\omega \tag{8-10}$$

and

$$\nu \nabla^2 \omega = \frac{\partial \psi}{\partial y}\frac{\partial \omega}{\partial x} - \frac{\partial \psi}{\partial x}\frac{\partial \omega}{\partial y} = -J(\psi, \omega)\,, \tag{8-11}$$

where $J(\cdot, \cdot)$ is the Jacobian. The pressure gradients no longer appear in Eqs. (8-10) and (8-11). A procedure for solving Eqs. (8-10) and (8-11) is described in Section 11-3.

8-2*d* Biharmonic-Streamfunction Formulation

Toward eliminating the field variables, we can still take one step further than the ψ-ω formulation. Upon substitution of Eq. (8-10) into Eq. (8-11), we are able to eliminate ω and concentrate our effort in solving the ψ field [19]. Clearly, the price is paid at the expense of having to solve a fourth-order differential equation and to meet, in the finite-element method, stringent elemental-continuity requirements.

Example 8-2 Consider again the shear-driven recirculating flow whose grid system is shown in Fig. 8-2*b* (p. 307). Derive a discretized form of the biharmonic-streamfunction equation by means of the central-difference scheme. Then write a computer program to solve for ψ on an 11×11 grid. Take Re = 10 and $h = 0.1$.

Solution: The biharmonic-streamfunction equation takes the following form:

$$\frac{\partial \psi}{\partial Y}\frac{\partial}{\partial X}(\nabla^2 \psi) - \frac{\partial \psi}{\partial X}\frac{\partial}{\partial Y}(\nabla^2 \psi) = \frac{1}{\text{Re}}\nabla^4 \psi\,, \tag{a}$$

where ∇^4 is the biharmonic operator, defined as

$$\nabla^4 = \frac{\partial^4}{\partial X^4} - 2\frac{\partial^4}{\partial X^2\,\partial Y^2} + \frac{\partial^4}{\partial Y^4}\,. \tag{b}$$

Discretization of a few terms in Eq. (*a*) can be demonstrated as

$$\frac{\partial^4 \psi}{\partial X^4} \approx \frac{1}{h^4}(\psi_{WW} - 4\psi_W + 6\psi_j - 4\psi_E + \psi_{EE})\,, \tag{c}$$

$$\frac{\partial^4 \psi}{\partial X^2\,\partial Y^2} \approx \frac{1}{h^4}(\psi_{NE} - 2\psi_E + \psi_{SE} - 2\psi_N + 4\psi_j - 2\psi_S + \psi_{NW} - 2\psi_W + \psi_{SW})\,, \tag{d}$$

$$\frac{\partial}{\partial X}\nabla^2 \psi \approx \frac{1}{2h^3}(\psi_{EE} + \psi_{SE} + \psi_{NE} - 4\psi_E - \psi_{WW} - \psi_{SW} - \psi_{NW} + 4\psi_W)\,, \tag{e}$$

and

$$\frac{\partial \psi}{\partial X} \approx \frac{1}{2h}(\bar{\psi}_E - \bar{\psi}_W)\,, \tag{f}$$

where the overbar denotes "previously-iterated." Substituting Eqs. (*c*)-(*f*) and their counterparts in the *y*-direction into Eq. (*a*), we finally obtain

$$\begin{aligned}20\psi_j = &-(Q_1 + 1)\psi_{WW} - (Q_2 + 1)\psi_{SS} + (Q_1 - 1)\psi_{EE} + (Q_2 - 1)\psi_{NN}\\ &+ (4Q_1 + 8)\psi_W + (4Q_2 + 8)\psi_S + (-4Q_1 + 8)\psi_E + (-4Q_2 + 8)\psi_N\\ &+ (-Q_1 + Q_2 - 2)\psi_{NW} - (Q_1 + Q_2 + 2)\psi_{SW} + (Q_1 - Q_2 - 2)\psi_{SE}\\ &+ (Q_1 + Q_2 - 2)\psi_{NE}\,,\end{aligned} \tag{g}$$

where

$$Q_1 = \frac{\text{Re}}{4}(\bar{\psi}_N - \bar{\psi}_S) \quad \text{and} \quad Q_2 = \frac{\text{Re}}{4}(\bar{\psi}_W - \bar{\psi}_E)\,.$$

At the grid points adjacent to the moving boundary (say, grid point *E*), the fictitious value at the grid point east to grid point *EE* can be specified by

$$V_{\text{wall}} = 1 \approx \frac{1}{2h}(\psi_E - \psi_{\text{fictitious}}) \tag{h}$$

or $$\psi_{\text{fictitious}} = \psi_E - 2h \, .$$

The program listing and the computed result with Re = 10 are shown in Appendix B-2.

8-2*e* Pressure Gradient Method

Since only the pressure gradients $\partial p/\partial x$ and $\partial p/\partial y$, and not the pressure itself, appear in Eqs. (8-2) and (8-3), it may be possible and advantageous to treat them as new dependent variables and forget the pressure term temporarily. In light of this, we change Eqs. (8-1)–(8-3) to

$$\frac{\partial U}{\partial X} + \frac{\partial V}{\partial Y} = 0 \, , \tag{8-12}$$

$$U \frac{\partial U}{\partial X} + V \frac{\partial U}{\partial Y} = \frac{1}{\text{Re}} \left(\frac{\partial^2 U}{\partial X^2} + \frac{\partial^2 U}{\partial Y^2} \right) - \alpha \, , \tag{8-13}$$

and

$$U \frac{\partial V}{\partial X} + V \frac{\partial V}{\partial Y} = \frac{1}{\text{Re}} \left(\frac{\partial^2 V}{\partial X^2} + \frac{\partial^2 V}{\partial Y^2} \right) - \beta \, , \tag{8-14}$$

along with

$$\frac{\partial \alpha}{\partial Y} = \frac{\partial \beta}{\partial X} \, , \tag{8-15}$$

where the velocity and the coordinates are normalized to a characteristic velocity v_∞ and a characteristic length L, respectively. The new variables α and β are defined as

$$\alpha = \frac{L}{\rho v_\infty^2} \frac{\partial p}{\partial x} = \frac{\partial P}{\partial X} \tag{8-16a}$$

and

$$\beta = \frac{L}{\rho v_\infty^2} \frac{\partial p}{\partial y} = \frac{\partial P}{\partial Y} \, . \tag{8-16b}$$

There are now four unknowns that must satisfy the four equations (8-12)–(8-15). The new system is closed provided proper boundary conditions are given.

The noticeable feature of the system of Eqs. (8-12)–(8-15) is the presence of both α and β and their first derivatives. When Eqs. (8-12)–(8-15) are discretized by using the central-difference scheme, all the nodal values of α and β at the alternating grids as well as the adjacent grids will be linked. Consequently, we do not need to use the staggered grid for α and β, but must pay the price of including one more unknown than in the case involving the pressure itself. Using the central difference scheme and a grid system of uniform interval size $\Delta x = \Delta y = h$, we can discretize Eqs. (8-12)–(8-15) into

$$U_E - U_W + V_N - V_S = 0 \, , \tag{8-17}$$

$$\left(\frac{2}{h\,\mathrm{Re}}+\bar{U}_j\right)U_W+\left(\frac{2}{h\,\mathrm{Re}}+\bar{V}_j\right)U_S+\left(\frac{2}{h\,\mathrm{Re}}-\bar{U}_j\right)U_E+\left(\frac{2}{h\,\mathrm{Re}}-\bar{V}_j\right)U_N$$
$$-\frac{8}{h\,\mathrm{Re}}U_j-2h\alpha_j=0\,, \tag{8-18}$$

$$\left(\frac{2}{h\,\mathrm{Re}}+\bar{U}_j\right)V_W+\left(\frac{2}{h\,\mathrm{Re}}+\bar{V}_j\right)V_S+\left(\frac{2}{h\,\mathrm{Re}}-\bar{U}_j\right)V_E+\left(\frac{2}{h\,\mathrm{Re}}-\bar{V}_j\right)V_N$$
$$-\frac{8}{h\,\mathrm{Re}}V_j-2h\beta_j=0\,, \tag{8-19}$$

and
$$\alpha_N-\alpha_S-\beta_E+\beta_W=0\,. \tag{8-20}$$

The boundary conditions of α_j and β_j must be specified with care. For illustration, we again refer to Fig. 8-4. Now suppose we wish to find the boundary condition of β at grid point 10. Near the sliding wall, since

$$U(1,Y)=0 \quad \text{and} \quad \left(\frac{\partial V}{\partial Y}\right)_{X=1}=0\,, \tag{8-21}$$

Eq. (8-14) reduces to

$$\beta_{10}\approx\frac{1}{\mathrm{Re}}\left(\frac{\partial^2 V}{\partial X^2}\right)_{10}. \tag{8-22}$$

To approximate the second derivative $\partial^2 V/\partial X^2$ near the sliding wall, we may assume a parabolic profile for $V(X, 2h)$, $X\in[X_1, X_{10}]$, to be

$$V(t,2h)\approx(1-3t+2t^2)V_1+4(t-t^2)V_3+(2t^2-t)V_{10}\,, \tag{8-23}$$

where

$$t=\frac{X-X_1}{2h}\,.$$

Therefore,

$$\left(\frac{\partial^2 V}{\partial X^2}\right)_{10}\approx\left(\frac{\partial^2 V}{\partial t^2}\right)_{t=1}\left(\frac{dt}{dX}\right)^2=\frac{V_1-2V_3+V_{10}}{h^2}\,. \tag{8-24}$$

Of course, better approximations may also be used. Substituting Eq. (8-24) into Eq. (8-22) yields

$$\beta_{10}\approx\frac{V_1-2V_3+V_{10}}{\mathrm{Re}\,h^2}\,. \tag{8-25}$$

The other boundary conditions for α and β can be derived in a similar fashion. Once these boundary conditions are specified, Eqs. (8-17)–(8-20) can be readily solved by an iteration procedure, which is described as follows:

1. Guess an initial distribution of U and V, denoted by $\bar{U}$ and $\bar{V}$.
2. Evaluate the boundary conditions of α and β based on Eq. (8-25) and other similar equations.

3. Solve Eqs. (8-17)–(8-20) simultaneously for U, V, α, and β.
4. Substitute these new values of U and V for $\bar{U}$ and $\bar{V}$.
5. Return to step 2 and iterate.
6. Terminate the iteration when a certain predetermined convergence criterion is satisfied.

Example 8-3 Choose $h = \frac{1}{3}$ and $\mathrm{Re} = 6$ and assume that the initial velocity distribution is uniformly zero. Solve Eqs. (8-17)–(8-20) for the first iteration cycle.

Solution: Let

$$\bar{U}_j = \bar{V}_j = 0\,, \quad j = 1, 2, 3, 4\,. \tag{a}$$

Using Eq. (8-25), we obtain

$$\beta_9 = \beta_{10} = \tfrac{3}{2}\,. \tag{b}$$

To carry out step 3, we list all the algebraic equations for $j = 1, 2, 3, 4$ in Table 8-2. There are 16 equations with 16 unknowns U_j, V_j, α_j, and β_j, $j = 1, 2, 3, 4$. The computed values are listed in Table 8-3. They can be substituted into Eqs. (8-17)–(8-20) for the purpose of debugging the computer program. Interestingly, at the first iteration the velocity solution already exhibits an expected qualitative trend. At the 23rd iteration, convergence (within ±0.1%) is achieved and the converged result is $U_1 = -V_4 = -0.0622$, $U_2 = V_3 = 0.0812$, $U_3 = V_2 = -0.0739$, and $U_4 = -V_1 = 0.0691$.

Table 8-2 Sixteen algebraic equations for U_j, V_j, α_j, and β_j at $j = 1, 2, 3, 4$

Grid point	Algebraic equation
$j = 1$	$U_3 - V_2 = 0$
	$U_2 + U_3 - 4U_1 - \frac{2}{3}\alpha_1 = 0$
	$V_2 + V_3 - 4V_1 - \frac{2}{3}\beta_1 = 0$
	$\alpha_2 + \beta_3 = 0$
$j = 2$	$U_4 + V_1 = 0$
	$U_4 + U_1 - 4U_2 - \frac{2}{3}\alpha_2 = 0$
	$U_4 + V_1 - 4V_2 - \frac{2}{3}\beta_2 = 0$
	$\alpha_1 - \beta_4 = 0$
$j = 3$	$U_1 + V_4 = 0$
	$U_1 + U_4 - 4U_3 - \frac{2}{3}\alpha_3 = 0$
	$V_1 + V_4 + 1 - 4V_3 - \frac{2}{3}\beta_3 = 0$
	$-\alpha_4 - \frac{3}{2} + \beta_1 = 0$
$j = 4$	$-U_2 + V_3 = 0$
	$U_2 + U_3 - 4U_4 - \frac{2}{3}\alpha_4 = 0$
	$V_2 + 1 + V_3 - 4V_4 - \frac{2}{3}\beta_4 = 0$
	$\alpha_3 - \frac{3}{2} + \beta_2 = 0$

Table 8-3 Nodal values of U_j, V_j, α_j, and β_j for the first iteration cycle

Grid point	$U_j^{(0)}$	$V_j^{(0)}$	$\alpha_j^{(1)}$	$\beta_j^{(1)}$	$U_j^{(1)}$	$V_j^{(1)}$
1	0	0	$\frac{3}{4}$	$\frac{3}{4}$	$-\frac{1}{8}$	$-\frac{1}{8}$
2	0	0	$-\frac{3}{4}$	$\frac{3}{4}$	$\frac{1}{8}$	$-\frac{1}{8}$
3	0	0	$\frac{3}{4}$	$\frac{3}{4}$	$-\frac{1}{8}$	$\frac{1}{8}$
4	0	0	$-\frac{3}{4}$	$\frac{3}{4}$	$\frac{1}{8}$	$\frac{1}{8}$
5	0	0	0	0	0	0
6	0	0	0	0	0	0
7	0	0	0	0	0	0
8	0	0	0	0	0	0
9	0	1	0	$\frac{3}{2}$	0	1
10	0	1	0	$\frac{3}{2}$	0	1
11	0	0	0	0	0	0
12	0	0	0	0	0	0

8-3 NUMERICAL STABILITY (ONE-DIMENSIONAL FLOWS)

In the preceding section nothing was mentioned about numerical stability. We took it for granted that the numerical scheme described by Eqs. (8-17)-(8-20) is stable without restriction on the Re value. When Re becomes large, however, numerical stability is no longer guaranteed. For example, at Re = 60, Eq. (8-18) can be rewritten as

$$(0.1+\bar{U}_j)U_W+(0.1+\bar{V}_j)U_S+(0.1-\bar{U}_j)U_E+(0.1-\bar{V}_j)U_N-0.4U_j-\tfrac{2}{3}\alpha_j=0\,. \tag{8-26}$$

During the iteration, if the values of $\bar{U}_j$ and $\bar{V}_j$ exceed 0.1, the coefficients of U_E and U_N become negative. In light of the extremal principle (see Section 5-3*a*), we know that Eq. (8-26) may become unstable. Therefore, some better numerical schemes must be employed as Re becomes large. For the purpose of clear illustration, we will start with a one-dimensional energy transport problem, since its governing equation takes a relatively simple form.

It has been reported [20, 21] that when convection and diffusion take place simultaneously in the same direction, there may exist a numerical instability problem due to the discretization of the differential equation. To show how such an instability appears, we will first introduce a simple physical problem.

Consider a one-dimensional steady-state flow moving inside a channel. The differential equation governing energy conservation can be derived, referring to Fig. 8-5, as

$$L\phi=-\frac{d^2\phi}{dx^2}+\frac{u}{\alpha}\frac{d\phi}{dx}=0\,, \tag{8-27}$$

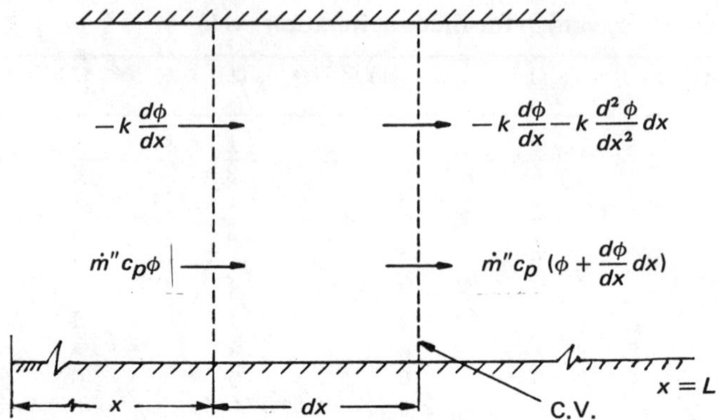

Figure 8-5 Conservation of ϕ in the elemental control volume with unit cross-sectional area.

where the mass flux $\dot{m}''$ has been assumed to be uniform. For simplicity, the two Dirichlet boundary conditions

$$\phi(0) = 0 \quad \text{and} \quad \phi(L) = 1 \tag{8-28a}$$

are prescribed. The convective boundary condition

$$a\left(\frac{d\phi}{dx}\right)_{x=L} = \phi(L) \tag{8-28b}$$

will be considered in Section 8-4. The analytical solution of Eqs. (8-27) and (8-28*a*) can be obtained as

$$\phi(x) = \frac{\exp(ux/\alpha) - 1}{\exp(uL/\alpha) - 1}, \tag{8-29}$$

which can be compared with the approximate solutions to be computed in the following section. Also note that ϕ may represent quantities other than the temperature.

8-3*a* Finite-Difference Method

Let us now use the finite-difference method to solve Eq. (8-27) with the central-difference approximation for the convective term. Thus, Eq. (8-27) can be discretized to

$$-\frac{\phi_{j-1} - 2\phi_j + \phi_{j+1}}{h^2} + \frac{u}{\alpha}\frac{\phi_{j+1} - \phi_{j-1}}{2h} - \frac{h^2}{6}\left(\frac{\partial^3\phi}{\partial x^3} - \frac{1}{2}\frac{\partial^4\phi}{\partial x^4}\right) + \cdots = 0. \tag{8-30}$$

Truncating the higher-order terms, we obtain

$$-\left(1 + \frac{\text{Pe}}{2}\right)\phi_{j-1} + 2\phi_j - \left(1 - \frac{\text{Pe}}{2}\right)\phi_{j+1} = 0, \tag{8-31}$$

where Pe is the local Peclet number, defined as uh/α. The value of Pe here will be assumed to be always positive. The difference scheme expressed in Eq. (8-31) will be shown later in Example 8-4 to be unstable if $\text{Pe} > 2$. To eliminate this instability, researchers [21-33] proposed an upwind-difference scheme in which $\partial\phi/\partial x$ is approximated by $(\phi_j - \phi_{j-1})/h$. Consequently, the discretized three-point equation corresponding to Eq. (8-27) becomes

$$-(1+\text{Pe})\phi_{j-1} + (2+\text{Pe})\phi_j - \phi_{j+1} = 0 \tag{8-32}$$

with the truncated higher-order terms

$$-\frac{h}{2}\frac{d\phi^2}{dx^2} + \frac{h^2}{6}\left(\frac{\partial^3\phi}{\partial x^3} - \frac{1}{2}\frac{\partial^4\phi}{\partial x^4}\right).$$

The finite-difference scheme represented by Eq. (8-32) will be shown in Example 8-5 to be unconditionally stable. However, since its truncation error is $\mathbf{O}(h)$, in comparison with $\mathbf{O}(h^2)$ introduced in the central-difference scheme, this scheme has lower accuracy than the latter one.

Example 8-4 Find the stability criterion for the central-difference scheme Eq. (8-31).

Solution: Let us write a general expression for the three-point algebraic equation as

$$A\phi_{j-1} + B\phi_j + C\phi_{j+1} = 0 \tag{a}$$

subject to $\phi_0 = 0$ and $\phi_J = 1$. For a discretization scheme similar to Eq. (*a*), the coefficients A, B, and C should satisfy

$$A + B + C = 0 \tag{b}$$

because Eq. (*a*) should also hold for the special case $\phi_{j-1} = \phi_j = \phi_{j+1}$. Now, we propose a solution

$$\phi_j = k_1 + k_2\left(\frac{A}{C}\right)^j \tag{c}$$

which, after substitution into Eq. (*a*), can be shown to be the finite-difference solution of Eq. (*a*). The two constants k_1 and k_2 are to satisfy the two boundary conditions. Based on Eq. (*c*), the stability criterion dictates that the value of A/C be positive, otherwise the sign of $(A/C)^j$ will be alternating and the value of ϕ_j oscillatory. Referring to Eq. (8-31), we require

$$\frac{1+\text{Pe}/2}{1-\text{Pe}/2} > 0 \quad \text{or} \quad \text{Pe} < 2. \tag{d}$$

This is the stability criterion for the central-difference scheme.

Example 8-5 Prove that the upwind-difference scheme, Eq. (8-32), is always stable.

Solution: Based on the argument presented in Example 8-4, the stability criterion dictates

$$1 + \mathrm{Pe} > 0 . \tag{a}$$

Since Eq. (*a*) is always true as long as $\mathrm{Pe} \geqslant 0$, the upwind-difference scheme, Eq. (8-32), is always stable.

8-3*b* Galerkin Finite-Element Method

The Galerkin finite-element method is an alternative for solving Eq. (8-27) numerically. If a symmetric pyramid weighting function is chosen, then the final discretized algebraic equation becomes identical to Eq. (8-31). The scheme thus becomes unstable when $\mathrm{Pe} > 2$. It has been proposed [34–38] that a modified Galerkin method with an asymmetric weighting function can remove the instability difficulty. The details are as follows.

We select an asymmetric weighting function

$$W_j(x) = N_j(x) + \mu n_j(x) \qquad \text{if } x \in [x_{j-1}, x_j] \tag{8-33a}$$

and

$$W_j(x) = N_j(x) - \mu n_j(x) \qquad \text{if } x \in [x_j, x_{j+1}] , \tag{8-33b}$$

where μ is a constant to be optimally determined and

$$n_j(x) = -\frac{3(x - x_{j-1})(x - x_j)}{h_j^2} .$$

Figure 8-6 shows the basis function $N_j(x)$ and the asymmetric weighting function $W_j(x)$. The fact that the left curve is fuller than the right one indicates that the flow moves eastward (see Section 8-3*d* for explanation). Using this asymmetric weighting function, we obtain the weak form of Eq. (8-27) as

$$(L\tilde{\phi}, W_j) = -\int_0^L W_j \frac{d^2\tilde{\phi}}{dx^2}\, dx + \frac{u}{\alpha}\int_0^L W_j \frac{d\tilde{\phi}}{dx}\, dx = 0 . \tag{8-34}$$

Using integration by parts yields

$$\int_0^L \left(\frac{dW_j}{dx} + \frac{uW_j}{\alpha}\right)\frac{d\tilde{\phi}}{dx}\, dx = 0 . \tag{8-35}$$

With the linear piecewise approximation for $\tilde{\phi}(x)$, $\Sigma_{j=0}^{J} N_j(x)\phi_j$, we can derive

$$A\phi_{j-1} + B\phi_j + C\phi_{j+1} = 0 , \tag{8-36}$$

where

$$A = \int_{x_{j-1}}^{x_j} \left(\frac{dW_j}{dx}\right)\left(\frac{dN_{j-1}}{dx}\right) dx + \frac{u}{\alpha}\int_{x_{j-1}}^{x_j} W_j \left(\frac{dN_{j-1}}{dx}\right) dx ,$$

$$B = \int_{x_{j-1}}^{x_{j+1}} \left(\frac{dW_j}{dx}\right)\left(\frac{dN_j}{dx}\right) dx + \frac{u}{\alpha}\int_{x_{j-1}}^{x_{j+1}} W_j \left(\frac{dN_j}{dx}\right) dx ,$$

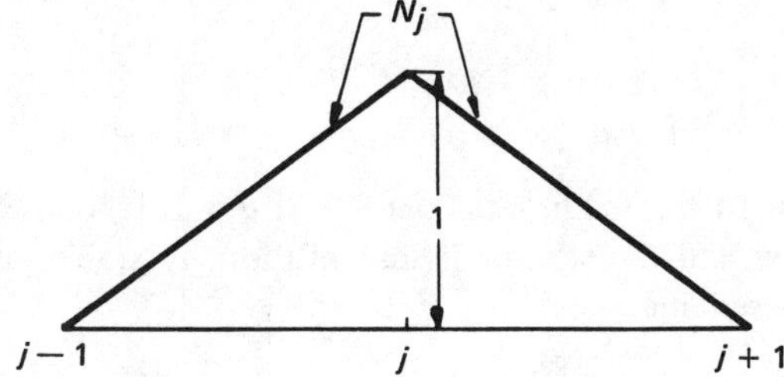

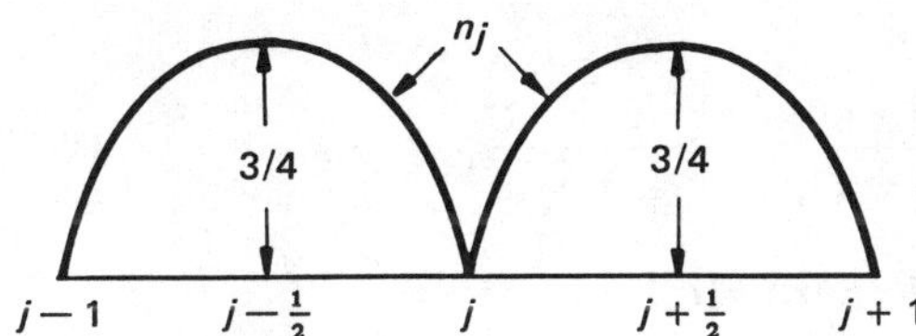

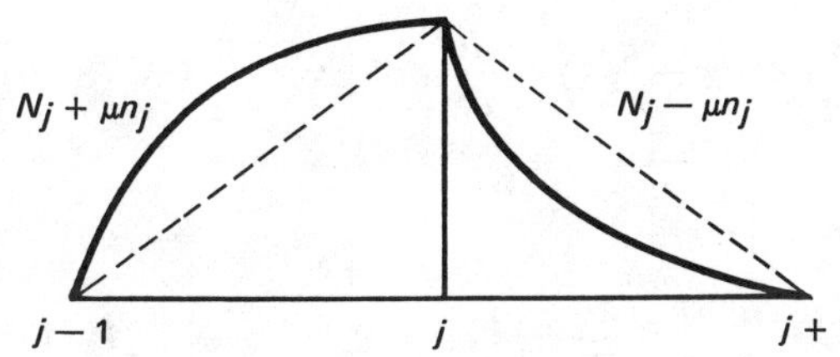

Figure 8-6 The basis function and the parabolic asymmetric weighting function on $[\omega_{j-1}, \omega_{j+1}]$.

and

$$C = \int_{x_j}^{x_{j+1}} \left(\frac{dW_j}{dx}\right)\left(\frac{dN_{j+1}}{dx}\right) dx + \frac{u}{\alpha} \int_{x_j}^{x_{j+1}} W_j \left(\frac{dN_{j+1}}{dx}\right) dx\,.$$

Finally, with the aid of Eqs. (8-33*a*) and (8-33*b*) and considerable algebra, we obtain

$$\left[1 + \frac{\text{Pe}}{2}(\mu + 1)\right]\phi_{j-1} - (2 + \text{Pe}\,\mu)\phi_j + \left[1 + \frac{\text{Pe}}{2}(\mu - 1)\right]\phi_{j+1} = 0\,. \quad (8\text{-}37)$$

It is interesting to note that Eq. (8-37) reduces to Eqs. (8-31) and (8-32), respectively, if $\mu = 0$ and $\mu = 1$.

Example 8-6 Find the stability criterion for the discretization scheme Eq. (8-37).

Solution: Equation (8-37) is stable if

$$\frac{1 + (\text{Pe}/2)(\mu + 1)}{1 + (\text{Pe}/2)(\mu - 1)} > 0 \qquad (a)$$

or

$$\text{Pe} < \frac{2}{1-\mu}. \tag{b}$$

If $\mu = 0$, Eq. (*b*) becomes identical to Eq. (*d*) in Example 8-2. If $\mu = 1$, it follows that Pe can be any positive number and the scheme is unconditionally stable, as expected for the upwind-difference scheme.

Although we may choose any value of μ between 0 and 1 for the computation, a guideline has been established by Christie et al. [39]. They showed that the error of the difference equation is minimized if

$$\mu = \frac{\exp(\text{Pe}/2) + \exp(-\text{Pe}/2)}{\exp(\text{Pe}/2) - \exp(-\text{Pe}/2)} - \frac{2}{\text{Pe}}. \tag{8-38}$$

8-3*c* Adjoint Variational Finite-Element Method

We can also employ the adjoint variational finite-element method to derive Eq. (8-36). The inner product

$$(L\phi, \delta\phi^*) = \int_0^L \frac{d^2\phi}{dx^2}\,\delta\phi^*\,dx - \frac{u}{\alpha}\int_0^L \frac{d\phi}{dx}\,\delta\phi^*\,dx \tag{8-39}$$

can be integrated by parts to yield

$$\int_0^L \frac{d\phi}{dx}\,\delta\left(\frac{d\phi^*}{dx}\right)dx + \frac{u}{\alpha}\int_0^L \frac{d\phi}{dx}\,\delta\phi^*\,dx = 0\,, \tag{8-40}$$

where $\delta\phi^*$ is the variation of the adjoint function ϕ^* that satisfies the adjoint equation

$$L^*\phi^* = -\frac{d^2\phi^*}{dx^2} - \frac{u}{\alpha}\frac{d\phi^*}{dx} = 0\,. \tag{8-41}$$

Equation (8-41) represents the conservation of ϕ in the control volume shown in Fig. 8-5 with convective flow in the negative x direction. Similarly, the inner product

$$(L^*\phi^*, \delta\phi) = -\int_0^L \frac{d^2\phi^*}{dx^2}\,\delta\phi\,dx - \frac{u}{\alpha}\int_0^L \frac{d\phi^*}{dx}\,\delta\phi\,dx \tag{8-42}$$

can be rewritten, after using integration by parts for both terms, as

$$\int_0^L \frac{d\phi^*}{dx}\,\delta\left(\frac{d\phi}{dx}\right)dx + \frac{u}{\alpha}\int_0^L \phi^*\delta\left(\frac{d\phi}{dx}\right)dx = 0\,. \tag{8-43}$$

Now we are ready to obtain a functional of ϕ and ϕ^* by adding Eqs. (8-40) and (8-43). This addition permits the variation sign δ to be brought out from the integrands, i.e.,

$$\delta I(\phi, \phi^*) = 0\,, \tag{8-44}$$

where

$$I = \int_0^L \frac{d\phi}{dx}\frac{d\phi^*}{dx}\,dx + \frac{u}{\alpha}\int_0^L \phi^* \frac{d\phi}{dx}\,dx\,. \tag{8-45}$$

Then we approximate the adjoint function ϕ^* by

$$\tilde{\phi}^*(x) = \sum_{j=0}^{J} W_j(x)\phi_j^*\,, \tag{8-46}$$

where $W_j(x)$ is defined in Eqs. (8-33*a*) and (8-33*b*). This approximation indicates that, on the interval $[x_{j-1}, x_j]$, more of the adjoint function is contributed by ϕ_j^*.

Example 8-7 Based on the approximation Eq. (8-46), what are the percentages of $\phi^*(x_{j-1/2})$ contributed by ϕ_{j-1}^* and ϕ_j^*? Assume $\mu = 1$.

Solution: On the interval $[x_{j-1}, x_j]$, the adjoint function is approximated by

$$\tilde{\phi}^*(x) = W_{j-1}(x)\phi_{j-1}^* + W_j(x)\phi_j^*$$

or

$$\tilde{\phi}^*(t) = \left[1 - t - \frac{\mu}{3}t(1-t)\right]\phi_{j-1}^* + \left[t + \frac{\mu}{3}t(1-t)\right]\phi_j^*\,, \tag{a}$$

where

$$t = \frac{x - x_{j-1}}{h}\,.$$

For $\mu = 1$ and $t = \frac{1}{2}$, Eq. (*a*) becomes

$$\tilde{\phi}^*(\tfrac{1}{2}) = \tfrac{5}{12}\phi_{j-1}^* + \tfrac{7}{12}\phi_j^* = (41.67\%)\phi_{j-1}^* + (58.33\%)\phi_j^*\,, \tag{b}$$

which confirms the fact that, in the adjoint problem governed by Eq. (8-41), the flow moves in the negative x direction.

The linear algebraic equation for ϕ_j can be obtained by taking the partial derivative of Eq. (8-45) with respect to ϕ_j^* as

$$\frac{\partial I}{\partial \phi_j^*} = \int_0^L \frac{d\tilde{\phi}}{dx}\frac{dW_j}{dx}\,dx + \frac{u}{\alpha}\int_0^L W_j \frac{d\tilde{\phi}}{dx}\,dx = 0\,,$$

which is identical to Eq. (8-35) obtained by the modified Galerkin method.

8-3*d* Selection of an Asymmetric Weighting Function

The choice of the asymmetric weighting function given in Eqs. (8-33*a*) and (8-33*b*) is not unique. In fact, Eqs. (8-33*a*) and (8-33*b*) are specified with a certain degree of arbitrariness. It is possible to derive an asymmetric weighting function more rigorously based on the adjoint character of the primary governing equation.

Integrating Eq. (8-35) by parts once again, we obtain

$$
\begin{aligned}
(L\tilde{\phi}, W_j) &= -\int_0^L \tilde{\phi}\,\frac{d^2 W_j}{dx^2}\,dx - \frac{u}{\alpha}\int_0^L \tilde{\phi}\,\frac{dW_j}{dx}\,dx + \text{boundary-condition terms} \\
&= (L^* W_j, \tilde{\phi})\,, \\
&= 0\,,
\end{aligned}
\tag{8-47}
$$

where

$$
L^* = -\frac{d^2}{dx^2} - \frac{u}{\alpha}\frac{d}{dx}\,.
$$

The boundary-condition terms vanish if our analysis is limited to consideration of interior grid points. According to Eq. (8-47), an appropriate asymmetric weighting function on $[x_{j-1}, x_j]$ must satisfy

$$
\frac{d^2 W_j}{dx^2} + \frac{u}{\alpha}\frac{dW_j}{dx} = 0 \tag{8-48}
$$

subject to

$$
W_j(x_{j-1}) = 0 \quad \text{and} \quad W_j(x_j) = 1\,. \tag{8-49}
$$

The solution to Eqs. (8-48) and (8-49) can be obtained as

$$
W_j(t) = \frac{e^{-\text{Pe}\,t} - 1}{e^{-\text{Pe}} - 1}\,, \tag{8-50}
$$

where

$$
t = \frac{x - x_{j-1}}{h} \qquad x \in [x_{j-1}, x_j]\,.
$$

On the interval $[x_j, x_{j+1}]$, Eq. (8-49) is replaced by

$$
W_j(x_j) = 1 \quad \text{and} \quad W_j(x_{j+1}) = 0\,. \tag{8-51}
$$

The corresponding solution is

$$
W_j(t) = \frac{e^{-\text{Pe}\,t} - e^{-\text{Pe}}}{1 - e^{-\text{Pe}}}\,. \tag{8-52}
$$

This asymmetric weighting function given by Eqs. (8-50) and (8-52) is plotted with a solid line in Fig. 8-7 for Pe = 1. The linear basis function is drawn with a dashed line for comparison. From the figure, we establish that the west curve of the asymmetric

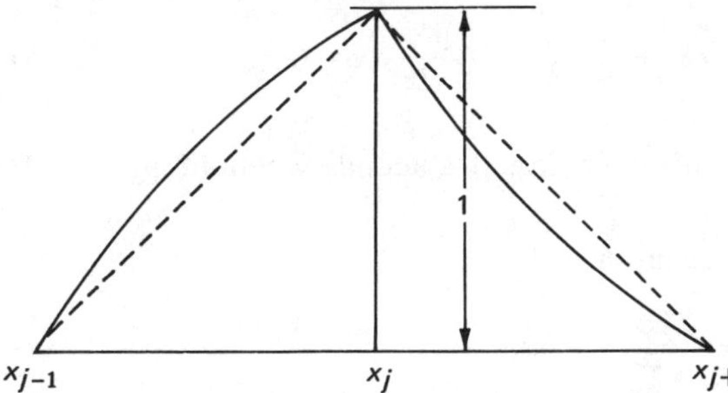

Figure 8-7 Exponential asymmetric weighting function.

weighting function should be fuller than the east one for a streamwise diffusion problem with eastbound convective flow. If other asymmetric weighting functions, such as Eqs. (8-33*a*) and (8-33*b*), are chosen, they should exhibit similar shapes.

It is left to the reader as an exercise (Problem 8-4) to prove that the Galerkin discretization scheme using Eqs. (8-50) and (8-52) will eventually lead Eqs. (8-27) and (8-28*a*) to Eq. (8-37) with μ specified according to Eq. (8-38).

8-4 CONVECTIVE BOUNDARY CONDITION

So far, in all the Galerkin and variational formulations we have assumed that the boundary conditions are of the Dirichlet type. Under such conditions, the boundary-condition terms in the weak forms generated by integration by parts disappear. In practice, however, Neumann and mixed boundary conditions often exist or sometimes are used intentionally to suppress the numerical oscillation [40, 41]. In those cases, we can no longer take it for granted that the boundary-condition terms automatically vanish. In this section, we will demonstrate how to deal with mixed boundary conditions by using the variational formulation.

Consider Eq. (8-27) subject to

$$\phi(0) = 1 \quad \text{and} \quad a\phi(L) = \left(\frac{d\phi}{dx}\right)_{x=L}, \tag{8-53}$$

where a is a constant. With the Galerkin method, the weak form of Eq. (8-27) becomes, after integration by parts,

$$\int_0^L \frac{dW_j}{dx}\frac{d\tilde{\phi}}{dx}\,dx - W_j \frac{d\tilde{\phi}}{dx}\bigg|_0^L + \frac{u}{\alpha}\int_0^L W_j \frac{d\tilde{\phi}}{dx}\,dx = 0\,. \tag{8-54}$$

The nodal unknowns now include $\phi_1, \phi_2, \ldots, \phi_J$, where ϕ_J is the floating unknown boundary condition. For $j = 1, 2, \ldots, J-1$, the boundary-condition terms vanish and Eq. (8-54) reduces to Eq. (8-35) since the weighting functions $W_1, W_2, \ldots, W_{J-1}$ equal zero at $x = 0, L$. For $j = J$, the boundary-condition term remains since $W_J(1) = 1$. Therefore, using the convective boundary condition prescribed in Eq. (8-53), we obtain

$$\int_0^L \frac{dW_J}{dx}\frac{d\tilde{\phi}}{dx}\,dx - a\phi_J + \frac{u}{\alpha}\int_0^L W_J\frac{d\tilde{\phi}}{dx}\,dx = 0\,. \tag{8-55}$$

This is the very equation that accounts for the convective boundary condition.

Example 8-8 Consider the simple equation

$$-\frac{d^2\phi}{dx^2} + 1 = 0 \tag{a}$$

subject to

$$\phi(0) = 0 \qquad \text{and} \qquad \phi'(1) = 4\phi(1)\,. \tag{b}$$

Use the Galerkin finite-element method with $h = \frac{1}{3}$ to find the nodal unknowns ϕ_1, ϕ_2, and ϕ_3.

Solution: We approximate $\phi(x)$ by

$$\tilde{\phi}(x) = N_1(x)\phi_1 + N_2(x)\phi_2 + N_3(x)\phi_3\,, \qquad x \in [0, 1]\,. \tag{c}$$

Following the Galerkin formulation leads to

$$\int_0^1 \frac{d\tilde{\phi}}{dx}\frac{dN_j}{dx}\,dx + \int_0^1 N_j\,dx - \frac{d\tilde{\phi}}{dx}N_j\Big|_0^1 = 0\,, \qquad j = 1, 2, 3\,. \tag{d}$$

Using the results derived in Table 1-4, we obtain the following simultaneous algebraic equations for ϕ_1, ϕ_2, and ϕ_3:

$$2\phi_1 - \phi_2 = -\tfrac{1}{9}\,, \tag{e}$$

$$-\phi_1 + 2\phi_2 - \phi_3 = -\tfrac{1}{9}\,, \tag{f}$$

and

$$\phi_2 + \tfrac{1}{3}\phi_2 = \tfrac{1}{18}\,, \tag{g}$$

which are solved to yield

$$\phi_1 = -\tfrac{1}{18}\,, \qquad \phi_2 = 0\,, \qquad \text{and} \qquad \phi_3 = \tfrac{1}{6}\,. \tag{h}$$

The analytical solution is

$$\phi(x) = \frac{x^2}{2} - \frac{x}{3} \tag{i}$$

also generates Eq. (h).

8-5 NUMERICAL STABILITY (TWO-DIMENSIONAL FLOWS)

The numerical methods applicable to problems of two-dimensional streamwise diffusion flow are similar to, but slightly more complicated than, those described in Section

8-3 for the one-dimensional counterparts. For a large local Peclet number, numerical instability also exists. In this section, we will direct our efforts toward solving a simple Poisson equation that governs a two-dimensional streamwise diffusion flow.

Consider the transport equation

$$\frac{\partial^2 \phi}{\partial x^2} + \frac{\partial^2 \phi}{\partial y^2} - \frac{u}{\alpha}\frac{\partial \phi}{\partial x} = 0 \tag{8-56}$$

subject to the boundary conditions

$$\phi(0, y) = \sin \pi y\,, \qquad \phi(1, y) = \phi(x, 0) = \phi(x, 1) = 0\,. \tag{8-57}$$

The exact solution of Eqs. (8-56) and (8-57) is

$$\phi(x, y) = \frac{\sin \pi y}{e^a - e^b}\left(e^{a+bx} - e^{b+ax}\right), \tag{8-58}$$

where

$$a = \frac{u}{2\alpha} + \left(\frac{u^2}{4\alpha^2} + \pi^2\right)^{1/2} \quad \text{and} \quad b = \frac{u}{2\alpha} - \left(\frac{u^2}{4\alpha^2} + \pi^2\right)^{1/2}.$$

This exact solution can be compared with those obtained by numerical methods in the following sections.

8-5*a* Finite-Difference Method

Let us first discretize Eq. (8-56) using the central-difference scheme. Figure 8-8 shows the five-point stencil. For simplicity, a uniform mesh size is used. With the approximation $\partial\phi/\partial x \approx (\phi_E - \phi_W)/2h$, Eq. (8-56) can be discretized to

$$-4\phi_j + \phi_N + \left(1 + \frac{\text{Pe}}{2}\right)\phi_W + \phi_S + \left(1 - \frac{\text{Pe}}{2}\right)\phi_E = 0\,, \tag{8-59}$$

where $\text{Pe} = uh/\alpha$. If $\text{Pe} > 2$, Eq. (8-59) becomes unstable according to the extremal principle (see Section 5-3*a*). Using the upwind difference $\partial\phi/\partial x \approx (\phi_j - \phi_W)/h$, we also obtain

$$-(\text{Pe} + 4)\phi_j + \phi_N + (1 + \text{Pe})\phi_W + \phi_S + \phi_E = 0\,. \tag{8-60}$$

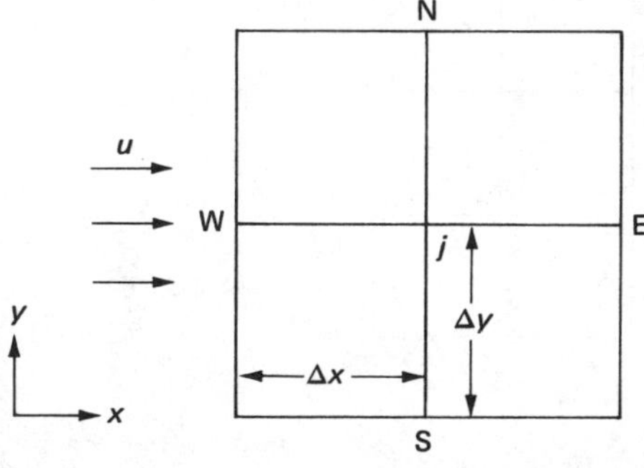

Figure 8-8 Five-point stencil used for the finite-difference approximation, $\Delta x = \Delta y = h$.

Equation (8-60) is unconditionally stable, but has lower accuracy than Eq. (8-59). Successively applying Eq. (8-59) or (8-60) to all the nodal points, we generate a banded matrix system of N algebraic equations, where N is the total number of grid points. If N is a large number, a direct matrix solver to solve for the N nodal unknowns may be inefficient. Alternatively, many indirect methods are available for solving a large matrix system iteratively [42, 43]. As an illustration, we will apply one of the most widely used methods, called successive line over-relaxation (SLOR) [44], to solve the system represented by Eq. (8-60). The procedures are as follows:

Step 1: Assign initially guessed values to the N nodal unknowns. On the basis of the boundary conditions given in Eq. (8-57), zero values may be appropriate.

Step 2: Start the computation at line $i = 1$ as shown in Fig. 8-9. For convenience, Eq. (8-60) is rearranged to

$$\phi_N - (\mathrm{Pe} + 4)\phi_j + \phi_S = -(1 + \mathrm{Pe})\phi_W - \bar{\phi}_E\,, \qquad j = 1, 2, \ldots, J-1\,, \tag{8-61}$$

where $\bar{\phi}_E$ represents the last iterated value. At this particular moment ($i = 1$), for example, $\bar{\phi}_E$ is equal to zero and ϕ_W is the boundary value at $x = 0$.

Step 3: Use the Thomas algorithm (see Section 5-1b) to solve $J-1$ simultaneous algebraic equations for $J-1$ nodal unknowns at line $i = 1$.

Step 4: Carry on the computation by sweeping the domain from $i = 2$ to $i = I - 1$. At each line location Eq. (8-61) and step 3 are used repeatedly.

Step 5: Over-relax N computed values according to

$$\phi^{[k+1]} = \phi^{[k]} + \lambda(\phi^{[k+1]} - \phi^{[k]})\,, \tag{8-62}$$

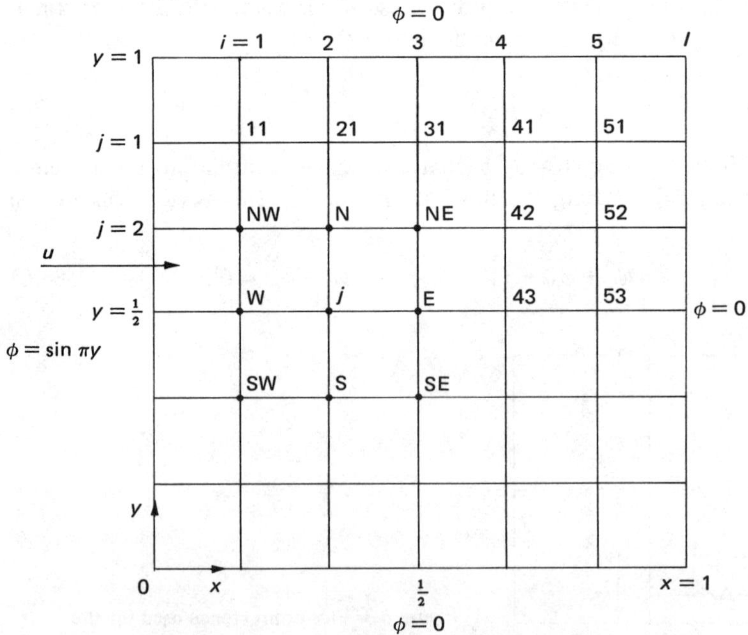

Figure 8-9 Network of the flow field described by Eqs. (8-56) and (8-57).

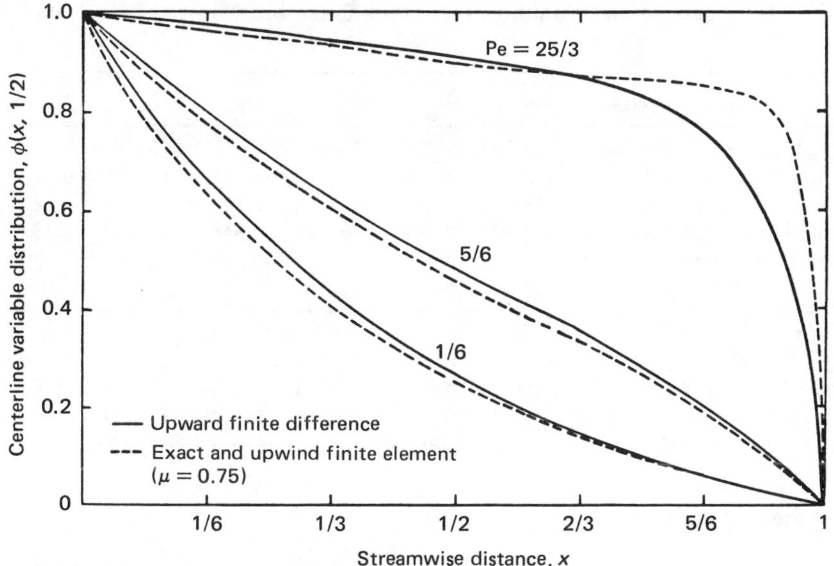

Figure 8-10 Centerline distributions of $\phi(x, y)$ for various values of Pe. The upwind finite element solution is indistinguishable from the exact solution.

where k denotes the kth iteration and λ is the relaxation parameter, which normally ranges from 1 to 2.† The choice of λ strongly influences the rate of convergence for a given scheme [45, 46].

Step 6: Repeat steps 2–5 until the solution converges, e.g.,

$$\frac{\phi^{[k+1]} - \phi^{[k]}}{\phi^{[k+1]}} \leqslant 10^{-6} .$$

Following steps 1–6, we solve Eq. (8-60) by SLOR iteration and list the iterated values for the first two cycles in Table 8-4. For a mesh system comprising 25 nodal unknowns, as shown in Fig. 8-9, the computed solutions along the centerline are shown in Fig. 8-10 at various values of Pe. The exact solutions (dashed lines) are also shown in Fig. 8-10 for comparison. The large discrepancy at $\mathrm{Pe} = \frac{25}{3}$ suggests that finer grid spacing is required near the exit $x = 1$ for larger Pe values.

8-5*b* Galerkin Method with Asymmetric Weighting Functions

The analysis described in Section 8-3*b* can be extended to two-dimensional flows governed by Eqs. (8-56) and (8-57). Following the Galerkin method, the weak form

$$(W_j, L\tilde{\phi}) = \iint_{\Omega} \left(W_j \frac{\partial^2 \tilde{\phi}}{\partial x^2} + W_j \frac{\partial^2 \tilde{\phi}}{\partial y^2} - \frac{u}{\alpha} W_j \frac{\partial \tilde{\phi}}{\partial x} \right) dx\, dy \, , \qquad x, y \in \Omega \quad (8\text{-}63)$$

†If the solution newly computed differs considerably from that previously computed, as in Example 8-1 solving the Navier-Stokes equations, under-relaxation should replace over-relaxation.

Table 8-4 Iterated values of nodal unknowns for two-dimensional flow at Pe = $\frac{5}{6}$ and $\lambda = 1.85$

	ϕ_{11}	ϕ_{12}	ϕ_{13}	ϕ_{21}	ϕ_{22}	ϕ_{23}	ϕ_{31}	ϕ_{32}
Assumed initial distribution	0	0	0	0	0	0	0	0
First iteration	0.2956	0.5119	0.5911	0.1763	0.3054	0.3526	0.1052	0.1822
Second iteration	0.3524	0.6104	0.7048	0.2441	0.4228	0.4883	0.1659	0.2873

	ϕ_{33}	ϕ_{41}	ϕ_{42}	ϕ_{43}	ϕ_{51}	ϕ_{52}	ϕ_{53}
Assumed initial distribution	0	0	0	0	0	0	0
First iteration	0.2104	0.0627	0.1087	0.1255	0.0374	0.0648	0.0749
Second iteration	0.3317	0.1110	0.1923	0.2220	0.0662	0.1147	0.1324

can be integrated by parts to

$$\iint_{\Omega} \left(\frac{\partial W_j}{\partial x} \frac{\partial \tilde{\phi}}{\partial x} + \frac{\partial W_j}{\partial y} \frac{\partial \tilde{\phi}}{\partial y} - \frac{u}{\alpha} \tilde{\phi} \frac{\partial W_j}{\partial x} \right) dx\, dy = 0 \,. \tag{8-64}$$

Before proceeding to discretize Eq. (8-64), we must approximate $\tilde{\phi}(x, y)$ in terms of the nodal unknowns and the basis functions within a typical element. Figure 8-11 shows a square element with vertices a, b, c, and d. The unknown function $\tilde{\phi}(x, y)$ inside e_{abcd} can be approximated by

$$\tilde{\phi}(x, y) = [N_a \quad N_b \quad N_c \quad N_d] \begin{Bmatrix} \phi_a \\ \phi_b \\ \phi_c \\ \phi_d \end{Bmatrix} \,. \tag{8-65}$$

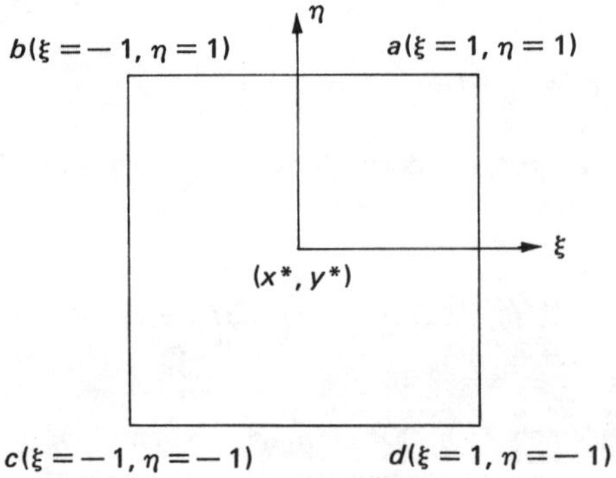

Figure 8-11 Square isoparametric element.

To determine the basis functions N_a, N_b, N_c, and N_d, we first assume

$$\tilde{\phi}(x,y) = p + qx + ry + sxy = [1 \quad x \quad y \quad xy] \begin{Bmatrix} p \\ q \\ r \\ s \end{Bmatrix} . \tag{8-66}$$

Therefore,

$$\begin{Bmatrix} \phi_a \\ \phi_b \\ \phi_c \\ \phi_d \end{Bmatrix} = [\mathrm{c}] \begin{Bmatrix} p \\ q \\ r \\ s \end{Bmatrix}, \quad \text{where } [\mathrm{c}] = \begin{bmatrix} 1 & x_a & y_a & x_a y_a \\ 1 & x_b & y_b & x_b y_b \\ 1 & x_c & y_c & x_c y_c \\ 1 & x_d & y_d & x_d y_d \end{bmatrix} \tag{8-67a}$$

or

$$\begin{Bmatrix} p \\ q \\ r \\ s \end{Bmatrix} = [\mathrm{c}]^{-1} \begin{Bmatrix} \phi_a \\ \phi_b \\ \phi_c \\ \phi_d \end{Bmatrix} . \tag{8-67b}$$

Substituting Eq. (8-67*b*) into Eq. (8-66) yields

$$\tilde{\phi}(x,y) = [1 \quad x \quad y \quad xy]\,[\mathrm{c}]^{-1} \begin{Bmatrix} \phi_a \\ \phi_b \\ \phi_c \\ \phi_d \end{Bmatrix} . \tag{8-68}$$

Then comparing Eq. (8-68) with Eq. (8-65) leads to

$$[N_a \quad N_b \quad N_c \quad N_d] = [1 \quad x \quad y \quad xy]\,[\mathrm{c}]^{-1} . \tag{8-69}$$

The final results for these basis functions are

$$N_a(\xi, \eta) = \tfrac{1}{4}(1 + \xi)(1 + \eta) , \tag{8-70a}$$

$$N_b(\xi, \eta) = \tfrac{1}{4}(1 - \xi)(1 + \eta) , \tag{8-70b}$$

$$N_c(\xi, \eta) = \tfrac{1}{4}(1 - \xi)(1 - \eta) , \tag{8-70c}$$

and

$$N_d(\xi, \eta) = \tfrac{1}{4}(1 + \xi)(1 - \eta) , \tag{8-70d}$$

where

$$\xi = \frac{2(x - x^*)}{h} \quad \text{and} \quad \eta = \frac{2(y - y^*)}{h}.$$

As shown in Sections 5-2 and 5-3 and in later sections, Eq. (8-70*a*)-(8-70*d*) are very useful and convenient in the finite-element formulation when square elements are used.

Now we are ready to seek an expression for the weighting function that incorporates the convection effect. Let

$$N_j(\xi, \eta) = \bar{N}_j(\xi)\bar{N}_j(\eta), \tag{8-71}$$

where

$$\bar{N}_a(\xi) = \bar{N}_d(\xi) = \frac{1+\xi}{2}, \quad \bar{N}_b(\xi) = \bar{N}_c(\xi) = \frac{1-\xi}{2},$$

$$\bar{N}_a(\eta) = \bar{N}_b(\eta) = \frac{1+\eta}{2},$$

and

$$\bar{N}_c(\eta) = \bar{N}_d(\eta) = \frac{1-\eta}{2}.$$

As proposed for the one-dimensional case, we define an asymmetric weighting function

$$W_j(\xi, \eta) = [\bar{N}_j(\xi) + \mu n(\xi)]\bar{N}_j(\eta) = N_j(\xi, \eta) + \mu n(\xi)\bar{N}_j(\eta) \tag{8-72a}$$

if the element is located to the left of grid point j, and

$$W_j(\xi, \eta) = [\bar{N}_j(\xi) - \mu n(\xi)]\bar{N}_j(\eta) = N_j(\xi, \eta) - \mu n(\xi)\bar{N}_j(\eta) \tag{8-72b}$$

if the element is located to the right of grid point j. Here $n(\xi) = \frac{3}{4}(1 - \xi^2)$ and μ is a constant to be determined such that the final discretized algebraic equation is both stable and accurate. In view of Fig. 8-9, we should use Eq. (8-72*a*) for elements e_{NW} and e_{SW} and Eq. (8-72*b*) for elements e_{SE} and e_{NE}. Outside these four elements, which surround grid point j, the weighting function $W_j(\xi, \eta)$ vanishes. With Eqs. (8-65), (8-70*a*)-(8-70*d*), (8-72*a*), and (8-72*b*), the discretization procedure can now be developed. Let us start with element e_{NW}. On comparing e_{NW} with the isoparametric element e_{abcd}, it is found that the vertex a corresponds to N, b to NW, c to W, and d to j. Therefore, leaving out the arguments ξ and η of $N_j(\xi, \eta)$ for clarity, we obtain

$$\iint_{e_{NW}} \left(\frac{\partial W_j}{\partial x}\frac{\partial \tilde{\phi}}{\partial x} + \frac{\partial W_j}{\partial y}\frac{\partial \tilde{\phi}}{\partial y} - \frac{u}{\alpha}\tilde{\phi}\frac{\partial W_j}{\partial x} \right) dx\, dy$$

$$= \phi_j \int_{-1}^{1}\int_{-1}^{1} \left(\frac{\partial W_d}{\partial \xi}\frac{\partial N_d}{\partial \xi} + \frac{\partial W_d}{\partial \eta}\frac{\partial N_d}{\partial \eta} - \frac{\text{Pe}}{2} N_d \frac{\partial W_d}{\partial \xi} \right) d\xi\, d\eta$$

$$+ \phi_N \int_{-1}^{1}\int_{-1}^{1} \left(\frac{\partial W_d}{\partial \xi}\frac{\partial N_a}{\partial \xi} + \frac{\partial W_d}{\partial \eta}\frac{\partial N_a}{\partial \eta} - \frac{\text{Pe}}{2} N_a \frac{\partial W_d}{\partial \xi} \right) d\xi\, d\eta$$

$$+ \phi_{NW} \int_{-1}^{1}\int_{-1}^{1} \left(\frac{\partial W_d}{\partial \xi}\frac{\partial N_b}{\partial \xi} + \frac{\partial W_d}{\partial \eta}\frac{\partial N_b}{\partial \eta} - \frac{\text{Pe}}{2} N_b \frac{\partial W_d}{\partial \xi} \right) d\xi\, d\eta$$

$$+ \phi_W \int_{-1}^{1}\int_{-1}^{1} \left(\frac{\partial W_d}{\partial \xi}\frac{\partial N_c}{\partial \xi} + \frac{\partial W_d}{\partial \eta}\frac{\partial N_c}{\partial \eta} - \frac{\text{Pe}}{2} N_c \frac{\partial W_d}{\partial \xi} \right) d\xi\, d\eta\,. \qquad (8\text{-}73)$$

Using the definition of $W_j(\xi, \eta)$, i.e., Eqs. (8-72*a*) and (8-72*b*), we can further simplify Eq. (8-73) to

$$\iint_{e_{NW}} = \phi_j \left[A_{dd} - \frac{\text{Pe}}{2} B_{d'd} + \mu \left(E_{dd'} + F_{dd} - \frac{\text{Pe}}{2} E_{dd} \right) \right]$$

$$+ \phi_N \left[A_{da} - \frac{\text{Pe}}{2} B_{d'a} + \mu \left(E_{da'} + F_{da} - \frac{\text{Pe}}{2} E_{da} \right) \right]$$

$$+ \phi_{NW} \left[A_{db} - \frac{\text{Pe}}{2} B_{d'b} + \mu \left(E_{db'} + F_{db} - \frac{\text{Pe}}{2} E_{db} \right) \right]$$

$$+ \phi_W \left[A_{dc} - \frac{\text{Pe}}{2} B_{d'c} + \mu \left(E_{dc'} + F_{dc} - \frac{\text{Pe}}{2} E_{dc} \right) \right], \qquad (8\text{-}74a)$$

where

$$A_{pq} = \int_{-1}^{1}\int_{-1}^{1} \left(\frac{\partial N_p}{\partial \xi}\frac{\partial N_q}{\partial \xi} + \frac{\partial N_p}{\partial \eta}\frac{\partial N_q}{\partial \eta} \right) d\xi\, d\eta\,,$$

$$B_{p'q} = \int_{-1}^{1}\int_{-1}^{1} \frac{\partial N_p}{\partial \xi} N_q\, d\xi\, d\eta\,,$$

$$E_{pq'} = \int_{-1}^{1} \int_{-1}^{1} \frac{dn}{d\xi} \bar{N}_p(\eta) \frac{\partial N_q}{\partial \xi} \, d\xi \, d\eta \, ,$$

$$F_{pq} = \int_{-1}^{1} \int_{-1}^{1} n \frac{d\bar{N}_p(\eta)}{d\eta} \frac{\partial N_q}{\partial \eta} \, d\xi \, d\eta \, ,$$

$$E_{pq} = \int_{-1}^{1} \int_{-1}^{1} n \frac{dn}{d\xi} \bar{N}_p(\eta) N_q \, d\xi \, d\eta \, . \tag{8-75}$$

It should be noted that N_p and $\bar{N}_p$ are different, and they are defined, respectively, by Eqs. (8-70*a*)-(8-70*d*) and Eq. (8-71). Similarly, comparing the other three elements e_{SW}, e_{SE}, and e_{NE} with the isoparametric element e_{abcd}, we derive

$$\begin{aligned} \iint_{e_{SW}} &= \phi_j \left[A_{aa} - \frac{\text{Pe}}{2} B_{a'a} + \mu \left(E_{aa'} + F_{aa} - \frac{\text{Pe}}{2} E_{aa} \right) \right] \\ &+ \phi_W \left[A_{ab} - \frac{\text{Pe}}{2} B_{a'b} + \mu \left(E_{ab'} + F_{ab} - \frac{\text{Pe}}{2} E_{ab} \right) \right] \\ &+ \phi_S \left[A_{ad} - \frac{\text{Pe}}{2} B_{a'd} + \mu \left(E_{ad'} + F_{ad} - \frac{\text{Pe}}{2} E_{ad} \right) \right] \\ &+ \phi_{SW} \left[A_{ac} - \frac{\text{Pe}}{2} B_{a'c} + \mu \left(E_{ac'} + F_{ac} - \frac{\text{Pe}}{2} E_{ac} \right) \right] , \end{aligned} \tag{8-74b}$$

$$\begin{aligned} \iint_{e_{SE}} &= \phi_j \left[A_{bb} - \frac{\text{Pe}}{2} B_{b'b} - \mu \left(E_{bb'} + F_{bb} - \frac{\text{Pe}}{2} E_{bb} \right) \right] \\ &+ \phi_S \left[A_{bc} - \frac{\text{Pe}}{2} B_{b'c} - \mu \left(E_{bc'} + F_{bc} - \frac{\text{Pe}}{2} E_{bc} \right) \right] \\ &+ \phi_{SE} \left[A_{bd} - \frac{\text{Pe}}{2} B_{b'd} - \mu \left(E_{bd'} + F_{bd} - \frac{\text{Pe}}{2} E_{bd} \right) \right] \\ &+ \phi_E \left[A_{ba} - \frac{\text{Pe}}{2} B_{b'a} - \mu \left(E_{ba'} + F_{ba} - \frac{\text{Pe}}{2} E_{ba} \right) \right] , \end{aligned} \tag{8-74c}$$

and

$$\iint_{e_{NE}} = \phi_j \left[A_{cc} - \frac{\text{Pe}}{2} B_{c'c} - \mu \left(E_{cc'} + F_{cc} - \frac{\text{Pe}}{2} E_{cc} \right) \right]$$

$$+ \phi_E \left[A_{cd} - \frac{\text{Pe}}{2} B_{c'd} - \mu \left(E_{cd'} + F_{cd} - \frac{\text{Pe}}{2} E_{cd} \right) \right]$$

$$+ \phi_{NE} \left[A_{ca} - \frac{\text{Pe}}{2} B_{c'a} - \mu \left(E_{ca'} + F_{ca} - \frac{\text{Pe}}{2} E_{ca} \right) \right]$$

$$+ \phi_N \left[A_{cb} - \frac{\text{Pe}}{2} B_{c'b} - \mu \left(E_{cb'} + F_{cb} - \frac{\text{Pe}}{2} E_{cb} \right) \right] . \qquad (8\text{-}74d)$$

Adding up Eqs. (8-74*a*)-(8-74*d*) yields a discretized algebraic equation for Eq. (8-64) at grid point j as

$$H_j \phi_j + H_N \phi_N + H_W \phi_W + H_S \phi_S + H_E \phi_E + H_{NW} \phi_{NW} + H_{SW} \phi_{SW} + H_{SE} \phi_{SE}$$
$$+ H_{NE} \phi_{NE} = 0 , \qquad (8\text{-}76)$$

where the definitions of all the coefficients H are listed in Table 8-5. Consulting Table 8-6 for the numerical values of the integrals A_{pq}, B_{pq}, etc., we finally simplify Eq. (8-76) to

$$(8 + 2\mu\,\text{Pe})\phi_j - \left(1 - \frac{\mu}{2}\,\text{Pe}\right)(\phi_N + \phi_S) - \left[1 + \text{Pe} - \frac{\mu}{2}(3 - 2\,\text{Pe})\right]\phi_W$$

$$- \left[1 - \text{Pe} + \frac{\mu}{2}(3 + 2\,\text{Pe})\right]\phi_E - (\phi_{NW} + \phi_{SW})\left[1 + \frac{\text{Pe}}{4} + \frac{\mu}{4}(3 + \text{Pe})\right]$$

$$- (\phi_{NE} + \phi_{SE})\left[1 - \frac{\text{Pe}}{4} - \frac{\mu}{4}(3 - \text{Pe})\right] = 0 . \qquad (8\text{-}77)$$

With a proper choice of μ, Eq. (8-77) can be both stable and accurate. After such an elaborate derivation, it is gratifying that the use of Eq. (8-77) is more advantageous than that of Eqs. (8-59) and (8-60), which were obtained rather easily by the finite-difference method. The numerical solution computed by using Eq. (8-77) with the SOR scheme is also plotted in Fig. 8-10. With $\mu = 0.75$, Eq. (8-77) performs impressively well.

If the symmetric pyramid weighting function is chosen, i.e., $\mu = 0$, Eq. (8-77) reduces to

$$8\phi_j - \phi_N - (1 + \text{Pe})\phi_W - \phi_S - (1 - \text{Pe})\phi_E - \left(1 + \frac{\text{Pe}}{4}\right)(\phi_{NW} + \phi_{SW})$$

$$- \left(1 - \frac{\text{Pe}}{4}\right)(\phi_{NE} + \phi_{SE}) = 0 . \qquad (8\text{-}78)$$

Table 8-5 Definitions of coefficients in Eq. (8-76)

Coefficient	Definition
H_j	$\sum_{k=a}^{d} A_{kk} - \frac{\text{Pe}}{2} \sum_{k=a}^{d} B_{k'k} + \mu \Big[E_{dd'} + E_{aa'} - E_{bb'} - E_{cc'} + F_{dd} + F_{aa} - F_{bb} - F_{cc} + \frac{\text{Pe}}{2} (E_{bb} + E_{cc} - E_{dd} - E_{aa}) \Big]$
H_N	$A_{da} + A_{cb} - \frac{\text{Pe}}{2} (B_{d'a} + B_{c'b}) + \mu \Big[E_{da'} - E_{cb'} + F_{da} - F_{cb} - \frac{\text{Pe}}{2} (E_{da} - E_{cb}) \Big]$
H_W	$A_{dc} + A_{ab} - \frac{\text{Pe}}{2} (B_{d'c} + B_{a'b}) + \mu \Big[E_{dc'} + E_{ab'} + F_{dc} + F_{ab} - \frac{\text{Pe}}{2} (E_{dc} + E_{ab}) \Big]$
H_S	$A_{ad} + A_{bc} - \frac{\text{Pe}}{2} (B_{a'd} + B_{b'c}) + \mu \Big[E_{ad'} - E_{bc'} + F_{ad} - F_{bc} - \frac{\text{Pe}}{2} (E_{ad} - E_{bc}) \Big]$
H_E	$A_{ba} + A_{cd} - \frac{\text{Pe}}{2} (B_{b'a} + B_{c'd}) - \mu \Big[E_{ba'} + E_{cd'} + F_{ba} + F_{cd} - \frac{\text{Pe}}{2} (E_{ba} + E_{cd}) \Big]$
H_{NW}	$A_{db} - \frac{\text{Pe}}{2} B_{d'b} + \mu \left(E_{db'} + F_{db} - \frac{\text{Pe}}{2} E_{db} \right)$
H_{SW}	$A_{ac} - \frac{\text{Pe}}{2} B_{a'c} + \mu \left(E_{ac'} + F_{ac} - \frac{\text{Pe}}{2} E_{ac} \right)$
H_{SE}	$A_{bd} - \frac{\text{Pe}}{2} B_{b'd} - \mu \left(E_{bd'} + F_{bd} - \frac{\text{Pe}}{2} E_{bd} \right)$
H_{NE}	$A_{ca} - \frac{\text{Pe}}{2} B_{c'a} - \mu \left(E_{ca'} + F_{ca} - \frac{\text{Pe}}{2} E_{ca} \right)$

We note that Eq. (8-78) takes a form different from that of Eq. (8-59), which was obtained by the central-difference scheme. This is unlike the one-dimensional case, in which the discretized algebraic equations derived from both the central-difference scheme and the Galerkin method with symmetric weighting functions are identical.

It is also interesting to note that, regardless of how lengthy the expressions for the coefficients shown in Eq. (8-77) and (8-78) are, their sum must be equal to zero. The reason lies in the fact that Eqs. (8-77) and (8-78) must also be applicable to flow systems that have uniform dependent variables throughout the flow field.

Table 8-6 Numerical values of some integrals needed for the finite-element method with square elements

Integral expression	Numerical value
$A_{ad} = \int_{-1}^{1} \int_{-1}^{1} \left(\frac{\partial N_a}{\partial \xi} \frac{\partial N_d}{\partial \xi} + \frac{\partial N_a}{\partial \eta} \frac{\partial N_d}{\partial \eta} \right) d\eta \, d\xi$	$-\frac{1}{6}$
or a–b, b–c, c–d (pairs on the same edge)	
$A_{aa} = \int_{-1}^{1} \int_{-1}^{1} \left[\left(\frac{\partial N_a}{\partial \xi} \right)^2 + \left(\frac{\partial N_a}{\partial \eta} \right)^2 \right] d\eta \, d\xi$	$\frac{2}{3}$
or replacing a with b, c, d	
$A_{ac} = \int_{-1}^{1} \int_{-1}^{1} \left(\frac{\partial N_a}{\partial \xi} \frac{\partial N_c}{\partial \xi} + \frac{\partial N_a}{\partial \eta} \frac{\partial N_c}{\partial \eta} \right) d\eta \, d\xi$	$-\frac{1}{3}$
or b–d (diagonally opposite pairs)	
$\int_{-1}^{1} \int_{-1}^{1} (N_a N_d) \, d\eta \, d\xi$	$\frac{2}{9}$
or a–b, b–c, c–d	
$\int_{-1}^{1} \int_{-1}^{1} N_a^2 \, d\eta \, d\xi$	$\frac{4}{9}$
or b, c, d	
$\int_{-1}^{1} \int_{-1}^{1} (N_a N_c) \, d\eta \, d\xi$	$\frac{1}{9}$
or b–d	
$\iint_{e_{sup}} N_j \, dx \, dy$	$(\Delta x)(\Delta y)$
$B_{a'a}, B_{d'd}, B_{a'b}, B_{d'c}$	$\frac{1}{3}$
$B_{b'b}, B_{c'c}, B_{b'a}, B_{c'd}$	$-\frac{1}{3}$
$B_{a'd}, B_{d'a}, B_{a'c}, B_{d'b}$	$\frac{1}{6}$
$B_{b'c}, B_{c'b}, B_{c'a}, B_{b'd}$	$-\frac{1}{6}$
All $E_{pq'}$	0

Table 8-6 Numerical values of some integrals needed for the finite-element method with square elements (*Continued*)

Integral expression	Numerical value
$F_{pp}, p = a, b, c, d$	$\frac{1}{4}$
$F_{ab}, F_{ba}, F_{cd}, F_{dc}$	$\frac{1}{4}$
$F_{ac}, F_{ca}, F_{bd}, F_{db}$ $F_{ad}, F_{da}, F_{bc}, F_{cb}$	$-\frac{1}{4}$
$E_{aa}, E_{dd}, E_{ba}, E_{cd}$	$-\frac{1}{3}$
$E_{bb}, E_{cc}, E_{ab}, E_{dc}$	$\frac{1}{3}$
$E_{da}, E_{ca}, E_{ad}, E_{bd}$	$-\frac{1}{6}$
$E_{cb}, E_{db}, E_{bc}, E_{ac}$	$\frac{1}{6}$

8-5*c* Variational Finite-Element Method

In Section 8-3*c*, for one-dimensional flows, we proved nonrigorously that the Galerkin method and the variational principle lead to the same discretized algebraic equation. In this section, we will use the variational principle to derive Eq. (8-64). First, let us write the inner product

$$(L\phi, \delta\phi^*) = \iint_\Omega \left(\frac{\partial^2 \phi}{\partial x^2}\, \delta\phi^* + \frac{\partial^2 \phi}{\partial y^2}\, \delta\phi^* - \frac{u}{\alpha}\, \frac{\partial \phi}{\partial x}\, \delta\phi^* \right) dx\, dy\,, \qquad x, y \in \Omega\,, \tag{8-79}$$

where $\delta\phi^*$ is the variation of the adjoint function ϕ^*, which satisfies the adjoint equation

$$L^*\phi^* = \frac{\partial^2 \phi^*}{\partial x^2} + \frac{\partial^2 \phi^*}{\partial y^2} + \frac{u}{\alpha}\, \frac{\partial \phi^*}{\partial x} = 0\,. \tag{8-80}$$

Equation (8-79) can be integrated by parts to

$$\iint_\Omega \left[\left(\frac{\partial \phi}{\partial x} \right) \delta \left(\frac{\partial \phi^*}{\partial x} \right) + \frac{\partial \phi}{\partial y}\, \delta \left(\frac{\partial \phi^*}{\partial y} \right) - \frac{u}{\alpha}\, \phi\, \delta \left(\frac{\partial \phi^*}{\partial x} \right) \right] dx\, dy = 0\,. \tag{8-81}$$

On the other hand, the inner product $(L\phi^*, \delta\phi)$ can also be integrated by parts to yield

$$\iint_\Omega \left[\frac{\partial \phi^*}{\partial x}\, \delta \left(\frac{\partial \phi}{\partial x} \right) + \frac{\partial \phi^*}{\partial y}\, \delta \left(\frac{\partial \phi}{\partial y} \right) - \frac{u}{\alpha}\, \frac{\partial \phi^*}{\partial x}\, \delta\phi \right] dx\, dy = 0\,. \tag{8-82}$$

Adding Eqs. (8-81) and (8-82) enables us to move the variation sign δ to the front of the integral, i.e.,

$$\delta I(\phi, \phi^*) = 0 \,, \tag{8-83}$$

where

$$I = \iint_{\Omega} \left[\left(\frac{\partial \phi}{\partial x}\right)\left(\frac{\partial \phi^*}{\partial x}\right) + \left(\frac{\partial \phi}{\partial y}\right)\left(\frac{\partial \phi^*}{\partial y}\right) - \frac{u}{\alpha}\,\phi\,\frac{\partial \phi^*}{\partial x} \right] dx\,dy \,. \tag{8-84}$$

If the function ϕ^* is approximated by $\Sigma_{j=1}^{N} W_j \phi_j^*$, where W_j is defined in Eqs. (8-72*a*) and (8-72*b*), we may take the partial derivative of Eq. (8-84) with respect to ϕ_j^*, change ϕ to $\tilde{\phi}$ since the space is enlarged, and obtain

$$\frac{\partial I}{\partial \phi_j^*} = \iint_{e_{\text{sup}}} \left[\left(\frac{\partial \tilde{\phi}}{\partial x}\right)\left(\frac{\partial W_j}{\partial x}\right) + \left(\frac{\partial \tilde{\phi}}{\partial y}\right)\left(\frac{\partial W_j}{\partial y}\right) - \frac{u}{\alpha}\,\tilde{\phi}\,\frac{\partial W_j}{\partial x} \right] dx\,dy = 0 \,, \tag{8-85}$$

where the integration domain is reduced to e_{sup} since only the integrals evaluated in e_{NW}, e_{SW}, e_{SE}, and e_{NE} as shown in Fig. 8-9 contain ϕ_j^*. Equation (8-85) is identical to Eq. (8-64) obtained by the Galerkin method.

SYMBOLS

h_j	mesh interval size $(= x_j - x_{j-1})$
I	functional of the function ϕ or functional of the function ϕ and its adjoint function ϕ^*
k	thermal conductivity, W/m·K
L	differential operator, or width of a square enclosure, or length of a tube
$\dot{m}''$	mass flux, g/cm^2·s
n_j	auxiliary weighting function $[= -3(x - x_{j-1})(x - x_j)/h_j^2$ or $\frac{3}{4}(1 - \xi^2)]$
N_j	bilinear basis function [Eqs. (8-70*a*)–(8-70*d*)]
$\bar{N}_j(\xi)$, $\bar{N}_j(\eta)$	component basis function [Eq. (8-71)]
P	normalized pressure $(= p/\rho v_\infty^2)$
Pe	Peclet number $(= u\,\Delta x/\alpha)$
Re	Reynolds number $(= v_\infty L/\nu)$
U	normalized velocity in x direction $(= u/v_\infty)$
V	normalized velocity in y direction $(= v/v_\infty)$
W_j	weighting function used in the Galerkin method [Eqs. (8-33*a*) and (8-33*b*), or Eqs. (8-72*a*) and (8-72*b*)]
α	thermal diffusivity $(= k/\rho c_p)$, m^2/s, or pressure gradient $[= (\partial p/\partial x)(L/\rho v_\infty^2)]$
β	pressure gradient $[= (\partial p/\partial y)(L/\rho v_\infty^2)]$
δ	operator denoting "variation of"
η	isoparametric coordinate $[= 2(y - y^*)/\Delta y]$
λ	relaxation parameter

μ constant to be optimized [Eqs. (8-33*a*), (8-33*b*), (8-72*a*), and (8-72*b*)]
ξ isoparametric coordinate [$= 2(x - x^*)/\Delta x$]
ψ stream function, m^2/s
ω vorticity ($= -\partial u/\partial y + \partial v/\partial x$), s^{-1}
Ω domain of the system

Subscripts

j grid point of current interest
sup support
W, S, E, N west, south, east, north
a, b, c, d index indicating four vertices of isoparametric element

Superscripts

$\bar{}$ previously iterated values (not applicable to N_j)
$*$ adjoint, or center of a square element
$[k]$ kth iteration
$'$ prime denoting correction

REFERENCES

1. O. R. Burggraf, Analytical and Numerical Studies of the Structure of Steady Separated Flows, *J. Fluid Mech.*, vol. 24, pp. 113–151, 1966.
2. T. M. Shih, A Literature Survey on Numerical Heat Transfer, *Numerical Heat Transfer*, vol. 5, pp. 369–420, 1982.
3. F. H. Harlow and J. E. Welch, Numerical Calculation of Time-dependent Viscous Incompressible Flow of Fluid with Free Surface, *Phys. Fluids*, vol. 8, pp. 2182–2189, 1965.
4. S. V. Patankar and D. B. Spalding, A Calculation Procedure for Heat, Mass and Momentum Transfer in Three-dimensional Parabolic Flows, *Int. J. Heat Mass Transfer*, vol. 15, pp. 1787–1806, 1972.
5. S. V. Patankar, *Numerical Heat Transfer and Fluid Flow*, chap. 6, Hemisphere, Washington, D.C., and McGraw-Hill, New York, 1980.
6. C. Taylor and P. Hood, A Numerical Solution of the Navier-Stokes Equations using Finite Element Technique, *Comput. Fluids*, vol. 1, pp. 73–100, 1973.
7. M. Kawahara, N. Yoshimura, K. Nakagawa, and H. Ohsaka, Steady and Unsteady Finite Element Analysis of Incompressible Viscous Fluid, *Int. J. Numer. Methods Eng.*, vol. 10, pp. 437–456, 1976.
8. D. K. Gartling, Convective Heat Transfer Analysis by the Finite Element Method, *Comput. Methods Appl. Mech. Eng.*, vol. 12, pp. 365–382, 1977.
9. G. E. Schneider, G. D. Raithby, and M. M. Yovanovich, Finite Element Solution Procedures for Solving the Incompressible, Navier-Stokes Equations using Equal Order Variable Interpolation, *Numer. Heat Transfer*, vol. 1, pp. 433–451, 1978.
10. P. S. Huyakorn, C. Taylor, R. L. Lee, and P. M. Gresho, A Comparison of Various Mixed-Interpolation Finite Elements in the Velocity-Pressure Formulation of the Navier-Stokes Equations, *Comput. Fluids*, vol. 6, pp. 25–35, 1978.
11. S. Y. Tuann and M. D. Olson, A Transient Finite Element Solution Method for the Navier-Stokes Equations, *Comput. Fluids*, vol. 6, pp. 141–152, 1978.

12. C. P. Jackson and K. A. Cliffe, Mixed Interpolation in Primitive Variable Finite Element Formulations for Incompressible Flow, *Int. J. Numer. Methods Eng.*, vol. 17, pp. 1659–1688, 1981.
13. R. S. Marshall, J. C. Heinrich, and O. C. Zienkiewicz, Natural Convection in a Square Enclosure by a Finite Element, Penalty Function Method using Primitive Fluid Variables, *Numer. Heat Transfer*, vol. 1, pp. 315–330, 1978.
14. T. J. R. Hughes, W. K. Liu, and A. Brooks, Finite Element Analysis of Incompressible Viscous Flows by the Penalty Function Formulation, *J. Comput. Phys.*, vol. 30, pp. 1–60, 1979.
15. M. Bercovier and M. Engelman, A Finite Element for the Numerical Solution of Viscous Incompressible Flows, *J. Comput. Phys.*, vol. 30, pp. 181–201, 1979.
16. J. C. Heinrich and R. S. Marshall, Penalty Function Solution of Steady-State Navier-Stokes Equations, *AIAA J.*, vol. 17, pp. 789–790, 1979.
17. J. N. Reddy and A. Satake, A Comparison of a Penalty Finite Element Model with the Stream Function-Vorticity Model of Natural Convection in Enclosures, *ASME J. Heat Transfer*, vol. 102, pp. 659–666, 1980.
18. J. C. Heinrich and R. S. Marshall, Viscous Incompressible Flow by a Penalty Function Finite Element Method, *Comput. Fluids*, vol. 9, pp. 73–83, 1981.
19. M. D. Olsen and S. Y. Tuann, New Finite Element Results for the Square Cavity, *Comput. Fluids*, vol. 7, pp. 123–135, 1979.
20. R. Courant, E. Issacson, and M. Rees, On the Solution of Nonlinear Hyperbolic Differential Equations by Finite Differences, *Commun. Pure Appl. Math.*, vol. 5, pp. 243–255, 1952.
21. K. E. Torrance and J. A. Rockett, Numerical Study of Natural Convection in an Enclosure with Localized Heating from Below, *J. Fluid Mech.*, vol. 36, pp. 33–54, 1969.
22. D. B. Spalding, A Novel Finite Difference Formulation for Differential Equations Involving Both First and Second Derivatives, *Int. J. Numer. Methods Eng.*, vol. 4, pp. 551–559, 1972.
23. A. K. Runchal, Convergence and Accuracy of Three Finite Difference Schemes for a Two-dimensional Conduction and Convection Problem, *Int. J. Numer. Methods Eng.*, vol. 4, pp. 541–550, 1972.
24. G. D. Raithby and K. E. Torrance, Upstream-weighted Differencing Schemes and their Application to Elliptic Problems Involving Fluid Flow, *Comput. Fluids*, vol. 2, pp. 191–206, 1974.
25. P. J. Roache, *Computational Fluid Dynamics*, p. 64, Hermosa, Albuquerque, N.M., 1976.
26. D. F. Griffiths, On the Approximation of Convection Problems in Fluid Dynamics, *Int. J. Numer. Methods Eng.*, vol. 11, pp. 1477–1483, 1977.
27. N. C. Steele and K. E. Barrett, A Second Order Numerical Method for Laminar Flow at Moderate to High Reynolds Numbers: Entrance Flow in a Duct, *Int. J. Numer. Methods Eng.*, vol. 12, pp. 405–414, 1978.
28. J. L. Siemieniuch and I. Gladwell, Analysis of Explicit Difference Methods for a Diffusion-Convection Equation, *Int. J. Numer. Methods Eng.*, vol. 12, pp. 899–916, 1978.
29. R. K. C. Chan, A Balanced Expansion Technique for Constructing Accurate Finite Difference Advection Scheme, *Int. J. Numer. Methods Eng.*, vol. 12, pp. 1131–1150, 1978.
30. K. W. Morton, Stability of Finite Difference Approximations to a Diffusion-Convection Equation, *Int. J. Numer. Methods Eng.*, vol. 15, pp. 677–683, 1980.
31. D. F. Griffiths, I. Christie, and A. R. Mitchell, Analysis of Error Growth for Explicit Difference Schemes in Conduction-Convection Problems, *Int. J. Numer. Methods Eng.*, vol. 15, pp. 1075–1081, 1980.
32. J. M. Varah, Stability Restrictions on Second Order, Three Level Finite Difference Schemes for Parabolic Equations, *SIAM J. Numer. Anal.*, vol. 17, pp. 300–309, 1980.
33. B. P. Leonard, A Convectively Stable, Third-Order Accurate Finite-Difference Method for Steady Two-dimensional Flow and Heat Transfer, in T. M. Shih (ed.), *Numerical Properties and Methodologies in Heat Transfer*, pp. 211–228, Hemisphere, Washington, D.C., and Springer-Verlag, Berlin, 1983.
34. J. C. Heinrich, P. S. Huyakorn, O. C. Zienkiewicz, and A. R. Mitchell, An Upwind Finite Element Scheme for Two-dimensional Convective Transport Equation, *Int. J. Numer. Methods Eng.*, vol. 11, pp. 131–143, 1977.

35. D. F. Griffiths and J. Lorentz, An Analysis of the Petrov-Galerkin Method Applied to a Model Problem, *Comput. Methods Appl. Mech. Eng.*, vol. 14, pp. 39–64, 1978.
36. J. C. Heinrich, On Quadratic Elements in Finite Element Solutions of Steady-State Convection-Diffusion Equation, *Int. J. Numer. Methods Eng.*, vol. 15, pp. 1041–1052, 1980.
37. J. W. Barrett and K. W. Morton, Optimal Finite Element Solutions to Diffusion-Convection Problems in One Dimension, *Int. J. Numer. Methods Eng.*, vol. 15, pp. 1457–1474, 1980.
38. D. W. Kelly, S. Nakazawa, O. C. Zienkiewicz, and J. C. Heinrich, A Note on Upwinding and Anisotropic Balancing Dissipation in Finite Element Approximations to Convective Diffusion Problems, *Int. J. Numer. Methods Eng.*, vol. 15, pp. 1705–1711, 1980.
39. I. Christie, D. F. Griffiths, and A. R. Mitchell, Finite Element Methods for Second Order Differential Equations with Significant First Derivatives, *Int. J. Numer. Methods Eng.*, vol. 10, pp. 1389–1396, 1976.
40. D. K. Gartling, Some Comments on the Paper by Heinrich, Huyakorn, Zienkiewicz and Mitchell, *Int. J. Numer. Methods Eng.*, vol. 12, pp. 187–190, 1978.
41. M. W. Chang and B. A. Finlayson, On the Proper Boundary Conditions for the Thermal Entry Problem, *Int. J. Numer. Methods Eng.*, vol. 15, pp. 935–942, 1980.
42. R. S. Verga, *Matric Iterative Analysis*, Prentice-Hall, Englewood Cliffs, N.J., 1962.
43. W. F. Ames, *Numerical Methods for Partial Differential Equations*, p. 104, Academic, New York, 1977.
44. D. M. Young, Iterative Methods for Solving Partial Difference Equations of Elliptic Types, *Trans. Am. Math. Soc.*, vol. 76, pp. 92–111, 1954.
45. G. Birkhoff, R. S. Verga, and D. M. Young, Alternating Direction Implicit Methods, in F. L. Alt and M. Rubinoff (eds.), *Advances in Computers*, pp. 189–273, Academic, New York, 1962.
46. A. Ostrowski, An Over and Under Relaxation in the Theory of the Cyclic Single Step Iteration, *Math. Tables Other Aids Comput.*, vol. 7, pp. 152–159, 1953.

PROBLEMS

8-1 Run the computer program listed in Appendix B-1 at Re = 100. What happens? Modify the program to remove the difficulty.

8-2 Insert a block of statements into Appendix B-2 to recover the flow velocities. How are the results compared with those printed in Appendix B-1?

8-3 Based on Eqs. (8-1)–(8-3), derive the pressure Poisson equation

$$-\nabla^2 P = 2\left(\frac{\partial V}{\partial X}\frac{\partial U}{\partial Y} - \frac{\partial U}{\partial X}\frac{\partial V}{\partial Y}\right),$$

or, another version,

$$-\nabla^2 P = \frac{\partial^2 (U^2)}{\partial X^2} + 2\,\frac{\partial^2 (UV)}{\partial X \partial Y} + \frac{\partial^2 (V^2)}{\partial Y^2}.$$

8-4 Prove that the Galerkin discretization scheme using the asymmetric weighting function given in Eqs. (8-50) and (8-52) will lead Eqs. (8-27)–(8-28*a*) to an unconditionally stable scheme:

$$-\phi_{j-1} + (1 + e^{-\mathrm{Pe}})\phi_j - e^{-\mathrm{Pe}}\phi_{j+1} = 0 .$$

8-5 Consider the control volume confined by $-1 \leqslant X \leqslant 1$ and $-1 \leqslant Y \leqslant 1$. The velocities U and V and the pressure gradients α and β can be interpolated in terms of the nodal unknowns at the four vertices, e.g.,

$$U(X, Y) = N_{NW}U_{NW} + N_{SW}U_{SW} + N_{SE}U_{SE} + U_{NE}U_{NE} .$$

Use the control volume method (integrating with respect to X, Y over the control volume) to discretize Eqs. (8-13) and (8-14). Compare your result with Eqs. (8-18) and (8-19).

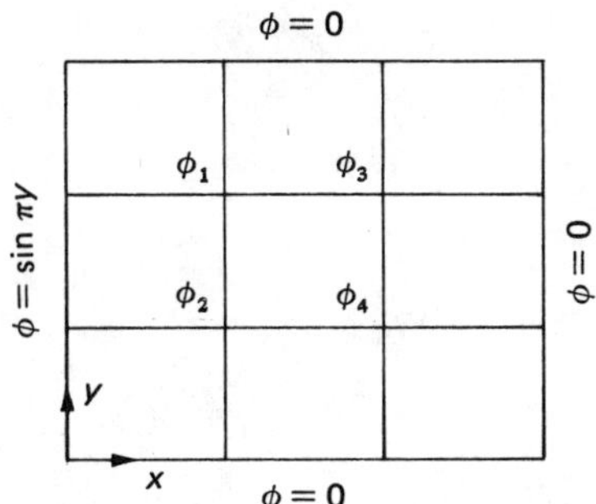

Figure 8-12 A 2 × 2 grid system.

8-6 (*a*) Use the central-difference scheme, Eq. (8-31), with $\Delta x = 0.2$ to solve ϕ_1, ϕ_2, ϕ_3, and ϕ_4 under the conditions $\phi_0 = 0$ and $\phi_5 = 1$. Choose Pe = 1 and 5. Do your solutions at Pe = 5 exhibit oscillatory behavior? (*b*) Repeat (*a*), using the upwind-difference scheme.

8-7 Consider the algebraic equation (8-37). Obtain the nodal unknowns ϕ_1, ϕ_2, ϕ_3, and ϕ_4 under the conditions $\phi_0 = 0$ and $\phi_5 = 1$. Let Pe be equal to 1 and choose the μ value (*a*) according to Eq. (8-38) and (*b*) arbitrarily to be 1. Are your solutions in (*a*) more accurate than those in (*b*)?

8-8 Derive the functional $I(\phi, \phi^*)$ of the governing equation

$$\frac{d^2\phi}{dx^2} + a(x)\frac{d\phi}{dx} + b(x)\phi + c(x) = 0$$

subject to $\phi(0) = 0$ and $\phi'(L) = \alpha\phi(L)$.

8-9 Consider Eqs. (8-56) and (8-57) and refer to Fig. 8-12. Find ϕ_1, ϕ_2, ϕ_3, and ϕ_4 with Pe = 1 and 10, using (*a*) the central-difference scheme and (*b*) the upwind-difference scheme.

8-10 Repeat Problem 8-5 using the modified Galerkin method–i.e., Eq. (8-77)–with $\mu = \frac{1}{2}$.

8-11 Derive Eq. (8-77) from Eq. (8-76) with the aid of Tables 8-5 and 8-6.

8-12 Referring to Fig. 8-12, find the discretized equation at $j = 4$ by the Galerkin method if the boundary condition on $x = 3h$ is $\partial\phi/\partial x = 0$.

CHAPTER

NINE

TRANSPORT OF ENERGY AND SPECIES

This chapter is presented with the understanding that the ultimate goal of tackling a heat convection problem is, in general, to solve the differential equation governing the conservation of energy, and not that governing the transport of momentum described in Chapters 6–8. In addition, we will assume here that the momentum equation can be decoupled from the energy equation and that the velocity distribution has already been computed and therefore is prescribed a priori. This assumption will be removed in Chapter 11, when we investigate free-convection phenomena in which the flow is induced by a thermal gradient.

In multispecies systems, energy transport is also influenced by the distributions of the species mass fractions, whose conservation equation is analogous in form to that of the energy. For this reason, we will concentrate on solving the energy equation, and only briefly discuss the effect of the coexistence of binary species on the boundary condition.

In Section 9-1, the governing equations describing energy transport in two-dimensional laminar flows are presented without derivation. How the Lewis number enters the problem and why its unit value simplifies the expression for the heat flux are demonstrated.

In Section 9-2, we consider heat convection in boundary-layer flow. The ordinary differential equation derived after similarity transformation and the partial differential equation derived after von Mises transformation are solved by using, respectively, the collocation method and the finite-difference method.

Section 9-3 presents the elliptic partial differential equation that governs energy transport near the leading edge of the boundary-layer flow. This equation is discretized by the Galerkin finite-element method. The difficulties associated with use of the least-squares finite-element method are discussed. Finally, with appropriate numerical data, attempts are made to examine the stability of the finite-difference scheme.

Finally, duct flows are considered in Section 9-4. Emphasis is placed on the finite-difference discretization at the centerline of the duct where the singularity exists.

9-1 GOVERNING EQUATIONS

When energy conservation is considered over an infinitesimal control volume $dx \cdot dy$ for a two-dimensional, steady-state, multispecies, laminar flow, the transport equation, neglecting radiation and dissipation, can be written as

$$\rho u \frac{\partial h}{\partial x} + \rho v \frac{\partial h}{\partial y} = -\nabla \cdot \dot{\mathbf{q}}'' , \tag{9-1}$$

where $\dot{\mathbf{q}}''$ is the heat flux vector, defined as

$$\dot{\mathbf{q}}'' = -k \, \nabla T + \sum_{k=1}^{n} \mathbf{j}_k'' h_k \tag{9-2}$$

and $\mathbf{j}_k''$ is the mass flux vector, defined as

$$\mathbf{j}_k'' = -\rho D \, \nabla Y_k \, .^{\dagger} \tag{9-3}$$

Here Y_k is the mass fraction of species k. We prefer to simplify Eq. (9-1) into an equation in which the enthalpy or the temperature of the mixture is the only dependent variable. For mixtures, we have the following relations:

$$h = \sum_{k=1}^{n} Y_k h_k \, ,$$

$$\nabla h_k = c_{p,k} \, \nabla T \, ,$$

and

$$c_p = \sum_{k=1}^{n} Y_k c_{p,k} \, .$$

Therefore,

$$\nabla h = \sum_{k=1}^{n} Y_k \, \nabla h_k + \sum_{k=1}^{n} h_k \, \nabla Y_k = c_p \, \nabla T + \sum_{k=1}^{n} h_k \, \nabla Y_k \, . \tag{9-4}$$

†If Soret diffusion is significant, the mass flux is additionally affected by the thermal gradient. See, for example, [1, 2].

Eliminating ∇T and $\mathbf{j}_k$ from Eqs. (9-2)–(9-4), we obtain

$$\dot{\mathbf{q}}'' = -\frac{k}{c_p}\,\nabla h + \frac{k}{c_p}\sum_{k=1}^{n}(1-\mathrm{Le})h_k\,\nabla Y_k\,, \tag{9-5}$$

where

$$\mathrm{Le} = \frac{D}{\alpha}$$

is called the Lewis number. If Le is nearly unity (for gases), then Eq. (9-5) conveniently reduces to

$$\dot{\mathbf{q}}'' = -\frac{k}{c_p}\,\nabla h\,. \tag{9-6}$$

Otherwise, the energy equation and the species equation are coupled. For ease of analysis, we will further assume that k/c_p is constant. With the assumptions of unity Le and constant k/c_p, Eq. (9-1) is simplified to

$$u\,\frac{\partial h}{\partial x} + v\,\frac{\partial h}{\partial y} = \alpha\left(\frac{\partial^2 h}{\partial x^2} + \frac{\partial^2 h}{\partial y^2}\right). \tag{9-7a}$$

For single-species systems or for multispecies systems with $c_{p,1} = c_{p,2} = \cdots = c_{p,n}$, Eq. (9-7*a*) reduces to

$$u\,\frac{\partial T}{\partial x} + v\,\frac{\partial T}{\partial y} = \alpha\left(\frac{\partial^2 T}{\partial x^2} + \frac{\partial^2 T}{\partial y^2}\right). \tag{9-7b}$$

Although in this chapter we assume that the velocity fields $u(x, y)$ and $v(x, y)$ are known, it should be realized that, for multispecies systems, specification of the boundary condition of $v(x, y)$ at the interface generally requires knowledge of the mass fraction of the evaporating species. For an air-water vapor binary system such as the boundary layer attached to a porous wet surface, the boundary condition of $v(x, y)$ at the interface ($y = 0$) is derived from the conservation of water vapor as

$$v(x, 0) = \frac{-D(\partial Y_{\mathrm{wa}}/\partial y)_{y=0}}{1 - Y_{\mathrm{wa}}(x, 0)}\,, \tag{9-8}$$

where the subscript "wa" denotes water vapor. In this system, momentum transport and species transport are coupled via Eq. (9-8), and the equation of species conservation

$$u\,\frac{\partial Y_k}{\partial x} + v\,\frac{\partial Y_k}{\partial y} = D\left(\frac{\partial^2 Y_k}{\partial x^2} + \frac{\partial^2 Y_k}{\partial y^2}\right) \tag{9-9}$$

must be solved simultaneously with the momentum equation.

9-2 BOUNDARY-LAYER FLOWS (PARABOLIC EQUATIONS)

In this section we will investigate the classical problem of heat transfer in the boundary-layer flow over an isothermal flat plate. According to Section 6-1, one of the most important boundary-layer simplifications

$$\frac{\partial^2 T}{\partial y^2} \gg \frac{\partial^2 T}{\partial x^2}$$

reduces Eq. (9-7b) to

$$u\frac{\partial T}{\partial x} + v\frac{\partial T}{\partial y} = \alpha\frac{\partial^2 T}{\partial y^2}, \tag{9-10}$$

which is a partial differential equation of parabolic type. Equation (9-10) can be solved in several ways.

9-2*a* Similarity (Exact) Solution

If the boundary conditions $T(x, 0)$, $T(x, y_\infty)$, and $T(0, y)$ are all constants, then the similarity solution of Eq. (9-10) exists. Using the similarity transformation described in Section 6-2, we can readily transform Eq. (9-10) to

$$L\theta = \frac{d^2\theta}{d\eta^2} + \frac{1}{2}\Pr f\frac{d\theta}{d\eta} = 0 \tag{9-11}$$

subject to

$$\theta(0) = 0 \quad \text{and} \quad \theta(\eta_\infty) = 1,$$

where

$$\eta = y\left(\frac{u_\infty}{\nu x}\right)^{1/2}, \quad \theta = \frac{T - T_w}{T_\infty - T_w}, \quad \text{and} \quad f = \psi(\nu x u_\infty)^{-1/2}. \tag{9-12}$$

Equation (9-11) can be analytically integrated to yield

$$\theta(\eta) = \frac{\int_0^\eta \left[\exp\left(-\frac{\Pr}{2}\int_0^\eta f(\xi)\,d\xi\right)\right]d\eta}{\int_0^\infty \left[\exp\left(-\frac{\Pr}{2}\int_0^\eta f(\xi)\,d\xi\right)\right]d\eta}. \tag{9-13}$$

Since this solution is presented in standard heat transfer textbooks [3-5] and is of little interest in the numerical analysis, we will skip the computation of Eq. (9-13) and plot the result in Fig. 9-1 with a solid line for $\Pr = 1$. Once $\theta(\eta)$ is obtained, the heat flux at the wall can be expressed by

$$\dot{q}''_w(x) = k(T_w - T_\infty)\left(\frac{u_\infty}{\nu x}\right)^{1/2}\left(\frac{d\theta}{d\eta}\right)_{\eta=0}, \tag{9-14}$$

which is a quantity of primary interest in heat transfer research.

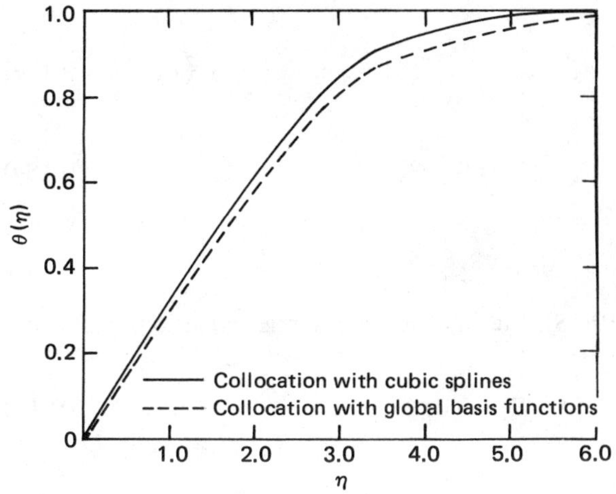

Figure 9-1 Temperature distributions calculated from Eq. (9-13), (9-15), and (9-27). The exact solution is indistinguishable from the solid line.

9-2*b* Collocation Method

In this section we will use the collocation method to solve Eq. (9-11). First, we let the piecewise cubic splines be the basis functions. For convenience, we also set $\text{Pr} = 1$ and $\eta_\infty = 6$, subdivide the domain into two intervals $[0, 3]$ and $[3, 6]$, and designate the locations at $\eta = 0$ and $\eta = 6$ as $j = 0$ and $j = 2$, respectively. According to Section 7-1, the approximate function $\tilde{\theta}(\eta)$ thus can be expressed by

$$\tilde{\theta}(\eta) = e_{-1}B_{-1}(\eta) + e_0 B_0(\eta) + e_1 B_1(\eta) + e_2 B_2(\eta) + e_3 B_3(\eta)\,, \tag{9-15}$$

where $B_j(\eta), j = -1, 0, \ldots, 3$, are piecewise cubic splines that satisfy

$$B_{j-2}(\eta_j) = B_{j+2}(\eta_j) = 0\,, \quad B_{j-1}(\eta_j) = B_{j+1}(\eta_j) = 1\,, \quad B_j(\eta_j) = 4\,, \tag{9-16a}$$

$$B'_{j-2}(\eta_j) = B'_{j+2}(\eta_j) = 0\,, \quad -B'_{j-1}(\eta_j) = B'_{j+1}(\eta_j) = \frac{3}{h}\,, \quad B'_j(\eta_j) = 0\,, \tag{9-16b}$$

and

$$B''_{j-2}(\eta_j) = B''_{j+2}(\eta_j) = 0\,, \quad B''_{j-1}(\eta_j) = B''_{j+1}(\eta_j) = \frac{6}{h^2}\,, \quad B''_j(\eta_j) = -\frac{12}{h^2}\,. \tag{9-16c}$$

Note that $h = 3$ in this problem. Since there are five undetermined coefficients in Eq. (9-15), we need to derive five algebraic equations; two of them can be obtained simply from the boundary conditions $\theta(0) = 0$ and $\theta(6) = 1$, and three can be derived by substituting Eq. (9-15) into Eq. (9-11) at three collocation points. In preparation, we find, at $\eta = 0$,

$$\tilde{\theta}(0) = e_{-1} + 4e_0 + e_1\,, \quad \tilde{\theta}'(0) = e_1 - e_{-1}\,, \quad \tilde{\theta}''(0) = \tfrac{2}{3}e_{-1} - \tfrac{4}{3}e_0 + \tfrac{2}{3}e_1\,, \tag{9-17a}$$

at $\eta = 3$,

$$\tilde{\theta}(3) = e_0 + 4e_1 + e_2\,, \quad \tilde{\theta}'(3) = e_2 - e_0\,, \quad \tilde{\theta}''(3) = \tfrac{2}{3}e_0 - \tfrac{4}{3}e_1 + \tfrac{2}{3}e_2\,, \tag{9-17b}$$

and at $\eta = 6$,

$$\tilde{\theta}(6) = e_1 + 4e_2 + e_3 , \quad \tilde{\theta}'(6) = e_3 - e_1 , \quad \tilde{\theta}''(6) = \tfrac{2}{3}e_1 - \tfrac{4}{3}e_2 + \tfrac{2}{3}e_3 . \tag{9-17c}$$

The five constraints mentioned before are

$$\tilde{\theta}(0) = 0 , \tag{9-18a}$$

$$L\tilde{\theta}_j = 0 , \quad j = 0, 1, 2 , \tag{9-18b}$$

and

$$\tilde{\theta}(6) = 1 . \tag{9-18c}$$

With the aid of Eqs. (9-17) and (9-18), the final set of algebraic equations can be obtained and written in matrix form as

$$[\mathbf{A}]\{\mathbf{E}\} = \{\mathbf{B}\} , \tag{9-19}$$

where

$$[\mathbf{A}] = \begin{bmatrix} 1 & 4 & 1 & 0 & 0 \\ 1 & -2 & 1 & 0 & 0 \\ 0 & 4-3f(3) & -8 & 4+3f(3) & 0 \\ 0 & 0 & 4-3f(6) & -8 & 4+3f(6) \\ 0 & 0 & 1 & 4 & 1 \end{bmatrix} , \tag{9-20a}$$

$$\{\mathbf{E}\}^T = \{e_{-1} \quad e_0 \quad e_1 \quad e_2 \quad e_3\} , \tag{9-20b}$$

and

$$\{\mathbf{B}\}^T = \{0 \quad 0 \quad 0 \quad 0 \quad 1\} . \tag{9-20c}$$

With $f(3) = 1.3968$ and $f(6) = 4.2796$ (from Appendix A), Eq. (9-19) can be solved:

$$e_{-1} = -e_1 = -0.1696 , \quad e_0 = 0 , \quad e_2 = 0.1657 , \quad \text{and} \quad e_3 = 0.1677 . \tag{9-21}$$

On the basis of this result, we plot Eq. (9-15) with a solid line in Fig. 9-1, along with the exact solution for comparison. The agreement is so good even with only two intervals considered that the two solutions are indistinguishable.

Alternatively, we may transform Eq. (9-11) into

$$\frac{d^2\theta}{dt^2} + 3f\frac{d\theta}{dt} = 0 \tag{9-22}$$

subject to

$$\theta(0) = 0 , \quad \theta(1) = 1 ,$$

where $t = \eta/6$. Using the global basis functions, we then obtain

$$\tilde{\theta}(t) = t + c_1 t(1-t) + c_2 t^2(1-t) + c_3 t^3(1-t) . \tag{9-23}$$

Differentiating Eq. (9-23) once and twice yields

$$\tilde{\theta}'(t) = 1 + c_1(1-2t) + c_2(2t - 3t^2) + c_3(3t^2 - 4t^3) \tag{9-24a}$$

and $$\tilde{\theta}''(t) = -2c_1 + c_2(2-6t) + c_3(6t-12t^2)\,. \tag{9-24b}$$

Substituting Eqs. (9-24*a*) and (9-24*b*) into Eq. (9-22) and letting t be $\frac{1}{4}$, $\frac{1}{2}$, and $\frac{3}{4}$, respectively, we obtain

$$[\mathbf{G}]\{\mathbf{C}\} = \{\mathbf{H}\}\,, \tag{9-25}$$

where

$$[\mathbf{G}] = \begin{bmatrix} -2+\frac{3}{2}f(\frac{1}{4}) & \frac{1}{2}+\frac{15}{16}f(\frac{1}{4}) & \frac{3}{4}+\frac{3}{8}f(\frac{1}{4}) \\ -2 & -1+\frac{3}{4}f(\frac{1}{2}) & \frac{3}{4}f(\frac{1}{2}) \\ -2-\frac{3}{2}f(\frac{3}{4}) & -\frac{5}{2}-\frac{9}{16}f(\frac{3}{4}) & -\frac{9}{4} \end{bmatrix}, \tag{9-26a}$$

$$\{\mathbf{C}\}^T = \{c_1 \quad c_2 \quad c_3\}\,, \tag{9-26b}$$

and $$\{\mathbf{H}\}^T = \{-3f(\tfrac{1}{4}) \quad -3f(\tfrac{1}{2}) \quad -3f(\tfrac{3}{4})\}\,. \tag{9-26c}$$

It is noteworthy that $[\mathbf{G}]$ given in Eq. (9-26*a*) is a full matrix, while $[\mathbf{A}]$ given in Eq. (9-20*a*) is a banded matrix. Therefore, the use of piecewise cubic splines results in a simpler computation than the use of global basis functions. To solve for the column vector $\{\mathbf{C}\}$, we first find from Appendix A

$$f(\tfrac{1}{4}) = 0.3701\,, \quad f(\tfrac{1}{2}) = 1.3968\,, \quad \text{and} \quad f(\tfrac{3}{4}) = 2.7903\,.$$

With these values, Eq. (9-25) is solved to yield

$$c_1 = 0.5819\,, \quad c_2 = 2.8411\,, \quad \text{and} \quad c_3 = -3.0181\,.$$

Therefore, this collocation solution is

$$\tilde{\theta}(\eta) = \frac{\eta}{6} + 0.582\,\frac{\eta}{6}\left(1-\frac{\eta}{6}\right) + 2.841\left(\frac{\eta}{6}\right)^2\left(1-\frac{\eta}{6}\right) - 3.018\left(\frac{\eta}{6}\right)^3\left(1-\frac{\eta}{6}\right), \tag{9-27}$$

which, along with the solution obtained by the cubic spline collocation method, is plotted in Fig. 9-1 in comparison with the exact solution.

9-2*c* Finite-Difference Method

Since the boundary-layer thickness is not known a priori and the transverse velocity v is generally of minor interest, it is convenient to transform Eq. (9-10) by the von Mises transformation (see Section 6-3) into

$$\frac{\partial\theta}{\partial x} + b\omega\,\frac{\partial\theta}{\partial\omega} = \frac{\partial}{\partial\omega}\left(\frac{c}{\Pr}\,\frac{\partial\theta}{\partial\omega}\right),$$

where

$$\omega = \int_0^y \frac{u\,dy}{\psi_E}\,.$$

This transformed equation is similar in form to Eq. (6-39) and therefore can be discretized in the manner shown in Section 6-3. Alternatively, Lapalce transform was used by Leveque [6, 7] to solve this equation. See also [8–10].

9-3 LEADING EDGE OF THE THERMAL BOUNDARY LAYER

The boundary-layer assumption $\partial^2 T/\partial y^2 \gg \partial^2 T/\partial x^2$ becomes invalid in the vicinity of the leading edge of the thermal boundary layer since the temperature of the fluid approaching the wall suddenly undergoes a finite change from the location $x = 0^-$ to 0^+. Clearly, the methods described in Section 9-2 can no longer be used for this problem. We will attempt to solve Eq. (9-7) by using the Galerkin finite-element method first. Figure 9-2 shows the mesh with numbered grid points. At the location $x = x_0$, where it is assumed that the hydrodynamic boundary layer is already developed, the normalized temperature of the wall drops from 1 to 0. The thermal boundary layer thus starts to develop at $x = x_0$. We will further assume that, by proper arrangement, grid point 21 lies on the edge of the thermal boundary layer. For such an elliptic system, the governing equation takes the form

$$L\theta = U\frac{\partial \theta}{\partial x} + V\frac{\partial \theta}{\partial y} - \frac{\alpha}{u_{25}}\left(\frac{\partial^2 \theta}{\partial x^2} + \frac{\partial^2 \theta}{\partial y^2}\right) = 0 \tag{9-28a}$$

subject to

$$\theta(0, y) = \theta(x, \delta) = 1\,, \quad \theta(x, 0) = 0\,, \quad \text{and} \quad \theta(L, y) = s(y)\,, \tag{9-28b}$$

where the velocities are normalized on u_{25}, a quantity obtained from the similarity solution. The value of u_{25}/α will be assumed to be sufficiently small that numerical convective instability does not arise (see Sections 8-3 and 8-5 for details). We will summarize all the assumptions as follows.

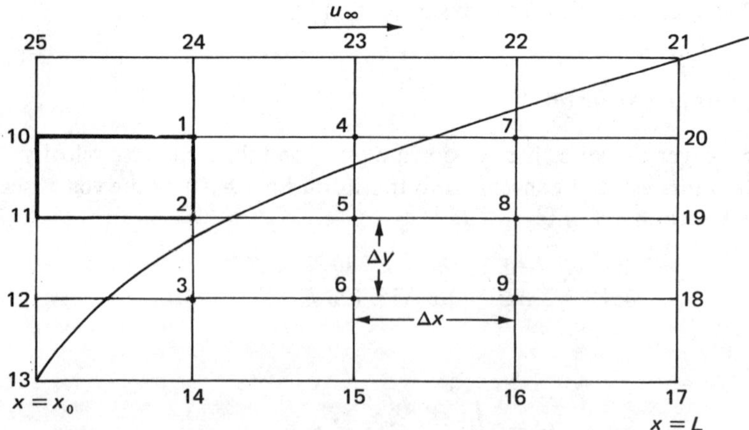

Figure 9-2 Mesh system of the thermal boundary layer near the leading edge.

1. The hydrodynamic boundary layer and similarity pattern have been developed.
2. The velocity distributions $U(x, y)$ and $V(x, y)$ are given a priori.
3. At the location $x = L$, the temperature profile $\theta(L, y)$ can be obtained from the similarity solution.
4. Grid point 21 lies on the edge of the thermal boundary layer. Consequently, $\theta_{21} = 1$.
5. The value of u_{25}/α is sufficiently small that numerical stability exists.

9-3*a* Galerkin Finite-Element Method

We now write the inner product

$$(L\tilde{\theta}, N_j) = \int_{x=0}^{L} \int_{y=0}^{\delta} (L\tilde{\theta}) N_j(x, y)\, dx\, dy \tag{9-29}$$

and seek the weak solution $\tilde{\theta}(x, y)$ that satisfies

$$\int_0^L \int_0^\delta \left(\tilde{U} N_j \frac{\partial \tilde{\theta}}{\partial x} + \tilde{V} N_j \frac{\partial \tilde{\theta}}{\partial y} + \frac{\alpha}{u_{25}} \frac{\partial N_j}{\partial x} \frac{\partial \tilde{\theta}}{\partial x} + \frac{\alpha}{u_{25}} \frac{\partial N_j}{\partial y} \frac{\partial \tilde{\theta}}{\partial y} \right) dx\, dy = 0\,, \tag{9-30}$$

where

$$\tilde{\theta}(x, y) = \sum_{j=1}^{25} N_j(x, y) \theta_j\,. \tag{9-31}$$

The local basis functions $N_j(x, y)$ of the C^0 square element are defined in Eqs. (5-47*a*)-(5-47*d*) in terms of isoparametric coordinates. Substituting Eq. (9-31) into Eq. (9-30), considering only the element e_{sup} within which $N_j(x, y)$ has local support, and matching e_{sup} with the isoparametric element e_{abcd}, we finally obtain

$$K_{NW} + K_{SW} + K_{SE} + K_{NE} = 0\,, \tag{9-32}$$

where $$K_{NW} = B_{dd}\theta_j + B_{ad}\theta_N + B_{bd}\theta_{NW} + B_{cd}\theta_W\,, \quad j = 1, 2, \ldots, 9\,, \tag{9-33a}$$

$$K_{SW} = B_{aa}\theta_j + B_{ba}\theta_W + B_{ca}\theta_{SW} + B_{da}\theta_S\,, \tag{9-33b}$$

$$K_{SE} = B_{bb}\theta_j + B_{cb}\theta_S + B_{db}\theta_{SE} + B_{ab}\theta_E\,, \tag{9-33c}$$

and $$K_{NE} = B_{cc}\theta_j + B_{dc}\theta_E + B_{ac}\theta_{NE} + B_{bc}\theta_N\,. \tag{9-33d}$$

All the coefficients B_{pq} take the form

$$B_{pq} = \int_{-1}^{1} \int_{-1}^{1} \left(\frac{\partial N_p}{\partial \xi} N_q \tilde{U} \gamma + \frac{\partial N_p}{\partial \eta} N_q \tilde{V} + \frac{1}{\text{Pe}} \frac{\partial N_p}{\partial \xi} \frac{\partial N_q}{\partial \xi} + \frac{\gamma^2}{\text{Pe}} \frac{\partial N_p}{\partial \eta} \frac{\partial N_q}{\partial \eta} \right) d\xi\, d\eta\,, \quad p, q = a, b, c, d\,, \tag{9-34}$$

where γ is the ratio of the mesh sizes, defined as

$$\gamma = \frac{\Delta x}{\Delta y},$$

whose value should be determined by numerical experiment since the assumption that grid point 25 lies on the edge of the thermal boundary layer was made a priori. Mesh sizes Δx and Δy are set to $L/4$ and $\delta(L)/4$, respectively, and the Peclet number is defined as

$$\text{Pe} = \frac{u_{25}\,\Delta y}{\alpha}.$$

In the following example we will present an evaluation procedure for an integral given in Eq. (9-34).

Example 9-1 Evaluate

$$A_1 = \int_{-1}^{1}\int_{-1}^{1} \frac{\partial N_d}{\partial \eta} N_d V \, d\xi \, d\eta . \tag{a}$$

Solution: From Eq. (5-47*d*) we have

$$N_d(\xi, \eta) = \tfrac{1}{4}(1+\xi)(1-\eta) \tag{b}$$

and

$$\frac{\partial N_d}{\partial \eta} = -\frac{1}{4}(1+\xi). \tag{c}$$

Routinely, we approximate the nondimensionalized transverse velocity V as

$$\tilde{V}(\xi, \eta) = \sum_{p=a}^{d} N_p(\xi, \eta) V_p . \tag{d}$$

Substitution of Eq. (*d*) into Eq. (*a*) will generate four double integrals, i.e.,

$$I_a = -\frac{1}{16}\int_{-1}^{1}\int_{-1}^{1} (1+\xi)^2(1-\eta) N_a(\xi, \eta)\, d\xi \, d\eta = -\frac{1}{12}, \tag{e}$$

$$I_b = -\tfrac{1}{36}, \quad I_c = -\tfrac{1}{18}, \quad \text{and} \quad I_d = -\tfrac{1}{6}. \tag{f–h}$$

Therefore,

$$A_1 = -\tfrac{1}{12} V_a - \tfrac{1}{36} V_b - \tfrac{1}{18} V_c - \tfrac{1}{6} V_d, \tag{i}$$

where the values of $V_p, p = a\text{–}d$, can be found from the similarity solution.

Repeated applications of Eq. (9-32) to the nine interior grid points will generate a set of nine linear algebraic equations. Although the numerical result for such a crude approximation may not be accurate, the mathematical formulation and computer programming can be readily extended to finer mesh systems.

9-3*b* Difficulties Associated with the Least-Squares Finite-Element Method

It is interesting to examine the applicability of the least-squares finite-element method to Eqs. (9-28*a*) and (9-28*b*). First, we recall that the least-squares functional takes the form

$$J = \int_0^L \int_0^\delta (L\tilde{\theta})^2 \, dx \, dy \, . \tag{9-35}$$

If $\tilde{\theta}(x, y)$ is interpolated as given in Eq. (9-31), the nine nodal temperatures θ_j should be determined such that

$$\frac{\partial J}{\partial \theta_j} = 0 \, , \quad j = 1, 2, \ldots, 9 \, ,$$

or

$$\int_0^L \int_0^\delta (L\tilde{\theta})(LN_j) \, dx \, dy = 0 \, , \tag{9-36}$$

Unfortunately, the differentiability of $\tilde{\theta}$ cannot be weakened by integrating Eq. (9-36) by parts. Consequently, since $N_j(x, y)$ is a bilinear function of x and y, it follows that

$$\frac{\partial^2 N_j}{\partial x^2} = \frac{\partial^2 N_j}{\partial y^2} = \frac{\partial^2 \tilde{\theta}}{\partial x^2} = \frac{\partial^2 \tilde{\theta}}{\partial y^2} = 0 \, , \tag{9-37}$$

and the problem degenerates to

$$L\tilde{\theta} = U \frac{\partial \tilde{\theta}}{\partial x} + V \frac{\partial \tilde{\theta}}{\partial y} = 0 \, . \tag{9-38}$$

This situation is, of course, undesirable. One way to remedy this difficulty is to introduce auxiliary functions $\tilde{s}$ and $\tilde{t}$ defined as

$$\tilde{s} = \frac{\partial \tilde{\theta}}{\partial x} \quad \text{and} \quad \tilde{t} = \frac{\partial \tilde{\theta}}{\partial y} \, , \tag{9-39a, b}$$

and consider, instead of Eq. (9-35), the penalty functional

$$J_p = \int_0^L \int_0^\delta \left[U\tilde{s} + V\tilde{t} - \frac{\alpha}{u_{25}} \left(\frac{\partial \tilde{s}}{\partial x} + \frac{\partial \tilde{t}}{\partial y} \right) \right]^2 dx \, dy$$
$$+ \lambda_1 \int_0^L \int_0^\delta \left(\tilde{s} - \frac{\partial \tilde{\theta}}{\partial x} \right)^2 dx \, dy + \lambda_2 \int_0^L \int_0^\delta \left(\tilde{t} - \frac{\partial \tilde{\theta}}{\partial y} \right)^2 dx \, dy \, , \tag{9-40}$$

where λ_1 and λ_2 are penalty constants to be optimized such that positive-definiteness of the resulting matrix remains. The functions $\tilde{s}$ and $\tilde{t}$ are expressed by

$$\tilde{s}(x,y) = \sum_{j=1}^{25} N_j(x,y)s_j \tag{9-41a}$$

and

$$\tilde{t}(x,y) = \sum_{j=1}^{25} N_j(x,y)t_j \,. \tag{9-41b}$$

There are now 9 X 3 nodal unknowns, which can be determined by solving the following 27 algebraic equations:

$$\frac{\partial J_p}{\partial \theta_j} = 0\,, \qquad \frac{\partial J_p}{\partial s_j} = 0\,, \qquad \text{and} \qquad \frac{\partial J_p}{\partial t_j} = 0\,, \qquad j = 1, 2, \ldots, 9\,. \tag{9-42}$$

Although the differentiability of the approximate solution $\tilde{\theta}$ is now weakened according to Eq. (9-40), another difficulty arises: the system now becomes unclosed. The reason is that the nodal parameters s_j and t_j, $j = 10, 11, \ldots, 25$, which are supposed to be boundary conditions, unfortunately are not given. Under such circumstances, instead of expressing $\tilde{s}(x, y)$ in terms of four s_j at the four vertices of the rectangular element, we may replace the unspecified boundary conditions of s_j by θ_j.

Example 9-2 Consider a rectangular element with four vertices 1, 10, 11, and 2, as shown in Fig. 9-2. Express $\tilde{s}(x, y)$ within this element in terms of s_1, s_2, θ_1, θ_2, θ_{10}, and θ_{11}.

Solution: We first assume

$$\tilde{s}(\xi, \eta) = a + b\xi + c\eta + d\xi\eta = \frac{2}{h}\frac{\partial \tilde{\theta}}{\partial \xi}, \tag{a}$$

where

$$\xi = \frac{2(x - x^*)}{h}, \qquad \eta = \frac{2(y - y^*)}{k}, \tag{b}$$

and x^* and y^* are the coordinates of the element center. At grid points 1 and 2, Eq. (a) yields two constraints

$$s_1 = a + b + c + d \tag{c}$$

and

$$s_2 = a + b - c - d\,. \tag{d}$$

Next, to seek the other two constraints, we integrate Eq. (a) with respect to ξ to obtain, at $\eta = -1$,

$$\frac{\theta_2 - \theta_{11}}{h} = a - c \tag{e}$$

and at $\eta = 1$,

$$\frac{\theta_1 - \theta_{10}}{h} = a + c . \qquad (f)$$

Solving the algebraic equations

$$\begin{bmatrix} 1 & 1 & 1 & 1 \\ 1 & 1 & -1 & -1 \\ 1 & 0 & -1 & 0 \\ 1 & 0 & 1 & 0 \end{bmatrix} \begin{Bmatrix} a \\ b \\ c \\ d \end{Bmatrix} = \begin{Bmatrix} s_1 \\ s_2 \\ \dfrac{(\theta_2 - \theta_{11})}{h} \\ \dfrac{(\theta_1 - \theta_{10})}{h} \end{Bmatrix} , \qquad (g)$$

we obtain

$$\begin{Bmatrix} a \\ b \\ c \\ d \end{Bmatrix} = \begin{Bmatrix} \dfrac{1}{2h}(\theta_1 + \theta_2 - \theta_{10} - \theta_{11}) \\ \dfrac{s_1 + s_2}{2} - \dfrac{1}{2h}(\theta_1 + \theta_2 - \theta_{10} - \theta_{11}) \\ \dfrac{1}{2h}(\theta_1 - \theta_2 - \theta_{10} + \theta_{11}) \\ \dfrac{s_1 - s_2}{2} - \dfrac{1}{2h}(\theta_1 - \theta_2 - \theta_{10} + \theta_{11}) \end{Bmatrix} , \qquad (h)$$

which can be readily substituted into Eq. (*a*) to yield an expression for $\tilde{s}(x, y)$ in terms of $s_1, s_2, \theta_1, \theta_2, \theta_{10}$, and θ_{11}.

9-3*c* Finite-Difference Method

One of the most straightforward numerical techniques for solving Eqs. (9-28*a*) and (9-28*b*) is the finite-difference method. By use of the second-order central-difference scheme, we can readily derive

$$\left(\frac{4}{\text{Pe}} + \frac{4\gamma^2}{\text{Pe}}\right)\theta_j - \left(U_j + \frac{2}{\text{Pe}}\right)\theta_W - \left(\gamma V_j + \frac{2\gamma^2}{\text{Pe}}\right)\theta_S - \left(\frac{2}{\text{Pe}} - U_j\right)\theta_E$$
$$- \left(\frac{2\gamma^2}{\text{Pe}} - \gamma V_j\right)\theta_N = 0 , \quad j = 1, 2, \ldots, 9 , \qquad (9\text{-}43)$$

Before action is taken to solve the nine simultaneous equations, it is always advisable to check the algebraic procedure leading to Eq. (9-43) and to examine the stability of the scheme. The former task can be partially accomplished by letting

$$\theta_j = \theta_W = \theta_S = \theta_E = \theta_N$$

and then showing that the left-hand side of Eq. (9-43) is indeed equal to zero. The latter task, if rigorous proof is desired, requires an intensive matrix stability analysis.

By means of the extremum principle (Section 5-3*a*), however, we can readily assert that the quantities in parentheses in front of θ_E and θ_N must be positive to ensure numerical stability, i.e.,

$$\mathrm{Pe} < \frac{2}{U_j}. \tag{9-44}$$

and

$$\mathrm{Pe} < \frac{2\gamma}{V_j}.$$

Since the maximum value of U_j is unity, we will choose Pe to be 2 for the following numerical calculation. Another parameter whose value needs to be chosen carefully is γ. With the number of interior grid points fixed at nine, the value of γ should be adjusted such that grid point 21 lies on the edge of the thermal boundary layer. To estimate γ reasonably well, we first adopt

$$x_0 = x_{25} = 5\ \mathrm{cm}\,, \quad y_{25} = 0.01\ \mathrm{cm}\,, \quad u_\infty = 1000\ \mathrm{cm/s}\,, \quad \nu = 0.15\ \mathrm{cm^2/s}\,,$$
$$\text{and}\quad \mathrm{Pe} = 2\,. \tag{9-45}$$

From Eq. (9-45), we compute

$$\eta_{25} = y_{25}\left(\frac{u_\infty}{\nu x_{25}}\right)^{1/2} = 0.365\,. \tag{9-46}$$

From the similarity solution of the boundary-layer flow, we find that, for such a small value of η, the streamwise velocity is nearly linear in η or y. Therefore, we may approximate

$$U_{12} = U_3 = U_6 = U_9 = U_{18} = 0.25\,, \tag{9-47a}$$

$$U_{11} = U_2 = U_5 = U_8 = U_{19} = 0.5\,, \tag{9-47b}$$

$$U_{10} = U_1 = U_4 = U_7 = U_{20} = 0.75\,, \tag{9-47c}$$

and

$$U_{25} = U_{24} = U_{23} = U_{22} = U_{21} = 1\,. \tag{9-47d}$$

The ratio of the transverse velocity to the streamwise velocity can be computed from

$$\frac{v}{u} = \frac{1}{2}\,\mathrm{Re}_x^{-1/2}\,\frac{\eta f' - f}{f'}\,. \tag{9-48}$$

At $\eta = \eta_{25}$, we compute, with the aid of Appendix A,

$$V_{25} = \frac{v_{25}}{u_{25}} = \frac{1}{2}\,\frac{(0.00548)(0.365 \times 0.1 - 0.017)}{0.1} = 0.00054\,.$$

With the approximation that the transverse velocity v is nearly linear in y near the wall, it follows that

$$V_3 = V_6 = V_9 = 1.35 \times 10^{-4}\,,$$

$$V_2 = V_5 = V_8 = 2.7 \times 10^{-4}\,,$$

and $$V_1 = V_4 = V_7 = 4.05 \times 10^{-4} ,$$

To estimate the nodal temperatures, we assume

$$\theta(x, y) = 1.5 \frac{y}{\delta(x)} - 0.5 \left[\frac{y}{\delta(x)} \right]^3 \tag{9-49}$$

at any streamwise location. The y coordinates of the grid points, in reference to Fig. 9-2, are

$$\frac{y_{18}}{\delta} \approx 0.25 , \quad \frac{y_8}{\delta} \approx 0.62 , \quad \frac{y_9}{\delta} \approx 0.31 ,$$

and

$$\frac{y_6}{\delta} \approx 0.37 . \tag{9-50}$$

Consequently, we obtain

$$\theta_{18} = 0.367 , \quad \theta_8 = 0.811 , \quad \theta_9 = 0.450 ,$$

and $$\theta_6 = 0.530 . \tag{9-51}$$

Next, let us substitute all the estimated data into Eq. (9-43) for $j = 9$ to obtain

$$(2 + 2\gamma^2)(0.45) - (0.25 + 1)(0.53) - 0 - (1 - 0.25)(0.367)$$
$$- (\gamma^2 - 1.35 \times 10^{-4} \gamma)(0.811) = 0$$

or, after simplification,

$$\gamma^2 + 1.23 \times 10^{-3} \gamma - 0.425 = 0 , \tag{9-52}$$

whose significant root is

$$\gamma = 0.652 . \tag{9-53}$$

For easy presentation, we will set γ to unity. Then it is found that, in the coefficients of θ_S and θ_N,

$$V_j \approx \mathbf{O}(10^{-4}) \ll \frac{2}{\mathrm{Pe}} = 1 . \tag{9-54}$$

Therefore, Eq. (9-43) may be reduced to

$$4\theta_j - (U_j + 1)\theta_W - \theta_S - (1 - U_j)\theta_E - \theta_N = 0 \tag{9-55}$$

subject to the boundary conditions

$$\theta_{10} = \theta_{11} = \theta_{12} = \theta_{22} = \theta_{23} = \theta_{24} = 1 , \tag{9-56a}$$

$$\theta_{14} = \theta_{15} = \theta_{16} = 0 , \tag{9-56b}$$

and $$\theta_{18} = 0.367 , \quad \theta_{19} = 0.6875 , \quad \theta_{20} = 0.914 . \tag{9-56c}$$

The last three boundary conditions, Eq. (9-56*c*), were found according to Eq. (9-49). Repetitive application of Eq. (9-55) to grid points $j = 1, 2, \ldots, 9$ generates the following 9×9 matrix system:

$$[\mathbf{A}]\{\theta\} = \{\mathbf{B}\}, \tag{9-57}$$

where

$$[\mathbf{A}] = \begin{bmatrix} 4 & -1 & 0 & -0.25 & 0 & 0 & 0 & 0 & 0 \\ -1 & 4 & -1 & 0 & -0.5 & 0 & 0 & 0 & 0 \\ 0 & -1 & 4 & 0 & 0 & -0.75 & 0 & 0 & 0 \\ -1.75 & 0 & 0 & 4 & -1 & 0 & -0.25 & 0 & 0 \\ 0 & -1.5 & 0 & -1 & 4 & -1 & 0 & -0.5 & 0 \\ 0 & 0 & -1.25 & 0 & -1 & 4 & 0 & 0 & -0.75 \\ 0 & 0 & 0 & -1.75 & 0 & 0 & 4 & -1 & 0 \\ 0 & 0 & 0 & 0 & -1.5 & 0 & -1 & 4 & -1 \\ 0 & 0 & 0 & 0 & 0 & -1.25 & 0 & -1 & 4 \end{bmatrix},$$

$$\{\theta\}^T = \{\theta_1 \quad \theta_2 \quad \cdots \quad \theta_9\},$$

and $$\{\mathbf{B}\}^T = \{2.75 \quad 1.5 \quad 1.25 \quad 1 \quad 0 \quad 0 \quad 1.229 \quad 0.3438 \quad 0.2753\}.$$

The final solution, along with the boundary condition, is presented in Table 9-1 in accordance with the position of the grid point in the mesh system. Since numerous assumptions have been made, this solution is not expected to be accurate. After numerical experiments, it is found that the solution at $\gamma = 0.5$, also shown in Table 9-1, appears to be most reasonable.

9-4 DUCT FLOWS

In this section we will focus on duct flows, which, although generally governed by simpler transport equations than external flows, are common phenomena relevant to heat transfer.

Table 9-1 Temperature distribution near the leading edge of a thermal boundary layer

$\gamma = 1.0$					$\gamma = 0.5$				
1	1	1	1	1	1	1	1	1	1
1	0.9599	0.9119	0.8763	0.9140	1	0.9941	0.9822	0.9616	0.9140
1	0.8616	0.7485	0.6806	0.6875	1	0.9586	0.9023	0.8262	0.6875
1	0.6122	0.4496	0.3795	0.3670	1	0.7870	0.6371	0.5113	0.3670
	0	0	0	0		0	0	0	0

For incompressible flows moving inside a tube with constant thermal properties, the energy transport equation in cylindrical coordinates can be written as

$$\frac{\partial}{\partial x}(uT) + \frac{1}{r}\frac{\partial}{\partial r}(rvT) = \frac{\alpha}{r}\frac{\partial}{\partial r}\left(r\frac{\partial T}{\partial r}\right) + \alpha\frac{\partial^2 T}{\partial x^2}, \qquad (9\text{-}58a)$$

which is sometimes referred to as the conservative form because it can be derived directly from the consideration of energy conservation. With the aid of the continuity equation

$$\frac{\partial u}{\partial x} + \frac{1}{r}\frac{\partial}{\partial r}(rv) = 0,$$

Eq. (9-58*a*) can be written alternatively as

$$u\frac{\partial T}{\partial x} + v\frac{\partial T}{\partial r} = \frac{\alpha}{r}\frac{\partial}{\partial r}\left(r\frac{\partial T}{\partial r}\right) + \alpha\frac{\partial^2 T}{\partial x^2}. \qquad (9\text{-}58b)$$

We will use the finite-difference method to discretize Eq. (9-58*a*) and, in particular, derive the discretization equation along the centerline of the tube $r = 0$.

9-4*a* Finite-Difference Method

At grid point (i, j) in a mesh of uniform Δx and Δr, where $x = i\,\Delta x$ and $r = j\,\Delta r$, adoption of the central difference leads to

$$\frac{\partial}{\partial x}(uT) = \frac{1}{2\,\Delta x}(u_{i+1,j}T_{i+1,j} - u_{i-1,j}T_{i-1,j}), \qquad (9\text{-}59a)$$

$$\frac{1}{r}\frac{\partial}{\partial r}(rvT) = \frac{1}{2(\Delta r)^2 j}(r_{i,j+1}v_{i,j+1}T_{i,j+1} - r_{i,j-1}v_{i,j-1}T_{i,j-1}), \quad j \neq 0, \qquad (9\text{-}59b)$$

$$\frac{\alpha}{r}\frac{\partial}{\partial r}\left(r\frac{\partial T}{\partial r}\right) = \alpha\left(\frac{\partial^2 T}{\partial r^2} + \frac{1}{r}\frac{\partial T}{\partial r}\right)$$

$$= \frac{\alpha}{(\Delta r)^2}\left[\left(1 - \frac{1}{2j}\right)T_{i,j-1} + \left(1 + \frac{1}{2j}\right)T_{i,j+1} - 2T_{i,j}\right], \quad j \neq 0, \qquad (9\text{-}59c)$$

and

$$\alpha\frac{\partial^2 T}{\partial x^2} = \frac{\alpha}{(\Delta x)^2}(T_{i-1,j} + T_{i+1,j} - 2T_{i,j}). \qquad (9\text{-}59d)$$

Substituting Eqs. (9-59*a*)-(9-59*d*) into Eq. (9-58*a*) and rearranging, we obtain

$$(2\gamma^2 + 2)T_{i,j} - \left(1 - \frac{u_{i+1,j}\Delta x}{2\alpha}\right)T_{i+1,j} - \left(1 + \frac{u_{i-1,j}\Delta x}{2\alpha}\right)T_{i-1,j}$$
$$- \left[\gamma^2\left(1 + \frac{1}{2j}\right) - \frac{\gamma^2 v_{i,j+1}r_{i,j+1}}{2\alpha j}\right]T_{i,j+1} - \left[\gamma^2\left(1 - \frac{1}{2j}\right)\right.$$
$$\left. + \frac{\gamma^2 v_{i,j-1}r_{i,j-1}}{2\alpha j}\right]T_{i,j-1} = 0, \quad j \neq 0, \qquad (9\text{-}60)$$

where $\gamma = \Delta x/\Delta r$. When the flow velocities $u_{i+1,j}$ and $v_{i,j+1}$ become large such that the coefficients of $T_{i+1,j}$ and $T_{i,j+1}$ become positive, Eq. (9-60) violates the extremum principle (see also Section 5-3*a*) and numerical instability arises. One way to ensure stability is to adopt the upwind scheme, which replaces Eq. (9-59*a*), for example, by

$$\frac{\partial}{\partial x}(uT) = \begin{cases} \dfrac{1}{\Delta x}(u_{i+1/2,j}T_{i,j} - u_{i-1/2,j}T_{i-1,j}) & \text{if } u > 0\,, \quad (9\text{-}61a) \\ \dfrac{1}{\Delta x}(u_{i+1/2,j}T_{i+1,j} - u_{i-1/2,j}T_{i,j}) & \text{if } u < 0\,. \quad (9\text{-}61b) \end{cases}$$

In Eq. (9-61*a*) it is postulated that, due to the influence of convection, the temperatures at $i+\frac{1}{2}$ and $i-\frac{1}{2}$ are taken to be the upstream temperatures at i and $i-1$, respectively. The first three terms in Eq. (9-60) then become

$$\left(2\gamma^2 + 2 + \frac{u_{i+1/2,j}\Delta x}{\alpha}\right) T_{i,j} - T_{i+1,j} - \left(1 + \frac{u_{i-1/2,j}\Delta x}{\alpha}\right) T_{i-1,j}\,,$$

where the coefficients of $T_{i+1,j}$ and $T_{i-1,j}$ are always negative if $u > 0$. Thus, numerical stability is ensured.

At $r = 0$ (or $j = 0$) Eq. (9-60) is no longer valid since not only is j in the denominator but also $T_{i,-1}$ does not exist. One way to remove the difficulty is to use the L'Hospital rule and introduce a fictitious temperature $T_{i,-1}$.

Example 9-3 Derive a finite-difference equation similar to Eq. (9-60) at the centerline $r = 0$.

Solution: The radial convective term in Eq. (9-58*a*) can be rewritten as

$$\frac{1}{r}\frac{\partial}{\partial r}(rvT) = T\frac{\partial v}{\partial r} + v\frac{\partial T}{\partial r} + \frac{vT}{r}\,. \tag{a}$$

Since any derivative $\partial\phi/\partial r$ should vanish at $r = 0$ due to symmetry, and since based on the L'Hospital rule

$$\lim_{r\to 0}\frac{vT}{r} = \lim_{r\to 0}\frac{\partial}{\partial r}(vT) = 0\,, \tag{b}$$

we conclude that

$$\frac{1}{r}\frac{\partial}{\partial r}(rvT) = 0\,. \tag{c}$$

Next, the radial conduction term

$$\frac{\alpha}{r}\frac{\partial}{\partial r}\left(r\frac{\partial T}{\partial r}\right)$$

can be changed to

$$\alpha\left(\frac{\partial^2 T}{\partial r^2}+\frac{1}{r}\frac{\partial T}{\partial r}\right)=2\alpha\frac{\partial^2 T}{\partial r^2}=\frac{2\alpha}{(\Delta r)^2}(T_{i,-1}+T_{i,1}-2T_{i,0}), \qquad (d)$$

where $T_{i,-1}$ is a fictitious quantity and is equal to $T_{i,1}$ due to symmetry. Thus,

$$\frac{\alpha}{r}\frac{\partial}{\partial r}\left(r\frac{\partial T}{\partial r}\right)=\frac{4\alpha}{(\Delta r)^2}(T_{i,1}-T_{i,0}), \qquad (e)$$

and Eq. (9-58*a*) can be discretized to

$$(4\gamma^2+2)T_{i,0}-\left(1-\frac{u_{i+1,0}\,\Delta x}{2\alpha}\right)T_{i+1,0}-\left(1+\frac{u_{i-1,0}\,\Delta x}{2\alpha}\right)T_{i-1,0}$$
$$-4\gamma^2 T_{i,1}=0. \qquad (f)$$

9-4*b* Control Volume Method

The validity of introducing the fictitious temperature $T_{i,-1}$ may be assessed by using a different approach to derive Eq. (*f*). This approach is presented in the following example.

Example 9-4 Derive the discretized form of Eq. (9-58*a*) at the centerline $r=0$ by the control volume method.

Solution: In reference to Fig. 9-3, we write all the components contributing to the energy balance as follows:

$$\text{Conduction across the bottom face}=k\,\frac{\pi(\Delta r)^2}{4}\left(\frac{T_{i-1,0}-T_{i,0}}{\Delta x}\right), \qquad (a)$$

$$\text{Conduction across the top face}=k\,\frac{\pi(\Delta r)^2}{4}\left(\frac{T_{i+1,0}-T_{i,0}}{\Delta x}\right), \qquad (b)$$

$$\text{Conduction across the circumferential surface}=k\pi\,\Delta r\,\Delta x\left(\frac{T_{i,1}-T_{i,0}}{\Delta r}\right), \qquad (c)$$

$$\text{Upstream convection}=\rho c_p\,\frac{\pi(\Delta r)^2}{4}(u_{i-1/2,0}T_{i-1/2,0}), \qquad (d)$$

and

$$\text{Downstream convection}=-\rho c_p\,\frac{\pi(\Delta r)^2}{4}(u_{i+1/2,0}T_{i+1/2,0}). \qquad (e)$$

Adding Eqs. (*a*)-(*e*) and setting the sum to zero, we obtain

$$\frac{k\pi(\Delta r)^2}{4\,\Delta x}(T_{i-1,0}+T_{i+1,0}-2T_{i,0})+k\pi\,\Delta x(T_{i,1}-T_{i,0})$$
$$+\rho c_p\,\frac{\pi(\Delta r)^2}{4}(u_{i-1/2,0}T_{i-1/2,0}-u_{i+1/2,0}T_{i+1/2,0})=0. \qquad (f)$$

If we further assume

$$u_{i-1/2,0}T_{i-1/2,0}=\tfrac{1}{2}(u_{i-1,0}T_{i-1,0}+u_{i,0}T_{i,0})$$

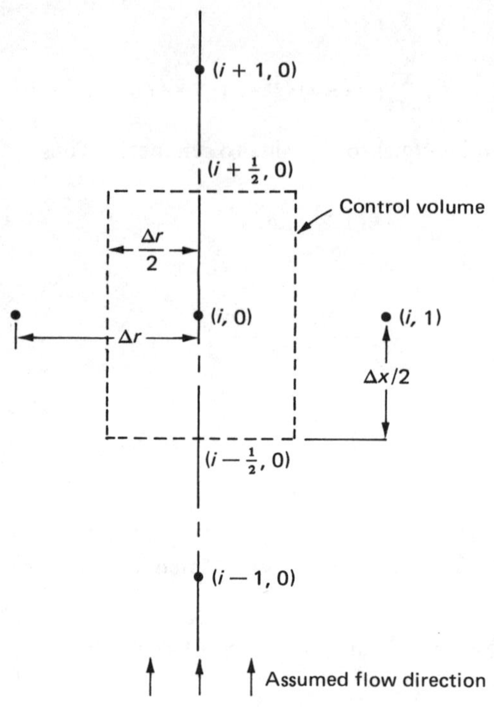

Figure 9-3 Cylindrical control volume enclosing centerline grid points $(i, 0)$.

and $$u_{i+1/2,0}T_{i+1/2,0} = \tfrac{1}{2}(u_{i,0}T_{i,0} + u_{i+1,0}T_{i+1,0}) \,,$$

then Eq. (*f*) can be simplified to

$$(4\gamma^2 + 2)T_{i,0} - \left(1 - \frac{u_{i+1,0}\,\Delta x}{2\alpha}\right)T_{i+1,0} - \left(1 + \frac{u_{i-1,0}\,\Delta x}{2\alpha}\right)T_{i-1,0}$$
$$-4\gamma^2 T_{i,1} = 0 \,, \qquad (g)$$

which is identical to Eq. (*f*) of Example 9-3.

For further information regarding the discretization formulation at the centerline, see, for example, [11-13].

Among the analyses of duct flows, much effort has been devoted to investigating the Graetz problem [14-27], i.e., flows that are hydrodynamically developed but thermally developing. Various numerical methods, such as finite difference [14-19], Galerkin [20-24], local similarity [25], and perturbation [26] methods, have been used.

SYMBOLS

$B_j(\eta)$ piecewise cubic splines at $\eta = \eta_j$

$c_{p,k}$ specific heat of species k, J/kg·K

$f(\eta)$	similarity function [Eq. (9-12)]
h	enthalpy, J/kg, or mesh size
j_k''	mass flux of species k, kg/m^2·s
J	least-squares functional
$N_j(x,y)$	basis function of C^0 square element [Eqs. (5-47*a*)-(5-47*d*)]
Pe	Peclet number ($= u_{25}\,\Delta y/\alpha$)
Pr	Prandtl number ($= \nu/\alpha$)
$\dot{\mathrm{q}}''$	conductive heat flux, W/m^2
s	$\partial\tilde{\theta}/\partial x$
t	$\partial\tilde{\theta}/\partial y$ or $\eta/6$
U	u/u_{25}
V	v/u_{25}
x^*, y^*	coordinates of the center of a typical square element
Y_k	mass fraction of species k
α	thermal diffusivity ($= k/\rho c_p$), m^2/s
γ	mesh-size ratio ($= \Delta x/\Delta y$ or $\Delta x/\Delta r$)
η	similarity variable [$= y(u_\infty/\nu x)^{1/2}$] or isoparametric coordinate [$= 2(y - y^*)/\Delta y$]
θ	normalized temperature [$= (T - T_w)/(T_\infty - T_w)$]
ξ	isoparametric coordinate [$= 2(x - x^*)/\Delta x$]
$\psi(x,y)$	stream function, m^2/s
ω	normalized stream function

Subscripts

p	penalty
wa	water

REFERENCES

1. R. Srivastava and D. E. Rosner, A New Approach to the Correlation of Boundary Layer Mass Transfer Rates with Thermal Diffusion and/or Variable Properties, *Int. J. Heat Mass Transfer*, vol. 22, pp. 1281–1294, 1979.
2. Y. Tambour, On Thermal-Diffusion Effects in Chemically Frozen Multicomponent Boundary Layer with Surface Catalytic Recombination behind a Strong Moving Shock, *Int. J. Heat Mass Transfer*, vol. 23, pp. 321–327, 1980.
3. W. M. Kays, *Convective Heat and Mass Transfer*, p. 207, McGraw-Hill, New York, 1966.
4. E. R. G. Eckert and R. M. Drake, Jr., *Heat and Mass Transfer*, p. 307, McGraw-Hill, New York, 1972.
5. B. Gebhart, *Heat Transfer*, p. 242, McGraw-Hill, New York, 1971.
6. A. Leveque, Transmission de Chaleur par Convection, *Annls. Mines*, vol. 13, pp. 283–290, 1928.
7. A. Leveque, Les Lois de la Transmission de Chaleur par Convection, *Annls. Mines*, vol. 13, pp. 201–239, 1928.
8. R. L. Mahajan and B. Gebhart, Higher-Order Boundary Layer Effects in Plane Horizontal Natural Convection Flows, *ASME J. Heat Transfer*, vol. 102, pp. 368–371, 1980.
9. F. N. Lin, Thermal Response Behavior of Boundary Layer Flows, *Int. J. Heat Mass Transfer*, vol. 21, pp. 683–690, 1978.

10. R. Ghez, Mass Transport and Surface Reactions in Leveque's Approximation, *Int. J. Heat Mass Transfer*, vol. 21, pp. 745–750, 1978.
11. P. C. Sukanek and R. P. Rhodes, Centerline Formulation in the Numerical Computation of Axisymmetric Flows, *AIAA J.*, vol. 16, pp. 1099–1102, 1978.
12. G. de Vahl Davis, A Note on a Mesh for Use with Polar Coordinates, *Numer. Heat Transfer*, vol. 2, pp. 261–266, 1979.
13. P. C. Sukanek, Conservation Errors in Axisymmetric Finite-Difference Equations, *AIAA J.*, vol. 17, pp. 99–101, 1979.
14. J. W. Ou and K. C. Cheng, Natural Convection Effects on Graetz Problem in Horizontal Isothermal Tubes, *Int. J. Heat Mass Transfer*, vol. 20, pp. 953–960, 1977.
15. A. F. Emery, P. K. Neighbors, and F. B. Gessner, Computational Procedure for Developing Turbulent Flow and Heat Transfer, *Numer. Heat Transfer*, vol. 2, pp. 399–416, 1979.
16. M. W. Collins, Finite Difference Analysis for Developing Laminar Flow in Circular Tubes Applied to Forced and Combined Convection, *Int. J. Numer. Methods Eng.*, vol. 15, pp. 381–404, 1980.
17. L. C. Chow, A. Campo, and C. L. Tien, Heat Transfer Characteristics for Laminar Flow between Parallel Plates with Suction, *Int. J. Heat Mass Transfer*, vol. 23, pp. 740–743, 1980.
18. R. Chilukuri and R. H. Pletcher, Numerical Solutions to the Partially Parabolized Navier-Stokes Equations for Developing Flow in a Channel, *Numer. Heat Transfer*, vol. 3, pp. 169–188, 1980.
19. S. J. Rhee and D. K. Edwards, Laminar Entrance Flow in a Flat Plate Duct with Asymmetric Suction and Heating, *Numer. Heat Transfer*, vol. 4, pp. 85–100, 1981.
20. S. Del Giudice, Step-by-Step Analysis of Flow Development in Ducts, *Numer. Heat Transfer*, vol. 2, pp. 291–302, 1979.
21. V. Javeri, Laminar Heat Transfer in a Rectangular Channel for the Temperature Boundary of the Third Kind, *Int. J. Heat Mass Transfer*, vol. 21, pp. 1029–1034, 1978.
22. M. Strada, S. Del Giudice, and G. Comini, Finite Element Solutions for Laminar Forced Convection in the Thermal Entrance Region of Ducts, *Numer. Heat Transfer*, vol. 1, pp. 471–488, 1978.
23. S. Del Giudice, M. Strada, and G. Comini, Laminar Heat Transfer in the Entrance Region of Ducts, *Numer. Heat Transfer*, vol. 2, pp. 487–496, 1979.
24. G. Comini, S. Del Giudice, and M. Strada, Finite Element Analysis of Laminar Flow in the Entrance Region of Ducts, *Int. J. Numer. Methods Eng.*, vol. 15, pp. 507–517, 1980.
25. H. T. Lin and Y. P. Shih, Unsteady Thermal Entrance Heat Transfer of Power-Law Fluids in Pipes and Plates Slits, *Int. J. Heat Mass Transfer*, vol. 24, pp. 1531–1539, 1981.
26. L. S. Yao, Free-forced Convection in the Entry Region of a Heated Straight Pipe, *ASME J. Heat Transfer*, vol. 100, pp. 212–219, 1978.

PROBLEMS

9-1 Even if the Lewis number is not equal to unity, prove that Eq. (9-5) can be reduced to Eq. (9-6) for a binary system provided $c_{p,1} = c_{p,2}$.

9-2 (*a*) Based on Eq. (9-13), use a crude approximation to check whether $\theta'(0)$ decreases as Pr decreases. Note that this trend is consistent with our physical intuition and that $\theta'(0) = f''(0) = 0.332$ if Pr is unity. [A more elaborate calculation yields $\theta'(0) = 0.292$ at Pr = 0.7.]

(*b*) Consulting Table 7-5, we obtain $f''(0) = 0.16495$ at $\lambda = v_w \mathrm{Re}_x^{1/2}/u_\infty = 0.25$. Calculate $f(\eta)$ by any integration scheme from $\eta = 0$ to $\eta = 6$.

(*c*) Use Eq. (9-13) to calculate $\theta(\eta)$ and $\theta'(0)$.

(*d*) Assume $\theta'(0) = a\lambda + b$ and determine the values of a and b.

(*e*) Dry air ($T_\infty = 38$°C, $\nu = 0.15$ cm²/s) flows at 3 m/s over a wetted flat plate [$T_w = 55$°C, $Y_{wa}(x, 0) = 0.1$]. Use Eq. (9-8) and the definition of the similarity variable η to find the mass transfer rate (g/s) from the leading 0.3 m of the plate. (Assume unit Lewis number so that $\nu/D = 0.7$.)

9-3 Use the quadratic spline collocation method to solve Eq. (9-11). The procedure is analogous to that described between Eqs. (9-15) and (9-21) except that now θ'' will be approximated by the 3-point relation.

9-4 (*a*) Derive the set of three nonlinear algebraic equations $\partial J_p/\partial \theta_j = 0$, $\partial J_p/\partial s_j = 0$, and $\partial J_p/\partial t_j = 0$, where J_p is given by Eq. (9-40).

(*b*) Prove that the solution obtained in (*a*) (however, do not actually obtain it) will yield a *minimum* value of J_p.

9-5 Derive two equations analogous to Eq. (9-57) for $\gamma = 0.3$ and 1.5, respectively. If a computer is accessible, solve for $\theta_j, j = 1, 2, \ldots, 9$. Does the solution appear qualitatively reasonable?

CHAPTER

TEN

RADIATION

Thermal radiation is a heat transfer mode in addition to conduction and convection, and it becomes especially important in high-temperature systems. It is different from the conduction and convection modes in that radiation can be transported through long distances and the evaluation of the radiation flux generally involves multiple integrals. A fundamental law governing the radiation phenomenon is Planck's spectral distribution of emissive power, given as a function of absolute temperature and wavelength by

$$\dot{q}''_\lambda = \frac{2\pi C_1}{\lambda^5 [\exp(C_2/\lambda T) - 1]} \text{ W/cm}^2 \cdot \mu\text{m} , \tag{10-1}$$

where $C_1 = hc_0^2$ and $C_2 = hc_0/k_p$, with h, c_0, and k_p being the Planck constant, speed of light in the vacuum, and Boltzmann constant, respectively. Upon integration of Eq. (10-1) with respect to λ, the radiative flux over the entire wavelength range can be evaluated. It is the evaluation of this integral that makes radiation a challenging subject.

In Section 10-1 we will briefly describe radiative exchange among black and diffusive surfaces. This section is brief since most of the material not only is intensively covered in standard radiation textbooks but also is remotely related to numerical analyses.

Section 10-2 describes the gray-gas radiative flux. When the gas is radiatively participating, the equation of transfer that associates the intensity with the gas temperature is needed in addition to the equation of energy conservation. The classical problem of a radiatively ideal medium (with constant absorption coefficient) between two infinite parallel black plates is analyzed. The analysis of this problem is the foundation for more advanced analyses of nongray gaseous radiation.

The nongray-gas radiative flux is discussed in Section 10-3. Total band absorptance is also introduced. By introducing this quantity, integrations with respect to wavelength and physical distance are avoided.

In Section 10-4, the integrodifferential equation governing combined conduction and radiation in a one-dimensional system is numerically solved by both the finite-difference method and the variational method.

In Section 10-5, we will describe an approach to solving the two-dimensional combined conduction and radiation problem without having to assume that the radiative flux is dominant in one direction. This approach is called the discretized-intensity method; it enables us to obtain a set of algebraic equations applicable to all the grid points in the computational domain.

Finally, in Section 10-6, we will solve the integrodifferential equation governing combined conductive, convective, and radiative boundary-layer flow. To simplify the computation, the radiative flux is assumed to be transversely dominant, i.e., $\partial\dot{q}_r''/\partial y \gg \partial\dot{q}_r''/\partial x$. The collocation method with cubic splines is adopted to investigate a thermally developing flow in the channel.

10-1 RADIATIVE EXCHANGE AMONG SURFACES

In reference to Fig. 10-1, the energy balance at surface A_1 can be written as

$$q_1'' = q_{o,1}'' - q_{i,1}'' , \tag{10-2}$$

where the subscripts o and i stand for outgoing and incoming, respectively. If A_2 is the only surface facing A_1, then the radiation arriving at A_1 is solely that leaving A_2, i.e.,

$$A_1\dot{q}_{i,1}'' = A_2\dot{q}_{o,2}''F_{2\text{-}1} ,$$

where $F_{2\text{-}1}$ is called the shape (or view, configuration) factor, defined as the ratio of the radiation arriving at surface A_1 to the total radiation emitted by surface A_2.

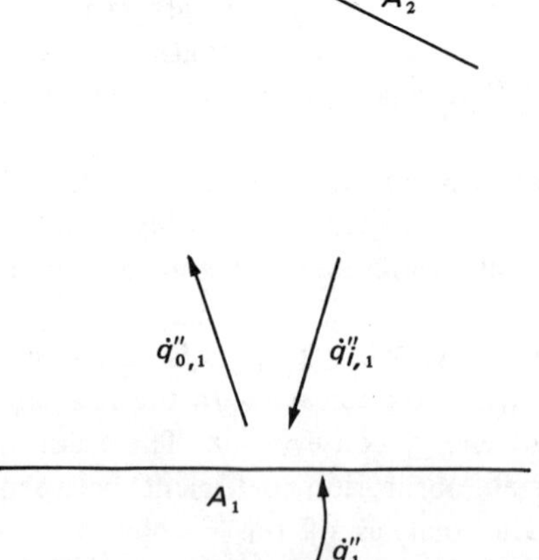

Figure 10-1 Radiation balance for surface A_1.

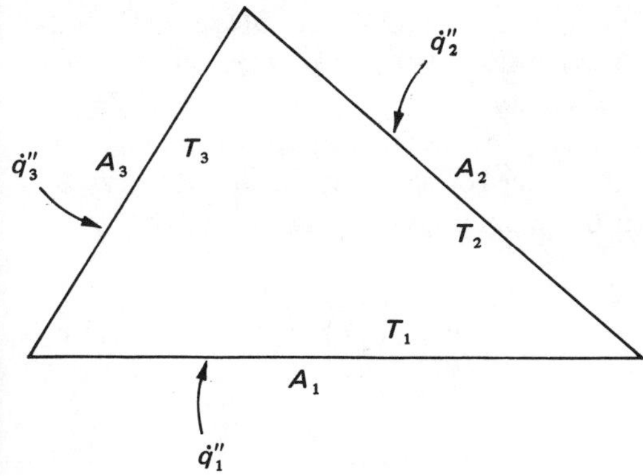

Figure 10-2 Radiative exchange within a black triangular enclosure.

Expressions for the shape factors for various configurations are given in standard radiation textbooks [1, pp. 748-791; 2, pp. 339-349]. For an infinitely long enclosure consisting of surfaces $A_1, A_2, \ldots, A_n$, it follows, from the definition of the shape factor and the reciprocal rule $A_jF_{j\text{-}k} = A_kF_{k\text{-}j}$, that

$$F_{1\text{-}1} + F_{1\text{-}2} + \cdots + F_{1\text{-}n} = 1\,. \tag{10-3}$$

10-1*a* Black Surfaces

A surface is idealized and said to be black if it absorbs all the incident radiation and emits its maximum radiation isotropically according to

$$\dot{q}_o'' = \int_0^\infty \dot{q}_\lambda''\, d\lambda = \sigma T^4\,, \tag{10-4}$$

where σ is the Stefan-Boltzman constant and has the numerical value 5.67×10^{-12} W/cm$^2 \cdot$K^4.

For convenience, let us consider radiation exchange among the three surfaces A_1, A_2, and A_3 of a triangular enclosure as shown in Fig. 10-2. Taking the energy balance over A_1 yields

$$\dot{q}_1 = A_1\sigma T_1^4 - A_2F_{2\text{-}1}\sigma T_2^4 - A_3F_{3\text{-}1}\sigma T_3^4\,, \tag{10-5}$$

where $\dot{q}_1$ (without the double prime) denotes the radiative flow rate (watts), and each surface temperature is assumed to be uniform. By using the reciprocal rule, Eq. (10-5) can be rewritten as

$$\dot{q}_1'' = \sigma T_1^4 - F_{1\text{-}2}\sigma T_2^4 - F_{1\text{-}3}\sigma T_3^4\,. \tag{10-6a}$$

Similarly,

$$\dot{q}_2'' = \sigma T_2^4 - F_{2\text{-}1}\sigma T_1^4 - F_{2\text{-}3}\sigma T_3^4 \tag{10-6b}$$

and

$$\dot{q}_3'' = \sigma T_3^4 - F_{3\text{-}1}\sigma T_1^4 - F_{3\text{-}2}\sigma T_2^4\,. \tag{10-6c}$$

If any three of the six quantities $\dot{q}_k''$ and T_k, $k = 1, 2, 3$, are prescribed, the set of Eqs. (10-6*a*)-(10-6*c*) is closed. Note also that the sum $\dot{q}_1'' + \dot{q}_2'' + \dot{q}_3''$ should vanish, since the enclosure is maintained in the steady state.

If the temperatures are not uniform along the surfaces as shown in Fig. 10-3, Eqs. (10-6*a*)-(10-6*c*) are no longer valid. Referring to Fig. 10-3, which shows a system of two parallel black plates of finite length L, we obtain, according to Eq. (10-2),

$$\dot{q}_1''(x_1) = \sigma T_1^4(x_1) - \int_0^L \sigma T_2^4(x_2) F(x_2, x_1)\, dx_2$$

and

$$\dot{q}_2''(x_2) = \sigma T_2^4(x_2) - \int_0^L \sigma T_1^4(x_1) F(x_1, x_2)\, dx_1$$

or, after nondimensionalization,

$$\psi_1(\xi_1) = \Theta_1(\xi_1) - \int_0^1 \Theta_2(\xi_2) F(\xi_2, \xi_1)\, d\xi_2 \tag{10-7a}$$

and

$$\psi_2(\xi_2) = \Theta_2(\xi_2) - \int_0^1 \Theta_1(\xi_1) F(\xi_1, \xi_2)\, d\xi_1\,, \tag{10-7b}$$

where $\psi = \dot{q}''/\sigma T_\infty^4$, $\Theta = T^4/T_\infty^4$, $\xi_1 = x_1/L$, $\xi_2 = x_2/L$, T_∞ is a known reference temperature, and $F(\xi_1, \xi_2)$ is the shape factor between the two strips dx_1 and dx_2, expressed by

$$F(\xi_1, \xi_2) = \frac{1}{2[(\xi_2 - \xi_1)^2 + 1]^{3/2}}\,.$$

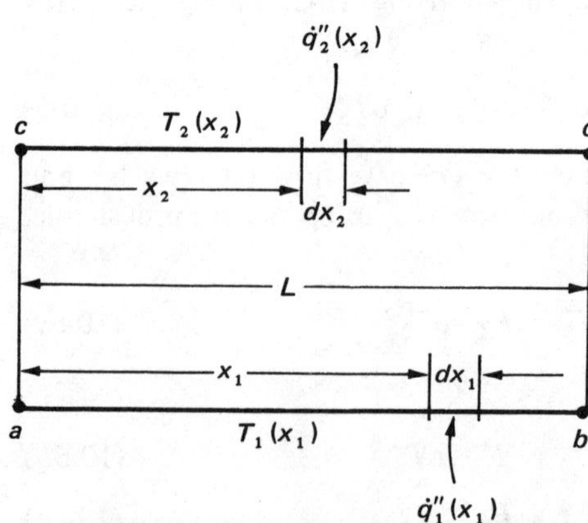

Figure 10-3 Radiative exchange between two parallel black plates with nonuniform temperature and heat flux distributions.

Once $\psi_1(\xi_1)$ and $\psi_2(\xi_2)$ are prescribed, Eqs. (10-7*a*) and (10-7*b*) can be solved either analytically or numerically.

Example 10-1 Discretize Eqs. (10-7*a*) and (10-7*b*) by taking the crude approximations

$$\Theta_1(\xi_1) = \Theta_a(1-\xi_1) + \Theta_b \xi_1 \tag{a}$$

and

$$\Theta_2(\xi_2) = \Theta_c(1-\xi_2) + \Theta_d \xi_2 \,. \tag{b}$$

The nondimensionalized heat fluxes $\psi_1(\xi_1)$ and $\psi_2(\xi_2)$ are assumed to be given.

Solution: With Eqs. (*a*) and (*b*) we have, at $\xi_1 = 0$,

$$\int_0^1 \Theta_2(\xi_2) F(\xi_2, 0)\, d\xi_2 = \frac{1}{2}\int_0^1 \frac{\Theta_c(1-\xi_2) + \Theta_d \xi_2}{(\xi_2^2+1)^{3/2}}\, d\xi_2 = e_1\Theta_c + e_2\Theta_d \,. \tag{c}$$

Similarly, at $\xi_2 = 0$,

$$\int_0^1 \Theta_1(\xi_1) F(\xi_1, 0)\, d\xi_2 = e_1\Theta_a + e_2\Theta_b \,. \tag{d}$$

At $\xi_1 = 1$ we also obtain

$$\int_0^1 \Theta_2(\xi_2) F(\xi_2, 1)\, d\xi_2 = \frac{1}{2}\int_0^1 \frac{\Theta_c(1-\xi_2) + \Theta_d \xi_2}{[(\xi_2-1)^2+1]^{3/2}}\, d\xi_2 = e_3\Theta_c + e_4\Theta_d \,, \tag{e}$$

and likewise at $\xi_2 = 1$,

$$\int_0^1 \Theta_1(\xi_1) F(\xi_1, 1)\, d\xi_1 = e_3\Theta_a + e_4\Theta_b \,. \tag{f}$$

Substituting Eqs. (*c*)–(*f*) into Eqs. (10-7*a*) and (10-7*b*) with proper rearrangement leads to

$$[\mathbf{A}]\,\{\mathbf{\Theta}\} = \{\mathbf{B}\}\,, \tag{g}$$

where

$$[\mathbf{A}] = \begin{bmatrix} 1 & 0 & -e_1 & -e_2 \\ -e_1 & -e_2 & 1 & 0 \\ 0 & 1 & -e_3 & -e_4 \\ -e_3 & -e_4 & 0 & 1 \end{bmatrix},$$

$$\{\Theta\}^T = \{\Theta_a \quad \Theta_b \quad \Theta_c \quad \Theta_d\} \quad \text{and} \quad \{\mathbf{B}\}^T = \{\psi_1(0) \quad \psi_2(0) \quad \psi_1(1) \quad \psi_2(1)\}.$$

Once $\{\Theta\}^T$ is solved, the temperature distribution can be obtained from Eqs. (*a*) and (*b*).

10-1*b* Diffuse-Gray Surfaces

A surface is said to be diffuse-gray if its emissivity ϵ and absorptivity α are independent of the inclination (emitting or absorbing) angle and wavelength. By Kirchhoff's law $\epsilon(T) = \alpha(T)$ and the fact that the outgoing radiative flux is the sum of emission and reflection, it follows that

$$\dot{q}''_o = \epsilon\sigma T^4 + (1-\epsilon)\dot{q}''_i\,. \tag{10-8}$$

If $\epsilon = 1$, Eq. (10-8) reduces to Eq. (10-4). Elimination of $\dot{q}''_i$ from Eqs. (10-2) and (10-8) gives

$$\dot{q}'' = \frac{\epsilon}{1-\epsilon}(\sigma T^4 - \dot{q}''_o)\,, \tag{10-9}$$

which is an important equation relating $\dot{q}''$ to $\dot{q}''_o$. In reference to Fig. 10-4, which shows a diffuse-gray triangular enclosure, the energy balance equations corresponding to Eqs. (10-6*a*)–(10-6*c*) for black surfaces are

$$\dot{q}''_1 = \dot{q}''_{o,1} - F_{1\text{-}2}\dot{q}''_{o,2} - F_{1\text{-}3}\dot{q}''_{o,3}\,, \tag{10-10a}$$

$$\dot{q}''_2 = \dot{q}''_{o,2} - F_{2\text{-}1}\dot{q}''_{o,1} - F_{2\text{-}3}\dot{q}''_{o,3}\,, \tag{10-10b}$$

and

$$\dot{q}''_3 = \dot{q}''_{o,3} - F_{3\text{-}1}\dot{q}''_{o,1} - F_{3\text{-}2}\dot{q}''_{o,2}\,. \tag{10-10c}$$

There are nine quantities $\dot{q}''_k$, $\dot{q}''_{o,k}$, and T_k, $k = 1, 2, 3$, satisfying only six equations, namely Eqs. (10-10*a*)–(10-10*c*) and Eq. (10-9) applied three times to three surfaces. If three of the nine quantities are given, the system is closed.

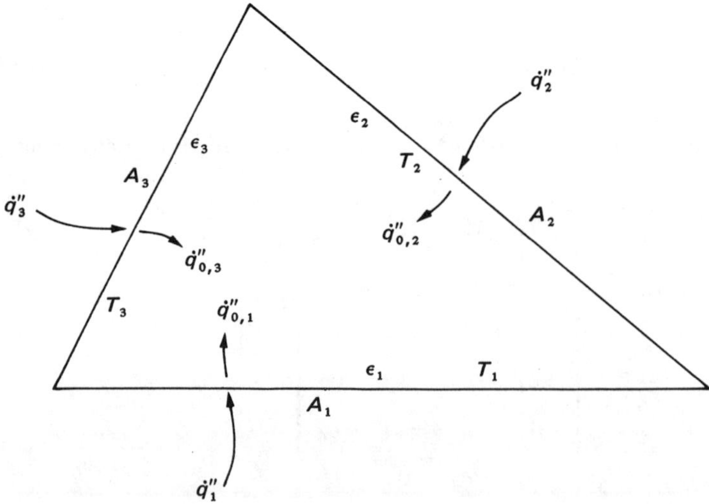

Figure 10-4 Radiative exchange within a diffuse-gray triangular enclosure.

10-2 GRAY–GAS RADIATION

The classical problem of radiation exchange among surfaces without the participation of gases is relatively simple and well understood. The emphasis in current radiation research has shifted to gaseous radiation because of the complicated and interesting phenomena associated with high-temperature industrial applications such as combustion, nuclear fission and fusion, and glass melting. Two major difficulties arise in the study of gaseous radiation. First, when a radiatively participating medium is present in a system, the radiation is absorbed and emitted not only at the system boundaries but also in the interior of the system. We therefore need to know the distributions of field variables such as temperature and species mass fraction. The mathematics required to seek this information is inevitably involved. Second, the absorption coefficients of the absorbing-emitting gases are in general strongly dependent on wavelength. The integrations that must be carried out with respect to wavelength and other independent variables to compute the radiative flux are formidable. Most of the effort in understanding gaseous radiation is devoted to the derivation of reasonable simplifications, which will be described in the following sections.

10-2*a* Equation of Transfer

In the analysis of gaseous radiation it is customary to introduce a convenient quantity called the intensity I_λ, which is defined as the radiative flux per projected surface area, per solid angle, and per wavelength. According to this definition, the radiative flux $\dot{q}''_\lambda$ can be related to the intensity I_λ by

$$\dot{q}''_\lambda = 2\pi \int_0^{\pi/2} I_\lambda(\beta) \sin\beta \cos\beta \, d\beta \,, \tag{10-11a}$$

where β is the inclination angle and circumferential symmetry has been assumed. If I_λ is independent of β, as in the black-surface case, then Eq. (10-11*a*) reduces to

$$\dot{q}''_\lambda = \pi I_\lambda \,. \tag{10-11b}$$

It has been assumed [3] that the variation of the intensity along one direction due to attenuation and emission is proportional to, respectively, the intensity itself and the black intensity that obeys Eqs. (10-1) and (10-11*b*). Mathematically, this assumption can be written as

$$dI_\lambda = -I_\lambda a_\lambda \, ds + I_{b\lambda} a_\lambda \, ds \,, \tag{10-12}$$

where a_λ is the spectral absorption coefficient and s is the physical distance traveled by the intensity. In this section, we will concentrate on an ideal case: gray gases. Dropping the subscript λ and integrating Eq. (10-12) with respect to s, we obtain

$$I(\tilde{\kappa}) = I(\tilde{\kappa}_0) \exp(\tilde{\kappa}_0 - \tilde{\kappa}) + \int_{\tilde{\kappa}_0}^{\tilde{\kappa}} I_b(\tilde{t}) \exp(\tilde{t} - \tilde{\kappa}) \, d\tilde{t} \,, \tag{10-13a}$$

where

$$\tilde{\kappa} = \int_{s_0}^{s} a\, ds \tag{10-14a}$$

and s_0 is the starting position of the intensity. In many one-dimensional problems it is preferable to use, instead of Eq. (10-13*a*), the form

$$I(\kappa) = I\left(\frac{\kappa_0}{\mu}\right) \exp\left(\frac{\kappa_0 - \kappa}{\mu}\right) + \int_{\kappa_0}^{\kappa} I_b(t)\mu^{-1} \exp\left(\frac{t-\kappa}{\mu}\right) dt\,, \tag{10-13b}$$

where

$$\kappa = \int_{x_0}^{x} a\, dx\,, \quad x = s\mu\,, \quad \text{and} \quad \mu = \cos\beta\,. \tag{10-14b}$$

The relation between $\tilde{\kappa}$ and κ is depicted in Fig. 10-5. Equation (10-13*a*) or (10-13*b*) will yield the distribution of the intensity if the temperature profile and the expression for the absorption coefficient are given.

Example 10-2 Consider a one-dimensional gas slab confined between two black plates, as shown in Fig. 10-5. The distributions of gas temperature and absorption coefficient are given, respectively, as

$$T = T_0\left(1 + \frac{x}{D}\right) \tag{a}$$

and

$$a = \frac{2T}{3T_0 D}\,. \tag{b}$$

Find the radiative flux $\dot{q}''$ emerging from a small hole at $x = D$.

Solution: Using Eqs. (10-11) and (10-13*b*), we derive, with $\kappa_0 = 0$,

$$\dot{q}'' = \dot{q}''_{\text{att}} + \dot{q}''_{\text{em}}\,, \tag{c}$$

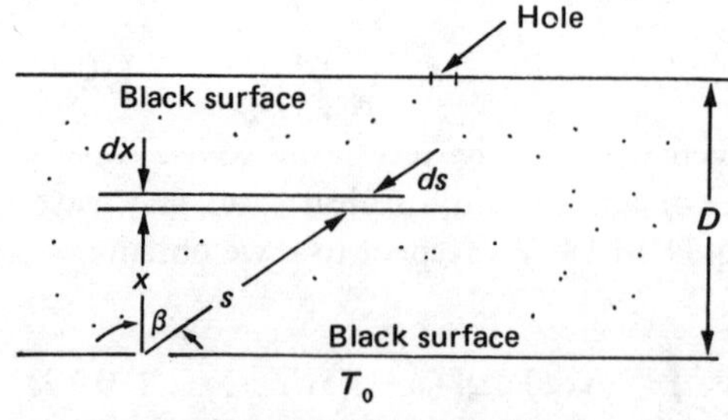

Figure 10-5 One-dimensional gas slab confined between two black plates.

where "att" and "em" denote attenuation and emission, and

$$\dot{q}''_{\mathrm{att}}(\kappa) = 2\pi I(0) \int_0^{\pi/2} \exp\left(-\frac{\kappa}{\mu}\right) \sin\beta \cos\beta \, d\beta \tag{d}$$

and

$$\dot{q}''_{\mathrm{em}}(\kappa) = 2 \int_0^{\pi/2} \int_0^{\kappa} \sigma T^4(t)\mu^{-1} \exp\left(\frac{t-\kappa}{\mu}\right) \sin\beta \cos\beta \, dt \, d\beta \, . \tag{e}$$

But

$$\kappa(x) = \int_0^x \left(\frac{2T}{3T_0 D}\right) dx = \frac{2}{3}\frac{x}{D} + \frac{1}{3}\left(\frac{x}{D}\right)^2 . \tag{f}$$

Therefore, changing the variable $\cos\beta$ to μ, we obtain

$$\dot{q}''_{\mathrm{att}}(\kappa_D) = 2\pi I(0) \int_0^1 \exp\left(-\frac{1}{\mu}\right) \mu \, d\mu \tag{g}$$

and

$$\dot{q}''_{\mathrm{em}}(\kappa_D) = 2 \int_0^1 \int_0^1 \sigma T^4(t) \exp\left(\frac{t-1}{\mu}\right) dt \, d\mu \, . \tag{h}$$

The integral in Eq. (*g*) is known to be the exponential integral $E_3(1)$, for which the general expression is

$$E_n(t) = \int_0^1 \mu^{n-2} \exp\left(-\frac{t}{\mu}\right) d\mu \, . \tag{i}$$

The numerical values of $E_1(t)$, $E_2(t)$, and $E_3(t)$ for $0 \leqslant t \leqslant 3.5$ are given in Table 10-1. To evaluate the double integral in Eq. (*h*), we first find from Eqs. (*a*) and (*f*) that

$$T^4(x) = T_0^4\left(1 + \frac{x}{D}\right)^4 = T_0^4(1 + 3\kappa)^2 \, , \tag{j}$$

and then use the trapezoidal rule to approximate

$$\int_0^1 \int_0^1 (1 + 3t)^2 \exp\left(\frac{t-1}{\mu}\right) dt \, d\mu = \int_0^1 (1 + 3t)^2 E_2(1-t) \, dt$$

$$\approx \frac{1}{4}\left[E_2(1) + 6.25E_2(0.5) + 16E_2(0)\right] . \tag{k}$$

Table 10-1 Values of exponential integrals $E_n(x)$

x	$E_1(x)$	$E_2(x)$	$E_3(x)$
0	∞	1.0000	0.5000
0.01	4.0379	0.9497	0.4903
0.02	3.3547	0.9131	0.4810
0.03	2.9591	0.8817	0.4720
0.04	2.6813	0.8535	0.4633
0.05	2.4679	0.8278	0.4549
0.06	2.2953	0.8040	0.4468
0.07	2.1508	0.7818	0.4388
0.08	2.0269	0.7610	0.4311
0.09	1.9187	0.7412	0.4236
0.10	1.8229	0.7225	0.4163
0.20	1.2227	0.5742	0.3519
0.30	0.9057	0.4691	0.3000
0.40	0.7024	0.3894	0.2573
0.50	0.5598	0.3266	0.2216
0.60	0.4544	0.2762	0.1916
0.70	0.3738	0.2349	0.1661
0.80	0.3106	0.2009	0.1443
0.90	0.2602	0.1724	0.1257
1.00	0.2194	0.1485	0.1097
1.25	0.1464	0.1035	0.0786
1.50	0.1000	0.0731	0.0567
1.75	0.0695	0.0522	0.0412
2.00	0.0489	0.0375	0.0301
2.25	0.0348	0.0272	0.0221
2.50	0.0249	0.0198	0.0163
2.75	0.0180	0.0145	0.0120
3.00	0.0130	0.0106	0.0089
3.25	0.0095	0.0078	0.0066
3.50	0.0070	0.0058	0.0049

Therefore,

$$\dot{q}''_{\text{em}}(\kappa_D) = 9.0949\sigma T_0^4 \,. \tag{l}$$

Since, according to Eqs. (*a*) and (10-11*b*), we have

$$I(0) = \frac{\sigma T_0^4}{\pi}\,, \tag{m}$$

it follows that

$$\dot{q}''(\kappa_D) = 9.3143\sigma T_0^4 \,. \tag{n}$$

From this example, we find that the calculation of the radiative flux is quite straightforward if the temperature distribution is prescribed. Under normal circumstances, however, this information is not available, and an additional equation governing the energy balance is needed to solve for $T(x)$ and $I(x)$ simultaneously.

10-2*b* One-dimensional Local Radiative Flux

Pure radiation (no conduction and convection) in a one-dimensional gray-gas slab has been intensively investigated [1, pp. 446-453; 2, pp. 255-267; 4]. In particular, for the case where the absorption coefficient of the gray gas is constant† the solution is well documented. To avoid repetition, we will only highlight the important procedures here.

In reference to Fig. 10-6, the northbound and southbound radiative fluxes emerging from a control volume of infinitesimal thickness dx can be written as

$$\dot{q}''_N(\kappa) = 2\pi \int_0^{\pi/2} I_N(\kappa, \beta) \cos\beta \sin\beta \, d\beta = 2\pi \int_0^1 I_N(\kappa, \mu)\mu \, d\mu \quad (10\text{-}15a)$$

and

$$\dot{q}''_S(\kappa) = 2\pi \int_0^{\pi/2} I_S(\kappa, \beta + \pi) \cos(\beta + \pi) \sin(\beta + \pi) \, d\beta$$

$$= 2\pi \int_0^1 I_S(\kappa, -\mu)\mu \, d\mu , \quad (10\text{-}15b)$$

where $\mu = \cos\beta$ and $\beta \in [0, \pi/2]$. From Eq. (10-13*b*), we have

$$I_N(\kappa, \mu) = I_N(0) \exp\left(-\frac{\kappa}{\mu}\right) + \int_0^{\kappa} \frac{\sigma T^4(t)}{\pi} \mu^{-1} \exp\left(\frac{t - \kappa}{\mu}\right) dt \quad (10\text{-}16a)$$

and

$$I_S(\kappa, -\mu) = I_S(D) \exp\left(\frac{\kappa - \kappa_D}{\mu}\right) - \int_{\kappa_D}^{\kappa} \frac{\sigma T^4(t)}{\pi} \mu^{-1} \exp\left(\frac{\kappa - t}{\mu}\right) dt . \quad (10\text{-}16b)$$

†Such gray gases are sometimes called radiatively ideal gases.

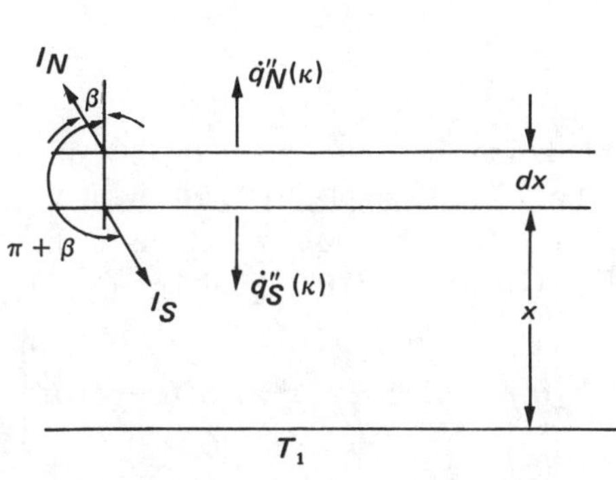

Figure 10-6 Consideration of radiative fluxes emerging from the infinitesimal slab.

The net local radiative flux in the north direction is therefore

$$\dot{q}'' = \dot{q}_N'' - \dot{q}_S'' = 2\pi \int_0^1 (I_N - I_S)\mu \, d\mu \,. \tag{10-17}$$

For steady-state pure radiation systems, the radiative flux remains constant along x. Therefore,

$$\frac{d\dot{q}''}{dx} = 0 \tag{10-18}$$

or

$$\int_0^1 \frac{dI_N}{d\kappa} \mu \, d\mu = \int_0^1 \frac{dI_S}{d\kappa} \mu \, d\mu \,. \tag{10-19}$$

Differentiating Eqs. (10-16*a*) and (10-16*b*) with respect to κ, realizing that the Leibnitz rule must be used, and substituting the result into Eq. (10-19), we finally derive

$$\Theta(\kappa) = \frac{1}{2} \int_0^1 \left[\exp\left(-\frac{\kappa}{\mu}\right) + \int_0^{\kappa} \Theta(t)\mu^{-1} \exp\left(\frac{t-\kappa}{\mu}\right) dt + \Theta_2 \exp\left(\frac{\kappa - \kappa_D}{\mu}\right) \right.$$
$$\left. + \int_{\kappa}^{\kappa_D} \Theta(t)\mu^{-1} \exp\left(\frac{\kappa - t}{\mu}\right) dt \right] d\mu \tag{10-20}$$

and

$$\psi(\kappa) = 2 \int_0^1 \left[\mu \exp\left(-\frac{\kappa}{\mu}\right) + \int_0^{\kappa} \Theta(t) \exp\left(\frac{t-\kappa}{\mu}\right) dt - \mu\Theta_2 \exp\left(\frac{\kappa - \kappa_D}{\mu}\right) \right.$$
$$\left. + \int_{\kappa_D}^{\kappa} \Theta(t) \exp\left(\frac{\kappa - t}{\mu}\right) dt \right] d\mu \,, \tag{10-21}$$

where

$$\Theta = \frac{T^4}{T_1^4} \quad \text{and} \quad \psi = \frac{\dot{q}''}{\sigma T_1^4} \,.$$

For the purpose of simplifying the computation, it is customary to introduce the exponential integral function defined by Eq. (*i*) in Example 10-2. Furthermore, since Eq. (10-18) implies that $\psi(\kappa)$ is uniform for all κ, the problem can be further simplified by letting $\psi(\kappa) = \psi(0)$. Therefore, Eqs. (10-20) and (10-21) reduce to

$$\Theta(\kappa) = \frac{1}{2} \left[E_2(\kappa) + \int_0^{\kappa} \Theta(t) E_1(\kappa - t) \, dt + \Theta_2 E_2(\kappa_D - \kappa) + \int_{\kappa}^{\kappa_D} \Theta(t) E_1(t - \kappa) \, dt \right] \tag{10-22}$$

and

$$\psi(0) = 1 - 2\left[\Theta_2 E_3(\kappa_D) + \int_0^{\kappa_D} \Theta(t) E_2(t)\,dt\right], \tag{10-23}$$

which is a set of linear integral equations. If the integrals are approximated–for example, by the trapezoidal rule–a set of linear algebraic equations will result.

Example 10-3 Solve Eqs. (10-22) and (10-23) numerically by taking only a single grid point at $\kappa = \frac{1}{2}$ with $\kappa_D = 1$ and $\Theta_2 = \frac{1}{2}$.

Solution: Let us designate $\Theta(\frac{1}{2})$ by Θ^*. Then

$$\Theta^* \approx \frac{1}{2}\left[E_2\left(\frac{1}{2}\right) + \frac{1}{2}(1+\Theta^*)\int_0^{1/2} E_1\left(\frac{1}{2}-t\right)dt + \frac{1}{2}E_2\left(\frac{1}{2}\right) + \frac{1}{2}\left(\Theta^* + \frac{1}{2}\right)\int_{1/2}^{1} E_1\left(t - \frac{1}{2}\right)dt\right] \tag{a}$$

and

$$\psi \approx 1 - 2\left[\frac{1}{2}E_3(1) + \Theta^* \int_0^1 E_2(t)\,dt\right]. \tag{b}$$

Consulting Table 10-1 and noting that

$$\int_{x_1}^{x_2} E_n(x)\,dx = E_{n+1}(x_1) - E_{n+1}(x_2), \tag{c}$$

we finally obtain

$$\Theta^* = \tfrac{3}{4} \quad \text{and} \quad \psi = 0.30485\,. \tag{d}$$

The accuracy of these crudely approximated solutions will be improved, of course, if more grid points are taken.

If radiation takes place simultaneously with conduction and convection, the equation of energy conservation (10-23) must be replaced by a nonlinear integro-differential equation and, naturally, the equality $\psi(\kappa) = \psi(0)$ no longer holds. This more complicated case will be considered in Section 10-6.

10-3 NONGRAY–GAS RADIATION

The gray gases mentioned in the preceding section seldom exist in reality. In realistic systems, the absorption coefficients of gases are strongly dependent on wavelength.

The evaluation of the radiative flux given in Eq. (10-11a) thus involves that of quadruple integrals with respect to (1) solid angle, (2) wavelength, (3) physical distance, and (4) optical path κ.

Example 10-4 Express the northbound local radiative flux in a gaseous slab in terms of a quadruple integral. Neglect the contribution due to attenuation.

Solution: Integrating the spectral local radiative flux with respect to wavelength yields

$$\dot{q}'' = \int_0^\infty \dot{q}''_\lambda \, d\lambda , \tag{a}$$

where $\dot{q}''_\lambda$ is given by Eq. (10-11a) as

$$\dot{q}''_\lambda = 2\pi \int_0^{\pi/2} I_\lambda \sin\beta \cos\beta \, d\beta , \tag{b}$$

where ω is the solid angle. If attenuation is neglected, the local intensity becomes, according to Eq. (10-13a),

$$I_\lambda = \int_{\tilde{\kappa}_0}^{\tilde{\kappa}} I_{b\lambda}(\tilde{t}) \exp(\tilde{t} - \tilde{\kappa}) \, d\tilde{t} . \tag{c}$$

But by definition we know that

$$\tilde{\kappa} = \int_{s_0}^{s} a \, ds . \tag{d}$$

Embedding Eqs. (a)-(d) into a single equation leads to

$$\dot{q}''(s) = \int_{\lambda=0}^{\infty} \int_{\beta=0}^{\pi/2} \int_{\tilde{t}=s_0}^{s} I_{b\lambda}(\tilde{t}) \exp\left[\tilde{t} - \int_{s_0}^{s} a(z)\, dz\right] \sin\beta \cos\beta \, d\tilde{t} \, d\beta \, d\lambda , \tag{e}$$

in which the quadruple integral appears. Since I_b and the absorption coefficient a are functions of the temperature, which subsequently is a function of the space coordinate s, the evaluation of $\dot{q}''(s)$ becomes a formidable task.

It is desirable that some simplifications be made for the computation of the nongray-gas radiative flux. At least two such simplifications are available in the literature.

10-3*a* First Simplification: Introduction of Total Band Absorptance

The total band absorptance for the ith band is defined as

$$A_i(x) = \int_0^\infty \left[1 - \exp\left(-\int_0^x a_\lambda(z)\,dz\right)\right] d\lambda\,. \qquad (10\text{-}24)$$

Seeking an explicit expression for $A_i(x)$ in terms of several band parameters for various bands and various gases can be separated from a given physical problem and thus is beyond the scope of this book. Interested readers can consult [5-13]. It is important, however, to know how to express $\dot{q}''(x)$ in terms of $A_i(x)$. See Example 10-5.

10-3*b* Second Simplification: Approximation of Exponential Integrals

It can be shown that the exponential integrals $E_2(t)$ and $E_3(t)$ can be closely approximated by $\exp(-2t)$ and $\frac{1}{2}\exp(-2t)$, respectively. Such an approximation, along with the adoption of $A_i(x)$, greatly simplifies the computation of the radiative flux.

Example 10-5 Express the local radiative emission flux, which is x-direction dominant, in terms of the total band absorptance $A_i(x)$.

Solution: Equation (e) in Example 10-4 can be rewritten as

$$\dot{q}''(x) = 2\pi \int_0^\infty \int_0^1 \int_{x_0}^x I_{b\lambda}(t) \exp\left[-\frac{1}{\mu}\int_t^x a_\lambda(z)\,dz\right] dt\,d\mu\,d\lambda\,. \quad (a)$$

Equation (a) can be written as

$$\dot{q}''(x) = 2\pi \int_0^\infty \int_{x_0}^x I_{b\lambda}(t) E_2\left[\int_t^x a_\lambda(z)\,dz\right] dt\,d\lambda\,. \qquad (b)$$

With the approximation

$$E_2(t) \approx \exp(-2t)\,,$$

Eq. (b) can be rearranged to

$$\dot{q}''(x) = 2\pi \int_{x_0}^x \int_0^\infty I_{b\lambda}(t)\,d\lambda\,dt$$

$$-2\pi \int_{x_0}^x \int_0^\infty I_{b\lambda}(t) \left\{1 - \exp\left[2\int_t^x a_\lambda(z)\,dz\right]\right\} d\lambda\,dt\,. \qquad (c)$$

But we recall that, for a blackbody,

$$\sigma T^4 = \int_0^\infty \pi I_{b\lambda}\, d\lambda \tag{d}$$

and that, by definition,

$$A_i(2x-2t) = \int_0^\infty \left\{1 - \exp\left[2\int_t^x a_\lambda(z)\, dz\right]\right\} d\lambda\,. \tag{e}$$

Substituting Eqs. (d) and (e) into Eq. (c) yields

$$\dot{q}''(x) = 2\sigma \int_{x_0}^{x} T^4(t)\, dt - 2\pi \sum_{i=1}^{n} \int_{x_0}^{x} I_{b\lambda_i}(t) A_i(2x-2t)\, dt\,, \tag{f}$$

where the black spectral intensity $I_{b\lambda_i}(t)$ can be evaluated by using Eq. (10-1) at $\lambda = \lambda_i$. It is seen that using the simplifications described in Sections 10-3a and 10-3b reduces the quadruple integrals to a single integral.

10-4 COMBINED CONDUCTION AND RADIATION: ONE-DIMENSIONAL RADIATIVE FLUX

The classical problem of combined conduction and radiation within a gray medium bounded by two infinite parallel black plates has received as much attention as that of pure radiation (Section 10-2b). Again, to avoid repetition, we will present this problem only briefly here.

For steady-state combined conductive and radiative systems, the conservation of energy over a control volume of infinitesimal thickness dx is governed by

$$\frac{d\dot{q}_r''}{dx} = k\frac{d^2T}{dx^2}\,, \tag{10-25}$$

which replaces Eq. (10-18) for the pure radiation case. Equation (10-17), however, remains valid. If the boundary conditions of Eq. (10-25) are

$$T(0) = T_1 \quad \text{and} \quad T(D) = T_2\,,$$

the substitution of Eq. (10-17) into Eq. (10-25) will eventually yield [14]

$$N_1 \frac{d^2\theta}{d\kappa^2} = \theta^4(\kappa) - \frac{1}{2}\left[E_2(\kappa) + \theta_2^4 E_2(\kappa_D - \kappa) + \int_0^{\kappa_D} \theta^4(t) E_1(|\kappa - t|)\, dt\right] \tag{10-26}$$

subject to

$$\theta(0) = 1 \quad \text{and} \quad \theta(\kappa_D) = \theta_2\,,$$

where

$$N_1 = \frac{ka}{4\sigma T_1^3}, \qquad \kappa = ax, \qquad \text{and} \qquad \theta = \frac{T}{T_1}.$$

To illustrate the solution procedures to be presented in the following sections, we will assign θ_2, N_1, and κ_D the values 0.1, 0.1, and 1, respectively. Thus, Eq. (10-26) reduces to

$$A\theta = -\frac{d^2\theta}{d\kappa^2} + 10\theta^4 - 5\left[E_2(\kappa) + \int_0^1 \theta^4(t)E_1(|\kappa - t|)\,dt\right] = 0, \tag{10-27}$$

where A is a nonlinear operator and θ_2^4, being small, was neglected. Equation (10-27) will be solved by the finite-difference method and the variational method.

10-4*a* Finite-Difference Method

The most straightforward numerical method for solving Eq. (10-27) is probably the finite-difference method, by which we approximate

$$\frac{d^2\theta}{d\kappa^2} = \frac{1}{h^2}(\theta_{j-1} - 2\theta_j + \theta_{j+1}), \tag{10-28}$$

where h is the uniform mesh size.

Example 10-6 Use Eq. (10-28) and the trapezoidal rule to approximate the second derivative and the integral, respectively, and then solve Eq. (10-27) for two grid points.

Solution: For the crude approximation of only three intervals, we write

$$\int_0^1 \theta^4(t)E_1(|\kappa - t|)\,dt = \int_0^{1/3} \theta^4(t)E_1\left(\frac{1}{3} - t\right)dt$$

$$+ \int_{1/3}^{2/3} \theta^4(t)E_1\left(t - \frac{1}{3}\right)dt + \int_{2/3}^{1} \theta^4(t)E_1\left(t - \frac{1}{3}\right)dt. \tag{a}$$

At $j = 1$, for example, the first integral can be approximated by

$$\int_0^{1/3} \theta^4(t)E_1\left(\frac{1}{3} - t\right)dt \approx \frac{1}{2}(1 + \theta_1^4)\int_0^{1/3} E_1\left(\frac{1}{3} - t\right)dt$$

$$= 0.285(1 + \theta_1^4). \tag{b}$$

Using Eq. (10-28) and Eqs. (*a*) and (*b*) and setting $h = \frac{1}{3}$, we finally obtain

$$9(-1 + 2\theta_1 - \theta_2) + 10\theta_1^4 - 5[0.43 + 0.285(1 + \theta_1^4) + 0.285(\theta_1^4 + \theta_2^4)$$

$$+ 0.09(\theta_2^4 + 0.0001)] = 0 \tag{c}$$

and

$$9(-\theta_1 + 2\theta_2 + 0.1) + 10\theta_2^4 - 5[0.248 + 0.09(1 + \theta_1^4) + 0.285(\theta_1^4 + \theta_2^4) + 0.285(\theta_2^4 + 0.0001)] = 0 , \qquad (d)$$

which is a set of nonlinear algebraic equations. Linearization of Eqs. (c) and (d) can be achieved by using

$$\theta^4 \approx -3\bar{\theta}^4 + 4\bar{\theta}^3 \theta , \qquad (e)$$

where the overbar denotes the previously iterated value. Initially, a reasonable guess can be

$$\bar{\theta}(\kappa) = 1 - 0.9\kappa , \qquad (f)$$

which yields $\bar{\theta}_1 = 0.7$ and $\bar{\theta}_2 = 0.4$ for the first iteration. The final iterated values are found to be

$$\theta_1 = 0.78 \quad \text{and} \quad \theta_2 = 0.5 . \qquad (g)$$

10-4*b* Variational Method with Global Basis Functions

Whenever the variational method is to be adopted, we must first determine whether the governing equation has a corresponding variational principle. In order to do this, we seek (see Section 1-2*c* or 2-1*b*) the following inner product:

$$(A\theta, \delta\theta) = -\int_0^1 \frac{d^2\theta}{d\kappa^2}\,\delta\theta\, d\kappa + 10\int_0^1 \theta^4\,\delta\theta\, d\kappa - 5\int_0^1 E_2(\kappa)\,\delta\theta\, d\kappa - 5\int_0^1\int_0^1 \theta^4(t)\,\delta\theta\, E_1(|\kappa - t|)\, dt\, d\kappa = 0 . \qquad (10\text{-}29)$$

Therefore, the variational principle I can be identified as

$$I = \frac{1}{2}\int_0^1 \left(\frac{d\theta}{d\kappa}\right)^2 d\kappa + 2\int_0^1 \theta^5\, d\kappa - 5\int_0^1 \theta E_2(\kappa)\, d\kappa - \int_0^1\int_0^1 \theta^5(\kappa) E_1(|\kappa - t|)\, dt\, d\kappa . \qquad (10\text{-}30)$$

Let us now choose the set of basis functions

$$\{1, \kappa, \kappa(1-\kappa), \kappa^2(1-\kappa), \ldots, \kappa^n(1-\kappa)\}$$

and assume

$$\tilde{\theta}(\kappa) = 1 - 0.9\kappa + c(\kappa - \kappa^2) . \qquad (10\text{-}31)$$

For easy presentation, only the leading term was taken. When Eq. (10-31) is substituted into Eq. (10-30), the variational principle I becomes a function of c. The value of c should be chosen such that

$$\frac{dI}{dc} = 0 \tag{10-32}$$

or

$$\int_0^1 (-2\kappa c + c - 0.9)(1 - 2\kappa)\, d\kappa + 10 \int_0^1 [1 - 0.9\kappa + c(\kappa - \kappa^2)]^4 (\kappa - \kappa^2)\, d\kappa$$

$$-5 \int_0^1 (\kappa - \kappa^2) E_2(\kappa)\, d\kappa$$

$$-5 \int_0^1 \int_0^1 [1 - 0.9\kappa + c(\kappa - \kappa^2)]^4 (\kappa - \kappa^2) E_1(|\kappa - t|)\, dt\, d\kappa = 0\,, \tag{10-33}$$

which can be reduced to a nonlinear algebraic equation by using the Gaussian quadrature formula

$$\int_0^1 f(\kappa)\, d\kappa \approx 0.2778 f(\kappa_1) + 0.4444 f(\kappa_2) + 0.2778 f(\kappa_3)\,. \tag{10-34}$$

These three Gaussian quadrature knots are found (see also Section 2-4c) to be located at

$$\kappa_1 = 0.1127\,, \qquad \kappa_2 = 0.5\,, \qquad \text{and} \qquad \kappa_3 = 0.8873\,.$$

According to Eq. (10-34), the second and third integrals in Eq. (10-33) then can be approximated, respectively, by

$$\int_0^1 [1 - 0.9\kappa + c(\kappa - \kappa^2)]^4 (\kappa - \kappa^2)\, d\kappa \approx 0.02778(0.8986 + 0.1c)^4$$

$$+ 0.1111(0.55 + 0.25c)^4 + 0.02778(0.2014 + 0.1c)^4$$

and, with the additional aid of Table 10-1,

$$\int_0^1 (\kappa - \kappa^2) E_2(\kappa)\, d\kappa \approx 0.2778(0.1 \times 0.72) + 0.4444(0.25 \times 0.3266)$$

$$+ 0.2778(0.1 \times 0.18) = 0.0613\,.$$

The fourth term in Eq. (10-33) is a double integral, of which the integral with respect to κ can be approximated by

$$\int_0^1 [1-0.9\kappa+c(\kappa-\kappa^2)]^4(\kappa-\kappa^2)E_1(|\kappa-t|)\,d\kappa$$

$$= 0.02788(0.8986+0.1c)^4E_1(|0.1127-t|)+0.1111(0.55+0.25c)^4E_1(|0.5-t|)$$
$$+0.02278(0.2014+0.1c)^4E_1(|0.8873-t|)\,.$$

With further approximations

$$\int_0^1 E_1(|0.1127-t|)\,dt \approx 1.0\,,$$

$$\int_0^1 E_1(|0.5-t|)\,dt \approx 1.8\,,$$

and

$$\int_0^1 E_1(|0.8873-t|)\,dt \approx 0.58,$$

Eq. (10-33) is finally reduced to

$$\frac{c}{3}+0.1389(0.8986+0.1c)^4+0.0985(0.55+0.25c)^4+0.1973(0.2014+0.1c)^4$$
$$-0.3065=0\,, \tag{10-35}$$

of which one possible root is

$$c=0.51\,.$$

Therefore, the approximate solution to Eq. (10-27) is found to be

$$\tilde{\theta}(\kappa)=1-0.39\kappa-0.51\kappa^2\,,$$

which agrees fairly well with the result of [14].

10-5 COMBINED CONDUCTION AND RADIATION (TWO-DIMENSIONAL RADIATIVE FLUX)

In most research involving radiatively participating media, the radiative flux has been assumed to be one-dimensional. Even in boundary-layer flow, which is definitely a two-dimensional system, the radiative transfer has been assumed negligible along the streamwise direction. Such an assumption has been widely, sometimes blindly, adopted because a more accurate computation involving two-dimensional radiative fluxes is a formidable task. There are two difficulties. First, triple integrals must be computed with respect to the physical distance, the optical thickness, and the solid angle in order to obtain the local radiative flux. Second, if the two-dimensional physical domain is represented by a square mesh, the discretized equation of energy transport at each grid point must be derived not only elaborately but also separately. There does not exist a discretized equation for a typical grid point j that can be applied to other grid points by conveniently changing the index j. The first difficulty is inherent in the radiative flux calculation, even for the one-dimensional case. The second is the major hindrance that discourages the adoption of a two-dimensional radiative flux model. An illustration may show this point more clearly.

Figure 10-7 shows a portion of a typical two-dimensional mesh system that encloses a radiatively participating gas. The temperature prescribed on the boundary is nonuniform, i.e.,

$$T_1 \neq T_2 \neq T_3 \neq T_4 \neq T_5 .$$

We wish to take the energy balance over the infinitesimal control volume enclosing a typical grid point j. The energy that flows into the control volume includes conduction and radiation. The radiation arriving at the north face of the control volume is contributed by intensities I_A and I_B as well as infinitely many other intensities. Radiation of intensity I_A starts at grid point 5 and intersects the grid lines at points 6, 7, and 8. Since the magnitude of the intensity depends on the temperature distribution along the path traveled by the radiation, it follows that

$$I_A = \text{function of } T_5, T_6, T_7, T_8, \ldots$$

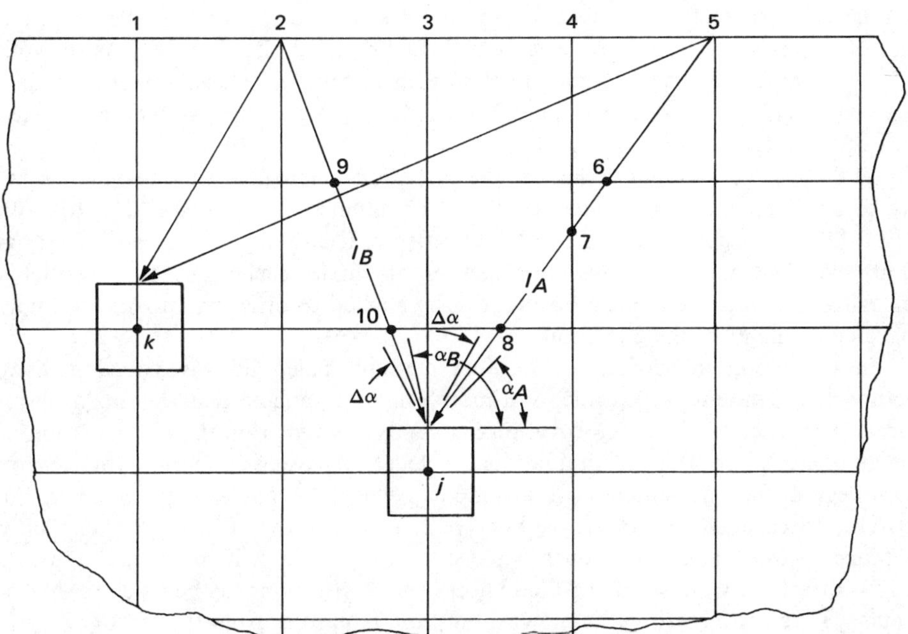

Figure 10-7 A portion of a two-dimensional discretized system that encloses a radiative-conductive medium.

and similarly that

$$I_B = \text{function of } T_2, T_9, T_{10}, \ldots .$$

But points 6, . . . , 10 do not coincide with the grid nodes, and the temperatures $T_6, \ldots, T_{10}$ must be interpolated from the nodal temperatures. This interpolation must be performed for all the components of radiation that may impinge on the control volume. After accomplishing this tedious interpolation, we still have to integrate the intensities with respect to the inclination angle α by approximating

$$\dot{q}_r'' = \int_0^{\pi} I(\alpha) \cos \alpha \, d\alpha \approx I_A \cos \alpha_A \, \Delta\alpha + I_B \cos \alpha_B \, \Delta\alpha + \cdots ,$$

in order to obtain the radiative flux entering the north face of the control volume. Furthermore, the final result obtained is valid only at grid point j. At another grid point k we must repeat the elaborate interpolation. The entire task is formidable. It is thus wise to tackle the problem by different approaches.

There exist two widely used methods for solving the multidimensional radiative and conductive (or/and convection) problems. They are the zone method [15-25] and the flux method [26-30].

In the zone method, the enclosure or the system domain is subdivided into a number of surface and volume zones that are sufficiently small to be assumed isothermal and of uniform properties. This method must rely heavily on knowledge and evaluation of the total exchange areas. For systems having irregular geometry or a large number of zones, specification of these exchange areas, used as the input data of the computation, may become a difficult task. Following the pioneer work [15], researchers applied the zone method to heat transfer calculations for furnaces and boilers [16, 18, 20–22] and pipe flows [24, 25]. Most of the systems considered were of the axisymmetric type.

The flux method approximates the radiation impinging on one-dimensional, two-dimensional, and three-dimensional control volumes as two fluxes [26, 30], four fluxes [28], and six fluxes [27, 29], respectively. It also introduces nodal fluxes as additional field variables. This introduction of fluxes enables us to construct a generalized discretization equation that can be applied to other grid points by simply changing the index of the grid point.

In the modified version of the flux method, called the discretized-intensity method [31], the intensity, which is a more primitive variable than the flux, is introduced. The discretized energy equation is directly derived from the consideration of energy conservation over a computational molecule enclosing grid point j; the governing integrodifferential equation is abandoned. Consequently, instead of using Eq. (10-11a) to calculate the radiative flux, we can express the radiative flux explicitly in terms of the intensities through a physical interpretation. If the nodal intensities in four directions west, south, east, and north (more directions can be taken for higher accuracy) are introduced as unknowns, only the intensities from the vicinity of grid point j, and not from more remote locations, will appear in the governing equation. These nodal intensities distributed in the entire domain are then related to one another through the equation of transfer in the discretized form.

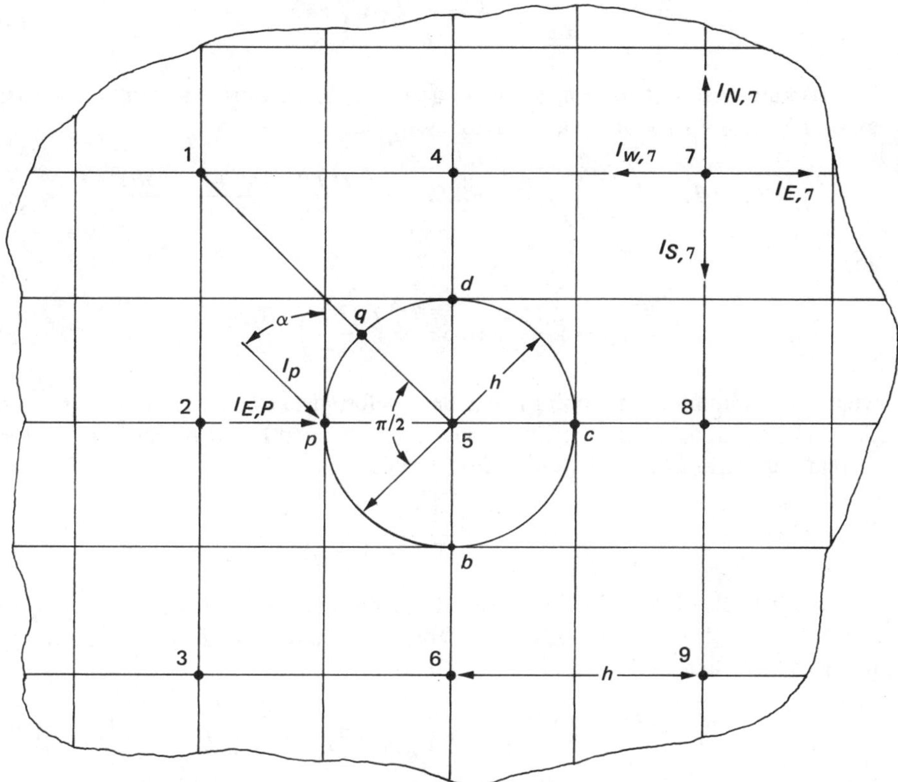

Figure 10-8 The circular computational molecule enclosing a typical grid point 5.

We will describe the discretized-intensity method in detail as follows. Consider a circular computational molecule enclosing grid point 5, as shown in Fig. 10-8. A circle is chosen as the shape of the computational cell because it can be used when more directions are considered. The conductive flow into the cell from grid point 2 can be approximated by

$$\dot{q}_{c,2\text{-}5} \approx k\left(\frac{T_2 - T_5}{h}\right)\left(\frac{\pi h}{4}\right). \tag{10-36}$$

Referring to Fig. 10-8, we find that the total conductive flow into the cell per unit time is

$$\dot{q}_c \approx \frac{\pi k}{4}\,(T_2 + T_4 + T_6 + T_8 - 4T_5)\,, \tag{10-37}$$

which is the familiar five-point relation derived by using a square computational cell. For higher accuracy, the nine-point relation can be derived with little additional effort.

Example 10-7 Derive a nine-point relation similar to Eq. (10-37).

Solution: The expression for the conductive flow entering the circular cell from grid point 2, as shown in Fig. 10-8, can be approximated by

$$\dot{q}_{c,2\text{-}5} \approx k\left(\frac{T_2 - T_5}{h}\right)\left(\frac{\pi h}{8}\right). \tag{a}$$

To estimate the contribution to the conductive flow from grid point 1, we first expand T_1 and T_5 about T_q in a Taylor's series as

$$T_1 \approx T_q - \left(\frac{dT}{ds}\right)_q (\sqrt{2}h - 0.5h) + \left(\frac{d^2T}{ds^2}\right)_q \frac{(\sqrt{2}h - 0.5h)^2}{2!} \tag{b}$$

and

$$T_5 \approx T_q + \left(\frac{dT}{ds}\right)_q (0.5h) + \left(\frac{d^2T}{ds^2}\right)_q \frac{(0.5h)^2}{2!}, \tag{c}$$

where the circumferential point q has been referred to as the origin $s = 0$, s being the distance along a straight line through grid points 1 and 5. Eliminating the second derivative from Eqs. (b) and (c) leads to

$$\left(\frac{dT}{ds}\right)_q \approx \frac{-\sigma^2 T_5 + (\sigma^2 - 1)T_q + T_1}{0.5h\sigma(1 + \sigma)}, \tag{d}$$

where $\sigma = (\sqrt{2} - 0.5)/0.5$ is the ratio of the distance between points 1 and q to the distance between q and 5. Furthermore, we assume that T_q can be expressed in terms of T_1 and T_5 by

$$T_q \approx \left(\frac{\sqrt{2} - 0.5}{\sqrt{2}}\right) T_5 + \left(\frac{0.5}{\sqrt{2}}\right) T_1 = \frac{\sigma T_5 + T_1}{1 + \sigma}. \tag{e}$$

Substituting Eq. (e) into Eq. (d), we obtain

$$\left(\frac{dT}{ds}\right)_q \approx \frac{T_1 - T_5}{\sqrt{2}h}, \tag{f}$$

as expected. Therefore,

$$\dot{q}_{c,1\text{-}5} \approx k\left(\frac{T_1 - T_5}{\sqrt{2}h}\right)\left(\frac{\pi h}{8}\right). \tag{g}$$

In reference to Fig. 10-8, the total conductive flow into the cell is

$$\dot{q}_c \approx \pi k \left[\frac{T_2 + T_4 + T_6 + T_8}{8} + \frac{T_1 + T_3 + T_7 + T_9}{8\sqrt{2}} - \frac{(\sqrt{2} + 1)T_5}{2\sqrt{2}}\right]. \tag{h}$$

Next, let us consider the radiation balance. Since the method should be applicable to analyses of two-dimensional slabs in which solid angle has little meaning, the intensity should be defined differently from the conventional way. Here we define the intensity as the heat flow rate per projected area per inclination angle α, i.e.,

$$I = \frac{d^2\dot{q}}{(dA \sin\alpha)\, d\alpha}, \tag{10-38}$$

where α is the inclination angle measured from the emitting (or receiving) surface. For the isotropic intensity emitted from a blackbody, Eq. (10-38) yields an important relation

$$I_b = \tfrac{1}{2}\sigma T_b^4 \tag{10-39}$$

between the temperature and the intensity.†

After defining the intensity appropriate for a two-dimensional slab, we will express the eastbound intensity arriving at point p (not a grid point) in terms of the eastbound intensity arriving at grid point 2 and the temperature at grid point p. Following Eq. (10-13a) and taking grid point 2 as the origin of emission, we derive

$$I_{E,p} = I_{E,2} \exp(-\kappa_p) + \int_0^{\kappa_p} I_b(\kappa^*) \exp(\kappa^* - \kappa_p)\, d\kappa^* , \tag{10-40a}$$

where the subscripts p and E denote "at point p" and "eastbound," respectively.

If the computational molecule is sufficiently small, it may be reasonable to assume that $I_b(\kappa^*)$ is fairly constant along the path from 0 to κ_p. Hence, Eq. (10-40a) can be reduced to

$$I_{E,p} = I_{E,2} \exp(-\kappa_p) + \tfrac{1}{2}\sigma T_p^4 [1 - \exp(-\kappa_p)] , \tag{10-40b}$$

where κ_p is equal to $ah/2$ if a is uniform. The temperature T_p is assumed to be the average of T_2 and T_5. However, it is noteworthy that the intuitive interpolation

$$I_{E,p} \simeq \tfrac{1}{2}(I_{E,2} + I_{E,5})$$

will yield erroneous results since $I_{E,p}$ is "upstream" with respect to $I_{E,5}$, and therefore should not be influenced by $I_{E,5}$.

Let us further assume that the intensities arriving at point p from the vicinity of grid point 2 are approximately of the same magnitude as $I_{E,p}$. This assumption, although rather severe, will greatly simplify our presentation here without altering the fundamental concept of the discretized-intensity method. Relaxation of this restriction is ultimately needed and is discussed in [31]. Under these assumptions, the eastbound radiative flux arriving at boundary point p can be approximated by

$$\vec{q}''_{r,p} = \int_0^{\pi} I_p \sin\alpha\, d\alpha \simeq 2I_{E,p} = 2I_{E,2} \exp(-\kappa_p) + \sigma T_p^4 [1 - \exp(-\kappa_p)] .$$

Similarly, the westbound radiative flux *emerging* at point p is

$$\overleftarrow{q}''_{r,p} \simeq 2I_{W,5} \exp(-\kappa_p) + \sigma T_p^4 [1 - \exp(-\kappa_p)] .$$

†It is worth noting that introduction of this definition does not alter the physics. Although the magnitude of I_b in Eq. (10-39) is larger than that of the conventional I_b by a factor of $\pi/2$, the magnitude of the radiative flux that is actually considered in the energy balance remains unchanged. See Problem 10-8.

Therefore, the *net* radiative flow (symbol without a double prime) entering the control volume through point p per unit time can be derived to be

$$\dot{q}_{r,p} = \frac{\pi h}{4} (\vec{\dot{q}}''_{r,p} - \overleftarrow{\dot{q}}''_{r,p}) = \frac{\pi h}{2} (I_{E,2} - I_{W,5}) \exp(-\kappa_p) . \tag{10-41a}$$

Likewise, we can derive

$$\dot{q}_{r,b} = \frac{\pi h}{2} (I_{N,6} - I_{S,5}) \exp(-\kappa_b) , \tag{10-41b}$$

$$\dot{q}_{r,c} = \frac{\pi h}{2} (I_{W,8} - I_{E,5}) \exp(-\kappa_c) , \tag{10-41c}$$

and

$$\dot{q}_{r,d} = \frac{\pi h}{2} (I_{S,4} - I_{N,5}) \exp(-\kappa_d) . \tag{10-41d}$$

Adding Eqs. (10-41*a*)-(10-41*d*) yields the net total radiative flow entering the control volume per unit time as

$$\dot{q}_r = \frac{\pi h}{2} \exp\left(-\frac{ah}{2}\right) (I_{E,2} + I_{N,6} + I_{W,8} + I_{S,4} - I_{W,5} - I_{S,5} - I_{E,5} - I_{N,5}) , \tag{10-42}$$

in which, for convenience in the presentation, the assumption that

$$\kappa_p = \kappa_b = \kappa_c = \kappa_d = \frac{ah}{2}$$

has been made. The fact that the conductive-radiative system is maintained in steady state dictates

$$\dot{q}_r + \dot{q}_c = 0$$

or

$$\xi(I^*_{E,2} + I^*_{N,6} + I^*_{W,8} + I^*_{S,4} - I^*_{W,5} - I^*_{S,5} - I^*_{E,5} - I^*_{N,5}) + \theta_2 + \theta_4 + \theta_6 + \theta_8 - 4\theta_5 = 0 , \tag{10-43}$$

where

$$\xi = 2h\sigma T_0^3 \frac{\exp(-ah/2)}{k} ,$$

$$I^* = \frac{I}{\sigma T_0^4} ,$$

and

$$\theta = \frac{T}{T_0} .$$

Equation (10-43) is the discretized energy equation at grid point 5. Nevertheless, since it was derived in a general manner, it can be applied to other grid points as well simply by changing 2, 4, 5, 6, and 8 to appropriate indices.

In addition to Eq. (10-43), we need to derive four algebraic equations relating the four intensities I_W, I_S, I_E, and I_N to the temperature. The equation of transfer in the differential form

$$dI = I_b\, d\kappa - I\, d\kappa \tag{10-44}$$

can be used. Consider, for example, the eastbound beam passing through the control volume. Equation (10-44) can be discretized by using the finite-difference approximation $[dI/d\kappa \approx (I_{E,c} - I_{E,p})/ah]$ to become

$$I_{E,c} - I_{E,p} = ah(\tfrac{1}{2}\sigma T_5^4 - T_{E,5}) \,.$$

Utilizing Eq. (10-40*b*) and assuming that T_p is the average of T_2 and T_5, we derive, after algebra,

$$(A_1 + ah)I_{E,5}^* = A_1 I_{E,2}^* - A_2[(\theta_5 + \theta_8)^4 - (\theta_2 + \theta_5)^4] + \frac{ah}{2}\theta_5^4 \,. \tag{10-45a}$$

Similarly, we can derive

$$(A_1 + ah)I_{W,5}^* = A_1 I_{W,8}^* - A_2[(\theta_5 + \theta_2)^4 - (\theta_8 + \theta_5)^4] + \frac{ah}{2}\theta_5^4 \,, \tag{10-45b}$$

$$(A_1 + ah)I_{S,5}^* = A_1 I_{S,4}^* - A_2[(\theta_5 + \theta_6)^4 - (\theta_4 + \theta_5)^4] + \frac{ah}{2}\theta_5^4 \,, \tag{10-45c}$$

and

$$(A_1 + ah)I_{N,5}^* = A_1 I_{N,6}^* - A_2[(\theta_5 + \theta_4)^4 - (\theta_6 + \theta_5)^4] + \frac{ah}{2}\theta_5^4 \,. \tag{10-45d}$$

Equations (10-43) and (10-45) constitute a set of algebraic equations at $j = 5$ that θ_5, $I_{W,5}^*$, $I_{S,5}^*$, $I_{E,5}^*$, and $I_{N,5}^*$ must satisfy. Repetitive application of this set of equations to all J interior grid points thus enables us to generate a set of $5 \times J$ algebraic equations that $5 \times J$ unknowns must satisfy.

Example 10-8 A conductive radiative medium with $\kappa = ah = 1$ is confined in a black enclosure of unit depth, as shown in Fig. 10-9. (Note that, if the enclosure is of infinite depth, the system becomes three-dimensional. Equation (10-38) then should be replaced by the definition of the conventional intensity. The final expression for the heat flux, however, remains the same regardless of the dimensionality, as stated on p. 393.) The system is squarely meshed and is subject to the following boundary conditions:

$$\theta_{10} = \theta_{11} = \theta_{12} = \theta_{18} = \cdots = \theta_{25} = 1$$

and

$$\theta_{14} = \theta_{15} = \theta_{16} = 2 \,. \tag{a}$$

Find the nodal temperature distribution for $\xi = 0$ and 1.

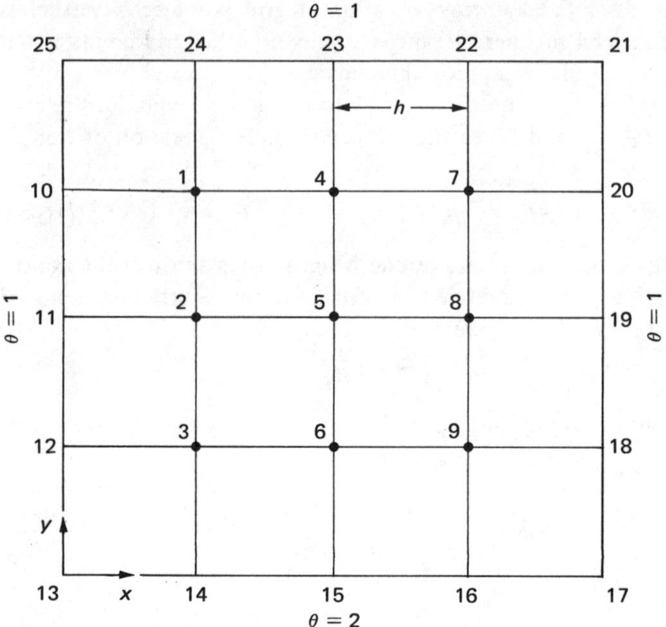

Figure 10-9 The 5 × 5 discretized conductive-radiative system.

Solution: First, let us examine whether the algebraic system is closed. When Eqs. (10-43) and (10-45) are applied to grid points $j = 1, 2, \ldots, 9$, the nodal unknowns involved are

$$\theta_1, \theta_2, \ldots, \theta_9,$$

$$I^*_{W,j}, I^*_{S,j}, I^*_{E,j}, I^*_{N,j}, \quad j = 1, 2, \ldots, 9$$

$$I^*_{W,k}, I^*_{E,k}, \quad k = 10, 11, 12, 18, 19, 20$$

$$I^*_{N,l}, I^*_{S,l}, \quad l = 14, 15, 16, 22, 23, 24\,.$$

The total number of unknowns is counted and found to be $9 + 4 \times 9 + 2 \times 6 + 2 \times 6 = 69$. On the other hand, the discretized equations include Eq. (10-43) and Eqs. (10-45*a*)-(10-45*d*) applied to $j = 1, 2, \ldots, 9$; the total number of equations is 45. Thus, it seems that 24 boundary conditions for the intensities are required. We will derive the boundary conditions for $I^*_{E,10}$, and simply discuss and and write down the remaining boundary conditions.

From Eq. (10-39), we immediately have

$$I^*_{E,10} = \tfrac{1}{2}\,. \tag{b}$$

Regarding the boundary conditions for $I^*_{W,10}$, we note that $I^*_{W,2}$ does not appear either in Eq. (10-42) or in Eq. (10-45*b*). This implies that, if we change the indices in these two equations to those corresponding to discretization at grid point 1, $I_{W,10}$ will not appear in the resulting equations. Consequently, the total number of unknowns is reduced to

$$9 + 4 \times 9 + 6 + 6 = 57 ,$$

and we only need $57 - 45 = 12$ boundary conditions. Following the above argument, we list these boundary conditions in Table 10-2. With these boundary conditions, the given set of algebraic equations is thus closed. The corresponding linearized matrix system can be solved by an iteration scheme. We tabulate the nodal temperature distribution in Table 10-3 in accordance with the mesh location for $\xi = 0$ and $\xi = 1$, respectively. The case of $\xi = 0$ corresponds to pure conduction. It can be seen that the temperature of the medium tends to increase throughout the field when the medium is radiatively participating. More results can be found in [31].

To assess the validity of the discretized-intensity method, we may analyze the two-dimensional problem for extremely elongated systems such as 5×41 mesh systems, and compare the solution with the classical result for the one-dimensional conductive-radiative system. The solution for the one-dimensional system and the temperature distributions along the vertical lines of symmetry for 5×5, 5×21, and 5×41 mesh systems are plotted against κ in Fig. 10-10 for comparison. As the system becomes more elongated, the temperature distribution becomes closer to that of the limiting one-dimensional case. The validity of the discretized-intensity method is partially demonstrated by this agreement.

10-6 COMBINED CONDUCTION, CONVECTION, AND RADIATION

In the last section of this chapter, we will investigate the phenomena involving all three heat transfer modes: conduction, convection, and radiation. Limited by the

Table 10-2 Twelve boundary conditions for the nodal intensities

Boundary point	Boundary conditions based on Eq. (10-39)	Intensities not involved in the analysis
$k = 10, 11, 12$	$I^*_{E,k} = \frac{1}{2}$	$I^*_{W,k}$
$k = 14, 15, 16$	$I^*_{N,k} = 8$	$I^*_{S,k}$
$k = 18, 19, 20$	$I^*_{W,k} = \frac{1}{2}$	$I^*_{E,k}$
$k = 22, 23, 24$	$I^*_{S,k} = \frac{1}{2}$	$I^*_{N,k}$

Table 10-3 Nodal temperatures of the 5 × 5 mesh system for $\xi = 0, 1.0$

		$\xi = 0$		
	1.0	1.0	1.0	
1.0	1.0717	1.0985	1.0717	1.0
1.0	1.1880	1.2504	1.1880	1.0
1.0	1.4291	1.5272	1.4290	1.0
	2.0	2.0	2.0	
		$\xi = 1.0$		
	1.0	1.0	1.0	
1.0	1.2682	1.2960	1.2682	1.0
1.0	1.4157	1.4476	1.4158	1.0
1.0	1.5674	1.5977	1.5675	1.0
	2.0	2.0	2.0	

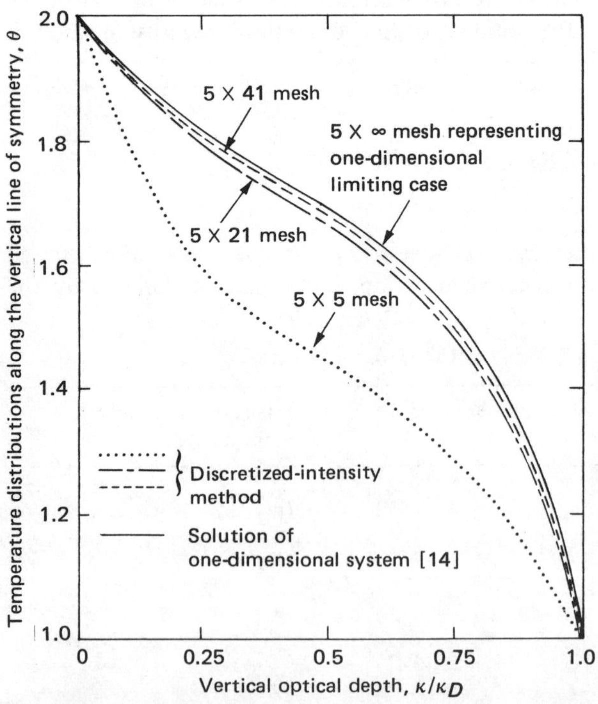

Figure 10-10 Temperature distributions along the vertical line of symmetry for 5 × 5, 5 × 21, and 5 × 41 mesh systems in comparison with the solution for one-dimensional systems [14].

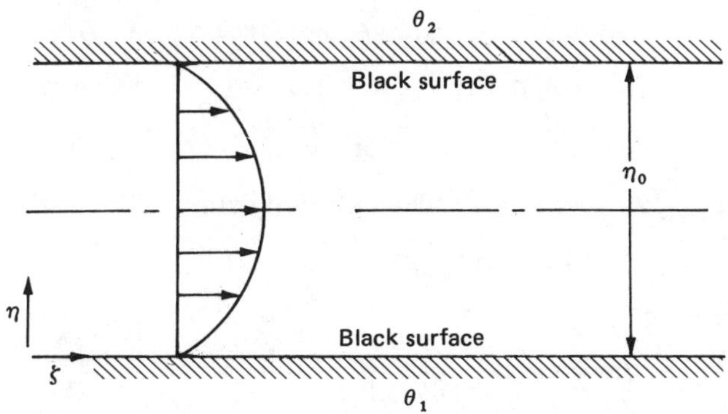

Figure 10-11 System schematic of a thermally developing flow between two parallel plates.

scope of this book, we will focus on the thermally developing flows of a radiatively participating fluid between two parallel infinite black plates, schematically shown in Fig. 10-11. For studies of other combined-modes heat transfer problems, [32-34] may be consulted. The energy transport equation, after nondimensionalization, can be written as

$$A\theta = \eta(\eta_0 - \eta)\frac{\partial \theta}{\partial \zeta} - \frac{\partial^2 \theta}{\partial \eta^2} + \frac{1}{4N_1}\frac{\partial q_r}{\partial \eta} = 0 \tag{10-46}$$

and

$$\frac{\partial q_r}{\partial \eta} = -2\theta_1^4 E_2(\eta) - 2\theta_2^4 E_2(\eta_0 - \eta) - \int_0^{\eta_0} \theta^4(\eta^*) E_1(|\eta - \eta^*|)\, d\eta^* + 4\theta^4(\eta)\,, \tag{10-47}$$

where A represents a nonlinear operator and ζ and η are dimensionless x and y coordinates, respectively. Equation (10-46) is a nonlinear integrodifferential equation; its derivation can be found in textbooks of thermal radiation heat transfer such as [1, 2, 32]. For simplicity, consider the following initial ($\zeta = 0$) and boundary conditions:

$$\theta(0, \eta) = \theta_0(\eta) = 0\,, \tag{10-48a}$$

$$\theta(\zeta, 0) = \theta_1 = 1\,, \tag{10-48b}$$

and

$$\theta(\zeta, \eta_0) = \theta_2 = 1\,. \tag{10-48c}$$

We will adopt the collocation method, using cubic splines as the basis functions, to solve Eqs. (10-46)-(10-48). At a given streamwise location ζ_0, we approximate $\theta(\zeta_0, \eta)$ by a linear combination of three cubic splines, i.e.,

$$\tilde{\theta}(\zeta_0, \eta) = c_0 B_0(\eta) + c_1 B_1(\eta) + c_2 B_2(\eta)\,. \tag{10-49}$$

The three undetermined constants can be computed by collocating Eq. (10-46) at one collocation point $\eta = \eta_0/2$

$$A\theta\left(\zeta_0, \frac{\eta_0}{2}\right) = 0\,, \tag{10-50a}$$

and by utilizing the two boundary conditions Eqs. (10-48*b*) and (10-48*c*)

$$\tilde{\theta}(\zeta_0, 0) = c_0 B_0(0) + c_1 B_1(0) + c_2 B_2(0) = 1 \tag{10-50b}$$

and

$$\tilde{\theta}(\zeta_0, 2) = c_0 B_0(2) + c_1 B_1(2) + c_2 B_2(2) = 1 \,. \tag{10-50c}$$

To carry out the computation, we specify the following set of data:

$$\eta_0 = 2 \,, \quad \zeta_0 = 0.1 \,, \quad \text{and} \quad N_1 = 0.5 \,.$$

With the upwind difference for $\partial\theta/\partial\zeta$

$$\frac{\partial\tilde{\theta}}{\partial\zeta} = \frac{\tilde{\theta}(0.1, \eta) - \theta_0(\eta)}{\Delta\zeta} = 10\tilde{\theta}(0.1, \eta) \,,$$

Eq. (10-50*a*) can be rewritten as

$$10\tilde{\theta}(0.1, 1) - \tilde{\theta}''(0.1, 1) + \frac{1}{2}\left[-4E_2(1) - \int_0^2 \theta^4(0.1, \eta^*)E_1(|1 - \eta^*|)\, d\eta^* + 4\theta^4(0.1, 1)\right] = 0 \,, \tag{10-51}$$

where the prime denotes differentiation with respect to η. Instead of solving Eqs. (10-50*b*), (10-50*c*), and (10-51) simultaneously, it is more convenient to first simplify Eqs. (10-50*b*) and (10-50*c*) (with the aid of Table 7-1) into

$$4c_0 + c_1 = 1$$

and

$$c_1 + 4c_2 = 1 \,,$$

which leads to

$$c_0 = c_2 = \frac{1 - c_1}{4} \,.$$

Then, on the basis of Eq. (10-49), we obtain

$$\tilde{\theta}(0.1, 1) = c_0 + 4c_1 + c_2 = 3.5c_1 + 0.5 \tag{10-52a}$$

and

$$\tilde{\theta}''(0.1, 1) = -15c_1 + 3 \,. \tag{10-52b}$$

Now the remaining task is to compute the integral in Eq. (10-51). For easy presentation, we will use the trapezoidal rule with a crude integration step size $\Delta\eta = 1$

$$\int_0^2 \tilde{\theta}^4(0.1, \eta^*)E_1(|1 - \eta^*|)\, d\eta^* \approx [\tilde{\theta}^4(0.1, 0)E_1(1) + \tilde{\theta}^4(0.1, 1)E_1(0.01)] \times \left(\frac{0.99}{2}\right) + [\tilde{\theta}^4(0.1, 1)E_1(0.01) + \tilde{\theta}^4(0.1, 2)E_1(1)] \times \left(\frac{1.01}{2}\right) , \tag{10-53a}$$

in which

$$\tilde{\theta}^4(0.1, 0.99) \approx \tilde{\theta}^4(0.1, 1)$$

has been assumed. Consulting Table 10-1, we simplify Eq. (10-53*a*) to

$$\int_0^2 \tilde{\theta}^4(0.1, \eta^*) E_1(|1-\eta^*|)\, d\eta^* = 0.2194 + 4.0379\tilde{\theta}^4(0.1, 1)\,. \qquad (10\text{-}53b)$$

Substituting Eq. (10-53*b*) into Eq. (10-51) and simplifying the result, we finally obtain the following nonlinear algebraic equation:

$$50c_1 + 1.5933 - 0.019(3.5c_1 + 0.5)^4 = 0\,. \qquad (10\text{-}54)$$

Among the four roots, the one that renders the present problem physically significant is

$$c_1 = -0.0319\,.$$

Therefore, the temperature profile of the channel flow at $\zeta = 0.1$ is given as

$$\theta(0.1, \eta) = 0.2580B_0(\eta) - 0.0319B_1(\eta) + 0.2580B_2(\eta)\,. \qquad (10\text{-}55)$$

If we are interested in the temperature at the centerline, it can be calculated and found to be

$$\theta(0.1, 1) = 0.3884\,.$$

Physically, this indicates that the flow initially at zero value, when traveling the distance from $\zeta = 0$ to $\zeta = 0.1$, receives energy from the channel walls via conduction and radiation. For more accurate numerical computations, see [39].

SYMBOLS

a_λ	spectral absorption coefficient, m^{-1}
A_i	total band absorptance for the ith band, μm
D	thickness of one-dimensional slab, m
$E_n(t)$	exponential integral [Eq. (*i*) of Example 10-2]
$F_{i\text{-}j}$	shape factor from surface i to surface j
I	intensity, $W/m^2 \cdot$angle [Eq. (10-11*a*) or (10-38)] or variational principle
N_1	conduction-radiation parameter ($= ka/4\sigma T_1^3$)
s	physical distance traveled by the intensity, m
t	dummy variable
T_0	temperature at $x = 0$ or reference temperature, K
α	inclination angle measured from surface
β	normal angle measured from normal of surface ($= \pi/2 - \alpha$)
ϵ	emissivity
θ	T/T_1
Θ	T^4/T_1^4 or T^4/T_∞^4
κ	optical path ($= \int_{x_0}^{x} a\, dx$)
$\tilde{\kappa}$	optical path ($= \int_{s_0}^{s} a\, ds$)
λ	wavelength, μm
μ	$\cos\beta$

ψ — $\dot{q}''/\sigma T_0^4$ or $\dot{q}''/\sigma T_1^4$
ω — solid angle
ξ — radiation-conduction parameter $[= 2h\sigma T_0^3 \exp(-ah/2)/k]$

Subscripts

att — attenuation
b — blackbody
em — emission
i — incoming
o — outgoing
r — radiation
1 — surface 1

Superscript

$\bar{}$ — overbar denoting previously iterated values

REFERENCES

1. R. Siegel and J. R. Howell, *Thermal Radiation Heat Transfer*, McGraw-Hill, New York, 1972.
2. E. M. Sparrow and R. D. Cess, *Radiation Heat Transfer*, Hemisphere, Washington, D.C., and McGraw-Hill, New York, 1978.
3. R. M. Goody, *Atmospheric Radiation, Theoretical Basis*, vol. 1, Clarendon, Oxford, 1964.
4. M. A. Heaslet and R. F. Warming, Radiative Transport and Wall Temperature Slip in an Absorbing Planar Medium, *Int. J. Heat Mass Transfer*, vol. 8, pp. 979–994, 1965.
5. D. K. Edwards, L. K. Glassen, W. C. Hauser, and J. S. Tuchscher, Radiation Heat Transfer in Nonisothermal Nongray Gases, *ASME J. Heat Transfer*, vol. 89, pp. 219–229, 1967.
6. C. L. Tien, Thermal Radiation Properties of Gases, *Adv. Heat Transfer*, vol. 5, pp. 253–324, 1968.
7. D. K. Edwards and A. Balakrishnan, Slab Band Absorptance for Molecular Gas Radiation, *J. Quant. Spectrosc. Radiat. Transfer*, vol. 12, pp. 1379–1387, 1972.
8. D. K. Edwards, Molecular Gas Band Radiation, *Adv. Heat Transfer*, vol. 12, pp. 115–193, 1976.
9. G. L. Hubbard and C. L. Tien, Infrared Mean Absorption Coefficients of Luminous Flames and Smoke, *ASME J. Heat Transfer*, vol. 100, pp. 235–239, 1978.
10. S. N. Tiwari and S. K. Gupta, Accurate Spectral Modeling for Infrared Radiation, *ASME J. Heat Transfer*, vol. 100, pp. 240–246, 1978.
11. D. A. Nelson, Band Radiation within Diffuse-walled Enclosures, Part I: Exact Solutions for Simple Enclosure; Part II: An Approximate Method Applied to Simple Enclosures, *ASME J. Heat Transfer*, vol. 101, pp. 81–89, 1979.
12. W. W. Yuen and C. L. Tien, A Successive Approximation Approach to Problems in Radiative Transfer with a Differential Formulation, *ASME J. Heat Transfer*, vol. 102, pp. 86–91, 1980.
13. C. N. Liu and T. M. Shih, Laminar, Mixed-Convection, Boundary-Layer, Nongray-Radiative, Diffusion Flames, *ASME J. Heat Transfer*, vol. 102, pp. 724–730, 1980.
14. R. Viskanta and R. J. Grosh, Heat Transfer by Simultaneous Conduction and Radiation in an Absorbing Medium, *ASME J. Heat Transfer*, vol. 84, pp. 63–72, 1962.
15. H. C. Hottel and E. S. Cohen, Radiant Heat Transfer in a Gas-filled Enclosure: Allowance for Non-uniformity of Gas Temperature, *AIChE J.*, vol. 4, pp. 3–14, 1958.

16. H. C. Hottel and A. F. Sarofim, The Effect of Gas Flow Patterns on Radiative Transfer in Cylindrical Furnaces, *Int. J. Heat Mass Transfer*, vol. 8, pp. 1153–1169, 1965.
17. H. C. Hottel and A. F. Sarofim, *Radiative Transfer*, pp. 83–106, McGraw-Hill, New York, 1967.
18. H. C. Hottel, First Estimates of Industrial Furnace Performance—The One-Gas-Zone Model Reexamined, in N. H. Afgan and J. M. Beer (eds.), *Heat Transfer in Flames*, pp. 5–28, Hemisphere, Washington, D.C., 1974.
19. J. M. Beer, Methods for Calculating Radiative Heat Transfer from Flames in Combustors and Furnaces, in N. H. Afgan and J. M. Beer (eds.), *Heat Transfer in Flames*, pp. 29–45, Hemisphere, Washington, D.C., 1974.
20. F. R. Steward and H. K. Guruz, Mathematical Simulation of an Industrial Boiler by the Zone Method of Analysis, in N. H. Afgan and J. M. Beer (eds.), *Heat Transfer in Flames*, pp. 47–71, Hemisphere, Washington, D.C., 1974.
21. K. A. Bueters, J. G. Logoli, and W. W. Habelt, Performance Predictions of Tangentially Fired Utility Furnaces by Computer Model, Fifteenth Symp. (Int'l) on Comb., The Comb. Inst., 1974, pp. 1245–1260.
22. A. Lowe, T. F. Wall, and J. M. Stewart, A Zoned Heat Transfer Model of a Large Tangentially Fired Pulverized Coal Boiler. Fifteenth Symp. (Int'l) on Comb., The Comb. Inst., 1974, pp. 1261–1270.
23. N. Selcuk, R. G. Siddall, and J. M. Beer, A Comparison of Mathematical Models of the Radiative Behavior of a Large-Scale Experimental Furnace, Sixteenth Symp. (Int'l) on Comb., The Comb. Inst., 1976, pp. 53–62.
24. N. K. Nahra and T. F. Smith, Combined Radiation-Convection for a Real Gas, *ASME J. Heat Transfer*, vol. 99, pp. 486–491, 1978.
25. C. W. Clausen and T. F. Smith, Radiative and Convective Transfer for Real Gas Flow through a Tube with Specified Wall Heat Flux, *ASME J. Heat Transfer*, vol. 100, pp. 376–378, 1979.
26. R. G. Siddall, The Flux Method of Furnace Heat Transfer Analysis, Fourth Symp. on Flames and Industry, Brit. Flame Res. Comm. and Inst. Fuel at Imperial College, 1972.
27. S. V. Patankar and D. B. Spalding, A Computer Model for Three-dimensional Flow in Furnaces, Fourteenth Symp. (Int'l) on Comb., The Comb. Inst., 1972, pp. 605–614.
28. A. D. Gosman and F. C. Lockwood, Incorporation of a Flux Model for Radiation into a Finite Difference Procedure for Furnace Calculations, Fourteenth Symp. (Int'l) on Comb., The Comb. Inst., 1972, pp. 661–671.
29. S. V. Patankar and D. B. Spalding, Simultaneous Predictions of Flow Patterns and Radiation for Three-dimensional Flames, in N. H. Afgan and J. M. Beer (eds.), *Heat Transfer in Flames*, pp. 73–94, Hemisphere, Washington, D.C., 1974.
30. N. Selcuk and R. G. Siddall, Two-Flux Spherical Harmonic Modeling of Two-dimensional Radiative Transfer in Furnaces, *Int. J. Heat Mass Transfer*, vol. 19, pp. 313–321, 1976.
31. T. M. Shih and Y. N. Chen, A Discretized-Intensity Method Proposed for Two-dimensional Systems Enclosing Radiative and Conductive Media, *Numer. Heat Transfer*, vol. 6, pp. 117–134, 1983.
32. M. N. Oziski, *Radiative Transfer and Interaction with Conduction and Convection*, Wiley-Interscience, New York, 1978.
33. J. D. Bankston, J. R. Lloyd, and J. L. Novotny, Radiative Convection Interaction in an Absorbing-emitting Liquid in Natural Convection Boundary Layer Flow, *ASME J. Heat Transfer*, vol. 99, pp. 125–127, 1977.
34. R. K. James and D. K. Edwards, Effect of Molecular Gas Radiation on a Planar, Two-dimensional Turbulent-Jet-Diffusion Flame, *ASME J. Heat Transfer*, vol. 99, pp. 221–226, 1977.
35. D. Y. Goswami and R. I. Vachon, Radiative Heat-Transfer Analysis Using an Effective Absorptivity for Absorption, Emission and Scattering, *Int. J. Heat Mass Transfer*, vol. 11, pp. 1233–1239, 1977.
36. J. N. Reddy and V. D. Murty, Finite Element Solution of Integral Equations Arising in Radiative Heat Transfer and Laminar Boundary Layer Theory, *Numer. Heat Transfer*, vol. 1, pp. 389–401, 1978.

37. C. W. Clausen and T. F. Smith, Radiative and Convective Transfer for Real Gas Flow through a Tube with Specified Wall Heat Flux, *ASME J. Heat Transfer*, vol. 101, pp. 376–379, 1979.
38. A. Balakrishnan and D. K. Edwards, Molecular Gas Radiation in the Thermal Entrance Region of a Duct, *ASME J. Heat Transfer*, vol. 101, pp. 489–495, 1979.
39. T. C. Chawla and S. H. Chan, Combined Radiation Convection in Thermally Developing Poiseuille Flow with Scattering, *ASME J. Heat Transfer*, vol. 102, pp. 297–302, 1980.
40. C. H. Liu and E. M. Sparrow, Convective-radiative Interaction in a Parallel Plate Channel–Applied to Air-operated Solar Collectors, *Int. J. Heat Mass Transfer*, vol. 23, pp. 1137–1146, 1980.
41. F. B. Cheung, S. H. Chan, T. C. Chawla, and D. H. Cho, Radiative Heat Transfer in a Heat Generating and Turbulently Convecting Fluid Layer, *Int. J. Heat Mass Transfer*, vol. 23, pp. 1313–1324, 1980.
42. T. C. Chawla and S. H. Chan, Spline Collocation Solution of Combined Radiation-Convection in Thermally Developing Flows with Scattering, *Numer. Heat Transfer*, vol. 3, pp. 47–65, 1980.
43. T. C. Chawla, W. J. Minkowycz, and G. Leaf, Spline-Collocation Solution of Integral Equations Occurring in Radiative Transfer and Laminar Boundary-Layer Problems, *Numer. Heat Transfer*, vol. 3, pp. 133–148, 1980.
44. F. H. Azad and M. F. Modest, Combined Radiation and Convection in Absorbing, Emitting and Anisotropically Scattering Gas-Particle Tube Flow, *Int. J. Heat Mass Transfer*, vol. 24, pp. 1681–1698, 1981.

PROBLEMS

10-1 The normalized radiative flux distributions are given as

$$\psi_1(\xi_1) = \xi_1 \qquad \text{and} \qquad \psi_2(\xi_2) = 1 - \xi_2^2$$

in Eqs. (10-7*a*) and (10-7*b*). Finish Example 10-1 by computing Θ_a, Θ_b, Θ_c, and Θ_d.

10-2 Consider an equilateral triangular enclosure with $\epsilon_1 = 1$, $\epsilon_2 = 0.8$, $\epsilon_3 = 0.6$, $T_1 = 1000$ K, $T_2 = 2000$ K, and $\dot{q}_3'' = 0$ (i.e., surface 3 is insulated). Find T_3. Will T_3 increase when ϵ_3 increases? Is it possible that T_3 will lie beyond the range 1000 to 2000 K?

10-3 Repeat Example 10-2 for an isothermal medium $T = T_0$. Is the result consistent with what is expected based on intuition?

10-4 Repeat Example 10-3 with two interior grid points. Compare the result with the exact solution given in [4].

10-5 Use Newton's method to solve Eq. (10-35). Let $c^{[0]} = 1$.

10-6 The equation of transfer for a radiatively ideal gas is given by

$$I(\kappa) = I(0)\exp(-\kappa) + \int_0^{\kappa} \frac{\sigma T^4(t)}{\pi} \exp(t - \kappa)\, dt\,, \tag{P10-7a}$$

where $\kappa = ax\,, \quad x \in [0, 1]\,.$

The linear temperature distribution of the gas is assumed to be

$$T(x) = (T_1 - T_0)x + T_0\,.$$

(*a*) Find the expressions for $I(0.5a)$ and $I(a)$.
(*b*) What are the percentage errors of $|I(0.5a) - \frac{1}{2}[I(0) + I(a)]|$ for $a = 0.1$, 1, and 5?

(*c*) At what value of p is

$$I_{E,p} = \tfrac{1}{2}(I_{E,2} + I_{E,5})$$

a reasonable approximation? (The subscripts p, 2, and 5 refer to Fig. 10-8.)

10-7 Derive the energy conservation equation analogous to Eq. (10-43) accounting for the α dependence of intensities.

10-8 Replace Eq. (10-39) with the conventional definition

$$J_b = \frac{\sigma T_b^4}{\pi} = \frac{2I_b}{\pi}.$$

Show that this definition will also lead to Eq. (10-41*a*).

PART

THREE

IMPORTANT HEAT TRANSFER PHENOMENA

CHAPTER

ELEVEN

LAMINAR FREE CONVECTION AND MIXED CONVECTION

In Chapter 9 we assumed that the momentum equation and the energy equation could be decoupled. This decoupling was possible because the flow was driven by the shear of a sliding lid. When a thermal gradient exists in the flow field and buoyancy plays a significant role in accelerating or retarding the flow, the momentum transport and energy transport become intrinsically coupled. This mixed-convection problem is more complicated than its forced-convection counterpart not only because the system of simultaneous governing equations is enlarged but also because a nonlinearity arises in the energy equation.

When the ambient fluid around the system investigated is stagnant, mixed convection reduces to pure free convection. Mathematically, the only quantity that must be changed is the boundary condition of the flow velocity. Therefore, in this chapter we will investigate both free convection and mixed convection at the same time.

In Section 11-1, the governing equations for two-dimensional laminar mixed-convection flow are introduced in terms of both the primitive variables u, v, and p, and the vorticity ω and stream function ψ.

In Section 11-2, the governing equations are reduced to those governing the boundary-layer flow. The similarity equations governing the pure free-convection flow over a vertical isothermal plate are derived and subsequently solved by using the parameterized residual method, while the nonsimilarity method and the discretization methods are briefly introduced. Most of the procedures used here were already applied to the forced-convection boundary-layer problems described in Sections 6-2, 7-5, and 9-2.

Section 11-3 concerns the free-convection flow in an enclosure with one side heated. The stream function vorticity equations are solved by using the central finite-difference scheme. The solution procedure is illustrated in detail for a coarse-meshed system with only four grid points.

11-1 GOVERNING EQUATIONS

The two-dimensional, laminar, mixed-convection flow exerted by the force of gravity along the negative x direction can be described by the following conservation equations:

$$\frac{\partial}{\partial x}(\rho u) + \frac{\partial}{\partial y}(\rho v) = 0 \,, \tag{11-1}$$

$$\rho u \frac{\partial u}{\partial x} + \rho v \frac{\partial u}{\partial y} = \frac{\partial}{\partial x}\left(\mu \frac{\partial u}{\partial x}\right) + \frac{\partial}{\partial y}\left(\mu \frac{\partial u}{\partial y}\right) - \frac{\partial p}{\partial x} - \rho g \,, \tag{11-2}$$

$$\rho u \frac{\partial v}{\partial x} + \rho v \frac{\partial v}{\partial y} = \frac{\partial}{\partial x}\left(\mu \frac{\partial v}{\partial x}\right) + \frac{\partial}{\partial y}\left(\mu \frac{\partial v}{\partial y}\right) - \frac{\partial p}{\partial y} \,, \tag{11-3}$$

$$\rho c_p u \frac{\partial T}{\partial x} + \rho c_p v \frac{\partial T}{\partial y} = \frac{\partial}{\partial x}\left(k \frac{\partial T}{\partial x}\right) + \frac{\partial}{\partial y}\left(k \frac{\partial T}{\partial y}\right) , \tag{11-4}$$

and the equation of state for the fluid

$$\beta = \frac{-(\partial \rho / \partial T)_p}{\rho} = \text{constant} \,. \tag{11-5}$$

Since there are five unknowns $u(x,y)$, $v(x,y)$, $p(x,y)$, $\rho(x,y)$, and $T(x,y)$, the system of Eqs. (11-1)–(11-5) is closed provided that proper boundary conditions of the two-dimensional domain are given. To further simplify this closed system, we will confine our interest to Boussinesq flow [1], i.e.,

$$\frac{\rho_{\max} - \rho_{\min}}{\rho} \ll 1 \,, \tag{11-6}$$

along with constant thermal properties. For Eq. (11-6) to remain valid, the temperature variation of the flow must be small. Thus, Eq. (11-6) may be considered a rather severe assumption. On the other hand, if the fluid can be treated as an ideal gas, the assumption

$$\rho T = \frac{p}{R} = \text{constant}^{\dagger} \tag{11-7}$$

is generally true for most heat transfer applications.

Example 11-1 The pressure field $p(x, y)$ that appears in Eqs. (11-2) and (11-3) is not constant. Is Eq. (11-7) true then for most gaseous systems of finite size?

†In most heat transfer applications, the flow is nonrigorously said to be incompressible if it satisfies Eq. (11-7).

Solution Let the height H of the system exposed to atmospheric conditions be 1 m and the pressure difference between the top and the bottom of the system be approximated by the static head as

$$\Delta p \approx \rho g H = (1\ \text{kg/m}^3)(9.8\ \text{m/s}^2)(1\ \text{m}) = 9.8\ \text{NT/m}^2\ , \tag{a}$$

Therefore,

$$\frac{\Delta p}{p} = \frac{9.8\ \text{NT/m}^2}{1.01 \times 10^5\ \text{NT/m}^2} \approx 10^{-4}\ . \tag{b}$$

According to Eq. (11-7), we obtain

$$\frac{\rho_b}{\rho_t} = \frac{p_b}{p_t}\frac{T_t}{T_b} = \frac{p_t + \Delta p}{p_t}\frac{T_t}{T_b} = (1 + 10^{-4})\frac{T_t}{T_b} \approx \frac{T_t}{T_b}, \tag{c}$$

where the subscripts b and t denote bottom and top, respectively. We have thus established that Eq. (11-7) is true for gaseous systems of moderate size. This static head Δp is small in comparison with atmospheric pressure, but is sufficiently large to drive the flow.

For steady-state two-dimensional incompressible flows, it is common practice [2-52] to eliminate the pressure variable by introducing the stream function $\bar{\psi}(x,y)$ defined as

$$u = \frac{\partial \bar{\psi}}{\partial y} \quad \text{and} \quad v = -\frac{\partial \bar{\psi}}{\partial x}, \tag{11-8a, b}$$

and the vorticity $\bar{\omega}(x,y)$ defined as

$$\bar{\omega} = \frac{\partial v}{\partial x} - \frac{\partial u}{\partial y}. \tag{11-9}$$

With Eqs. (11-8) and (11-9), we can transform Eqs. (11-1)-(11-4) to

$$\nabla^2 \psi = -\omega\ , \tag{11-10}$$

$$\frac{\partial \psi}{\partial Y}\frac{\partial \omega}{\partial X} - \frac{\partial \psi}{\partial X}\frac{\partial \omega}{\partial Y} = \Pr \nabla^2 \omega - \Pr \text{Ra}\frac{\partial \theta}{\partial Y}, \tag{11-11}$$

and

$$\frac{\partial \psi}{\partial Y}\frac{\partial \theta}{\partial X} - \frac{\partial \psi}{\partial X}\frac{\partial \theta}{\partial Y} = \nabla^2 \theta\ , \tag{11-12}$$

where

$$\psi = \frac{\bar{\psi}}{\alpha}, \qquad \omega = \frac{\bar{\omega} L^2}{\alpha}, \qquad \text{and} \qquad \theta = \frac{T - T_{\min}}{T_{\max} - T_{\min}}$$

are dimensionless dependent field variables (p disappeared),

$$X = \frac{x}{L} \quad \text{and} \quad Y = \frac{y}{L}$$

are normalized coordinates, and the parameters

$$\mathrm{Pr} = \frac{\nu}{\alpha} \quad \text{and} \quad \mathrm{Ra} = \frac{g\beta\,\Delta T L^3}{\nu\alpha}$$

are the Prandtl number and the Rayleigh number, respectively. We realize that the number of unknowns is now reduced to three and the system of Eqs. (11-10)-(11-12) remains closed provided the corresponding boundary conditions are prescribed.†

The derivation of the buoyancy term Pr Ra $(\partial\theta/\partial Y)$ in Eq. (11-11) deserves special attention. One way to transform the term involving density variation into one involving temperature variation is to first replace $-\rho g - \partial p/\partial x$ with $(\rho_\infty - \rho)g - \partial(p + \rho_\infty g x)/\partial x$ in Eq. (11-2), which is then uniformly divided by ρ and differentiated with respect to y. The density term eventually leads to Pr Ra $(\partial\theta/\partial Y)$, while the pressure term is dropped through cross-differentiation (see also Problem 11-1).

11-2 BOUNDARY-LAYER FLOWS (PARABOLIC EQUATIONS)

When the flow is of the boundary-layer type, the assumptions

$$\frac{\partial^2 u}{\partial y^2} \gg \frac{\partial^2 u}{\partial x^2}, \quad \frac{\partial^2 \theta}{\partial y^2} \gg \frac{\partial^2 \theta}{\partial x^2}, \quad \text{and} \quad \frac{\partial p}{\partial y} \approx 0$$

become valid. On the basis of the last assumption, it is customary to express the pressure gradient in terms of the density of the ambient fluid as

$$\frac{\partial p}{\partial x} \approx \frac{dp_\infty}{dx} = -\rho_\infty g\,. \tag{11-13}$$

Substituting this expression into Eq. (11-2) and assuming μ to be constant, we obtain

$$u\frac{\partial u}{\partial x} + v\frac{\partial u}{\partial y} = \nu\frac{\partial^2 u}{\partial y^2} + \frac{g(\rho_\infty - \rho)}{\rho}, \tag{11-14a}$$

along with the energy equation

$$u\frac{\partial T}{\partial x} + v\frac{\partial T}{\partial y} = \alpha\frac{\partial^2 T}{\partial y^2}\,. \tag{11-14b}$$

The buoyancy term in this case can be conveniently changed, according to Eq. (11-5), to

$$\frac{g(\rho_\infty - \rho)}{\rho} \approx g\beta(T - T_\infty)\,. \tag{11-15}$$

†We can also further eliminate ω by differentiating Eq. (11-10) once and twice and then substituting the result into Eq. (11-11). This procedure, however, raises the differential equation in ψ to fourth order, and correspondingly the discretization involves more nodal unknowns or requires smoother continuity across the elemental boundaries [53-61]. See Section 8-2*d*.

Since Eq. (11-3), being of a lower order of magnitude than Eq. (11-2), is not considered, the pressure variable disappears from the governing equations. It is therefore unnecessary to introduce the stream function and the vorticity for the purpose of avoiding calculation of the pressure field.

A typical mixed-convection boundary-layer flow is shown in Fig. 11-1. The vertical impermeable wall is maintained at a constant temperature slightly higher than the ambient temperature, and the directions of both the buoyancy force and the ambient free stream are upward. The boundary conditions of this system are

$$u(0,y) = u_\infty\ , \qquad T(0,y) = T_\infty\ , \tag{11-16a}$$

$$u(x,0) = 0\ , \qquad T(x,0) = T_w\ , \qquad v(x,0) = 0\ , \tag{11-16b}$$

and

$$u(x,\infty) = u_\infty\ , \qquad T(x,\infty) = T_\infty\ . \tag{11-16c}$$

Clearly, if the ambient flow velocity u_∞ equals zero, the given flow reduces to a pure free-convection case.

11-2*a* Similarity Method

As mentioned in Chapter 6, it is not a bad habit to examine the possibility that a similarity solution to the boundary-layer equation exists. To do this, we transform the x, y physical coordinates into x, η similarity coordinates, where

$$\eta = \frac{y}{A_1 \nu x^a} \tag{11-17}$$

is the similarity variable. The transformation is described by

$$\left(\frac{\partial}{\partial x}\right)_y = \left(\frac{\partial}{\partial x}\right)_\eta - \frac{a\eta}{x}\frac{\partial}{\partial \eta} \tag{11-18a}$$

and

$$\frac{\partial}{\partial y} = (A_1 \nu x^a)^{-1} \frac{\partial}{\partial \eta}\ . \tag{11-18b}$$

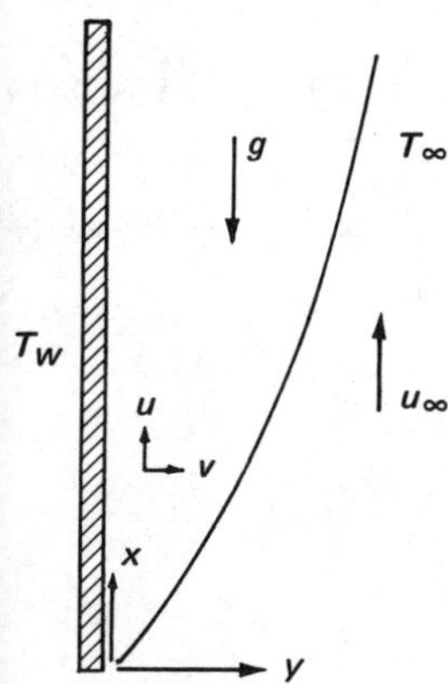

Figure 11-1 System schematic of a mixed-convection boundary-layer flow.

In addition, we introduce the similarity function $f(x, \eta)$ defined as

$$f(x, \eta) = \frac{\overline{\psi}(x, y)}{A_2 x^b} . \tag{11-19}$$

The constants A_1, A_2, a, and b are to be determined. Since, by definition,

$$u = \frac{\partial \overline{\psi}}{\partial y} \quad \text{and} \quad v = -\frac{\partial \overline{\psi}}{\partial x} ,$$

Eqs. (11-17)–(11-19) yield

$$u = U \frac{\partial f}{\partial \eta} \tag{11-20a}$$

and

$$v = A_2 \left(a\eta x^{b-1} \frac{\partial f}{\partial \eta} - bx^{b-1} f - x^b \frac{\partial f}{\partial x} \right) , \tag{11-20b}$$

where U is a reference velocity defined as

$$U = \frac{A_2 x^{b-a}}{A_1 \nu} . \tag{11-21}$$

Utilizing Eqs. (11-17)–(11-21), we can transform Eqs. (11-14a) and (11-14b) into

$$f''' - B_1(x) f'^2 + B_2(x) f f'' + B_3(x)\theta = B_4(x) \left(f' \frac{\partial f'}{\partial x} - f'' \frac{\partial f}{\partial x} \right) \tag{11-22}$$

and

$$\theta'' + \Pr B_2(x) f \theta' = B_4(x) \left(f' \frac{\partial \theta}{\partial x} - \theta' \frac{\partial f}{\partial x} \right) . \tag{11-23}$$

Here the prime denotes differentiation with respect to η; the functions $B_k(x)$, $k =$ 1, 2, 3, 4, are listed in Table 11-1 for bookkeeping. The boundary conditions corresponding to Eqs. (11-16a)–(11-16c) are

$$\frac{x}{b} \frac{\partial f(x, 0)}{\partial x} + f(x, 0) = 0 , \quad f'(x, 0) = 0 , \quad \theta(x, 0) = 1 , \tag{11-24a}$$

Table 11-1 Expressions for $B_k(x)$ in Eqs. (11-22) and (11-23)

$B_1(x)$	$A_1 A_2 (b - a) x^{a+b-1}$
$B_2(x)$	$A_1 A_2 b x^{a+b-1}$
$B_3(x)$	$\dfrac{(T_w - T_\infty) A_1^3 g \nu^2 x^{3a-b}}{T_\infty A_2}$
$B_4(x)$	$A_1 A_2 x^{a+b}$

and

$$f'(x,\eta_\infty)=\frac{u_\infty \nu x^{a-b}A_1}{A_2}, \qquad \theta(x,\eta_\infty)=0\,. \tag{11-24b}$$

There are at least two choices of a, b, A_1, and A_2 that can greatly simplify Eqs. (11-22)-(11-24), namely,

1. $a=\frac{1}{4}$, $b=\frac{3}{4}$, $A_1=\left[\frac{4T_\infty}{g\nu^2(T_w-T_\infty)}\right]^{1/4}$, and $A_2=\frac{4}{A_1}$.

and

2. $a=\frac{1}{2}$, $b=\frac{1}{2}$, $A_1=2(u_\infty \nu)^{-1/2}$, and $A_2=(u_\infty \nu)^{1/2}$.

For choice (1), Eqs. (11-22)-(11-24) reduce to

$$f'''-2f'^2+3ff''+\theta=4\xi\left(f'\frac{\partial f'}{\partial \xi}-f''\frac{\partial f}{\partial \xi}\right) \tag{11-25}$$

and

$$\theta''+3\,\mathrm{Pr}\,f\theta'=4\xi\left(f'\frac{\partial \theta}{\partial \xi}-\theta'\frac{\partial f}{\partial \xi}\right) \tag{11-26}$$

subject to the boundary conditions

$$\frac{4}{3}\,\xi\,\frac{\partial f(\xi,0)}{\partial \xi}+f(\xi,0)=0\,, \qquad f'(\xi,0)=0\,, \qquad \theta(\xi,0)=1\,, \tag{11-27a}$$

and

$$f'(\xi,\eta_\infty)=\tfrac{1}{2}\xi^{-1/2}\,, \qquad \theta(\xi,\eta_\infty)=0\,. \tag{11-27b}$$

We will devote our efforts to choice (1) and leave the derivation for choice (2) to the reader as an exercise (see Problem 11-3 and [62]). The parameter ξ indicates the strength of the buoyancy relative to the flow momentum, defined as

$$\xi=\frac{gx(T_w-T_\infty)}{u_\infty^2 T_\infty}=\frac{\mathrm{Gr}_x}{\mathrm{Re}_x^2}\,. \tag{11-28}$$

When the ambient fluid is stagnant ($u_\infty=0$), it follows that

$$f'(\xi,\eta_\infty)=0 \tag{11-29}$$

based on Eqs. (11-27b) and (11-28). At this point, observant readers may notice that the dependence of the similarity function $f(\xi,\eta_\infty)$ on ξ becomes very weak. In fact, if we boldly assume that f is entirely independent of ξ, Eqs. (11-25)-(11-27) are reduced to

$$f'''-2f'^2+3ff''+\theta=0 \tag{11-30}$$

and

$$\theta''+3\,\mathrm{Pr}\,f\theta'=0 \tag{11-31}$$

subject to

$$f(0)=0\,, \qquad f'(0)=0\,, \qquad \theta(0)=1\,, \tag{11-32a}$$

and
$$f'(\eta_\infty) = 0 , \quad \theta(\eta_\infty) = 0 . \tag{11-32b}$$

The parameter ξ is seen to disappear from the governing equations and the assumption above seems reasonable. This pure free-convection similarity problem—a special case of Eqs. (11-25)-(11-27)—is a classical problem [63] and is presented in heat transfer textbooks [64-66]. Since Eqs. (11-30)-(11-32) constitute a typical nonlinear two-point boundary-value problem involving two missing initial conditions $f''(0)$ and $\theta'(0)$, and since the solution procedure generally has not been emphasized in heat transfer textbooks, we will present the parameterized-residuals (PR) method [67] here (see also Section 7-5) to solve Eqs. (11-30)-(11-32).

Let us choose a polynomial†

$$\tilde{f}(\eta) = -\frac{\eta^3}{36} + \frac{7\eta^2}{24} , \tag{11-33}$$

which satisfies Eqs. (11-32*a*) and (11-32*b*) as well as an arbitrarily assigned condition $\tilde{f}'(1) = \frac{1}{2}$, as an approximate solution to $f(\eta)$. With Eqs. (11-31), (11-32*a*), and (11-32*b*), we can analytically obtain

$$\tilde{\theta}(\eta) = 1 - \frac{\int_0^{\eta} \exp(-3\tilde{f}\,\mathrm{Pr})\, d\eta}{\int_0^{\eta_\infty} \exp(-3\tilde{f}\,\mathrm{Pr})\, d\eta} . \tag{11-34}$$

Substituting Eqs. (11-33) and (11-34) into Eq. (11-30) generates a residual $R(\eta)$; embedding an artificial parameter $1-\lambda$ in front of the residual leads to

$$\tilde{f}''' + 3\tilde{f}\tilde{f}'' - 2\tilde{f}'^2 + \tilde{\theta} = (1-\lambda) R(\eta) . \tag{11-35}$$

It is clear that the guessed profiles Eqs. (11-33) and (11-34) are the solution to Eqs. (11-31), (11-32), and (11-35) at $\lambda = 0$ and that the solution sought is the one that satisfies the same equations at $\lambda = 1$. Linking the assumed solution at $\lambda = 0$ with the genuine solution at $\lambda = 1$ is our next task, and it can be performed by, for example, the parametric expansion method [68]. We first expand the functions at $\lambda + \Delta\lambda$ about those at λ in a Taylor's series as

$$\tilde{f}(\eta, \lambda + \Delta\lambda) = \tilde{f}(\eta, \lambda) + g(\eta, \lambda)\,\Delta\lambda + h(\eta, \lambda)\,\frac{(\Delta\lambda)^2}{2} + Q_1(\eta, \lambda) , \tag{11-36}$$

where

$$g = \frac{\partial \tilde{f}}{\partial \lambda} \quad \text{and} \quad h = \frac{\partial^2 \tilde{f}}{\partial \lambda^2} ;$$

$$\tilde{\theta}(\eta, \lambda + \Delta\lambda) = \tilde{\theta}(\eta, \lambda) + s(\eta, \lambda)\,\Delta\lambda + t(\eta, \lambda)\,\frac{(\Delta\lambda)^2}{2} + Q_2(\eta, \lambda) , \tag{11-37}$$

†Choice of other guessed profiles is also possible as long as they satisfy the boundary conditions. The final solutions obtained by the PR method have been found to be relatively insensitive to the initial guesses.

where

$$s = \frac{\partial \tilde{\theta}}{\partial \lambda} \quad \text{and} \quad t = \frac{\partial^2 \tilde{\theta}}{\partial \lambda^2} .$$

The higher-order terms $Q_1(\eta, \lambda)$ and $Q_2(\eta, \lambda)$ are to be estimated, whereas the functions g and s should satisfy

$$L_1 g + s + R = 0 \tag{11-38}$$

and

$$L_2 g + L_3 s = 0 \tag{11-39}$$

subject to

$$g(0) = 0 , \quad g'(0) = 0 , \quad g'(\eta_\infty) = 0 , \quad s(0) = 0 , \quad \text{and} \quad s(\eta_\infty) = 0 ; \tag{11-40}$$

the functions h and t should satisfy

$$L_1 h + t = -6gg'' + 4g'^2 \tag{11-41}$$

and

$$L_2 h + L_3 t = -6 \,\mathrm{Pr}\, gs' \tag{11-42}$$

subject to

$$h(0) = 0 , \quad h'(0) = 0 , \quad h'(\eta_\infty) = 0 , \quad t(0) = 0 , \quad \text{and} \quad t(\eta_\infty) = 0 . \tag{11-43}$$

The linear operators L_1, L_2, and L_3 are defined, respectively, as

$$L_1 = \frac{d^3}{d\eta^3} + 3\tilde{f}\frac{d^2}{d\eta^2} - 4\tilde{f}'\frac{d}{d\eta} + 3\tilde{f}'' , \tag{11-44a}$$

$$L_2 = 3 \,\mathrm{Pr}\, \tilde{\theta}' , \tag{11-44b}$$

and

$$L_3 = \frac{d^2}{d\eta^2} + 3 \,\mathrm{Pr}\, \tilde{f}\frac{d}{d\eta} . \tag{11-44c}$$

These linear two-point boundary-value problems Eqs. (11-38)–(11-43) can be transformed into initial-value problems by introducing superposition functions such that

$$g = \alpha_1 g_1 + \alpha_2 g_2 + g_3 , \qquad s = \alpha_1 s_1 + \alpha_2 s_2 + s_3 ,$$
$$h = \beta_1 h_1 + \beta_2 h_2 + h_3 , \qquad t = \beta_1 t_1 + \beta_2 t_2 + t_3 . \tag{11-45a–d}$$

The reason why two undetermined constants are needed is that now two initial conditions $g''(0)$ and $\theta'(0)$ are missing. These superposition functions g_k, h_k, s_k, and t_k, $k = 1, 2, 3$, should satisfy Eqs. (11-46) and (11-47) and the corresponding initial conditions listed in Table 11-2. By a forward integration scheme, such as the Runge-Kutta method, Eqs. (11-46*a*)–(11-46*c*) are integrated first to yield

$$\alpha_1 = \frac{g_2'(\eta_\infty)s_3(\eta_\infty) - g_3'(\eta_\infty)s_2(\eta_\infty)}{g_1'(\eta_\infty)s_2(\eta_\infty) - g_2'(\eta_\infty)s_1(\eta_\infty)} \tag{11-48a}$$

and

$$\alpha_2 = \frac{g_3'(\eta_\infty)s_1(\eta_\infty) - g_1'(\eta_\infty)s_3(\eta_\infty)}{g_1'(\eta_\infty)s_2(\eta_\infty) - g_2'(\eta_\infty)s_1(\eta_\infty)} . \tag{11-48b}$$

Table 11-2 Superposition equations and their corresponding initial conditions

Governing equation	Initial condition at $\eta = 0$	Equation number
$L_1 g_1 + s_1 = 0$ $L_2 g_1 + L_3 s_1 = 0$	$g_1 = 0, g_1' = 0, g_1'' = 1$ $s_1 = 0, s_1' = 0$	(11-46a)
$L_2 g_2 + s_2 = 0$ $L_2 g_2 + L_3 s_2 = 0$	$g_2 = 0, g_2' = 0, g_2'' = 0$ $s_2 = 0, s_2' = 1$	(11-46b)
$L_1 g_3 + s_3 = -R$ $L_2 g_3 + L_3 s_3 = 0$	$g_3 = 0, g_3' = 0, g_3'' = 0$ $s_3 = 0, s_3' = 0$	(11-46c)
$L_1 h_3 + t_3 = -6gg'' + 4g'^2$ $L_2 h_3 + L_3 t_3 = -6 \Pr gs'$	$h_3 = 0, h_3' = 0, h_3'' = 0$ $t_3 = 0, t_3' = 0$	(11-47)

After the values of α_1 and α_2 are computed, the functions g and s can be obtained from Eqs. (11-45*a*) and (11-45*b*) by superposition. Therefore, we are ready to integrate Eqs. (11-46*a*), (11-46*b*), and (11-47) to yield

$$\beta_1 = \frac{h_2'(\eta_\infty)t_3(\eta_\infty) - h_3'(\eta_\infty)t_2(\eta_\infty)}{h_1'(\eta_\infty)t_2(\eta_\infty) - h_2'(\eta_\infty)t_1(\eta_\infty)} \tag{11-49a}$$

and

$$\beta_2 = \frac{h_3'(\eta_\infty)t_1(\eta_\infty) - h_1'(\eta_\infty)t_3(\eta_\infty)}{h_1'(\eta_\infty)t_2(\eta_\infty) - h_2'(\eta_\infty)t_1(\eta_\infty)}. \tag{11-49b}$$

Similarly, h and t can be obtained from Eqs. (11-45*c*) and (11-45*d*). The next task is to estimate $Q_1''(0, \lambda)$ and $Q_2''(0, \lambda)$. For convenience, Eqs. (11-36) and (11-37) are rewritten as

$$\tilde{f}''(0, \lambda + \Delta\lambda) = \tilde{f}_s''(0, \lambda + \Delta\lambda) + Q_1''(0, \lambda) \tag{11-50}$$

and

$$\tilde{\theta}'(0, \lambda + \Delta\lambda) = \tilde{\theta}_s'(0, \lambda + \Delta\lambda) + Q_2'(0, \lambda), \tag{11-51}$$

where

$$\tilde{f}_s''(0, \lambda + \Delta\lambda) = \tilde{f}''(0, \lambda) + \alpha_1 \Delta\lambda + \beta_1 \frac{(\Delta\lambda)^2}{2} \tag{11-52}$$

with the subscript s representing second order, and

$$\tilde{\theta}_s'(0, \lambda + \Delta\lambda) = \tilde{\theta}'(0, \lambda) + \alpha_2 \Delta\lambda + \beta_2 \frac{(\Delta\lambda)^2}{2}. \tag{11-53}$$

Now we will assume that the two terminal values are single-valued functions of the two missing initial conditions, i.e.,

$$\tilde{f}'(\eta_\infty, \lambda + \Delta\lambda) = G_1(\kappa_1, \kappa_2) \tag{11-54}$$

and

$$\tilde{\theta}(\eta_\infty, \lambda + \Delta\lambda) = G_2(\kappa_1, \kappa_2), \tag{11-55}$$

where κ_1 and κ_2 are, respectively, the missing initial conditions $\tilde{f}''(0, \lambda + \Delta\lambda)$ and $\tilde{\theta}'(0, \lambda + \Delta\lambda)$; G_1 and G_2 are certain single-valued functions of κ_1 and κ_2. The two second-order terminal values can thus be expanded about the true ones in a Taylor's series as

$$\tilde{f}_s'(\eta_\infty, \lambda + \Delta\lambda) = 0 - Q_1''(0, \lambda)\left(\frac{\partial G_1}{\partial \kappa_1}\right)_{\kappa_2} - Q_2'(0, \lambda)\left(\frac{\partial G_1}{\partial \kappa_2}\right)_{\kappa_1} \tag{11-56}$$

and

$$\tilde{\theta}_s(\eta_\infty, \lambda + \Delta\lambda) = 0 - Q_1''(0, \lambda)\left(\frac{\partial G_2}{\partial \kappa_1}\right)_{\kappa_2} - Q_2'(0, \lambda)\left(\frac{\partial G_2}{\partial \kappa_2}\right)_{\kappa_1} . \tag{11-57}$$

The partial derivatives can be approximated by, for example,

$$\left(\frac{\partial G_1}{\partial \kappa_1}\right)_{\kappa_2} = \frac{G_1\,[\tilde{f}_s''(0,\lambda + \Delta\lambda) + \epsilon, \tilde{\theta}_s'(0, \lambda + \Delta\lambda)] - G_1\,[\tilde{f}_s''(0, \lambda + \Delta\lambda), \tilde{\theta}_s'(0, \lambda + \Delta\lambda)]}{\epsilon}, \tag{11-58}$$

where ϵ is chosen to be a small positive number $\mathrm{O}(10^{-3})$. It is extremely important to note that the missing condition $\tilde{\theta}_s'(0, \lambda + \Delta\lambda)$, i.e., κ_2, has been kept constant in evaluating Eq. (11-58). Similar care must be taken in approximating the other three derivatives. Equations (11-44*a*) and (11-44*b*) are then used to solve for Q_1'' and Q_2', which subsequently are added to $\tilde{f}_s''(0, \lambda + \Delta\lambda)$ and $\tilde{\theta}_s'(0, \lambda + \Delta\lambda)$ to yield the true missing initial conditions.

Example 11-2 Estimate Q_1'' and Q_2' if the following three sets of numerical results are given:

(*a*) $f_s''(0, 1) = 0.672, \theta_s'(0, 1) = 0.507, f_s'(7, 1) = 0.0028, \theta_s(7, 1) = 0.0016;$
(*b*) $f_b''(0, 1) = 0.673, \theta_b'(0, 1) = 0.507, f_b'(7, 1) = 0.0013, \theta_b(7, 1) = 0.0011;$
(*c*) $f_c''(0, 1) = 0.672, \theta_c'(0, 1) = -0.506, f_c'(7, 1) = 0.0015, \theta_c(7, 1) = 0.0012.$

The subscripts b and c represent computer run b and computer run c, respectively.

Solution According to Eq. (11-58), we compute

$$\left(\frac{\partial G_1}{\partial \kappa_1}\right)_{\kappa_2} \approx \frac{0.0013 - 0.0028}{0.001} = -1.5 .$$

Similarly,

$$\left(\frac{\partial G_1}{\partial \kappa_2}\right)_{\kappa_1} \approx \frac{0.0015 - 0.0028}{0.001} = -1.3 ,$$

$$\left(\frac{\partial G_2}{\partial \kappa_1}\right)_{\kappa_2} \approx \frac{0.0011 - 0.0016}{0.001} = -0.5 ,$$

and

$$\left(\frac{\partial G_2}{\partial \kappa_2}\right)_{\kappa_1} \approx \frac{0.0012 - 0.0016}{0.001} = -0.4 .$$

Therefore, Eqs. (11-56) and (11-57) can be rewritten as

$$0.0028 = 1.5Q_1'' + 1.3Q_2'$$

and

$$0.0016 = 0.5Q_1'' + 0.4Q_2' .$$

which are solved to yield

$$Q_1'' = 0.0192 \quad \text{and} \quad Q_2' = -0.02 .$$

The procedure described from Eq. (11-36) to Eq. (11-58) is repeated for $\lambda = 2\,\Delta\lambda$, $3\,\Delta\lambda, \ldots, 1$. The initial profile ($\lambda = 0$), intermediate profiles, and final profile ($\lambda = 1$) for $f'(\eta, \lambda)$ are plotted in Fig. 11-2 for Pr = 0.72. The profile at $\lambda = 1$ coincides with the reported result [63]. Once the solution at a certain value of Pr (0.72 in this case) becomes available, the PR method can be used repeatedly to generate solutions at other Pr values by replacing Eq. (11-36) with

$$\tilde{f}(\eta, \mathrm{Pr} + \Delta\,\mathrm{Pr}) = \tilde{f}(\eta, \mathrm{Pr}) + \left(\frac{\partial \tilde{f}}{\partial\,\mathrm{Pr}}\right)_{\mathrm{Pr}} \Delta\,\mathrm{Pr} + \left(\frac{\partial^2 \tilde{f}}{\partial\,\mathrm{Pr}^2}\right)_{\mathrm{Pr}} \frac{(\Delta\,\mathrm{Pr})^2}{2} + Q_1(\eta, \mathrm{Pr}) \quad (11\text{-}59)$$

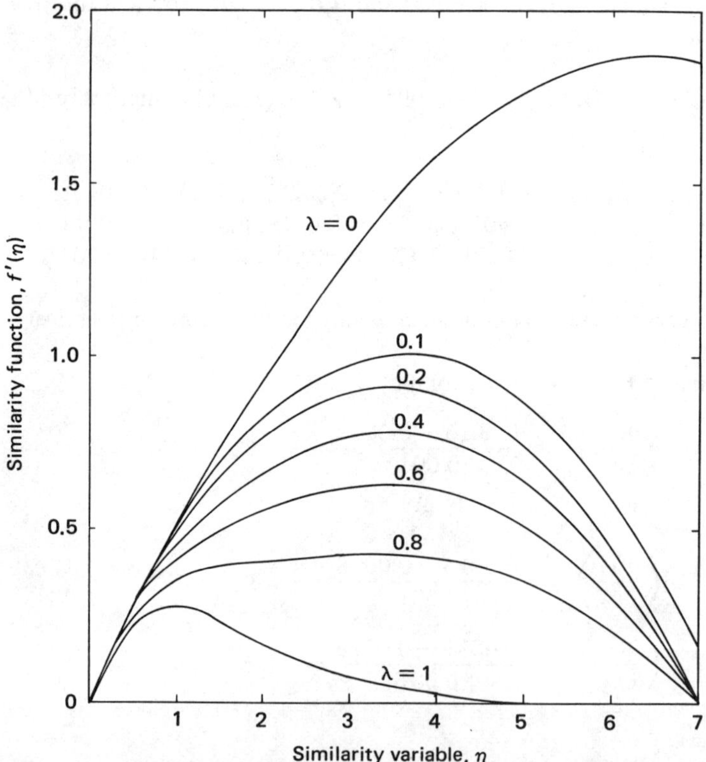

Figure 11-2 Similarity solutions for the parameterized free-convection similarity equations. The curve at $\lambda = 1$ coincides with the classic solution for Pr = 0.72.

and

$$\widetilde{\theta}(\eta,\mathrm{Pr}+\Delta\,\mathrm{Pr})=\widetilde{\theta}(\eta,\mathrm{Pr})+\left(\frac{\partial\widetilde{\theta}}{\partial\,\mathrm{Pr}}\right)_{\mathrm{Pr}}\Delta\,\mathrm{Pr}+\left(\frac{\partial^2\widetilde{\theta}}{\partial\,\mathrm{Pr}^2}\right)_{\mathrm{Pr}}\frac{(\Delta\,\mathrm{Pr})^2}{2}+Q_2(\eta,\mathrm{Pr})\,. \qquad (11\text{-}60)$$

Table 11-3 lists the dimensionless shear stresses and heat fluxes at the wall for various values of Pr. The absolute terminal values $f'(\eta_\infty,\ \mathrm{Pr})$ and $\theta(\eta_\infty,\ \mathrm{Pr})$ computed with these values are less than 0.001, more accurate than those computed with the literature values [69].

When the ambient fluid is moving ($u_\infty \neq 0$), the right-hand side of Eq. (11-27*b*) no longer vanishes. We are not able to justify the assumption that f and θ are weak functions of ξ and that a similarity solution may exist. In other words, the governing differential equations remain partial after the similarity transformation. This inconvenience motivated the development of the so-called local nonsimilarity method, which is presented in the following section.

11-2*b* Local Nonsimilarity Method

When a similarity solution does not exist for certain boundary-layer problems, such as that described by Eqs. (11-25)-(11-27), the local nonsimilarity method [70, 71] retains the streamwise derivatives and seeks the solutions of the auxiliary functions

$$F(\eta,\,\xi)=\frac{\partial f}{\partial \xi}\quad\text{and}\quad \Theta(\eta,\,\xi)=\frac{\partial \theta}{\partial \xi}\,. \qquad (11\text{-}61)$$

The first-order auxiliary equations are generated by differentiating Eqs. (11-25) and (11-26) with respect to ξ. Thus,

$$F'''-4f'F'+3(fF''+f''F)+\Theta=4(f'F'-f''F)+4\xi(F'^2-FF'')$$
$$+\text{ terms involving }\frac{\partial^2 f}{\partial \xi^2} \qquad (11\text{-}62)$$

and

$$\Theta''+3\,\mathrm{Pr}\,(f\Theta'+\theta'F)=4(f'\Theta-\theta'F)+4\xi(F'\Theta-\Theta'F)$$
$$+\text{ terms involving }\frac{\partial^2 f}{\partial \xi^2}\text{ and }\frac{\partial^2 \theta}{\partial \xi^2} \qquad (11\text{-}63)$$

subject to

$$\frac{7}{3}F(\xi,0)=\text{terms involving}\left(\frac{\partial^2 f}{\partial \xi^2}\right)_{\eta=0},\qquad F'(\xi,0)=0\,,\qquad \Theta(\xi,0)=0 \qquad (11\text{-}64a)$$

and

$$F'(\xi,\eta_\infty)=-\frac{\xi^{-3/2}}{\sqrt{2}}\,,\qquad \Theta(\xi,\eta_\infty)=0\,. \qquad (11\text{-}64b)$$

The primary functions $f(\xi,\,\eta)$ and $\theta(\xi,\,\eta)$ satisfy

$$f'''-2f'^2+3ff''+\theta=4\xi(f'F'-f''F) \qquad (11\text{-}65)$$

Table 11-3 Dimensionless shear stresses and heat fluxes at the wall for various values of Pr

	Pr	0	0.10	0.20	0.30	0.72	1.0	2.0	3.0	4.0	5.0
$f''(0)$	PR method	0.913685	0.843859	0.796095	0.761357	0.675875	0.642147	0.571241	0.530834	0.502901	0.481726
	[69]	–	0.8590	0.8006	0.7633	0.6760	0.6421	0.5713	0.5312	0.5036	0.4827
$-\theta'(0)$	PR method	0.142857	0.240721	0.310341	0.363045	0.504598	0.567113	0.716420	0.815468	0.891447	0.953860
	[69]	–	0.2326	0.3101	0.3641	0.5046	0.5671	0.7165	0.8145	0.8898	0.9517

and
$$\theta'' + 3\,\mathrm{Pr}\,f\theta' = 4\xi(f'\Theta - \theta' F) \tag{11-66}$$

subject to

$$\tfrac{4}{3}\xi F(\xi,0) + f(\xi,0) = 0\,, \quad f'(\xi,0) = 0\,, \quad \theta(\xi,0) = 1 \tag{11-67a}$$

and
$$f'(\xi,\eta_\infty) = \tfrac{1}{2}\xi^{-1/2}\,, \quad \theta(\xi,\eta_\infty) = 0\,. \tag{11-67b}$$

If we assume that terms involving second derivatives are small and can be deleted, the system of Eqs. (11-62)–(11-67) is then closed—four unknowns f, θ, F, and Θ satisfying four equations (11-62), (11-63), (11-65), and (11-66) which are of tenth differential order and subject to ten boundary conditions. If we wish to retain the kth-order derivatives for higher accuracy, Eqs. (11-62)–(11-64) can be further differentiated $k-1$ times with respect to ξ. In all cases, however, the $(k+1)$th-order derivatives must be deleted to ensure closure of the system. The resulting set of ordinary diferential equations remains to be solved by a numerical scheme for solving two-point boundary-value problems (see Section 3-4).

11-2*c* Discretization Methods

If the similarity transformation described by Eqs. (11-18)–(11-21) fails to transform the original partial differential equations into ordinary ones, its advantage is greatly reduced. On the other hand, although the local nonsimilarity method succeeds for nonsimilar boundary-layer problems, it involves considerable mathematical formulation and computer programming. Under such circumstances, discretization schemes such as the finite-difference method, finite-element method, or control volume method may be used. In Section 6-3 we mentioned the joint use of the von Mises transformation, control volume method, and upwind finite-difference scheme for solving forced-convection boundary-layer problems [72, 73]. These methods, of course, can also be applied to mixed-convection boundary-layer problems. The transformed streamwise momentum equation now takes the form

$$Au = \frac{\partial u}{\partial x} + (a + b\omega)\frac{\partial u}{\partial \omega} - \frac{\partial}{\partial \omega}\left(c\,\frac{\partial u}{\partial \omega}\right) - \frac{g(\rho_\infty - \rho)}{\rho u} = 0\,, \tag{11-68}$$

which can be discretized in the ω domain—for example, by using the control volume method and computing the integral

$$(A\tilde{u}, N_j) = \int_{\omega_{j-1/2}}^{\omega_{j+1/2}} A\tilde{u}\,d\omega\,,$$

where N_j is the unit step basis function (see Section 2-2), or the Galerkin finite-element method and computing the inner product

$$(A\tilde{u}, N_j) = \int_{\omega_{j-1}}^{\omega_{j+1}} (A\tilde{u})N_j\,d\omega\,,$$

where N_j is the piecewise linear pyramid function (see Section 1-2), or the finite-difference scheme

$$\left(\frac{\partial u}{\partial \omega}\right)_{\omega_j} = \frac{1}{2\,\Delta\omega}(\omega_{j+1} - \omega_{j-1})\,.$$

In any event, a singularity exists near the wall since

$$y_1 \propto \int_0^{\omega_1} \frac{d\omega}{u} \propto \int_0^{\omega_1} \frac{d\omega}{\omega} \to \infty\,, \tag{11-69a}$$

where the subscript 1 denotes the grid point adjacent to the wall. This singularity can be avoided by using the x-ω^2 transformation [74] (see also Section 6-3*b*), which replaces Eq. (11-69*a*) with

$$y_1 \propto \int_0^{\omega_1} \frac{d\omega}{\omega^{1/2}} = 2\omega_1^{1/2}\,, \tag{11-69b}$$

or the parameterized-residuals method [67], which does not employ any transformation. See Section 7-5 for details.

11-3 STREAMWISE DIFFUSION FLOWS (ELLIPTIC EQUATIONS)

Although boundary-layer formulations are quite important in engineering applications, many flow systems, such as flows inside enclosures or cavities and flows near the leading edge or the trailing edge of the boundary layer, cannot be idealized as boundary-layer flows. The differences between boundary-layer flows and streamwise diffusion flows were summarized at the beginning of Chapter 8. From the viewpoint of numerical analysis, the investigation of the latter appears more challenging. In this section we will use the central finite-difference scheme to solve Eqs. (11-10)–(11-12), which describe a buoyancy-driven flow inside an enclosure as shown in Fig. 11-3. The boundary conditions of the prescribed temperature T suggest that the flow will

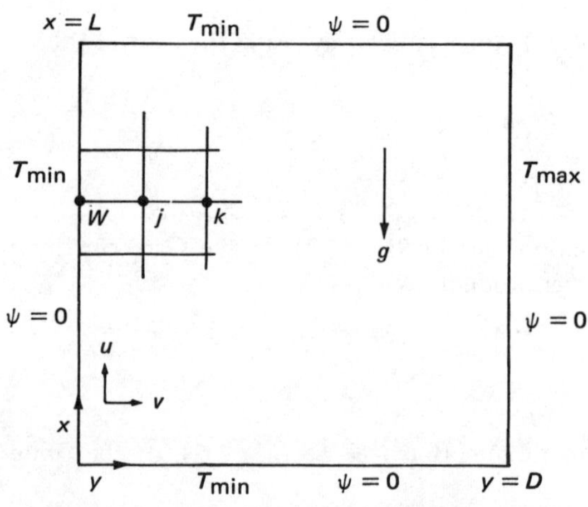

Figure 11-3 Free-convection flow in a rectangular enclosure. For consistency with conventional boundary-layer flow systems, the x direction is taken to be upward.

circulate counterclockwise. The zero boundary values of the stream function are arbitrary but most convenient.

Straightforward discretization of the energy equation (11-12), the vorticity equation (11-11), and the Poisson Eq. (11-10), respectively, leads to

$$16\theta_j - 4\sum_{p=W}^{N} \theta_p + (\psi_E - \psi_W)(\theta_N - \theta_S) - (\psi_N - \psi_S)(\theta_E - \theta_W) = 0\,, \tag{11-70}$$

$$16\omega_j - 4\sum_{p=W}^{N} \omega_p + \frac{(\psi_E - \psi_W)(\omega_N - \omega_S) - (\psi_N - \psi_S)(\omega_E - \omega_W)}{\mathrm{Pr}} = 2h\,\mathrm{Ra}\,(\theta_W - \theta_E)\,, \tag{11-71}$$

and

$$4\psi_j - \sum_{p=W}^{N} \psi_p = h^2\omega_j\,. \tag{11-72}$$

The boundary conditions for θ and ψ are given (Fig. 11-3) as

$$\theta(0, Y) = \frac{T(0, Y) - T_{\mathrm{min}}}{T_{\mathrm{max}} - T_{\mathrm{min}}} = \theta(X, 0) = \theta(X, 1) = 0\,, \tag{11-73a, b, c}$$

$$\theta\left(X, \frac{D}{L}\right) = 1\,, \tag{11-73d}$$

and

$$\psi(0, Y) = \psi(X, 0) = \psi(X, 1) = \psi\left(X, \frac{D}{L}\right) = 0\,. \tag{11-73e–h}$$

The specification of the boundary conditions for ω deserves some attention. From Eq. (11-10) we obtain, at the west grid point at the wall,

$$\omega_W = -\left(\frac{\partial^2\psi}{\partial Y^2}\right)_W\,. \tag{11-74}$$

But the stream function at grid point j adjacent to the wall can be expanded about ψ_W at the wall in a Taylor's series as

$$\psi_j = \psi_W + \left(\frac{\partial\psi}{\partial Y}\right)_W h + \left(\frac{\partial^2\psi}{\partial Y^2}\right)_W \frac{h^2}{2} + \cdots\,.$$

Since $(\partial\psi/\partial Y)_W$ must vanish because of the nonslip boundary condition, Eq. (11-74) yields the boundary condition

$$\omega_W = -\frac{2}{h^2}\,\psi_j\,. \tag{11-75}$$

The other boundary conditions for ω can be derived in a similar fashion.

Equation (11-75) suggests that $\psi(Y)$ near the wall (at a fixed X location) has been assumed to be a quadratic function of Y. If higher accurate is desired, it is possible to replace this quadratic function by a cubic function; i.e.,

$$\psi(Y) = a + bY + cY^2 + dY^3 . \tag{11-76}$$

The coefficients a, b, c, and d can be determined, in reference to Fig. 11-3, by the following four conditions:

$$\psi(0) = \psi_W , \quad \psi(h) = \psi_j , \quad \psi(2h) = \psi_k , \quad \text{and} \quad \left(\frac{\partial \psi}{\partial Y}\right)_{Y=0} = 0 .$$

After $\psi(Y)$ is specified, Eq. (11-74) can be used to compute ω_W.

The specification of the ω boundary condition can be considered one of the most intensively investigated subjects associated with the ω-ψ formulation. Interested readers may also consult [15, 17, 32] on this subject.

A conventional iterative procedure for solving the nonlinear algebraic system Eqs. (11-70)-(11-72) is summarized as follows:

1. Initially the stream function $\psi_j^{[0]}$ is assumed to be uniformly zero. This assumption makes Eq. (11-70) a pure conduction equation, which is solved to yield the initial temperature distribution $\theta_j^{[0]}$.
2. The $\theta_j^{[0]}$ distribution obtained is then substituted into Eq. (11-71) to generate the vorticity distribution $\omega_j^{[0]}$.
3. This $\omega_j^{[0]}$ distribution is substituted into Eq. (11-72) to generate an improved stream function distribution $\psi_j^{[1]}$.
4. With these improved $\psi_j^{[1]}$ values, we solve Eq. (11-70) to determine $\theta_j^{[1]}$, which are substituted into Eq. (11-71) to produce $\omega_j^{[1]}$.
5. The cycle of iteration is repeated until the values of the difference between $\omega_j^{[k]}$ and $\omega_j^{[k+1]}$ lie within certain predetermined limits.

Example 11-3 Consider the numbered mesh system shown in Fig. 11-4, in which the data

$$\frac{D}{L} = 1 , \quad h = \frac{1}{3} , \quad \text{Pr} = 1 , \quad \text{and} \quad \text{Ra} = 10^3 \tag{a}$$

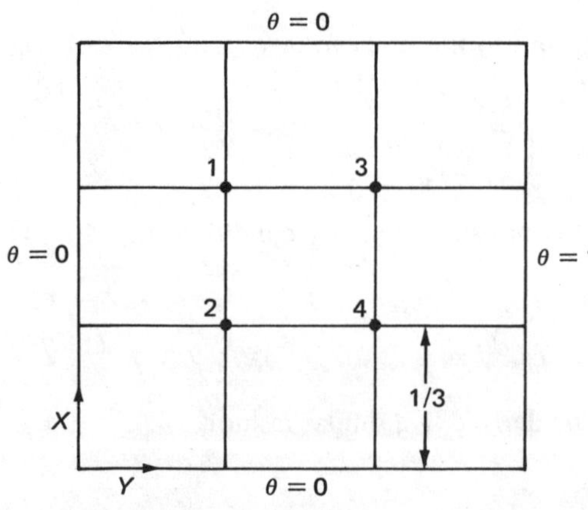

Figure 11-4 Two-by-two mesh system of a recirculating flow within a square enclosure. Pr = 1 and Ra = 10^3.

are chosen. Follow the iterative procedure outlined above to find $\psi_j^{[k]}$, $\theta_j^{[k]}$, and $\omega_j^{[0]}$, $k = 0, 1$ and $j = 1, 2, 3, 4$.

Solution Using the data given in Eq. (*a*), we reduce Eqs. (11-70)-(11-72) to

$$16\theta_j - 4\sum_{p=W}^{N} \theta_p + (\psi_E - \psi_W)(\theta_N - \theta_S) - (\psi_N - \psi_S)(\theta_E - \theta_W) = 0\,, \tag{b}$$

$$16\omega_j - 4\sum_{p=W}^{N} \omega_p + (\psi_E - \psi_W)(\omega_N - \omega_S) - (\psi_N - \psi_S)(\omega_E - \omega_W)$$

$$= \frac{2}{3} \times 10^3 (\theta_W - \theta_E)\,, \tag{c}$$

and

$$36\psi_j - 9\sum_{p=W}^{N} \psi_p = \omega_j\,. \tag{d}$$

For the pure conduction case $\psi_1^{[0]} = \psi_2^{[0]} = \psi_3^{[0]} = \psi_4^{[0]} = 0$, Eq. (*b*) is simplified to

$$[\mathbf{A}]\begin{Bmatrix} \theta_1^{[0]} \\ \theta_2^{[0]} \\ \theta_3^{[0]} \\ \theta_4^{[0]} \end{Bmatrix} = \begin{Bmatrix} 0 \\ 0 \\ 1 \\ 1 \end{Bmatrix}, \tag{e}$$

whose solution is listed in Table 11-4. The matrix $[\mathbf{A}]$ is defined as

$$[\mathbf{A}] = \begin{bmatrix} 4 & -1 & -1 & 0 \\ -1 & 4 & 0 & -1 \\ -1 & 0 & 4 & -1 \\ 0 & -1 & -1 & 4 \end{bmatrix}. \tag{f}$$

These $\theta_j^{[0]}$ values are substituted into Eq. (*c*) to yield

$$[\mathbf{A}]\begin{Bmatrix} \omega_1^{[0]} \\ \omega_2^{[0]} \\ \omega_3^{[0]} \\ \omega_4^{[0]} \end{Bmatrix} = \begin{Bmatrix} -62.5 \\ -62.5 \\ -145.83 \\ -145.83 \end{Bmatrix}. \tag{g}$$

Table 11-4 Nodal values of 2 × 2 mesh system for the first two iteration cycles

j	$\psi^{[0]}$	$\theta^{[0]}$	$\omega^{[0]}$	$\psi^{[1]}$	$\theta^{[1]}$
1	0	0.125	−41.67	−2.6042	0.2207
2	0	0.125	−41.67	−2.6042	0.1119
3	0	0.375	−62.50	−3.1829	0.5208
4	0	0.375	−62.50	−3.1829	0.1467

Note that the boundary conditions of ω_j, according to Eq. (11-75), are zero. The solution of Eq. (*g*) is also listed in Table 11-4. With these values of $\omega_j^{[0]}$, Eq. (*d*) can be rewritten as

$$[A]\begin{Bmatrix}\psi_1^{[1]}\\ \psi_2^{[1]}\\ \psi_3^{[1]}\\ \psi_4^{[1]}\end{Bmatrix}=\begin{Bmatrix}-4.630\\ -4.630\\ -6.9444\\ -6.9444\end{Bmatrix}. \qquad (h)$$

See Table 11-4 for the solution of Eq. (*h*). Finally, substitution of these $\psi_j^{[1]}$ into Eq. (*b*) yields

$$\begin{bmatrix}16 & -0.8171 & -6.6042 & 0\\ -7.1829 & 16 & 0 & -1.3958\\ -0.8171 & 0 & 16 & -6.6042\\ 0 & -7.1829 & -1.3958 & 16\end{bmatrix}\begin{Bmatrix}\theta_1^{[1]}\\ \theta_2^{[1]}\\ \theta_3^{[1]}\\ \theta_4^{[1]}\end{Bmatrix}=\begin{Bmatrix}0\\ 0\\ 7.1829\\ 0.8171\end{Bmatrix}. \qquad (i)$$

On comparing $\theta_j^{[1]}$ with $\theta_j^{[0]}$ in Table 11-4, we find that, even for such a coarse-meshed system, our result exhibits the expected qualitative trend. The warmest and coolest locations are at grid points 3 and 2, respectively, since the fluid that circulates counterclockwise receives heat along the right face of the enclosure and is cooled along the top and left faces. To continue the computation of $\omega_j^{[1]}$ using Eq. (*c*), we should remember that the boundary conditions of $\omega_j^{[1]}$, according to Eq. (11-75), are no longer zero because $\psi_j^{[1]}$ is nonzero. However, we will terminate our computation here, believing that interested readers will be able to proceed.

For systems with a finer mesh such as that shown in Fig. 11-5, the iteration can no longer be properly handled with desk calculators and should be performed on computers. In each computation of the matrices, an iterative scheme, such as the SOR method (see Sections 4-3*d* and 5-2*e*) or the Strongly Implicit Procedure (SIP) [75], can be adopted. Using the former, we repeat Example 11-2 for a 5 × 5 mesh system and list the convergent result in Table 11-5.

Care must be taken when the central finite-difference scheme is adopted for

Table 11-5 Nodal solution of the three field variables for natural convection within a square enclosure with 5 × 5 interior grids, Ra = 10^4 and Pr = 1

Stream function ψ						
0.0000	0.0000	0.0000	0.0000	0.0000	0.0000	0.0000
0.0000	−1.7145	−3.0627	−3.9393	−3.9095	−2.3843	0.0000
0.0000	−3.2754	−5.7305	−6.8211	−6.5077	−4.3446	0.0000
0.0000	−3.3032	−6.0941	−7.1329	−6.8179	−5.3523	0.0000
0.0000	−2.2479	−4.4638	−5.2764	−4.9964	−4.7436	0.0000
0.0000	−0.9465	−1.9550	−2.2801	−2.1130	−2.5775	0.0000
0.0000	0.0000	0.0000	0.0000	0.0000	0.0000	0.0000
Temperature θ						
0.0000	0.0000	0.0000	0.0000	0.0000	0.0000	1.0000
0.0000	0.1410	0.2864	0.4355	0.6011	0.7864	1.0000
0.0000	0.1848	0.3310	0.4034	0.4405	0.5620	1.0000
0.0000	0.1530	0.2639	0.2912	0.2793	0.3370	1.0000
0.0000	0.0969	0.1683	0.1844	0.1814	0.1451	1.0000
0.0000	0.0445	0.0772	0.0851	0.0951	0.0226	1.0000
0.0000	0.0000	0.0000	0.0000	0.0000	0.0000	1.0000
Vorticity ω						
0.00	123.45	220.51	283.63	281.49	171.67	0.00
123.45	−18.72	−31.19	−70.69	−101.05	−46.20	171.67
235.83	−84.72	−132.07	−143.07	−148.95	−112.82	312.81
237.83	−57.43	−134.86	−126.79	−118.16	−198.12	385.37
161.85	−10.01	−82.14	−80.36	−37.26	−217.73	341.54
68.15	15.01	−4.67	8.06	50.48	−124.33	185.58
0.00	68.15	140.76	164.17	152.13	185.58	0.00

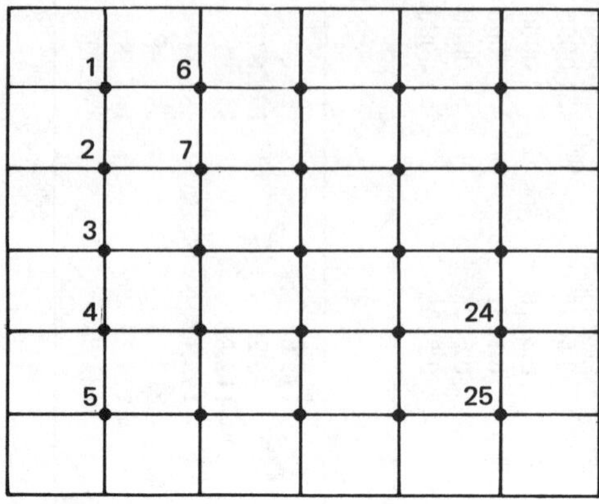

Figure 11-5 Five-by-five mesh system of a recirculating flow within a square enclosure. Pr = 1 and Ra = 10^4.

flow systems with large Rayleigh numbers (Ra $> 10^6$). Coupling Eqs. (11-11) and (11-12), we observe that, for large Ra, the "convective" term† $\partial\theta/\partial Y$ becomes dominant over the diffusive term $\nabla^2\theta$. The resulting matrix of coefficients is no longer diagonally dominant and a numerical instability may arise (see Sections 4-1 and 8-5). Under such circumstances, upwind finite-difference schemes or upwind finite-element schemes can remove the instability. Interested readers may consult the references cited in Chapter 8 for details.

Naturally, it is not mandatory to transform the primitive-variables equations (11-1)-(11-4) into the vorticity-stream function equations (11-10)-(11-12). In fact, for three-dimensional flow systems it becomes awkward to use the ω-ψ method. Another widely used technique is to solve directly for the primitive variables u, v, and p with a staggered grid. Interested readers may consult Section 8-2*a*.

SYMBOLS

D	width of enclosure, m
$f(\eta)$	similarity function [Eq. (11-19)]
F	$\partial f/\partial\xi$
g	$\partial f/\partial\lambda$ or gravitational acceleration, m/s^2
Gr_x	Grashof number [$= gx^3\beta(T_w - T_\infty)/\nu^2$]
h	$\partial^2 f/\partial\lambda^2$ or mesh size
L	length of system, m
L_1, L_2, L_3	linear differential operators [Eqs. (11-44*a*)-(11-44*c*)]
p	pressure, NT/m^2
Q_1, Q_2	high-order terms in Taylor's series expansions [Eqs. (11-36) and (11-37)]
R	gas constant, J/kg·K, or residual generated by substituting the approximate function into the governing equation
Ra	Rayleigh number ($= gL^3\beta\,\Delta T/\nu\alpha$)
s	$\partial\theta/\partial\lambda$

†Also called advective.

t	$\partial^2\theta/\partial\lambda^2$
U	reference velocity [Eq. (11-21)]
X, Y	dimensionless coordinates normalized on L
β	coefficient of thermal expansion, K^{-1}
$\alpha_1, \alpha_2, \beta_1, \beta_2$	superposition constants [Eqs. (11-45*a*)-(11-45*d*)]
η	similarity variable [Eq. (11-17)]
θ	$(T-T_{min})/(T_{max}-T_{min})$ or $(T-T_\infty)/(T_w-T_\infty)$
Θ	$\partial\theta/\partial\xi$
λ	artificial embedded parameter $0 \leqslant \lambda \leqslant 1$
ξ	mixed convection number $(= Gr_x/Re_x^2)$
ψ	dimensionless stream function $(= \bar{\psi}/\alpha)$ (notation adopted in this chapter for convenience)
$\bar{\psi}$	stream function, m^2/s
ω	dimensionless vorticity $(= \bar{\omega}L^2/\alpha)$ or dimensionless stream function [Eq. (6-37)]
$\bar{\omega}$	vorticity [Eq. (11-9)], s^{-1}

Subscript

s	second-order terms included
W	west of grid point j

REFERENCES

1. J. Boussinesq, Théorie Analytique de la Chaleur, Gauthier-Villars, Paris, vol. 2, p. 172, 1903.
2. A. E. Gill, The Boundary-Layer Regime for Convection in a Rectangular Cavity, *J. Fluid Mech.*, vol. 26, pp. 515–536, 1966.
3. G. de Vahl Davis, Laminar Natural Convection in an Enclosed Rectangular Cavity, *Int. J. Heat Mass Transfer*, vol. 11, pp. 1675–1693, 1968.
4. K. E. Torrance, Comparison of Finite-Difference Computations of Natural Convection, *J. Res. Natl. Bur. Stand.*, vol. 72B, p. 281, 1968.
5. A. D. Gosman, W. M. Pun, A. K. Runchal, D. B. Spalding, and M. Wolfshtein, *Heat and Mass Transfer in Recirculating Flows*, Academic, New York, 1969.
6. R. T. Cheng, Numerical Solution of the Navier-Stokes Equation by the Finite Element Method, *Phys. Fluids*, vol. 15, pp. 2098–2105, 1972.
7. A. J. Baker, Finite Element Solution Algorithm for Viscous Incompressible Fluid Dynamic, *Int. J. Numer. Methods Eng.*, vol. 6, pp. 89–101, 1973.
8. S. L. Smith and C. A. Brebbia, Finite Element Solution of Navier-Stokes Equations for Transient Two-dimensional Incompressible Flow, *J. Comput. Phys.*, vol. 17, pp. 235–245, 1975.
9. R. F. Boehm and D. Kamyab, Established Stripwise Laminar Natural Convection on a Horizontal Surface, *ASME J. Heat Transfer*, vol. 99, pp. 294–299, 1977.
10. C. Quon, Free Convection in an Enclosure Revisited, *ASME J. Heat Transfer*, vol. 99, pp. 340–342, 1977.
11. H. Ozoe, K. Yamamota, H. Sayama, and S. W. Churchill, Natural Convection Patterns in a Long Inclined Rectangular Box Heated from Below. Part II: Three-dimensional Numerical Results, *Int. J. Heat Mass Transfer*, vol. 20, pp. 131–139, 1977.
12. R. A. Wirtz, The Effect of Solute Layering on Lateral Heat Transfer in an Enclosure, *Int. J. Heat Mass Transfer*, vol. 20, pp. 841–846, 1977.

13. L. A. Clomburg, Convection in an Enclosure–Source and Sink Located along a Single Horizontal Boundary, *ASME J. Heat Transfer*, vol. 100, pp. 205–211, 1978.
14. B. Roux, J. C. Grondin, P. Bontoux, and B. Gilly, On a High-Order Accurate Method for the Numerical Study of Natural Convection in a Vertical Square Cavity, *Numer. Heat Transfer*, vol. 1, pp. 331–349, 1978.
15. L. C. Chow, Y. K. Cheung, and C. L. Tien, A New Finite-Difference Representation for the Vorticity at a Wall with Suction, *Numer. Heat Transfer*, vol. 1, pp. 417–423, 1978.
16. R. F. LeFeuvre, The Prediction of Two-dimensional Recirculating Flows Using a Simple Finite-Difference Grid for Non-rectangular Flow Fields, *Comput. Fluids*, vol. 6, pp. 203–218, 1978.
17. M. M. Gupta and R. P. Manohar, Boundary Approximations and Accuracy in Viscous Flow Computations, *J. Comput. Phys.*, vol. 31, pp. 265–288, 1979.
18. S. C. R. Dennis, D. B. Ingham, and R. N. Cook, Finite Difference Methods for Calculating Steady Incompressible Flows in Three Dimensions, *J. Comput. Phys.*, vol. 33, pp. 325–339, 1979.
19. A. S. Benjamin and V. E. Denny, On the Convergence of Numerical Solutions for 2-D Flows in a Cavity at Large Re, *J. Comput. Phys.*, vol. 33, pp. 340–358, 1979.
20. M. C. Charrier-Mojtabi, A. Mojtabi, and J. P. Caltagirone, Numerical Solution of a Flow due to Natural Convection in Horizontal Cylindrical Annulus, *ASME J. Heat Transfer*, vol. 101, pp. 171–173, 1979.
21. S. Bunditkul and W. J. Yang, Laminar Transport Phenomena in Parallel Channels with a Short Flow Construction, *ASME J. Heat Transfer*, vol. 101, pp. 217–221, 1979.
22. C. Quon, A Study of Penetrative Convection in Rotating Fluid, *ASME J. Heat Transfer*, vol. 101, pp. 261–264, 1979.
23. H. H. Wong and G. D. Raithby, Improved Finite Difference Methods based on a Critical Evaluation of the Approximation Errors, *Numer. Heat Transfer*, vol. 2, pp. 139–163, 1979.
24. J. T. Han, A Computational Method to Solve Nonlinear Elliptic Equations for Natural Convection in Enclosures, *Numer. Heat Transfer*, vol. 2, pp. 165–175, 1979.
25. I. P. Jones, A Numerical Study of Natural Convection in an Air-filled Cavity: Comparison with Experiment, *Numer. Heat Transfer*, vol. 2, pp. 193–213, 1979.
26. K. Kublbeck, G. P. Merker, and J. Straub, Advanced Numerical Computation of Two-dimensional Time-dependent Free Convection in Cavities, *Int. J. Heat Mass Transfer*, vol. 23, pp. 203–217, 1980.
27. N. Takemitsu, On a Finite-Difference Approximation for the Steady-State Navier-Stokes Equations, *J. Comput. Phys.*, vol. 36, pp. 236–248, 1980.
28. J. P. Caltagirone, M. Combarnous, and A. Mojtabi, Natural Convection between Two Concentric Spheres: Transition toward a Multicellular Flow, *Numer. Heat Transfer*, vol. 3, pp. 107–114, 1980.
29. M. M. Gupta and R. P. Manohar, On the Use of Central Difference Scheme for Navier-Stokes Equations, *Int. J. Numer. Methods Eng.*, vol. 15, pp. 557–573, 1980.
30. M. A. Yaghoubi and F. P. Incropera, Analysis of Natural Convection due to Localized Heating in a Shallow Water Layer, *Numer. Heat Transfer*, vol. 3, pp. 315–330, 1980.
31. J. Brandeis and J. Rom, Interactive Method for Computation of Viscous Flow with Recirculation, *J. Comput. Phys.*, vol. 40, pp. 396–410, 1981.
32. L. Quartapelle, Vorticity Conditioning in the Computation of Two-dimensional Viscous Flows, *J. Comput. Phys.*, vol. 40, pp. 453–477, 1981.
33. J. P. vanDoormaal, G. D. Raithby, and A. B. Strong, Prediction of Natural Convection in Nonrectangular Enclosures using Orthogonal Curvilinear Coordinates, *Numer. Heat Transfer*, vol. 4, pp. 21–38, 1981.
34. G. D. Raithby and H. H. Wong, Heat Transfer by Natural Convection across Vertical Air Layers, *Numer. Heat Transfer*, vol. 4, pp. 447–457, 1981.
35. H. Inaba, N. Seki, S. Fukusako, and K. Kanayama, Natural Convection Heat Transfer in a Shallow Rectangular Cavity with Different End Temperatures, *Numer. Heat Transfer*, vol. 4, pp. 459–468, 1981.

36. N. Ramachandran, J. P. Gupta, and Y. Jaluria, Two-dimensional Solidification with Natural Convection in the Melt and Convective and Radiative Boundary Conditions, *Numer. Heat Transfer*, vol. 4, pp. 469–484, 1981.
37. H. Ozoe and T. Shibata, Natural Convection in an Inclined Circular Cylindrical Annulus Heated and Cooled on its End Plates, *Int. J. Heat Mass Transfer*, vol. 24, pp. 727–737, 1981.
38. B. Gilly, P. Bontoux, and B. Roux, Influence of Thermal Wall Conditions on the Natural Convection in a Vertical Rectangular Differentially Heated Cavity, *Int. J. Heat Mass Transfer*, vol. 24, pp. 829–841, 1981.
39. G. S. Shiralkar and C. L. Tien, A Numerical Study of Laminar Natural Convection in Shallow Cavities, *ASME J. Heat Transfer*, vol. 103, pp. 226–231, 1981.
40. M. Kawahara, Periodic Galerkin Finite Element Method of Unsteady Periodic Flow of Viscous Fluid, *Int. J. Numer. Methods Eng.*, vol. 11, pp. 1093–1105, 1977.
41. B. Tabarrok and R. C. Lin, Finite Element Analysis of Free Convection Flows, *Int. J. Heat Mass Transfer*, vol. 20, pp. 945–952, 1977.
42. A. Campion-Renson and M. J. Crochet, On the Stream Function-Vorticity Finite Element Solutions of Navier-Stokes Equations, *Int. J. Numer. Methods Eng.*, vol. 14, pp. 1809–1818, 1978.
43. S. I. Abdel-Khalik, H. W. Li, and K. R. Randall, Natural Convection in Compound Parabolic Concentrators–A Finite Element Solution, *ASME J. Heat Transfer*, vol. 100, pp. 199–204, 1978.
44. K. E. Barrett, A Variational Principle for the Streamfunction-Vorticity Formulation of the Navier-Stokes Equations Incorporating No-Slip Conditions, *J. Comput. Phys.*, vol. 26, pp. 153–161, 1978.
45. A. Moult, D. Burley, and H. Rawson, The Numerical Solution of Two-dimensional, Steady Flow Problems by the Finite Element Method, *Int. J. Numer. Methods Eng.*, vol. 14, pp. 11–35, 1979.
46. M. Ikegawa, A New Finite Element Technique for the Analysis of Steady Viscous Flow Problems, *Int. J. Numer. Methods Eng.*, vol. 14, pp. 103–113, 1979.
47. D. W. Pepper and R. E. Cooper, Numerical Solution of Recirculating Flow by a Simple Finite Element Recursion Relation, *Comput. Fluids*, vol. 8, pp. 213–223, 1980.
48. G. Dhatt, B. K. Fomo, and C. Bourque, A ψ-ω Finite Element Formulation for the Navier-Stokes Equations, *Int. J. Numer. Methods Eng.*, vol. 17, pp. 199–212, 1981.
49. S. Y. Tuann and M. D. Olson, Review of Computing Methods for Recirculating Flows, *J. Comput. Phys.*, vol. 29, pp. 1–19, 1978.
50. G. P. Merker and L. G. Leal, Natural Convection in a Shallow Annular Cavity, *Int. J. Heat Mass Transfer*, vol. 23, pp. 677–686, 1980.
51. J. S. Vrentas, R. Narayanan, and S. S. Agrawal, Free Surface Convection in a Bounded Cylinder Geometry, *Int. J. Heat Mass Transfer*, vol. 24, pp. 1513–1529, 1981.
52. C. J. Chen, H. Naseri-Neshat, and K. S. Ho, Finite-Analytic Numerical Solution of Heat Transfer in Two-dimensional Cavity Flow, *Numer. Heat Transfer*, vol. 4, pp. 179–197, 1981.
53. D. L. Young, J. A. Ligget, and R. H. Gallagher, Steady Stratified Circulation in a Cavity, *ASCE J. Eng. Mech.*, vol. 102, pp. 1–17, 1976.
54. S. N. Singh, Free Convection from a Sphere in a Slightly-Thermally Stratified Fluid, *Int. J. Heat Mass Transfer*, vol. 20, pp. 1155–1160, 1977.
55. E. J. Shaughnessy and R. W. Douglass, The Effect of Stable Stratification on the Motion in a Rotating Spherical Annulus, *Int. J. Heat Mass Transfer*, vol. 21, pp. 1251–1259, 1978.
56. T. Kumada, R. Ishiguro, T. Sato, and T. Abe, Natural Evaporation of Sodium with Mist Formation, *ASME J. Heat Transfer*, vol. 101, pp. 306–312, 1979.
57. S. N. Singh and J. M. Elliott, Free Convection between Horizontal Concentric Cylinders in a Slightly-Thermally Stratified Fluid, *Int. J. Heat Mass Transfer*, vol. 22, pp. 639–646, 1979.
58. R. M. Turian and W. Aung, Convection and Frictional Heating in a Cone and Plate System, *Int. J. Heat Mass Transfer*, vol. 21, pp. 1087–1097, 1978.
59. M. D. Olson and S. Y. Tuann, New Finite Element Results for the Square Cavity, *Comput. Fluids*, vol. 7, pp. 123–135, 1979.

60. L. Iyican, Y. Bayazitoglu, and L. C. Witte, An Analytical Study of Natural Convective Heat Transfer within a Trapezoidal Enclosure, *ASME J. Heat Transfer*, vol. 102, pp. 640–647, 1980.
61. S. N. Singh and J. M. Elliott, Natural Convection between Concentric Spheres in a Slightly-Thermally Stratified Medium, *Int. J. Heat Mass Transfer*, vol. 24, pp. 395–406, 1981.
62. R. Hunt and G. Wilks, Continuous Transformation Computation of Boundary Layer Equations between Similarity Regions, *J. Comput. Phys.*, vol. 40, pp. 478–490, 1981.
63. S. Ostrach, An Analysis of Laminar-Free-Convection Flow, NACA Tech. Rep. 1111, 1953.
64. W. M. Rohsenow and H. Y. Choi, *Heat, Mass and Momentum Transfer*, chap. 7, Prentice-Hall, Englewood Cliffs, N.J., 1961.
65. J. P. Holman, *Heat Transfer*, chap. 7, McGraw-Hill, New York, 1976.
66. B. Gebhart, *Heat Transfer*, chap. 8, McGraw-Hill, New York, 1970.
67. T. M. Shih and H. J. Huang, A Method of Solving Nonlinear Differential Equations for Boundary-Layer Flows, *Numer. Heat Transfer*, vol. 4, pp. 159–178, 1981.
68. T. M. Shih, A Method to Solve Two-Point Boundary-Value Problems in Boundary-Layer Flows or Flames, *Numer. Heat Transfer*, vol. 2, pp. 177–191, 1979.
69. T. Y. Na and I. S. Habib, Solution of the Natural Convection Problem by Parametric Differentiation, *Int. J. Heat Mass Transfer*, vol. 17, pp. 457–459, 1974.
70. E. M. Sparrow, H. Quack, and C. J. Boerner, Local Nonsimilarity Boundary-Layer Solutions, *AIAA J.*, vol. 8, pp. 1936–1942, 1970.
71. E. M. Sparrow and H. S. Yu, Local Nonsimilarity Thermal Boundary-Layer Solutions, *ASME J. Heat Transfer*, vol. 93, pp. 328–334, 1971.
72. S. V. Patankar and D. B. Spalding, *Heat and Mass Transfer in Boundary Layers*, Intertext, London, 1970.
73. D. B. Spalding, *GENMIX: A General Computer Program for Two-dimensional Parabolic Phenomena*, vol. 1, Pergamon, Elmsford, N.Y., 1977.
74. V. E. Denny, A. F. Mills, and V. J. Jusionis, Laminar Film Condensation from a Stream-Air Mixture Undergoing Forced Flow down a Vertical Surface, *ASME J. Heat Transfer*, vol. 93, pp. 297–304, 1971.
75. H. L. Stone, Iterative Solution of Implicit Approximation of Multidimensional Partial Differential Equations, *SIAM J. Numer. Anal.*, vol. 5, pp. 530–558, 1968.

PROBLEMS

11-1 Derive Eqs. (11-10) and (11-11) from Eqs. (11-1)–(11-3).

11-2 Utilize Eqs. (11-17)–(11-19) to derive Eqs. (11-20*b*), (11-22), and (11-23).

11-3 Equations (11-22)–(11-24) can be greatly simplified if the coefficients are chosen properly. Let $a = \frac{1}{2}$, $b = \frac{1}{2}$, $A_1 = 2(u_\infty \nu)^{-1/2}$, and $A_2 = (u_\infty \nu)^{1/2}$ and perform the simplification. At $\xi = 0$, is the similarity equation reduced to the Blasius equation?

11-4 (*a*) Plot $\widetilde{\theta}(\eta)$ versus η according to Eq. (11-34) for Pr = 0.72. Then plot the residual $R(\eta)$ versus η according to Eqs. (11-33) and (11-35).

(*b*) Approximate $\widetilde{\theta}(\eta)$ by a third-degree polynomial. What is the polynomial expression for $R(\eta)$?

(*c*) Write the two-point boundary-value problem generated by procedure (*b*). What is the exact solution?

11-5 Use the Euler method (forward difference) to find the distributions of g_k and s_k, $k = 1, 2, 3$, by solving Eqs. (11-46*a*)–(11-46*c*) in Table 11-2. Take $\Delta\eta = 2$ and $\eta_\infty = 6$. What are the values of α_1 and α_2? For convenience, take the values of $\widetilde{\theta}(\eta)$ and $R(\eta)$ from Problem 11-4*a*.

11-6 Derive Eqs. (11-48*a*) and (11-48*b*) from Eqs. (11-45*a*)–(11-45*d*).

11-7 Suppose that the boundary at $y = 0$ shown in Fig. 11-3 is moving at velocity u_∞. Modify Eq. (11-75) accordingly.

11-8 Extend Example 11-2 by performing one more iteration cycle. Tabulate $\omega^{[1]}$, $\psi^{[2]}$, and $\theta^{[2]}$.

11-9 Repeat Example 11-2 for Ra = 10^8. Does any numerical instability occur?

CHAPTER

TWELVE

INTRODUCTION TO TURBULENT FLOWS

Taylor and von Karman [1] define turbulence as "an irregular motion which in general makes its appearance in fluids, gaseous or liquids, when they flow past solid surfaces or even when neighboring streams of the same fluid flow past or over one another."

For more insight into the physical description of turbulent flows, the reader may consult [1-5]. The main purpose of this chapter is to show how some physical problems involving turbulent flows can be solved numerically. The emphasis here is placed on the use of the numerical methods, not on numerical properties of the schemes such as stability and convergence. This is because the governing equations for turbulent flows are in general highly nonlinear and the numerical properties of nonlinear differential equations are not yet well understood.

For completeness, a brief review of turbulent flows will be presented in Section 12-1. With the information given in that section, readers may find the remaining sections of this chapter easier to comprehend.

In Section 12-2, the governing equations for a two-dimensional (although turbulence is generally three-dimensional) turbulent flow are presented. Because of the presence of Reynolds stresses, the system of equations is not closed.

To close the system of transport equations, we introduce in Section 12-3 the zero-equation and one-equation turbulence models, which greatly simplify the analysis but may be inadequate for elliptic-type flows. Using this model, we are able to derive an expression for the Nusselt number in terms of the Reynolds number and the Prandtl number for pipe flows.

Being aware of the inadequacy of the zero-equation or one-equation turbulence models, researchers developed the two-equation turbulence model described in Section 12-4. The widely used K-ϵ turbulence model is used in conjunction with the integral method to predict some qualitative characteristics of a circular turbulent jet.

Finally, higher-order closure theories are briefly mentioned in Section 12-5. This section may help to inform the readers of the state of the art of some advanced turbulence analyses.

12-1 BRIEF REVIEW OF TURBULENT FLOWS

In turbulent flows the instantaneous field variables ϕ, such as the flow velocity, density, pressure, temperature, and species mass fractions, can be approximated by

$$\phi = \Phi + \phi' , \tag{12-1}$$

where Φ is an average (in a certain sense) value of ϕ and ϕ' is the fluctuation. The two most important averages are the time average and the ensemble average. In mathematical form, we define the former as

$$\Phi = \frac{1}{2\,\Delta t} \int_{t-\Delta t}^{t+\Delta t} \phi(t)\, dt , \tag{12-2a}$$

where Δt must be sufficiently large in comparison with the time scale of the turbulence and sufficiently small in comparison with the period of slow variations of the averaged quantities in the flow field. The ensemble average is defined as

$$\Phi = \frac{1}{N} \sum_{k=1}^{N} \phi_k , \tag{12-2b}$$

where ϕ_k are the values of N identical experiments. For quasi-steady systems, i.e., $\partial\phi/\partial t \neq 0$ and $\partial\Phi/\partial t = 0$, it is safe to assume that the two averaging procedures lead to the same result. This assumption is known as the ergodic hypothesis.

Substituting Eq. (12-1) into the transport equations and taking the average, denoted by an overbar, we will generate terms such as $\overline{u'v'}$. For example

$$\frac{\partial}{\partial y} uv = \frac{\partial}{\partial y} [(U + u')(V + v')] = \frac{\partial}{\partial y} UV + u' \frac{\partial V}{\partial y} + v' \frac{\partial U}{\partial y} + \frac{\partial}{\partial y} u'v' ,$$

and then taking the time average yields

$$\frac{\partial}{\partial y} \overline{uv} = \frac{\partial}{\partial y} UV + \frac{\partial}{\partial y} \overline{u'v'} .$$

The terms $\overline{u'v'}$ and others are called Reynolds stresses. Since these terms are products of two fluctuations, they are sometimes also called second-order correlations. Similarly, terms such as $\overline{u'^2 v'}$ are called third-order correlations. If the turbulence does not have preference in any direction, i.e., $\overline{u'^2} = \overline{v'^2} = \overline{w'^2}$, it is said to be isotropic. In this case, the kinetic energy of the turbulence is reduced to

$$K = \tfrac{1}{2}(\overline{u'^2} + \overline{v'^2} + \overline{w'^2}) = \tfrac{3}{2}\overline{u'^2} . \tag{12-3}$$

The essence of the analysis for turbulent flows lies in either expressing the correlations in terms of the existing unknowns or solving the additional transport equations which the correlations must satisfy. Among the former strategies, one classical model [6] is to introduce the so-called Prandtl mixing length l, defined as

$$l = \frac{|u'|}{|\partial U/\partial y|} \quad \text{or} \quad \frac{|v'|}{|\partial U/\partial y|} .$$

The mixing length in this model is prescribed a priori based on some physical interpretation. See Section 12-3 for more details. In the latter strategies, two additional transport equations that $\overline{u'v'}$ and $\overline{u'^2}$ satisfy are derived. Therefore, the distributions of $\overline{u'v'}$ and $\overline{u'^2}$ are solved simultaneously with U and V and not specified artificially.

12-2 GOVERNING EQUATIONS

The Navier-Stokes equations and the energy transport equation governing laminar flows remain valid for turbulent flows. The only difference between the two sets of equations is that the dependent variables ϕ (representing u, v, p, and T) for turbulent flows become instantaneous quantities (if a time average is taken), which can be broken into two parts as shown in Eq. (12-1). We assume here that the time averages and ensemble averages are equivalent (ergodic hypothesis). Then substitution of Eq. (12-1) into Eqs. (8-1)-(8-3), including the time derivative, yields

$$\frac{\partial U}{\partial x} + \frac{\partial V}{\partial y} = 0 , \tag{12-4}$$

$$\frac{DU}{Dt} = \nabla\cdot(\nu \nabla U) - \frac{1}{\rho}\frac{\partial P}{\partial x} - \frac{\partial}{\partial x}\overline{u'^2} - \frac{\partial}{\partial y}\overline{u'v'} , \tag{12-5}$$

$$\frac{DV}{Dt} = \nabla\cdot(\nu \nabla V) - \frac{1}{\rho}\frac{\partial P}{\partial y} - \frac{\partial}{\partial x}\overline{u'v'} - \frac{\partial}{\partial y}\overline{v'^2} , \tag{12-6}$$

and

$$\frac{DT}{Dt} = \nabla\cdot(\alpha \nabla T) - \frac{\partial}{\partial x}\overline{T'u'} - \frac{\partial}{\partial y}\overline{T'v'} . \tag{12-7}$$

We will take it for granted that T stands for the average temperature. In deriving Eqs. (12-4)-(12-7), the effect of the density variation is completely ignored. For rigorous analyses that account for this effect, see, for example, [7, 8]. If the Reynolds stresses and the second-order correlations vanish, Eqs. (12-4)-(12-7) reduce to those governing laminar flows. These stresses and correlations are the quantities that characterize turbulent flows. Specifying or computing them with reasonable accuracy is a crucial task in the investigation of turbulent flows.

If we do not make any assumptions to relate the correlations to the average quantities or do not produce additional equations that the correlations must satisfy, Eqs. (12-4)-(12-7) are unclosed. Several models have been developed to close them, including zero-equation [6, 9] and one-equation turbulence models [10, 11], two-equation turbulence models [11-14], and higher-order closure theories [15-17], which are described in the following sections.

12-3 ZERO-EQUATION AND ONE-EQUATION TURBULENCE MODELS

The zero-equation model was developed mainly for the investigation of flows in which turbulent transport is significant only along the transverse direction. It is called "zero-equation" because no additional differential equation is generated to close the set of equations.

The governing equations for steady-state boundary-layer flow can be derived by simplifying Eqs. (12-4)–(12-7) to

$$\frac{\partial U}{\partial x} + \frac{\partial V}{\partial y} = 0 , \tag{12-8}$$

$$U \frac{\partial U}{\partial x} + V \frac{\partial U}{\partial y} = \nu \frac{\partial^2 U}{\partial y^2} - \frac{\partial}{\partial y} \overline{u'v'} , \tag{12-9}$$

and

$$U \frac{\partial T}{\partial x} + V \frac{\partial T}{\partial y} = \alpha \frac{\partial^2 T}{\partial y^2} - \frac{\partial}{\partial y} \overline{T'v'} . \tag{12-10}$$

In comparing the Reynolds stresses with the laminar stresses caused by viscosity effects in Eq. (12-9), it is tempting to assume that the Reynolds stresses behave like viscous stresses. It was further suggested that these turbulent stresses are directly proportional to the average velocity gradient $\partial u/\partial y$. Prandtl, who originated the zero-equation model, assumed that

$$|u'| = |v'| = l \left| \frac{\partial U}{\partial y} \right| . \tag{12-11}$$

Consequently,

$$-\overline{u'v'} = \nu_t \frac{\partial U}{\partial y} , \tag{12-12}$$

where the minus sign accounts for the fact that the value of u' is generally associated with a value of v' of the opposite sign; ν_t corresponds to the molecular diffusivity ν in laminar flow and is therefore called the "apparent" or "eddy" diffusivity; it is defined as

$$\nu_t = l^2 \left| \frac{\partial U}{\partial y} \right| . \tag{12-13}$$

Substituting Eq. (12-13) into Eq. (12-9) yields

$$U \frac{\partial U}{\partial x} + V \frac{\partial U}{\partial y} = \frac{\partial}{\partial y} \left[(\nu + \nu_t) \frac{\partial U}{\partial y} \right] . \tag{12-14}$$

Sometimes the sum of the molecular and turbulent diffusivities is denoted by ν_{eff}. Equations (12-8), (12-13), and (12-14) thus constitute a closed system with three unknowns u, v, and ν_t. We may use, for example, the control volume method to perform the discretization of these partial differential equations with respect to y. On the interval $[y_{j-1/2}, y_{j+1/2}]$, the dependent variables can be approximated by

$$\tilde{\Phi} = \begin{cases} N_{j-1}(y)\Phi_{j-1} + N_j(y)\Phi_j\,, & y \in [y_{j-1/2}, y_j] \\ N_j(y)\Phi_j + N_{j+1}(y)\Phi_{j+1}\,, & y \in [y_j, y_{j+1/2}]\,, \end{cases} \tag{12-15}$$

where $N_k(y)$, $k = j-1, j, j+1$, are the piecewise linear functions given by Eq. (1-16), and the Φ_k denote U_k, V_k, or $\nu_{t,k}$. Then Eqs. (12-8), (12-13), and (12-14) can be integrated with respect to y over the interval $[y_{j-1/2}, y_{j+1/2}]$. A set of nonlinear, first-order, ordinary differential equation will result.

Example 12-1 Use the control volume method to discretize Eq. (12-14) in the y coordinate.

Solution: Integrating Eq. (12-14) from $y_{j-1/2}$ to $y_{j+1/2}$ yields

$$\int_{y_{j-1/2}}^{y_{j+1/2}} U\frac{\partial U}{\partial x}\,dy + \int_{y_{j-1/2}}^{y_{j+1/2}} V\frac{\partial U}{\partial y}\,dy = \nu_{\text{eff},j+1/2}\left(\frac{\partial U}{\partial y}\right)_{j+1/2} - \nu_{\text{eff},j-1/2}\left(\frac{\partial U}{\partial y}\right)_{j-1/2}. \tag{a}$$

The first integral on the left side of Eq. (a) can be evaluated as

$$I_1 = \int_{y_{j-1/2}}^{y_{j+1/2}} U\frac{\partial U}{\partial x}\,dy = \int_{y_{j-1/2}}^{y_j} (N_{j-1}U_{j-1} + N_jU_j)\left(N_{j-1}\frac{dU_{j-1}}{dx} + N_j\frac{dU_j}{dx}\right)dy + \int_{y_j}^{y_{j+1/2}} (N_jU_j + N_{j+1}U_{j+1})\left(N_j\frac{dU_j}{dx} + N_{j+1}\frac{dU_{j+1}}{dx}\right)dy\,. \tag{b}$$

After algebra, we obtain

$$I_1 = \sum_k \Lambda_{k,k}U_k\frac{dU_k}{dx} + \sum_k \Lambda_{j,k}\frac{d}{dx}(U_jU_k) + 2\Lambda_{j,j}U_j\frac{dU_j}{dx}\,, \qquad k = j-1, j+1 \tag{c}$$

where, for example,

$$\Lambda_{j-1,j} = \int_{y_{j-1/2}}^{y_j} N_{j-1}N_j\,dy = \frac{y_j - y_{j-1/2}}{6}\,. \tag{d}$$

Likewise, for the second integral I_2 on the left side of Eq. (*a*), we can derive

$$I_2 = \sum_k \Lambda_{k,dk} U_k V_k + \sum_k \Lambda_{dj,k} U_j V_k + \sum_k \Lambda_{j,dk} U_k V_j + 2\Lambda_{j,dj} U_j V_j ,$$
$$k = j-1, j+1 \qquad (e)$$

where, for example,

$$\Lambda_{j,d(j+1)} = \int_{y_j}^{y_{j+1/2}} N_j \frac{dN_{j+1}}{dy} dy = \frac{3}{8} . \qquad (f)$$

Substituting Eqs. (*c*) and (*e*) into Eq. (*a*) and rearranging the result, we finally obtain

$$h(I_1 + I_2) - \nu_{\text{eff},j+1/2}(U_{j+1} - U_j) + \nu_{\text{eff},j-1/2}(U_j - U_{j-1}) = 0 . \qquad (g)$$

To actually solve Eq. (*g*) along with the discretized forms of Eq. (12-8) and (12-13) we must further integrate them with respect to x and specify the Prandtl mixing length l. We will omit these procedures here.

When the wall over which the boundary layer is formed is impermeable, it is reasonable to assume that, in the vicinity of the wall,

$$\frac{\partial U}{\partial x} \approx 0 \quad \text{and} \quad V \approx 0. \qquad (12\text{-}16)$$

Under this assumption, Eq. (12-14) reduces to

$$(\nu + \nu_t) \frac{\partial U}{\partial y} = \frac{\tau_w}{\rho} , \qquad (12\text{-}17)$$

where τ_w, representing the shear stress at the wall, is supposed to be an integration constant. In the literature describing turbulent flows Eq. (12-17) is commonly non-dimensionalized to

$$\left(1 + \frac{\nu_t}{\nu}\right) \frac{dU^+}{dy^+} = 1 \qquad (12\text{-}18a)$$

or

$$\left[1 + l^{+2} \left(\frac{dU^+}{dy^+}\right)\right] \frac{dU^+}{dy^+} = 1 , \qquad (12\text{-}18b)$$

where

$$y^+ = \frac{y u_\tau}{\nu}, \quad l^+ = \frac{l u_\tau}{\nu}, \quad u_\tau = \left(\frac{\tau_w}{\rho}\right)^{1/2} ,$$

and

$$U^+ = \frac{U}{u_\tau}.$$

Equation (12-18) is far simpler than Eq. (12-14). Once l^+ is specified, it is possible to integrate Eq. (12-18) numerically (sometimes even analytically). In the van Driest model [9] it is assumed, based on some physical phenomena, that

$$l^+ = ky^+\left[1 - \exp\left(-\frac{y^+}{A}\right)\right], \tag{12-19}$$

where k and A are empirically found to be 0.4 and 26, respectively. Substituting Eq. (12-19) into Eq. (12-18) and solving the parabolic equation for dU^+/dy^+ yields

$$U^+ = 2\int_0^{y^+}\left\{1 + \left[1 + 4k^2y^{+2}\left(1 - \exp\left(-\frac{y^+}{A}\right)\right)^2\right]^{1/2}\right\}^{-1} dy^+, \tag{12-20}$$

which can be readily integrated, e.g., by using the trapezoidal rule. The solution is listed in Table 12-1. From this $U^+(y^+)$ profile, the shear stress at the wall can be determined. If we wish to evaluate ν_t/ν, the rearranged form of Eq. (12-18*a*)

$$\frac{\nu_t}{\nu} = \frac{1}{dU^+/dy^+} - 1 \tag{12-21}$$

can be used.

Table 12-1 Normalized velocity and temperature distributions in a one-dimensional turbulent flow[a]

y^+	dU^+/dy^+	U^+	$T^+ - T_w^+$
0	1	0	0
2	0.996	1.996	19.7
4	0.952	3.945	35.8
6	0.831	5.73	45.9
8	0.673	7.23	50.9
10	0.534	8.44	53.6
15	0.314	10.6	57.2
20	0.207	11.9	58.8
25	0.149	12.7	60.0
30	0.114	13.4	60.7
40	0.076	14.3	61.2
50	0.056	15.0	61.9
70	0.038	15.9	62.8
100	0.025	16.8	63.8

[a]Calculation is based on Eq. (12-20) and Eq. (*b*) of Example 12-4. $\mathrm{Re}_D = 16{,}000$ and $\mathrm{Pr} = 10$.

Example 12-2 Show that the friction coefficient C_f for pipe flows is a function of Re_D only. For $\mathrm{Re}_D = 16{,}000$, what is the value of C_f?

Solution: From the definition of the mean flow velocity

$$U_m = \frac{2}{R^2} \int_0^R rU\,dr , \tag{a}$$

we derive

$$U_m^+ = \frac{2}{R^{+2}} \int_0^{R^+} (R^+ - y^+)U^+\,dy^+ = \text{function of } R^+ \text{ only} , \tag{b}$$

where the length and the velocity have been normalized on ν/u_τ and u_τ, respectively, and

$$y^+ = R^+ - r^+ .$$

The frictional coefficient C_f is defined in the literature as

$$C_f = \frac{2\tau_w}{\rho U_m^2} = \frac{2(dU^+/dy^+)_{y^+=0}}{U_m^{+2}} . \tag{c}$$

Since, from Table 12-1,

$$\left(\frac{dU^+}{dy^+}\right)_{y^+=0} = 1 ,$$

we obtain

$$C_f = \frac{2}{U_m^{+2}} = \text{function of } R^+ \text{ only} . \tag{d}$$

But R^+ is not a familiar term to us. We prefer to express C_f in terms of Re_D defined as

$$\mathrm{Re}_D = \frac{U_m(2R)}{\nu} = 2R^+U_m^+ . \tag{e}$$

Based on Eqs. (*b*) and (*e*), we find that Re_D is a function of R^+ only and thereby, based on Eq. (*d*), C_f is a function of Re_D only.

When Re_D is given, we guess a value of R^+, evaluate U_m^+ from Eq. (*b*), and then check whether these values satisfy Eq. (*e*). If not, iteration must be performed. At $\mathrm{Re}_D = 16{,}000$, we first guess $R^+ = 500$. Numerical integration of Eq. (*b*) leads to $U_m^+ = 16$, and Eq. (*e*) is satisfied. Therefore,

$$C_f = 0.0078 , \tag{f}$$

which is in good agreement with the value reported in [3, p. 73].

A similar treatment can be applied to the energy equation, Eq. (12-10). We may assume

$$-\overline{v'T'} = \alpha_t \frac{\partial T}{\partial y} \tag{12-22}$$

in an analogous form to Eq. (12-12).

Example 12-3 Consider the following normalized energy transport equation of parabolic type:

$$U^* \frac{\partial \theta}{\partial X} + V^* \frac{\partial \theta}{\partial Y} = \frac{1}{\text{Re}_1} \frac{\partial}{\partial Y} \left[\left(\frac{1}{\text{Pr}} + \frac{\alpha_t}{\nu} \right) \frac{\partial \theta}{\partial Y} \right], \tag{a}$$

where

$$U^* = \frac{U}{U_1}, \qquad V^* = \frac{V}{U_1}, \qquad X = \frac{x}{\Delta x},$$

$$Y = \frac{y}{\Delta x}, \qquad \gamma = \frac{\Delta y}{\Delta x}, \qquad \theta = \frac{T - T_w}{T_a - T_w},$$

and

$$\text{Re}_1 = \frac{U_1 \, \Delta x}{\nu}.$$

The subscripts 1 and a denote the grid points of the mesh system shown in Fig. 12-1. We will assume that θ_1 is the only unknown and that the upstream $\theta(Y)$ distribution is linear, with the relevant data given as $\gamma = 0.05$, $\text{Re}_1 = \gamma^{-2}$, $V_1/U_1 = 0.02$, $\theta_c = 0.9$, and $\text{Pr} = 1$. Use the upwind scheme and the central-difference scheme to discretize the X derivative and the Y derivatives, respectively. Find θ_1 for the laminar flow ($\alpha_t = 0$) and for a turbulent flow [$\alpha_{t,c}/\nu = 4$, with $\alpha_t(Y)$ linear in Y].

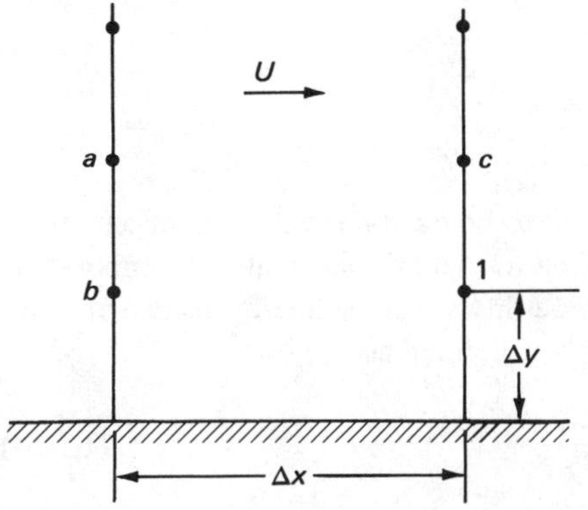

Figure 12-1 The mesh system of a turbulent flow near the wall.

Solution: At grid point 1, Eq. (*a*) can be discretized to

$$\theta_1 - \theta_b + \frac{V_1}{U_1}\frac{\theta_c - \theta_w}{2\gamma} = \frac{1}{\mathrm{Re}_1\,\gamma^2}\left[\left(\frac{1}{\mathrm{Pr}} + \frac{\alpha_{t,1.5}}{\nu}\right)(\theta_c - \theta_1) - \left(\frac{1}{\mathrm{Pr}} + \frac{\alpha_{t,0.5}}{\nu}\right)(\theta_1 - \theta_w)\right]. \qquad (b)$$

With $\gamma = 0.05$, $V_1/U_1 = 0.02$, and $\mathrm{Pr} = 1$, Eq. (*b*) can be rearranged to

$$\left(3 + \frac{\alpha_{t,1.5} + \alpha_{t,0.5}}{\nu}\right)\theta_1 = \theta_b + \left(0.8 + \frac{\alpha_{t,1.5}}{\nu}\right)\theta_c + \left(1.2 + \frac{\alpha_{t,0.5}}{\nu}\right)\theta_w . \qquad (c)$$

For laminar flows, $\alpha_t(Y)$ vanishes for all Y. With $\theta_b = 0.5$, $\theta_c = 0.9$, and $\theta_w = 0$, we compute

$$\theta_1 = 0.4067 . \qquad (d)$$

For turbulent flow with $\alpha_{t,0.5}/\nu = 1$ and $\alpha_{t,1.5}/\nu = 3$, we also obtain

$$\theta_1 = 0.56 . \qquad (e)$$

From the simple analysis above, we find that the nodal value θ_1 in turbulent flow tends to be closer to θ_c than in laminar flow, suggesting that turbulent mixing plays an important role in smoothing the field variable distribution and increasing the wall gradient.

As in the analysis of turbulent momentum transport, we may assume

$$\frac{\partial T}{\partial x} \approx 0 \quad \text{and} \quad V \approx 0$$

near the impermeable wall. Then Eq. (12-10) can be reduced to

$$\left(\frac{1}{\mathrm{Pr}} + \frac{\alpha_t}{\nu}\right)\frac{dT^+}{dy^+} = 1 , \qquad (12\text{-}23)$$

where

$$T^+ = \frac{T\rho c_p u_\tau}{\dot{q}''_w} .$$

If the thermal turbulent diffusivity α_t is assumed to be nearly equal to the momentum counterpart ν_t, Eq. (12-23) can also be integrated to yield the normalized temperature distribution. When turbulent heat transfer in pipe flows is considered, this distribution can be used to deduce the familiar Nusselt number, defined as

$$\mathrm{Nu}_D = \frac{h_c D}{k} , \qquad (12\text{-}24)$$

where h_c is the heat transfer coefficient, defined as†

$$h_c = \frac{\dot{q}''_w}{T_m - T_w} \tag{12-25}$$

and the mean temperature is defined as

$$T_m = \frac{\int_0^R TU(R-y)\,dy}{\int_0^R U(R-y)\,dy}. \tag{12-26}$$

Example 12-4 Show that the Nusselt number Nu_D is a function of Re_D and Pr. For $\mathrm{Re}_D = 16{,}000$ and $\mathrm{Pr} = 10$, what is the value of Nu_D?

Solution: From Eq. (12-26), we derive

$$T_m^+ - T_w^+ = \frac{2}{R^{+2} U_m^+} \int_0^{R^+} (R^+ - y^+) U^+ (T^+ - T_w^+)\, dy^+ . \tag{a}$$

To evaluate this integral, we need to know the distribution of $T^+(y^+)$. First, Eq. (12-21) can be substituted into Eq. (12-23) to yield (assuming $\nu_t \approx \alpha_t$)

$$T^+ - T_w^+ = \int_0^{y^+} \frac{dy^+}{1/\mathrm{Pr} - 1 + dy^+/dU^+} = \text{function of Pr only} . \tag{b}$$

Equations (*a*) and (*b*) imply that $T_m^+ - T_w^+$ is a function of Re_D and Pr. Next, in light of Eqs. (12-24) and (12-25), we derive

$$\mathrm{Nu}_D = \frac{2R^+ \mathrm{Pr}}{T_m^+ - T_w^+} . \tag{c}$$

Since $T_m^+ - T_w^+$ is a function of Re_D and Pr, we established that Nu_D is also a function of Re_D and Pr.

At $\mathrm{Pr} = 10$ Eq. (*b*) is evaluated, and the $T^+(y^+)$ distribution is listed in Table 12-1. This distribution is substituted into Eq. (*a*) to yield, after numerical integration,

$$T_m^+ - T_w^+ \approx 61.0 . \tag{d}$$

Therefore, with $R^+ = 500$ from Example 12-2, we obtain

$$\mathrm{Nu}_D = \frac{2 \times 500 \times 10}{61} = 164 , \tag{e}$$

which agrees well with the value reported in [3, pp. 170–171].

†Since we have used the symbol h to denote the interval size throughout this book, it is less confusing to assign a new symbol h_c for the heat transfer coefficient that is conventionally denoted by h.

Various turbulent flow problems have been numerically solved by adopting the Prandtl mixing length [18-21], the van Driest mixing length [22-33], the Cebeci-Smith model [34-36], and one-equation models [37-39]. Interested readers may consult these papers for details.

12-4 TWO-EQUATION TURBULENCE MODELS

The zero-equation and one-equation turbulence models described in the previous section greatly simplify the analysis of turbulent flows. For predictions of the frictional coefficient C_f and the Nusselt number Nu_D for turbulent flows in pipes or over flat walls, these models serve as useful approximations. If the flow patterns become more complicated, such that turbulent transport is important in both the x and y directions, the validity of a predetermined mixing length $l(y)$ certainly no longer exists. A typical example is turbulent flow in a two-dimensional enclosure. Before the analysis of this flow, information on $l(x,y)$ is not available. Under such circumstances the zero-equation turbulence model must be replaced by other more sophisticated models, such as the two-equation turbulence model described below.

Instead of expressing the Reynolds stresses in terms of average quantities, the two-equation turbulence model seeks to derive additional transport equations of the Navier-Stokes type which the Reynolds stresses must satisfy. If isotropic turbulence is assumed, it is possible to derive two transport equations of the following form:

$$\frac{D}{Dt}\left(\frac{\overline{u'^2}}{2}\right) = \nabla\cdot\left(\nu\nabla\frac{\overline{u'^2}}{2}\right) + S_1(\text{third-order correlations}) \qquad (12\text{-}27)$$

and

$$\frac{D}{Dt}\overline{u'v'} = \nabla\cdot(\nu\nabla\overline{u'v'}) + S_2(\text{third-order correlations}) . \qquad (12\text{-}28)$$

These two transport equations are to be solved simultaneously with Eqs. (12-4)-(12-6). If S_1 and S_2 can somehow be approximated in terms of second-order correlations and average quantities, there are five unknowns u, v, p, $\overline{u'^2}$, and $\overline{u'v'}$ satisfying five equations (12-4)-(12-6), (12-27), and (12-28); the system is closed (under the isotropy assumption, $\overline{u'^2} = \overline{v'^2}$).

Example 12-5 Show how a third-order correlation appears in Eq. (12-27).

Solution: We multiply the x-component Navier-Stokes equation

$$\frac{\partial u}{\partial t} + u\frac{\partial u}{\partial x} + v\frac{\partial u}{\partial y} = \nu\nabla^2 u - \frac{1}{\rho}\frac{\partial p}{\partial x} \qquad (a)$$

by u to yield

$$\frac{\partial}{\partial t}\left(\frac{u^2}{2}\right) + u\frac{\partial}{\partial x}\left(\frac{u^2}{2}\right) + v\frac{\partial}{\partial y}\left(\frac{u^2}{2}\right) = \nu u\nabla^2 u - \frac{u}{\rho}\frac{\partial p}{\partial x} . \qquad (b)$$

On substituting

$$u = U + u'$$

into Eq. (*b*), the second term on the left-hand side becomes

$$u \frac{\partial}{\partial x}\left(\frac{u^2}{2}\right) = (U + u')\frac{\partial}{\partial x}\left[\frac{1}{2}(U + u')^2\right] = u' \frac{\partial}{\partial x}\left(\frac{u'^2}{2}\right) + \cdots. \qquad (c)$$

After taking the time average over Eq. (*c*), we therefore obtain a third-order term such as $\overline{u' \, \partial(u'^2/2)/\partial x}$ in Eq. (12-27).

One of the two-equation turbulence models is the so-called K-ϵ model [11, 12]. In this model, the turbulent diffusivity ν_t is expressed in terms of the kinetic energy of turbulence K and the dissipation rate of kinetic energy of turbulence as

$$\nu_t = \frac{C_\mu K^2}{\epsilon}, \qquad (12\text{-}29)$$

while K and ϵ satisfy the following two transport equations in Cartesian ($y = r, n = 0$) or cylindrical ($n = 1$) coordinates:

$$\frac{DK}{Dt} = \frac{1}{r^n}\frac{\partial}{\partial r}\left(\frac{\nu + \nu_t}{\sigma_k} r^n \frac{\partial K}{\partial r}\right) + \frac{\partial}{\partial x}\left(\frac{\nu + \nu_t}{\sigma_k}\frac{\partial K}{\partial x}\right) + G - \epsilon \qquad (12\text{-}30)$$

and

$$\frac{D\epsilon}{Dt} = \frac{1}{r^n}\frac{\partial}{\partial r}\left(\frac{\nu + \nu_t}{\sigma_\epsilon} r^n \frac{\partial \epsilon}{\partial r}\right) + \frac{\partial}{\partial x}\left(\frac{\nu + \nu_t}{\sigma_\epsilon}\frac{\partial \epsilon}{\partial x}\right) + C_1 \frac{\epsilon G}{K} - C_2 \frac{\epsilon^2}{K}, \qquad (12\text{-}31)$$

where

$$G = \nu_t \left[2\left(\frac{\partial U}{\partial x}\right)^2 + 2\left(\frac{\partial V}{\partial r}\right)^2 + \left(\frac{\partial U}{\partial r} + \frac{\partial V}{\partial x}\right)^2\right].$$

For isotropic turbulence, K and ϵ are defined, respectively, as

$$K = \tfrac{3}{2}\overline{u'^2} \qquad (12\text{-}32a)$$

and

$$\epsilon = 15\nu \overline{\left(\frac{\partial u'}{\partial x}\right)^2}. \qquad (12\text{-}32b)$$

As far as numerical analysis is concerned, the definitions in Eqs. (12-32*a*) and (12-32*b*) do not play roles since the closed system now contains five unknowns u, v, p, K, and ϵ.

To demonstrate the discretization of Eq. (12-30), we consider its simplified version of parabolic type

$$u \frac{\partial K}{\partial x} + v \frac{\partial K}{\partial y} - \frac{\partial}{\partial y}\left(\nu_t \frac{\partial K}{\partial y}\right) - G + \epsilon = 0, \qquad (12\text{-}33)$$

in which $\nu_t \gg \nu$ has been assumed. Using the upwind difference for $\partial/\partial x$ and the central difference for $\partial/\partial y$, we write

$$\frac{\partial K}{\partial x} = \frac{1}{\Delta x}(K_j - K_W),$$

$$\frac{\partial K}{\partial y} = \frac{1}{2\,\Delta y}(K_N - K_S),$$

and

$$\frac{\partial}{\partial y}\left(\nu_t \frac{\partial K}{\partial y}\right) = \frac{1}{\Delta y}\left(\frac{\nu_{t,N} + \nu_{t,j}}{2}\,\frac{K_N - K_j}{\Delta y} - \frac{\nu_{t,j} + \nu_{t,S}}{2}\,\frac{K_j - K_S}{\Delta y}\right).$$

Substituting these discretized expressions into Eq. (12-33), with some rearrangement, eventually yields

$$a_j K_j - a_W K_W - a_S K_S - a_N K_N - G_j + \epsilon_j = 0\,, \tag{12-34}$$

where

$$a_j = \frac{\nu_{t,S} + \nu_{t,N} + 2\nu_{t,j}}{2(\Delta y)^2} + \frac{u_j}{\Delta x},$$

$$a_W = \frac{u_j}{\Delta x},$$

$$a_S = \frac{\nu_{t,S} + \nu_{t,j}}{2(\Delta y)^2} + \frac{v_j}{2\,\Delta y},$$

and

$$a_N = \frac{\nu_{t,N} + \nu_{t,j}}{2(\Delta y)^2} - \frac{v_j}{2\,\Delta y}.$$

Since deriving a discretized form such as Eq. (12-34) from a nonlinear differential equation usually requires lengthy algebra, it is advisable to use a simple way to check the derived algebraic result. We can first assign some simple functions to the field variables. These functions are designed so that their finite-difference expressions are exactly equal to the derivatives, i.e., the truncation errors vanish. We then substitute these simple functions into the original differential equation, generate a residual, and evaluate the residual at a given grid point. Finally, we evaluate nodal variables according to the assigned functions at the given grid point and check whether the residuals obtained from the differential equation and the difference equation are the same.

Example 12-5 Let $K = xy$, $u = y$, $v = 1$, and $\epsilon = 2$. Substitute these expressions into Eq. (12-33) and evaluate the result at $x = y = 2$. Then take $\Delta x = \Delta y = 1$ with grid point j at $x = y = 2$ and evaluate the nodal values according to the assumed expressions. Check whether the results obtained from Eq. (12-33) and Eq. (12-34) are the same.

Solution: The residual due to substitution of the assumed expressions into Eq. (12-33) is

$$R_1(x,y) = y^2 + x - C_\mu x^3 y - \frac{C_\mu x^2 y^2}{2} + 2 . \tag{a}$$

At $x = y = 2$, we obtain

$$R_1(2,2) = 8 - 24C_\mu . \tag{b}$$

Next we compute the nodal values of K and u at various nodal points and obtain

$$\begin{aligned} R_2(x,y) = & \left[\tfrac{1}{2}(2C_\mu + 18C_\mu + 2 \times 8C_\mu) + 2\right](4) - 2(2) - \left[\tfrac{1}{2}(2C_\mu + 8C_\mu) + \tfrac{1}{2}\right](2) \\ & - \left[\tfrac{1}{2}(18C_\mu + 8C_\mu) - \tfrac{1}{2}\right](6) - 8C_\mu + 2 = 8 - 24C_\mu . \end{aligned} \tag{c}$$

The fact that the results in Eqs. (b) and (c) are the same suggests that the algebraic procedure leading to Eq. (12-33) is probably correct.

The K-ϵ model has been extensively used [40-55]. The values of the empirical constants C_μ, C_1, C_2, σ_K, and σ_ϵ were chosen in each problem to best fit the experimental data. A typical set of these values is

$$C_\mu = 0.09 , \quad C_1 = 1.45 , \quad C_2 = 0.18 , \quad \sigma_K = 1.0 , \quad \text{and} \quad \sigma_\epsilon = 1.0 .$$

We will now adopt the integral method to illustrate a procedure for solving the K-ϵ equations governing a turbulent cylindrical free jet. Use of the integral method is allowed because the jet flow exhibits similar variable distributions at various streamwise locations, and the transport equations become the parabolic type. Consider a disk control volume representing an infinitesimal section of the circular jet between x and $x + dx$ as shown in Fig. 12-2. The streamwise momentum entering this control volume should be conserved since

$$\left(\frac{\partial U}{\partial r}\right)_{r=\delta} = 0 .$$

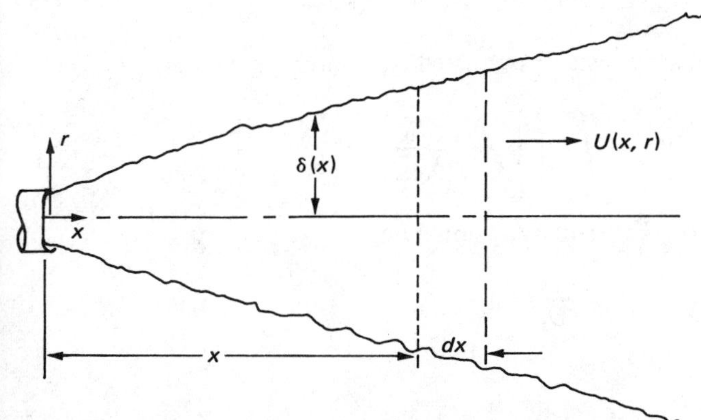

Figure 12-2 System schematic of a circular turbulent jet.

Consequently, at any streamwise location,

$$\frac{d}{dx}\int_0^\delta rU^2\,dr = 0\,. \tag{12-35}$$

The integral form of Eqs. (12-30) and (12-31) for the circular jet of the boundary-layer type can be derived as

$$\frac{d}{dx}\int_0^\delta rUK\,dr = C_\mu \int_0^\delta r\frac{K^2}{\epsilon}\left(\frac{\partial U}{\partial r}\right)^2 dr - \int_0^\delta r\epsilon\,dr \tag{12-36}$$

and

$$\frac{d}{dx}\int_0^\delta rU\epsilon\,dr = C_1 C_\mu \int_0^\delta rK\left(\frac{\partial U}{\partial r}\right)^2 dr - C_2\int_0^\delta r\frac{\epsilon^2}{K}\,dr\,. \tag{12-37}$$

For the integral method to be useful, we must be able to assume that $U(x, r)/U_c(x)$, $K(x, r)/K_c(x)$, and $\epsilon(x, r)/\epsilon_c(x)$ are functions of the dimensionless variable ξ only, where

$$\xi = \frac{r}{\delta(x)}\,.$$

If this assumption holds, Eqs. (12-35)-(12-37) can be reduced, respectively, to

$$\frac{d}{dx}(\delta^2 U_c^2) = 0\,, \tag{12-38}$$

$$a_1\frac{d}{dx}(\delta^2 U_c K_c) = C_\mu a_2\frac{K_c^2 U_c^2}{\epsilon_c} - a_3\delta^2\epsilon_c\,, \tag{12-39}$$

and

$$a_4\frac{d}{dx}(\delta^2 U_c\epsilon_c) = C_1 C_\mu a_5 K_c U_c^2 - C_2 a_6\delta^2\frac{\epsilon_c^2}{K_c}\,, \tag{12-40}$$

where a_k, $k = 1, 2, \ldots, 6$, are constants evaluated from integrals such as

$$a_1 = \int_0^1 \xi\left(\frac{U}{U_c}\right)\left(\frac{K}{K_c}\right)d\xi\,.$$

Typical assumed profiles for U/U_c and K/K_c could be

$$\frac{U}{U_c} = 1 - \xi^2$$

and

$$\frac{K}{K_c} = \exp(-1.15\xi^2)\,.$$

Thus there are four unknowns δ, U_c, K_c, and ϵ_c in Eqs. (12-38)-(12-40). A fourth ordinary equation can be derived by taking the momentum balance along the centerline where the transverse velocity V is zero due to symmetry. Hence, the momentum equation in cylindrical coordinates corresponding to Eq. (12-14) reduces to

$$U_c \frac{dU_c}{dx} = \left[(\nu + \nu_t)\frac{\partial^2 U}{\partial r^2}\right]_{r=0} + \left[\frac{(\nu + \nu_t)(\partial U/\partial r)}{r}\right]_{r=0}, \tag{12-41}$$

where the assumption that the r derivative of $\nu + \nu_t$ vanishes at the centerline has been incorporated. The second indeterminant term on the right-hand side of Eq. (12-41) can be evaluated as

$$\left[\frac{(\nu + \nu_t)(\partial U/\partial r)}{r}\right]_{r=0} = a_7\left(\nu + \frac{C_\mu K_c^2}{\epsilon_c}\right)\left(\frac{U_c}{\delta^2}\right),$$

where

$$a_7 = \lim_{\xi \to 0}\left(-\frac{2\xi}{\xi}\right) = -2 .$$

Therefore, Eq. (12-41)—the fourth equation—becomes

$$U_c \frac{dU_c}{dx} = -4\left(\nu + \frac{C_\mu K_c^2}{\epsilon_c}\right)\left(\frac{U_c}{\delta^2}\right). \tag{12-42}$$

Equations (12-38)-(12-40) and (12-42) are now closed provided some proper initial conditions are given. They constitute a set of nonlinear first-order ordinary differential equations, which can be integrated straightforwardly. No singularities seem to exist since all the values in the denominators are nonzero.

Sometimes, with luck, the unknowns in the ordinary differential equations can be assumed to be proportional to a certain power of x. This assumption is valid if, after the substitution of these expressions into the ordinary differential equations, all the x terms can be uniformly canceled.

Example 12-6 Consider the set of ordinary differential equations (12-38)-(12-40) and (12-42) deduced by using the integral method. Assume that δ, U_c, K_c, and ϵ_c are proportional to certain powers of x. Determine these powers for the four unknowns.

Solution: Let us assume

$$\delta \propto x^j, \quad U_c \propto x^k, \quad K_c \propto x^m, \quad \text{and} \quad \epsilon_c \propto x^n . \tag{a}$$

Substituting Eq. (*a*) into Eqs. (12-38)-(12-40) and (12-42) yields

$$2j + 2k = 0, \tag{b}$$

$$2j + k + m - 1 = 2m + 2k - n = 2j + n, \tag{c,d}$$

$$2j + k + n - 1 = m + 2k = 2j + 2n - m, \tag{e,f}$$

$$2k - 1 = k - 2j + 2m - n \,, \qquad (g)$$

and

$$2m - n = 0 \,. \qquad (h)$$

Ignoring the fact that there are three redundant equations, we solve Eqs. (*b*)-(*h*) to obtain

$$j = 1 \,, \quad k = -1 \,, \quad m = -2 \,, \quad \text{and} \quad n = -4 \,.$$

Once these powers are determined and x terms are dropped from the ordinary differential equations, a set of algebraic equations in which the proportionality constants are unknowns will result.

Two-equation models other than the K-ϵ model include K and l [39], K and the turbulent pseudovorticity [56], and K and the mean square variance of concentration [57]. Among the turbulence models studied and compared [58], the K-l model has been found to yield the best overall performance.

12-5 HIGHER-ORDER CLOSURE THEORIES

In Example 12-5 we showed how a third-order correlation appears in Eq. (12-27). In order to close the system of equations, these correlations must be artificially expressed in terms of the existing unknowns. In doing so, however, it is likely that the artificial interpolation will invite some uncertainties [Eqs. (12-19) and (12-29) are good examples]. Alternatively, the artificial interpretation can be avoided with the so-called higher-order closure theories [15-17]. In these theories, additional transport equations which the kth-order correlations must satisfy are derived. In these kth-order transport equations, $(k + 1)$th-order correlations will appear. If the system is to be closed at the kth order, the $(k + 1)$th-order correlations must be expressed in terms of kth- and lower-order correlations. For example, at $k = 4$ the closed system contains as many as 30 simultaneous equations. Undoubtedly such a high-order closure yields high accuracy, but at the same time it requires a great deal of computational work.

SYMBOLS

A	empirical constant [Eq. (12-19)], $A = 26$
C_f	frictional coefficient $(= 2\tau_w/\rho U_m^2)$
D	diameter of pipe, m
h_c	convective heat transfer coefficient, $\mathrm{W/m^2 \cdot K}$
k	empirical constant [Eq. (12-19)], $k = 0.4$
K	kinetic energy of turbulence, J/kg
l	Prandtl mixing length, m
Nu_D	Nusselt number based on pipe diameter $(= h_c D/k)$
Re_D	Reynolds number $(= U_m D/\nu)$
u_τ	reference velocity $[= (\tau_w/\rho)^{1/2}]$, m/s

U	time average of instantaneous velocity u, m/s
$\delta(x)$	thickness of circular turbulent jet, m
ϵ	dissipation rate of kinetic energy of turbulence, m^2/s^3
θ	normalized temperature $[= (T - T_w)/(T_a - T_w)]$
ν_t	turbulent diffusivity, m^2/s
ξ	dimensionless radius $[= r/\delta(x)]$
τ_w	shear stress at the wall, NT/m^2
Φ	time average of ϕ

Subscripts

c	centerline
eff	effective
m	mean
t	turbulent
w	wall
$1, a, b, c$	locations in Fig. 12-1

Superscripts

*	normalized on U_1
$-$	time average
$'$	fluctuation
+	normalization on ν/u_τ for lengths, u_τ for velocities, and $\dot{q}''_w/\rho c_p u_\tau$ for temperatures, respectively

REFERENCES

1. J. O. Hinze, *Turbulence*, p. 2, McGraw-Hill, New York, 1975.
2. G. K. Batchelor, *The Theory of Homogeneous Turbulence*, Cambridge Univ. Press, New York, 1953.
3. W. M. Kays, *Convective Heat and Mass Transfer*, chaps. 9 and 11, McGraw-Hill, New York, 1966.
4. H. Schlichting, *Boundary-Layer Theory*, part D, McGraw-Hill, New York, 1968.
5. P. Bradshaw, *An Introduction to Turbulence and Its Measurement*, Pergamon, London, 1971.
6. L. Prandtl, Über ein neues formelsystem für der ausgebildeten turbulenz, *Nachr. Akad. Wiss. Goettingen*, pp. 6–19, 1945.
7. A. Favre, Équations des Gaz Turbulents Compressibles I., *J. Mécanique*, vol. 4, pp. 361–390, 1965.
8. A. Favre, Équations des Gaz Turbulents Compressibles II., *J. Mécanique*, vol. 4, pp. 391–421, 1965.
9. E. R. van Driest, On Turbulent Flow near a Wall, *J. Aeronaut. Sci.*, vol. 23, pp. 1007–1011, 1956.
10. A. N. Kolmogorov, Equation of Turbulent Motion of an Incompressible Fluid, *Izv. Akad. Nauk SSSR Ser. Fiz.*, pp. 1–2, 1942.
11. B. E. Launder and D. B. Spalding, *Mathematical Models of Turbulence*, Academic, London, 1972.
12. W. P. Jones and B. E. Launder, The Prediction of Laminarization with a Two-Equation Model of Turbulence, *Int. J. Heat Mass Transfer*, vol. 15, pp. 301–314, 1972.

13. K. H. Ng and D. B. Spalding, A Turbulence Model for Boundary Layers near Walls, *Phys. Fluids*, vol. 15, pp. 20–30, 1972.
14. B. E. Launder and D. B. Spalding, The Numerical Computation of Turbulent Flows, *Comput. Math. Appl. Mech. Eng.*, vol. 3, pp. 269–289, 1974.
15. J. Rotta, Statistische Theorie Nichthomogener Turbulenz, *Z. Phys.*, vol. 129, pp. 547–572. 1951, and vol. 131, pp. 51–77, 1951.
16. B. E. Launder, G. J. Reece, and W. Rodi, Progress in the Development of a Reynolds-Stress Turbulence Closure, *J. Fluid Mech.*, vol. 68, pp. 537–566, 1975.
17. K. Hanjalic and B. E. Launder, Contribution towards a Reynolds-Stress Closure for Low-Reynolds-Number Turbulence, *J. Fluid Mech.*, vol. 74, pp. 593–610, 1976.
18. I. K. Madni and R. H. Pletcher, Prediction of Turbulent Forced Plumes Issuing Vertically into Stratified or Uniform Ambients, *ASME J. Heat Transfer*, vol. 99, pp. 99–104, 1977.
19. A. T. Wassel and A. F. Mills, Calculation of Variable Property Turbulent Friction and Heat Transfer in Rough Pipes, *ASME J. Heat Transfer*, vol. 101, pp. 469–474, 1979.
20. A. F. Emery, P. K. Neighbors, and F. B. Gessner, The Numerical Prediction of Developing Turbulent Flow and Heat Transfer in a Square Duct, *ASME J. Heat Transfer*, vol. 102, pp. 51–57, 1980.
21. S. Levy and J. M. Healzer, Application of Mixing Length Theory to Wavy Turbulent Liquid–Gas Interface, *ASME J. Heat Transfer*, vol. 103, pp. 492–500, 1981.
22. N. M. Schnurr, Numerical Predictions of Heat Transfer to Supercritical Helium in Turbulent Flow through Circular Tubes, *ASME J. Heat Transfer*, vol. 99, pp. 580–585, 1977.
23. A. T. Wassel and I. Catton, Diffusion from a Line Source in a Neutral or Stably Stratified Atmospheric Surface Layer, *Int. J. Heat Mass Transfer*, vol. 20, pp. 383–391, 1977.
24. R. Greif, An Experimental and Theoretical Study of Heat Transfer in Vertical Tube Flows, ASME J. Heat Transfer, vol. 100, pp. 86–91, 1978.
25. L. Thomas, A Simple Integral Approach to Turbulent Thermal Boundary Layer Flow, *ASME J. Heat Transfer*, vol. 100, pp. 744–746, 1978.
26. J. A. Liburdy, E. G. Groff, and G. M. Faeth, Structure of a Turbulent Thermal Plume Rising along an Isothermal Wall, *ASME J. Heat Transfer*, vol. 101, pp. 249–255, 1979.
27. A. M. Cary, Jr., D. M. Bushnell, and J. N. Hefner, Predicted Effects of Tangential Slot Injection on Turbulent Boundary Layer Flow over a Wide Speed Range, *ASME J. Heat Transfer*, vol. 101, pp. 699–704, 1979.
28. P. E. Pickett, M. F. Taylor, and D. M. McEligot, Heated Turbulent Flow of Helium-Argon Mixtures in Tubes, *Int. J. Heat Mass Transfer*, vol. 22, pp. 705–719, 1979.
29. C. C. Chieng and B. E. Launder, On the Calculation of Turbulent Transport in Flow through an Asymmetrically Heated Pipe, *Numer. Heat Transfer*, vol. 2, pp. 359–371, 1979.
30. M. Nakajima, K. Fukui, H. Ueda, and T. Mizushina, Buoyancy Effects on Turbulent Transport in Combined Free and Forced Convection between Vertical Parallel Plates, *Int. J. Heat Mass Transfer*, vol. 23, pp. 1325–1336, 1980.
31. C. E. Jobe and W. L. Hankey, Turbulent Boundary Layer Calculations in Adverse Pressure Gradient Flows, *AIAA J.*, vol. 18, pp. 1394–1397, 1980.
32. C. J. Chen and J. S. Chiou, Laminar and Turbulent Heat Transfer in the Pipe Entrance Region for Liquid Metals, *Int. J. Heat Mass Transfer*, vol. 24, pp. 1179–1189, 1981.
33. M. J. Adams and H. Krier, Unsteady Internal Boundary Layer Analysis Applied to Gun Barrel Wall Heat Transfer, *Int. J. Heat Mass Transfer*, vol. 24, pp. 1925–1935, 1981.
34. A. Polak and M. J. Werle, Interacting Turbulent Boundary Layer over a Wavy Wall, *ASME J. Heat Transfer*, vol. 100, pp. 678–683, 1978.
35. T. Cebeci, F. Thiele, P. G. Williams, and K. Stewartson, On the Calculation of Symmetric Wakes: 1, Two-dimensional Flows, *Numer. Heat Transfer*, vol. 2, pp. 35–60, 1979.
36. R. K. Ahluwalia and K. H. Im, Combined Conduction, Convection, Gas Radiation and Particle Radiation in MHD Diffusers, *Int. J. Heat Mass Transfer*, vol. 24, pp. 1421–1430, 1981.
37. C. Taylor, T. G. Hughes, and K. Morgan, Finite Element Solution of One-Equation Models of Turbulent Flow, *J. Comput. Phys.*, vol. 29, pp. 163–172, 1978.

38. F. Gori, M. A. ElHadidy, and D. B. Spalding, Numerical Predictions of Heat Transfer to Low-Prandtl-Number Fluids, *Numer. Heat Transfer*, vol. 2, pp. 441–454, 1979.
39. M. Kaviany and R. A. Seban, Transient Turbulent Thermal Convection in a Pool of Water, *Int. J. Heat Mass Transfer*, vol. 24, pp. 1742–1746, 1981.
40. O. A. Plumb and L. A. Kennedy, Application of a K-ϵ Turbulence Model to Natural Convection from a Vertical Isothermal Surface, *ASME J. Heat Transfer*, vol. 99, pp. 79–85, 1977.
41. M. R. F. Heikal, P. J. Walklate, and A. P. Hatton, The Effect of Free Stream Turbulence Level on the Flow and Heat Transfer in the Entrance Region of an Annulus, *Int. J. Heat Mass Transfer*, vol. 20, pp. 763–771, 1977.
42. M. D. Deshpande and D. P. Giddens, Direct Solution of Two Linear Systems of Equations Forming Coupled Tridiagonal-Type Matrices, *Int. J. Numer. Methods Eng.*, vol. 13, pp. 1049–1052, 1977.
43. S. V. Patankar, E. M. Sparrow, and M. Ivanovic, Thermal Interactions among the Confining Walls of a Turbulent Recirculating Flow, *Int. J. Heat Mass Transfer*, vol. 21, pp. 269–274, 1978.
44. S. J. Lin and S. W. Churchill, Turbulent Free Convection from a Vertical, Isothermal Plate, *Numer. Heat Transfer*, vol. 1, pp. 129–145, 1978.
45. N. C. G. Markatos, Transient Flow and Heat Transfer of Liquid Sodium Coolant in the Outlet Plenum of a Fast Nuclear Reactor, *Int. J. Heat Mass Transfer*, vol. 21, pp. 1565–1579, 1978.
46. G. Bergels, A. D. Gosman, and B. E. Launder, The Turbulent Jet in a Cross Stream at Low Injection Rates: A Three-dimensional Numerical Treatment, *Numer. Heat Transfer*, vol. 1, pp. 217–242, 1978.
47. J. C. Wu and A. Sugavanam, Method for the Numerical Solution of Turbulent Flow Problems, *AIAA J.*, vol. 16, pp. 948–955, 1978.
48. P. C. Sukanek and R. P. Rhodes, Centerline Formulation in the Numerical Computation of Axisymmetric Flows, *AIAA J.*, vol. 16, pp. 1099–1102, 1978.
49. J. G. Bartzis and N. E. Todreas, Turbulence Modeling of Axial Flow in a Bare Rod Bundle, *ASME J. Heat Transfer*, vol. 101, pp. 628–634, 1979.
50. G. D. Raithby and G. E. Schneider, The Prediction of Surface Discharge Jets by a Three-dimensional Finite Difference Method, *ASME J. Heat Transfer*, vol. 102, pp. 138–145, 1980.
51. C. C. Chieng and B. E. Launder, On the Calculation of Turbulent Heat Transport Downstream from an Abrupt Pipe Expansion, *Numer. Heat Transfer*, vol. 3, pp. 189–207, 1980.
52. G. Bergels, A. D. Gosman, and B. E. Launder, The Prediction of Three-dimensional Discrete-Hole Cooling Processes, Part 2: Turbulent Flow, *ASME J. Heat Transfer*, vol. 103, pp. 141–145, 1981.
53. M. Ljuboja and W. Rodi, Prediction of Horizontal and Vertical Turbulent Buoyant Wall Jets, *ASME J. Heat Transfer*, vol. 103, pp. 343–349, 1981.
54. S. Byggstoyl and W. Kollmann, Closure Model for Intermittent Turbulent Flows, *Int. J. Heat Mass Transfer*, vol. 24, pp. 1811–1822, 1981.
55. B. H. Hjertager and B. F. Magnussen, Calculation of Turbulent Three-dimensional Jet Induced Flow in Rectangular Enclosures, *Comput. Fluids*, vol. 9, pp. 395–407, 1981.
56. B. Sunden, A Theoretical Investigation of the Effect of Free Stream Turbulence on Skin Friction and Heat Transfer for a Bluff Body, *Int. J. Heat Mass Transfer*, vol. 22, pp. 1125–1135, 1979.
57. S. E. Tahry and A. D. Gosman, The Two- and Three-dimensional Dispersal of a Passive Scalar in a Turbulent Boundary Layer, *Int. J. Heat Mass Transfer*, vol. 24, pp. 35–46, 1981.
58. M. R. Malik and R. H. Pletcher, A Study of Some Turbulence Models for Flow and Heat Transfer in Ducts of Annular Cross-Section, *ASME J. Heat Transfer*, vol. 103, pp. 146–152, 1981.

PROBLEMS

12-1 Following the procedure described in Example 12-2, find values of R^+ and C_f if $\mathrm{Re}_D =$ 10,000. Compare your result with that in [3, p. 73].

12-2 Following the procedure described in Example 12-4, find the value of Nu_D if $Re_D = 10{,}000$ and $Pr = 10$. Compare your result with that in [3, pp. 170–171].

12-3 Derive K-ϵ Eqs. (12-30) and (12-31).

12-4 Based on Example 12-6, it may be assumed that

$$\delta = Ax\ , \qquad U_c = Bx^{-1}\ , \qquad K_c = Cx^{-2}\ , \qquad \text{and} \qquad \epsilon_c = Dx^{-4}\ . \qquad \text{(P12-4}a\text{–}d\text{)}$$

Substitute Eqs. (P12-4*a–d*) into Eqs. (12-38)–(12-40) and (12-42) and find the expressions for δ, U_c, K_c, and ϵ_c in terms of the empirical constants.

CHAPTER

THIRTEEN

INTRODUCTION TO COMBUSTION PHENOMENA

Combustion is a complicated phenomenon in which chemical reactions couple heat and mass transfer, often in a turbulent fashion. The numerical investigation of combustion problems is therefore a challenging task. We tend to feel contented if, by means of a certain numerical method, we are able to obtain a solution that appears physically reasonable or agrees with the available experimental data. No further attempt will be made in performing the numerical analysis to examine the stability and convergence of the scheme or to choose schemes that are better in terms of accuracy and efficiency. For this reason, numerical analysis has not received sufficient attention in the field of combustion research. The finite-difference method has been the technique predominantly adopted by combustion researchers; the numerical properties of the finite-difference schemes have been rarely discussed.

It is difficult to change this traditional attitude toward numerical combustion research. This is because, generally, no fruitful result can be obtained if the numerical analysis is applied to a complicated physical problem in which our major efforts must be devoted to fighting with the physical complexities. Since the use and numerical properties of various numerical methods have already been described in Part I and Part II of this book, we will focus our attention here on the following questions: (1) What do we have to know about combustion in order not to be regarded as a layperson? (2) What are the major numerical difficulties associated with combustion problems and the possible approaches to removing these difficulties? (3) What is the typical numerical solution for an idealized combustion system?

In Section 13-1, combustion phenomena are briefly reviewed. Some important combustion terminology used in this chapter is defined.

In Section 13-2, the general set of conservation equations that govern a two-dimensional gaseous combustion system is introduced. It can be readily reduced to simpler sets for various more idealized combustion systems.

In Section 13-3, the difficulties in solving the governing equations and the possible approaches to removing these difficulties are discussed.

Finally, the boundary-layer, laminar, radiative flame is considered in Section 13-4. An example is presented to illustrate the procedure for recovering the physical field variables from the Shvab-Zeldovich variables. A brief note on the existing scheme for investigating the boundary-layer turbulent flame is also given.

13-1 BRIEF REVIEW OF COMBUSTION PHENOMENA

In this section we will briefly describe the fundamentals of combustion phenomena and define some important terminology where appropriate. Several quantitative examples will also be presented for illustration. For more comprehensive material see, for example, [1-5].

Combustion is a phenomenon in which chemical reactions occur exothermally (in the overall sense) and energy and species are transported. A typical chemical reaction can be written as

$$\nu_f A + \nu_{ox} B \rightarrow \nu_p C \,. \tag{13-1}$$

The species A, B, and C stand for the fuel, oxidant (not necessarily oxygen), and product of combustion, respectively. The stoichiometric coefficients ν_f, ν_{ox}, and ν_p should be determined such that the number of each atom is conserved.

The law of mass action can be stated as

$$-\frac{\dot{m}_f'''}{\nu_f M_f} = -\frac{\dot{m}_{ox}'''}{\nu_{ox} M_{ox}} = \frac{\dot{m}_p'''}{\nu_p M_p} \,, \tag{13-2}$$

where $\dot{m}_i'''$, $i = r$, ox, p, denotes the increase in volumetric rate of the mass fraction of species i and M_i represents the molecular weight of species i. For the simple single-step reaction shown in Eq. (13-1), the reaction rates can be expressed as, for example,

$$m_f''' = \frac{dY_f}{dt} = -k_2 Y_f Y_{ox} \,, \tag{13-3}$$

where k_2 is the rate constant and the subscript 2 means second order. The order (or overall order) of a reaction is determined by the sum of the exponents of the reactant mass fractions that express the reaction rate.

Example 13-1 The orders of the following reactions:

$$N_2O_5 \rightarrow N_2O_4 + \tfrac{1}{2}O_2 \tag{a}$$

and

$$2CO + O_2 \rightarrow 2CO_2 \tag{b}$$

are first and third, respectively.

The rate constant k for a single-step reaction is empirically given by the so-called Arrhenius expression

$$k = A \exp\left(-\frac{E}{\tilde{R}T}\right), \tag{13-4}$$

where A and E are the frequency (or preexponential) factor and the activation energy, respectively, and their values are determined experimentally; $\tilde{R}$ is the universal gas constant. Generally, E is a constant but A depends weakly on T.

Example 13-2 Find the rate constant k_2 at $T = 2000$ K for the biomolecular reaction

$$O + H_2 \rightarrow OH + H \tag{a}$$

with

$$A = 2.1 \times 10^{-11}\ \text{cm}^3/\text{molecule}\cdot\text{s} \quad \text{and} \quad E = 9.40\ \text{kcal/gmol [6]}. \tag{b}$$

Solution: From Eq. (13-4), we first calculate

$$\exp\left(-\frac{E}{\tilde{R}T}\right) = \exp\left(\frac{-9.4 \times 10^3}{1.986 \times 2000}\right) = 0.0938\,. \tag{c}$$

Therefore,

$$k_2 = 1.97 \times 10^{-12}\ \text{cm}^3/\text{molecule}\cdot\text{s}\,. \tag{d}$$

Incidentally, at a given instant, if

$$Y_{\text{ox}} = 10^{15}\ \text{molecule/cm}^3 \quad \text{and} \quad Y_{H_2} = 10^{16}\ \text{molecule/cm}^3\,,$$

then

$$\dot{m}'''_{H_2} = -1.97 \times 10^{19}\ \text{molecule/cm}^3\cdot\text{s}\,.$$

Combining Eqs. (13-3) and (13-4), we can write an overall reaction rate as

$$-\frac{d(\text{fuel})}{dt} = A(\text{fuel})^b(\text{oxidant})^c \exp\left(-\frac{E}{\tilde{R}T}\right), \tag{13-5}$$

which neglects the detailed reaction kinetics and can be used in conjunction with the transport equations [7-10]. In reality, most chemical reactions occur in a chain sequence composed of four basic processes: initiation, branching, propagation, and termination of the radicals. For example, Eq. (13-1) may be composed of

$$C + O_2 \rightarrow CO + O \quad \text{(initiation)},$$

$$CO + O_3 \rightarrow CO_2 + O + O \quad \text{(branching)},$$

$$C_2O + CO \rightarrow C_3O_2 \quad \text{(propagation)},$$

and

$$CO + O + M \rightarrow CO_2 + M \quad \text{(termination)}.$$

In the last reaction, the catalyst M is a substance that affects the reaction rate but does not change in property or mass fraction during the reaction. While the investigation of multistep reactions is essential for an understanding of the characteristics of combustion, it has been customarily excluded from heat transfer research.

Even if a multistep chemical reaction can be idealized and described by Eq. (13-5), much uncertainty remains in determining the values of A, b, and c. Furthermore, from the computational viewpoint, this nonlinear term inevitably causes difficulties or inconvenience when present in the equations governing transport of energy and species. For these reasons, it will be desirable to eliminate this term, if possible. One classical way to achieve this elimination is to introduce the so-called Shvab-Zeldovich variables, defined as

$$\beta = \frac{Y_f}{\nu_f M_f} - \frac{Y_{ox}}{\nu_{ox} M_{ox}} \tag{13-6a}$$

and

$$\gamma = -\frac{h}{q_{ox}\nu_{ox}M_{ox}} - \frac{Y_{ox}}{\nu_{ox}M_{ox}}, \tag{13-6b}$$

where M_f and M_{ox} are the molecular weights of fuel and oxidant, respectively, and q_{ox} is the heat of reaction per kilogram of oxygen. More details of this formulation are presented in Section 13-4.

The Shvab-Zeldovich formulation has been widely adopted in the investigation of diffusion (or diffusion-controlled) flames exemplified by classical Burke-Schumann problem [11] and several analyses [12-18]. A diffusion flame may be defined as the combustion phenomenon in which the fuel and the oxidant are initially separated and the chemical reaction rate is much higher than the flow diffusion rate. Consequently, diffusion flames generally exhibit steep temperature and species gradients. Flames of this class are commonly seen in daily life and are to be distinguished from reaction-controlled flames. In the latter, the reactants are generally well mixed because there is sufficient time for the flow or diffusion processes to smooth out the spatial nonuniformities of species and temperature. In other words, the reaction rate is very low in comparison with the diffusion rate.

To quantitatively determine which mechanism controls a given flame, we introduce the Damkohler number, defined as

$$\mathrm{Da} = \frac{t_{residence}}{t_{reaction}}, \tag{13-7}$$

where $t_{residence}$ is the time required for the reactants to reside in the combustion system. This time can be estimated from the diffusion process in stagnant systems, i.e.,

$$t_{residence} \approx \frac{L^2}{D}, \tag{13-8a}$$

where D and L are the diffusion coefficient and characteristic length of the system, respectively. Alternatively, for flow (nonstagnant) systems, we may define

$$t_{residence} \approx \frac{L}{u}, \tag{13-8b}$$

where u is the characteristic flow velocity. The reaction time may be approximated as the reciprocal of the rate constant. Thus, according to Eq. (13-4),

$$t_{\text{reaction}} \approx \frac{1}{A \exp(-E/\tilde{R}T)} . \tag{13-9}$$

When the flame is diffusion-controlled, Da approaches infinity; when it is reaction-controlled, Da tends to vanish.

Flames can also be classified according to the phase of the combustion medium. A flame is said to be homogeneous if it takes place in a pure gaseous, liquid, or solid phase and is said to be heterogeneous if it takes place at the interface of at least two phases.

Since (1) most combustion phenomena take place in the gaseous phase, (2) gaseous combustion usually interacts with heat and mass transfer, and (3) in reaction-controlled flames the distributions of field variables are, in general, uniform (unchallenging to us) and the details of the reaction kinetics are needed (troublesome to us), we will concentrate on gaseous diffusion flames.

13-2 GOVERNING EQUATIONS

The general conservation equations that govern a two-dimensional gaseous diffusion flame can be written in Cartesian ($n = 0$, $r = y$) or cylindrical ($n = 1$) coordinates as follows.

Continuity:

$$\frac{\partial \rho}{\partial t} + \frac{\partial}{\partial x}(\rho u) + \frac{1}{r^n}\frac{\partial}{\partial r}(r^n \rho v) = 0 , \tag{13-10}$$

x-direction momentum:

$$\frac{\partial}{\partial t}(\rho u) + \frac{\partial}{\partial x}(\rho u^2) + \frac{1}{r^n}\frac{\partial}{\partial r}(r^n \rho v u) = \frac{\partial}{\partial x}\left(\mu \frac{\partial u}{\partial x}\right) + \frac{1}{r^n}\frac{\partial}{\partial r}\left(\mu r^n \frac{\partial u}{\partial r}\right) - \frac{\partial p}{\partial x} - \rho g , \tag{13-11}$$

r-direction momentum:

$$\frac{\partial}{\partial t}(\rho v) + \frac{\partial}{\partial x}(\rho u v) + \frac{1}{r^n}\frac{\partial}{\partial r}(r^n \rho v^2) = \frac{\partial}{\partial x}\left(\mu \frac{\partial v}{\partial x}\right) + \frac{1}{r^n}\frac{\partial}{\partial r}\left(\mu r^n \frac{\partial v}{\partial r}\right) - n\frac{\mu v}{r^2} - \frac{\partial p}{\partial r} , \tag{13-12}$$

Energy:

$$\frac{\partial}{\partial t}(\rho e) + \frac{\partial}{\partial x}(\rho u h) + \frac{1}{r^n}\frac{\partial}{\partial r}(r^n \rho v h) = -\frac{\partial \dot{q}''}{\partial x} - \frac{1}{r^n}\frac{\partial}{\partial r}(r^n \dot{q}'') + \dot{q}_c''' + \dot{q}_r''' , \tag{13-13}$$

Species:

$$\frac{\partial}{\partial t}(\rho Y_i) + \frac{\partial}{\partial x}(\rho u Y_i) + \frac{1}{r^n}\frac{\partial}{\partial r}(r^n \rho v Y_i) = \frac{\partial}{\partial x}\left(\rho D_i \frac{\partial Y_i}{\partial x}\right) + \frac{1}{r^n}\frac{\partial}{\partial r}\left(\rho D_i r^n \frac{\partial Y_i}{\partial r}\right) + \dot{m}_i''' , \tag{13-14}$$

$i =$ fuel, oxidant, and products of chemical reaction .

Gravity acts along the negative x direction; the heat flux $\dot{q}''$ was given in Eq. (9-5) as

$$\dot{q}'' = -\frac{k}{c_p} \nabla h + \frac{k}{c_p} \sum_{k=1}^{n} (1 - \mathrm{Le}) h_k \nabla Y_k . \tag{13-15}$$

The symbols $\dot{q}_c'''$ and $\dot{q}_r'''$ denote volumetric heat of combustion and volumetric radiation, respectively. The enthalpy h is related to the temperature by

$$h = \int_{T_0}^{T} c_p \, dT , \tag{13-16}$$

where, for a mixture of n species,

$$c_p = \sum_{i=1}^{n} Y_i c_{p,i} .$$

If all the species are assumed to be ideal gases, we also have the equation of state

$$p = \rho R T , \tag{13-17}$$

where

$$R = \sum_{i=1}^{n} Y_i R_i .$$

To obtain the mass fraction distributions $Y_i(x, r)$ of n species, we need to solve only $n - 1$ species transport equations since the mass fraction of the nth species can be obtained by

$$\sum_{i=1}^{n} Y_i = 1 . \tag{13-18}$$

Finally, if $\dot{q}_c'''$, $\dot{q}_r'''$, and all the $\dot{m}_i'''$ can be prescribed in terms of the existing field variables, then there are $n + 7$ unknowns, namely

$$u, v, \rho, p, h, T, \dot{q}'', Y_1, Y_2, \ldots, \text{and } Y_n ,$$

that should satisfy $n + 7$ governing equations, namely Eqs. (13-10)–(13-13), $n - 1$ times Eq. (13-14), and Eqs. (13-15)–(13-18). In principle, therefore, the set of equations is closed provided proper boundary conditions and initial conditions are given. Several examples of combustion systems that are governed by Eqs. (13-10)–(13-18) or the simplified versions of these equations are schematically shown in Fig. 13-1*a–f*. Of course, setting up the governing equations is one task, but actually solving them is another.

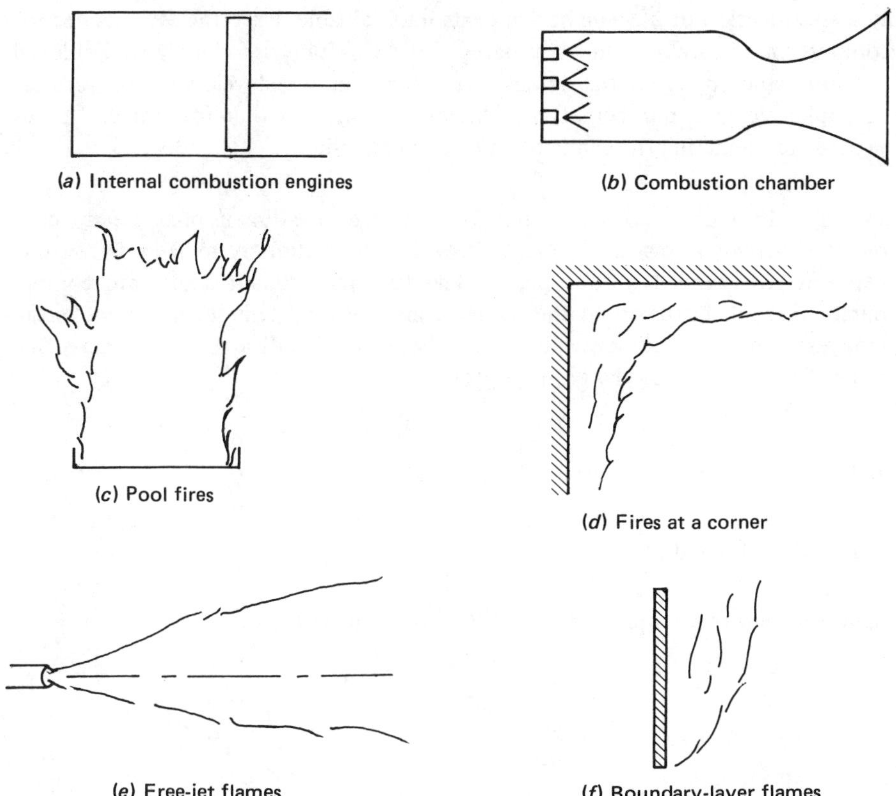

Figure 13-1 Six combustion systems that are governed by Eqs. (13-10)–(13-18).

13-3 ASSOCIATED DIFFICULTIES AND PROPOSED SOLUTIONS

In this section we will describe the difficulties that may arise when Eqs. (13-10)–(13-18) are to be solved.

13-3a Specification of $\dot{q}_c'''$ and $\dot{m}_i'''$

The volumetric rate of heat of combustion $\dot{q}_c'''$ and the volumetric rate of species generation $\dot{m}_i'''$ are known to be complicated functions of (at least) the temperature and the species mass fractions. Even if the idealized Arrhenius expression Eq. (13-5) is applicable to single-step chemical reactions, uncertainties in the values of A, b, c, and E remain. Unfortunately, real chemical reactions are generally multistep reactions and cannot be adequately described by the Arrhenius formulation. The detailed kinetics of multistep reactions depends on chemical, physical, and thermodynamic properties of the flame and is not fully understood. Without this input, a reliable and accurate specification of $\dot{q}_c'''$ and $\dot{m}_i'''$ appears unobtainable. Nevertheless, it is possible for combustion research to make progress. For this purpose the Shvab-Zeldovich formulation, which bypasses the details of the reaction, was developed. This formulation is applicable to a number of combustion systems and therefore much progress can be made in predicting the characteristics of combustion.

Example 13-3 Consider an extremely idealized one-dimensional steady combustion system, shown in Fig. 13-2, in which the stationary mixture is confined between two parallel porous plates. The fuel and oxidant come into contact, purely through diffusion, at the flame location $y = y_{\mathrm{fl}}$, where they react instantaneously and stoichiometrically. If the boundary conditions for the mass fractions of the fuel and oxidant are, respectively,

$$Y_f(0) = Y_{fw}\ , \qquad Y_{\mathrm{ox}}(0) = 0\ , \tag{a}$$

and

$$Y_f(L) = 0\ , \qquad Y_{\mathrm{ox}}(L) = Y_{\mathrm{ox}\infty}\ , \tag{b}$$

determine the flame location y_{fl} and the reaction rate $\dot{m}_f'''$. Assume a constant diffusion coefficient D.

Solution: For this simple system, Eq. (13-14) reduces to

$$\rho D \frac{d^2 Y_f}{dy^2} + \dot{m}_f''' = 0 \tag{c}$$

and

$$\rho D \frac{d^2 Y_{\mathrm{ox}}}{dy^2} + \dot{m}_{\mathrm{ox}}''' = 0\ . \tag{d}$$

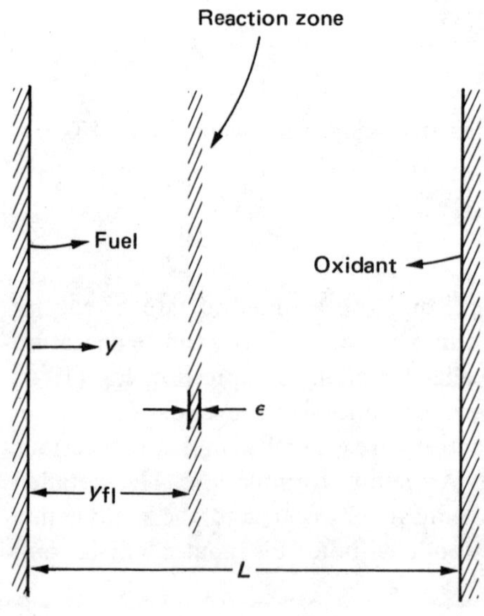

Figure 13-2 An idealized one-dimensional steady-state combustion system.

According to the law of mass action, Eq. (13-2), we deduce

$$\frac{d^2\beta}{dy^2} = 0 \,, \tag{e}$$

where β is the Shvab-Zeldovich variable defined by Eq. (13-6*a*). The two boundary conditions of $\beta(y)$, based on Eqs. (*a*) and (*b*), are

$$\beta(0) = \frac{Y_{fw}}{\nu_f M_f} \tag{f}$$

and

$$\beta(L) = -\frac{Y_{\text{ox}\infty}}{\nu_{\text{ox}} M_{\text{ox}}} \,. \tag{g}$$

The solution of Eqs. (*e*)-(*g*) is

$$\beta(y) = \frac{Y_{\text{ox}\infty}}{\nu_{\text{ox}} M_{\text{ox}}} \left[r - \frac{y}{L}(1+r) \right] , \tag{h}$$

where

$$r = \frac{Y_{fw}\nu_{\text{ox}} M_{\text{ox}}}{Y_{\text{ox}\infty}\nu_f M_f} \,. \tag{i}$$

Once the reactants meet, they react instantaneously, and hence $Y_f(y)$ and $Y_{\text{ox}}(y)$ should both vanish at $y = y_{\text{fl}}$. Therefore,

$$\beta(y_{\text{fl}}) = 0 \qquad \text{at } y = y_{\text{fl}} \,. \tag{j}$$

From Eq. (*h*), we obtain

$$y_{\text{fl}} = \frac{Lr}{1+r} \,. \tag{k}$$

To recover $Y_f(y)$ from $\beta(y)$, we know that $Y_{\text{ox}}(y)$ should vanish for all $y \leqslant y_{\text{fl}}$. Therefore, from Eq. (*h*) again,

$$\frac{Y_f(y)}{\nu_f M_f} = \frac{Y_{\text{ox}\infty}}{\nu_{\text{ox}} M_{\text{ox}}} \left[r - \frac{y}{L}(1+r) \right] , \qquad y \leqslant y_{\text{fl}} \,, \tag{l}$$

or

$$Y_f(y) = Y_{fw} \left[1 - \frac{y}{L}\left(\frac{1+r}{r}\right) \right] , \tag{m}$$

which is the mass fraction distribution of the fuel. The linearity of Eq. (*m*) suggests that $\dot{m}_f'''$ is zero everywhere except at the flame sheet, where the fuel diffusion flux enters the infinitesimal control volume dy through the left face but nothing emerges from the right face. Consequently, Eq. (*c*) can be integrated with respect to y from y_{fl} to $y_{\text{fl}} + \epsilon$ to yield

$$\rho D\left(\frac{dY_f}{dy}\right)_{y_{fl}+\epsilon} - \rho D\left(\frac{dY_f}{dy}\right)_{y_{fl}} + \int_{y_{fl}}^{y_{fl}+\epsilon} \dot{m}_f''' \, dy = 0 \,, \tag{n}$$

where ϵ is the thickness of the reaction zone. With the assumption that $\dot{m}_f'''$ is constant inside the reaction zone and the knowledge that

$$\rho D\left(\frac{dY_f}{dy}\right)_{y_{fl}+\epsilon} = 0 \,, \tag{o}$$

we estimate the reaction rate per surface area to be

$$\epsilon \dot{m}_f''' = \rho D\left(\frac{dY_f}{dy}\right)_{y_{fl}} = -\frac{\rho D Y_{fw}}{L}\left(\frac{1+r}{r}\right) . \tag{p}$$

13-3*b* Strong Coupling and High Nonlinearity via Variable Thermal Properties

For Boussinesq inert flows in which the temperature and density variations are small, it is a good approximation to assume constant thermal properties throughout the transport equations except for the buoyancy term—the only term that couples momentum and energy. For combustion systems, in which the temperature distribution is highly nonuniform (ranging from $O(10°\text{C})$ to $O(1000°\text{C})$), thermal properties such as μ, c_p, k, and D and especially ρ vary appreciably with temperature. Such a temperature dependence inevitably causes high nonlinearity and strong coupling among the governing equations and therefore increases the computational difficulty. One way to reduce this inconvenience is to introduce a new independent variable that is proportional to the product of the density and the transverse coordinate y.

Example 13-4 Seek a transformation to remove the density from the continuity and streamwise momentum equations governing a compressible boundary-layer flow.

Solution: Let

$$z = \int_0^y \frac{\rho}{\rho_\infty} \, dy \, . \tag{a}$$

Then the transformation from x, y coordinates to x, z coordinates is dictated by

$$\left(\frac{\partial}{\partial x}\right)_y = \left(\frac{\partial}{\partial x}\right)_z + \left(\frac{1}{\rho_\infty}\int_0^y \frac{\partial \rho}{\partial x} \, dy\right)\frac{\partial}{\partial z} \tag{b}$$

and

$$\frac{\partial}{\partial y} = \frac{\rho}{\rho_\infty}\frac{\partial}{\partial z} \,, \tag{c}$$

which is called the Howarth-Dorodnitzyn transformation [19, 20]. With Eqs. (*a*)-(*c*), the continuity and x-momentum equations are transformed into

$$\frac{\partial^2 \psi}{\partial x\, \partial y}\frac{\partial \psi}{\partial y} - \frac{\partial^2 \psi}{\partial y^2}\frac{\partial \psi}{\partial x} = \nu \frac{\partial^3 \psi}{\partial y^3} \qquad (d)$$

or

$$u\frac{\partial u}{\partial x} + w\frac{\partial u}{\partial z} = \nu_\infty \frac{\partial^2 u}{\partial z^2}, \qquad (e)$$

where

$$\psi = \int_0^y \frac{\rho u}{\rho_\infty}\, dy \quad \text{and} \quad w = \frac{u}{\rho_\infty}\int_0^y \frac{\partial \rho}{\partial x}\, dy + \frac{\rho v}{\rho_\infty} \qquad (f)$$

and the approximation $\rho\mu \approx \rho_\infty \mu_\infty$ has been made. It can be seen that the density disappears from Eqs. (*d*) and (*e*). Of course, when the solutions of $u(x, z)$ and $w(x, z)$ are obtained, the computational task remains incomplete; $u(x, y)$ and $v(x, y)$ must be recovered from $u(x, z)$ and $w(x, z)$. This is the drawback of the transformation.

13-3*c* Radiative Participation of Gases

Combustion phenomena generally take place at high temperatures, at which radiative transfer becomes significant. Since gases such as CO_2, H_2O, OH^-, H^+, and other hydrocarbons that are commonly generated during the chemical reactions are radiatively participating, both gaseous radiation and combustion coexist in the flame. Heat transfer involving nongray gaseous radiation alone is a complicated phenomenon, so it is clear that the coupling of radiation and combustion becomes exceedingly intricate. Analyses of such coupling can be much simplified if both the Shvab-Zeldovich formulation and the approximation of a dominantly one-dimensional radiative flux are adopted.

13-3*d* Unsteady and Irregular Boundary

The boundaries of some combustion systems, such as those shown in Fig. 13-1*c*-*f*, are not defined. If, for example, the edge of the plume where $u = 0$ is taken to be the boundary, it is not prescribed a priori and is constantly changing with time. Clearly, the computational task is more difficult with such an unsteady system boundary than with a stationary boundary. Furthermore, the boundary of the plume of large-scale fires is expected from observations to be highly irregular. This irregularity makes the computation even more inconvenient.

We recall that the von Mises transformation was used for the analysis of boundary-layer flow (Section 6-4) partly because of the irregularity of the boundary-layer domain. Extending this idea, it is also possible to introduce a dimensionless scaling variable to transform an unsteady and irregular boundary into a steady and regular one for fire plumes.

Example 13-5 Transform Eqs. (13-10)-(13-12) into a new set of equations where the domain becomes steady and rectangular.

Solution: Let us introduce a new scaling variable η defined as

$$\eta = \frac{r}{R(x,t)}, \tag{a}$$

where $R(x, t)$, shown in Fig. 13-3, is the thickness of the hydrodynamic plume such that $u = 0$ at $r = R(x, t)$. The transformation from the original independent variables x, r, t to x, η, t is

$$\left(\frac{\partial}{\partial x}\right)_{r,t} = \left(\frac{\partial}{\partial x}\right)_{\eta,t} - \eta A \frac{\partial}{\partial \eta}, \tag{b}$$

$$\left(\frac{\partial}{\partial r}\right)_{x,t} = \frac{1}{R(x,t)} \frac{\partial}{\partial \eta}, \tag{c}$$

and

$$\left(\frac{\partial}{\partial t}\right)_{x,r} = \left(\frac{\partial}{\partial t}\right)_{x,\eta} - \eta B \frac{\partial}{\partial \eta}, \tag{d}$$

where $A = (1/R)\,\partial R/\partial x$ and $B = (1/R)\,\partial R/\partial t$. With the aid of the transformation, Eqs. (13-10)-(13-12) in the nonconservative form can be changed, after algebra, to

$$\frac{\partial \rho}{\partial t} - \eta B \frac{\partial \rho}{\partial \eta} + \frac{\partial}{\partial x}(\rho u) - \eta A \frac{\partial}{\partial \eta}(\rho u) + \frac{1}{R\eta}\frac{\partial}{\partial \eta}(\eta \rho v) = 0\,, \tag{e}$$

$$\rho \frac{\partial u}{\partial t} + \Gamma_1 \frac{\partial u}{\partial x} + \Gamma_2 \frac{\partial u}{\partial \eta} = \Gamma_3 \mu \frac{\partial^2 u}{\partial \eta^2} + \mu \frac{\partial^2 u}{\partial x^2} - 2\eta A \mu \frac{\partial^2 u}{\partial x\, \partial \eta} - \frac{\partial P}{\partial x} + \eta A \frac{\partial P}{\partial \eta} - \rho g\,, \tag{f}$$

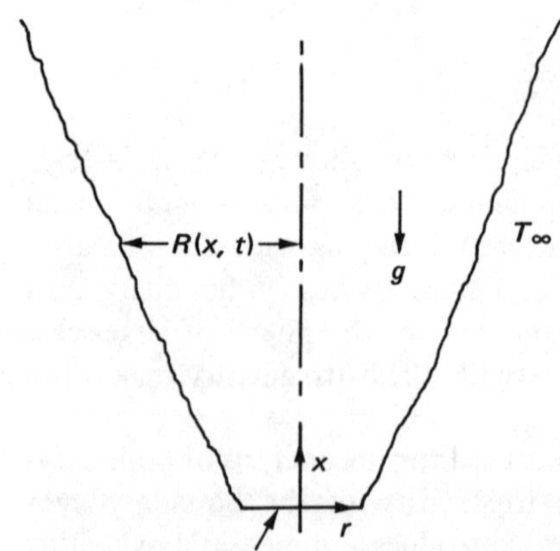

Figure 13-3 System schematic of an unsteady hydrodynamic plume.

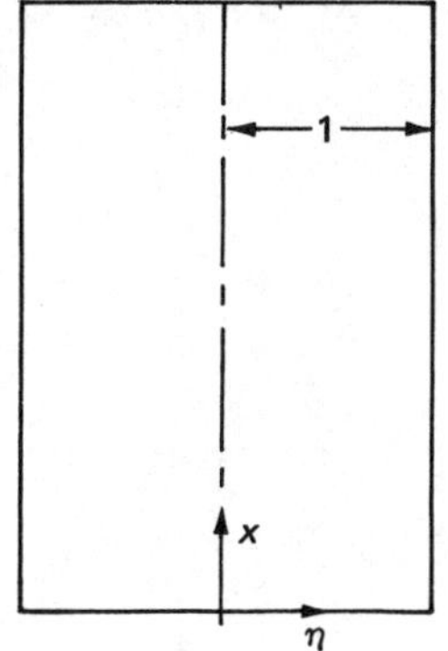

Figure 13-4 The transformed domain of Fig. 13-3.

and

$$\rho\frac{\partial v}{\partial t}+\Gamma_1\frac{\partial v}{\partial x}+\Gamma_2\frac{\partial v}{\partial \eta}=\Gamma_3\mu\frac{\partial^2 v}{\partial \eta^2}+\mu\frac{\partial^2 v}{\partial x^2}-2\eta A\mu\frac{\partial^2 v}{\partial x\,\partial \eta}-\frac{1}{R}\frac{\partial P}{\partial \eta}, \quad (g)$$

where

$$\Gamma_1=\rho u-\frac{\partial \mu}{\partial x}+\eta A\frac{\partial \mu}{\partial \eta},$$

$$\Gamma_2=\frac{\rho v}{R}-\rho u\eta A-\rho\eta B+\eta\frac{\partial}{\partial x}(\mu A)-\left(\eta A^2+\frac{1}{R^2\eta}\right)\frac{\partial}{\partial \eta}(\mu\eta),$$

and

$$\Gamma_3=\eta^2A^2+\frac{1}{R^2}.$$

For steady states, $\partial/\partial t=0$ and $B=0$. The transformed domain is shown in Fig. 13-4. It is not only steady but also of rectangular geometry. Thus a numerical method, such as the finite-difference method, can be readily used. The drawback of this transformation is that more terms appear in the governing equations and $R(x, t)$ must be prescribed a priori.

13-4 BOUNDARY-LAYER FLAMES

A typical boundary-layer flame is shown schematically in Fig. 13-5. The ambient air is driven upward by the buoyancy force generated by combustion. The oxygen and the pyrolyzed fuel come in contact inside the boundary layer, where they react at a rate much faster than the diffusion rate or the flow rate. Consequently, the reaction zone becomes thinner than the hydrodynamic (or approximately thermal) boundary layer, as shown, and a steep temperature gradient exists. The heat generated by combustion is conducted, convected, and radiated away from the reaction zone such that no heat is accumulated and such that a portion of the heat is absorbed by the fuel slab to

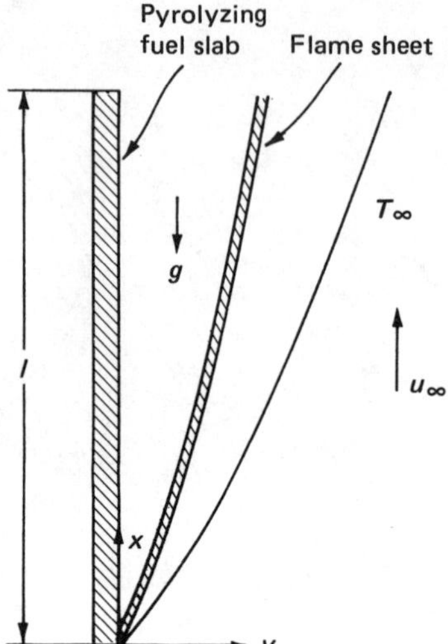

Figure 13-5 System schematic of a two-dimensional, laminar, steady-state, nongray-radiative, boundary-layer flame adjacent to a pyrolyzing fuel slab.

sustain the pyrolysis and combustion in a steady state. For simplification, in addition to the usual boundary-layer assumptions

$$u \gg v \quad \text{and} \quad \frac{\partial^2 u}{\partial y^2} \gg \frac{\partial^2 u}{\partial x^2},$$

it is assumed that the radiative transport is negligible along the streamwise direction, i.e.,

$$\frac{\partial \dot{q}_r''}{\partial y} \gg \frac{\partial \dot{q}_r''}{\partial x}.$$

This idealized problem has been solved numerically [12, 16]. Although many assumptions must be made to obtain a final solution, the nature of combustion phenomena is still more or less preserved. We will present the solution procedure for laminar flames in detail in the following section.

13-4*a* Laminar Flames

In contrast to boundary-layer inert flows, for which the critical values of the Reynolds number (for forced convection) and the Rayleigh number (for free convection) that determine the transition from laminar flows to turbulent flows have been well known for decades, we have not yet identified the parameters and their critical values that determine the transition from laminar flames to turbulent flames. It was experimentally observed, however, that the boundary-layer flame remains relatively laminar when the length of a fuel slab does not exceed $O(10 \text{ cm})$. For a steady, two-dimensional, radiative, boundary-layer flame with unit Lewis number, Eqs. (13-10)–(13-14) reduce to

$$\frac{\partial \rho u}{\partial x} + \frac{\partial \rho v}{\partial y} = 0, \tag{13-19}$$

$$\rho u \frac{\partial u}{\partial x} + \rho v \frac{\partial u}{\partial y} = \frac{\partial}{\partial y}\left(\mu \frac{\partial u}{\partial y}\right) + g(\rho_\infty - \rho)\,, \tag{13-20}$$

$$\rho u \frac{\partial h}{\partial x} + \rho v \frac{\partial h}{\partial y} = \frac{\partial}{\partial y}\left(\frac{k}{c_p}\frac{\partial h}{\partial y}\right) + \dot{q}_c''' + \dot{q}_r'''\,, \tag{13-21}$$

and

$$\rho u \frac{\partial Y_i}{\partial x} + \rho v \frac{\partial Y_i}{\partial y} = \frac{\partial}{\partial y}\left(\rho D \frac{\partial Y_i}{\partial y}\right) + \dot{m}_i'''\,. \tag{13-22}$$

The boundary conditions of the system shown in Fig. 13-5 are

$$\text{At } y = 0, \quad u = 0\,, \quad h = c_p(T_w - T_\infty)\,, \quad Y_f = Y_{fw}\,, \quad Y_N = Y_{Nw}\,,$$

$$v = -\frac{D(\partial Y_f/\partial y)_{y=0}}{1 - Y_{fw}}\,, \tag{13-23a}$$

$$\text{At } y = y_{\text{fl}}, \quad Y_f = Y_{\text{ox}} = 0\,, \tag{13-23b}$$

$$\text{At } y \to \infty, \quad u = u_\infty\,, \quad h = 0\,, \quad Y_{\text{ox}} = Y_{\text{ox}\infty}\,, \quad Y_N = Y_{N\infty} \tag{13-23c}$$

and at $x = 0$, all six dependent variables are prescribed as functions of y.

We note that the boundary condition for the transverse velocity v does not exist in Eq. (13-23*c*). This is because the set of Eqs. (13-19)–(13-22) is only of first order in v. The mass fraction of the combustion product Y_p can be simply obtained as

$$Y_p = 1 - Y_f - Y_{\text{ox}} - Y_N\,. \tag{13-24}$$

Since nitrogen remains relatively inert during combustion, we have

$$\dot{m}_N''' = 0\,.$$

Since $\dot{q}_c'''$ is a complicated function of several field variables, as mentioned in Section 13-3*a*, it is customary to introduce the Shvab-Zeldovich variables to eliminate $\dot{q}_c'''$ if possible. The two major assumptions associated with this formulation are that the Lewis number for all species is equal to unity and that a single-step overall chemical reaction takes place stoichiometrically. For laminar flames that are diffusion-controlled, it can be further proposed that the reaction zone is so thin or sheetlike that the reactants do not coexist. In other words, the mass fraction profiles of the fuel and oxygen do not overlap. Therefore, the computed Shvab-Zeldovich variable solutions can be used to recover the primitive variables Y_f, Y_{ox}, and h. For reaction-controlled flames, these primitive variables cannot be recovered from the Shvab-Zeldovich variables since they do coexist throughout the system.

On transformation of Eqs. (13-21) and (13-22) into Shvab-Zeldovich expressions and further normalization, we derive

$$\rho u \frac{\partial J_1}{\partial x} + \rho v \frac{\partial J_1}{\partial y} = \frac{\partial}{\partial y}\left(\frac{k}{c_p}\frac{\partial J_1}{\partial y}\right) + \frac{1}{Q_p Y_{\text{ox}\infty} - h_w}\frac{\partial \dot{q}_r''}{\partial y} \tag{13-25}$$

and

$$\rho u \frac{\partial J_2}{\partial x} + \rho v \frac{\partial J_2}{\partial y} = \frac{\partial}{\partial y}\left(\rho D \frac{\partial J_2}{\partial y}\right) \tag{13-26}$$

subject to the transformed boundary conditions

$$J_1 = J_2 = 1 \,, \quad \text{at } y = 0$$

and

$$J_1 = J_2 = 0 \,, \quad \text{at } y \to \infty \,,$$

where J_1 and J_2 are defined in the Symbols. For nonradiative flames, i.e., $\dot{q}_r'' = 0$, it is clear that J_1 and J_2 are identical.

Example 13-6 Derive Eq. (13-25).

Solution: Dividing Eq. (13-21) by $-q_{ox}\nu_{ox}M_{ox}$ and Eq. (13-22) by $-\nu_{ox}M_{ox}$ (let i denote oxygen) and then adding the results, we obtain, with $k/c_p = \rho D$,

$$\rho u \frac{\partial \gamma}{\partial x} + \rho v \frac{\partial \gamma}{\partial y} = \frac{\partial}{\partial y}\left(\frac{k}{c_p}\frac{\partial \gamma}{\partial y}\right) - \frac{\dot{q}_c'''}{q_{ox}\nu_{ox}M_{ox}} - \frac{\dot{q}_r'''}{q_{ox}\nu_{ox}M_{ox}} - \frac{\dot{m}_{ox}'''}{\nu_{ox}M_{ox}} \,. \qquad (a)$$

But by definition we have

$$\dot{q}_c''' = -\dot{m}_{ox}''' q_{ox} \,. \qquad (b)$$

Equation (a) reduces to

$$\rho u \frac{\partial \gamma}{\partial x} + \rho v \frac{\partial \gamma}{\partial y} = \frac{\partial}{\partial y}\left(\frac{k}{c_p}\frac{\partial \gamma}{\partial y}\right) - \frac{\dot{q}_r'''}{q_{ox}\nu_{ox}M_{ox}} \,. \qquad (c)$$

Dividing Eq. (c) by $\gamma_w - \gamma_\infty$ yields

$$\rho u \frac{\partial J_1}{\partial x} + \rho v \frac{\partial J_1}{\partial y} = \frac{\partial}{\partial y}\left(\frac{k}{c_p}\frac{\partial J_1}{\partial y}\right) - \frac{\dot{q}_r'''}{q_{ox}\nu_{ox}M_{ox}(\gamma_w - \gamma_\infty)} \,. \qquad (d)$$

But by definition

$$\gamma_w = -\frac{h_w}{q_{ox}\nu_{ox}M_{ox}} \qquad (e)$$

and

$$\gamma_\infty = -\frac{Y_{ox\infty}}{\nu_{ox}M_{ox}} \,. \qquad (f)$$

Substituting Eqs. (e) and (f) into Eq. (d) and setting $\dot{q}_r''' = -\partial \dot{q}''/\partial y$ therefore leads to Eq. (13-25).

The last boundary condition in Eq. (13-23a) must be expressed as well in terms of the Shvab-Zeldovich variable as

$$v(0) = \frac{Y_{fw} + Y_{ox\infty}\nu_f M_f/\nu_{ox}M_{ox}}{1 - Y_{fw}}\left(-D\frac{\partial J_2}{\partial y}\right)_{y=0} \,, \qquad (13\text{-}27)$$

while Eqs. (13-19) and (13-20) remain unchanged except that the buoyancy term now takes the new form

$$g\left(\frac{\rho_\infty}{\rho}-1\right)=g\left(\frac{1}{\theta_r}-1\right)[D_3-J_1(D_3-1)]\,, \quad y\leqslant y_{\mathrm{fl}}$$

$$=g\left(\frac{1}{\theta_r}-1\right)\left[D_3(1+r)\frac{J_2}{r}-(D_3-1)J_2\right]\,, \quad y>y_{\mathrm{fl}}\,, \quad (13\text{-}28a,b)$$

where

$$\theta_r=\frac{T_\infty}{T_w} \quad \text{and} \quad D_3=\frac{q_{\mathrm{ox}}Y_{\mathrm{ox}\infty}}{h_w}\,.$$

The laminar, boundary-layer, diffusion flame with negligible radiation has been a classical heat transfer problem. Investigation of the system without the buoyancy force was initiated by Emmons [21], who introduced the similarity variable and used the Shvab-Zeldovich formulation. The governing equation of momentum (or Shvab-Zeldovich variables because $\nu=\alpha=D$ was assumed) transport reduced to

$$f'''+ff''=0 \qquad (13\text{-}29a)$$

subject to the two-point boundary conditions

$$f'(0)=0\,, \quad f'(\eta_\infty)=2\,, \quad f(0)=Bf''(0)\,, \qquad (13\text{-}29b)$$

where the similarity variable η and the similarity function $f(\eta)$ were defined as

$$\eta=\frac{\mathrm{Re}_x^{1/2}}{2x}\int_0^y \frac{\rho}{\rho_\infty}\,dy$$

and

$$f'(\eta)=\frac{2u}{u_\infty}\,.$$

The similarity function is related to the stream function by

$$\psi=f(\eta)(\nu_\infty x u_\infty)^{1/2}\,.$$

The mass transfer number B is defined as

$$B=\frac{Y_{fw}+Y_{\mathrm{ox}\infty}\nu_f M_f/\nu_{\mathrm{ox}}M_{\mathrm{ox}}}{1-Y_{fw}}\,.$$

If the flame is quenched and the fuel ceases to pyrolyze, then $f(0)$ becomes zero and Eqs. (13-29a) and (13-29b) reduce to the Blasius equation.

Example 13-7 Derive the third boundary condition of Eq. (13-29b).

Solution: Based on the definition of the stream function, we derive

$$v=-\frac{\partial\psi}{\partial x}=\frac{1}{2}\left(\frac{\nu u_\infty}{x}\right)^{1/2}[\eta f'(\eta)-f(\eta)]\,. \qquad (a)$$

Utilizing Eq. (*a*) and Eq. (13-27) leads to

$$\frac{1}{2}\left(\frac{\nu u_\infty}{x}\right)^{1/2} f(0) = \frac{Y_{fw} + Y_{\text{ox}\infty}\nu_f M_f/\nu_{\text{ox}} M_{\text{ox}}}{1 - Y_{fw}} \left(D\,\frac{\partial J_2}{\partial y}\right)_{y=0},$$

which can be further simplified to

$$f(0) = \left(\frac{D_w \rho_w}{\nu_\infty \rho_\infty}\right)\left(\frac{Y_{fw} + Y_{\text{ox}\infty}\nu_f M_f/\nu_{\text{ox}} M_{\text{ox}}}{1 - Y_{fw}}\right) J_2'(0)\,. \tag{b}$$

If unit Schmidt number and constant $\rho\nu$ are assumed, Eq. (*b*) can be readily changed to the third boundary condition of Eq. (13-29*b*).

The purely free-convection, laminar, boundary-layer, diffusion flame with negligible radiation was also investigated, using the similarity and Shvab-Zeldovich assumptions [22-24]. The transformed nonlinear, fifth-order, ordinary differential equations constitute a challenging two-point boundary-value problem. The numerical solution was found [14] to be much more sensitive to the values of missing initial conditions $f''(0)$ and $J'(0)$ than in the inert free-convection case.

To specify the local radiative flux $\dot{q}_r''$, we first assume that the absorption coefficients of all the gases are constant and of the same value. Thus, according to [25, 26] (also see Chapter 10), we write

$$\dot{q}_r''(x, y) = 2\epsilon_w e_{bw} E_3(\kappa) + 4(1 - \epsilon_w) E_3(\kappa) \int_0^{\kappa_\delta} e_b(t) E_2(t)\, dt$$
$$+ 2\int_0^{\kappa} e_b(t) E_2(\kappa - t)\, dt - 2\int_{\kappa}^{\kappa_\delta} e_b(t) E_2(t - \kappa)\, dt\,, \tag{13-30}$$

where $\kappa = ay$ and $\kappa_\delta = a\delta$.

Finally, we need to find a relation between J_1 and T to close the set of Eqs. (13-19), (13-20), (13-25), (13-26), (13-28), and (13-30). Through straightforward algebra and use of definitions, the following expressions can be derived:

$$\frac{h}{h_w} = D_3 - J_1(D_3 - 1) \tag{13-31a}$$
$$\frac{Y_f}{Y_{fw}} = (1 + r)J_2 - r \tag{13-32a}$$
$$\Bigg\}\quad y \leqslant y_{\text{fl}}$$

and

$$\frac{h}{h_w} = D_3(1 + r)\frac{J_2}{r} - (D_3 - 1)J_1 \tag{13-31b}$$
$$\frac{Y_{\text{ox}}}{Y_{\text{ox}\infty}} = 1 - (1 + r)\frac{J_2}{r} \tag{13-32b}$$
$$\Bigg\}\quad y > y_{\text{fl}}\,,$$

where y_{fl} is determined by the assumption that Y_f and Y_{ox} vanish simultaneously at the flame sheet, i.e.,

$$J_2(y_{fl}) = \frac{r}{1+r}. \tag{13-33}$$

Since the mass burning rate is equal to the diffusion rate of the fuel vapor into the reaction zone, we obtain

$$\dot{m}_f'' = -\left(\rho D \frac{\partial Y_f}{\partial y}\right)_{y=y_{fl}} = -\rho D Y_{fw}(1+r)\left(\frac{\partial J_2}{\partial y}\right)_{y=y_{fl}}. \tag{13-34}$$

One iterative procedure for solving Eqs. (13-19), (13-20), (13-25), (13-26), (13-28), and (13-30) is to first solve for $u(x_k, y)$, $v(x_k, y)$, $J_1(x_k, y)$, and $J_2(x_k, y)$ at the location $x = x_k$, using the upstream solution $T(x_k - \Delta x, y)$ to compute $\dot{q}_r''(x_k, y)$. Substituting the solutions $J_1(x_k, y)$ and $J_2(x_k, y)$ into Eqs. (13-31a) and (13-31b), we recover $T(x_k, y)$ and then update $\dot{q}_r''(x_k, y)$. Nitrogen is an inert gas and therefore its normalized mass fraction $(Y_N - Y_{N\infty})/(Y_{Nw} - Y_{N\infty})$ also satisfies the J_2 transport Eq. (13-26) if Y_{Nw} is assumed constant. Consequently,

$$Y_N(x,y) = J_2(x,y)(Y_{Nw} - Y_{N\infty}) + Y_{N\infty}, \tag{13-35}$$

where

$$Y_{N\infty} = 1 - Y_{ox\infty}.$$

Once $Y_N(x, y)$ is computed from Eq. (13-35), the assumption that the fuel and the oxygen do not coexist yields

$$Y_p = \begin{cases} 1 - Y_f - Y_N, & y \leqslant y_{fl} \\ 1 - Y_{ox} - Y_N, & y > y_{fl}. \end{cases} \tag{13-36}$$

Example 13-8 The data on burning of a typical Plexiglas slab are given in Table 13-1. Assume now that, by a certain finite-difference scheme, the distributions of u, v, J_1, and J_2 are obtained at the streamwise location $x = x^*$. These computed values at four transverse locations are listed in Table 13-2. Also let $y_{fl} = 0.04$ m and $\delta = 0.08$ m. Find the nodal values for T, Y_f, Y_{ox}, Y_N, Y_p, $\dot{q}_r''$, and $\dot{m}_f''$.

Table 13-1 Data on burning of Plexiglas slab used in Example 13-7

Parameter	Value
u_∞, m/s	1
ν_∞, m²/s	1.5×10^{-5}
l, m	0.2
T_w, K	720
T_∞, K	300
Y_{ox}	0.230
Y_{fw}	0.593
$\nu_f M_f/\nu_{ox} M_{ox}$	0.516
q_{ox}, cal/kg	3.16×10^6
k_∞, cal/m·s·K	7.54×10^{-3}
c_p, cal/kg·K	310
ϵ_w	0.5
a, m⁻¹	5

Table 13-2 Nodal values of various dependent variables computed in Example 13-8

y	0	$0.5y_{fl}$	y_{fl}	δ	Equation used
u, m/s	0	0.60	1.3	1	Given
v, m/s	0.0049	0.01	0	0.009	Given
J_1	1	0.72	0.3655	0	Given
J_2	1	0.54	0.1667	0	Given
T, K	720	1260	1944	300	(13-31*a*) and (13-31*b*)
$\dot{q}''_r/\sigma T_w^4$	−4.8061	−6.8759	0.1813	13.0678	(13-30)
Y_f	0.288	0.129	0	0	(13-32*a*)
$\dot{m}''_f/\rho_\infty D_\infty$	0	0	41.8	0	(13-34)
Y_{ox}	0	0	0	0.23	(13-32*b*)
Y_N	0.518	0.6339	0.7280	0.77	(13-35)
Y_p	0.194	0.2371	0.272	0	(13-36)

Solution: We will only present the procedure for obtaining the rows for $\dot{q}''_r/\sigma T_w^4$ and $\dot{m}''_f/\rho_\infty D_\infty$. The remaining information can be obtained by straightforward substitution of the given data into the appropriate equations with $D_3 = 5.596$ and $r = 0.2$. From Eq. (13-30) we can obtain, at $y = 0$,

$$\frac{\dot{q}''(x,0)}{e_{bw}} = E_3(0) - \int_0^{\kappa\delta} \frac{e_b(t)}{e_{bw}} E_2(t)\,dt\,. \tag{a}$$

To calculate the integral in Eq. (*a*) and the radiative fluxes at $y = 0.5y_{fl}, y_{fl}, \delta$, we first evaluate, by the trapezoidal rule,

$$\int_0^{\kappa\delta} \frac{e_b(t)}{e_{bw}} E_2(t)\,dt \approx \frac{0.1}{2}\left[E_2(0) + \frac{e_b(0.1)}{e_{bw}} E_2(0.1)\right]$$
$$+ \frac{0.1}{2}\left[\frac{e_b(0.1)}{e_{bw}} E_2(0.1) + \frac{e_b(0.2)}{e_{bw}} E_2(0.2)\right]$$
$$+ \frac{0.2}{2}\left[\frac{e_b(0.2)}{e_{bw}} E_2(0.2) + \frac{e_b(0.4)}{e_{bw}} E_2(0.4)\right]$$
$$= 0.05(7.776) + 0.05(37.292) + 0.1(30.527)$$
$$= 5.3061\,. \tag{b}$$

Therefore,

$$\frac{\dot{q}''(x,0)}{e_{bw}} = -4.8061\,.$$

The negative value indicates that the net radiative flux impinges on the wall.

Regarding the calculation of $\dot{m}''_f$, we will assume

$$\dot{m}''_f = -\left(\rho D \frac{\partial Y_f}{\partial y}\right)_{y=y_{fl}} \approx \frac{-\rho_{fl} D_{fl}}{0.5 y_{fl}}(0 - 0.129)\,. \tag{c}$$

But for ideal gases it is reasonable to assume

$$\frac{\rho_{fl} D_{fl}}{\rho_\infty D_\infty} \approx \frac{T_{fl}}{T_\infty} . \tag{d}$$

Therefore,

$$\frac{\dot{m}_f''}{\rho_\infty D_\infty} = 41.8 . \tag{e}$$

13-4*b* Turbulent Flames

It has been experimentally observed [27-29] that the reaction zone in a turbulent boundary-layer flame is generally of finite thickness (sometimes called a flame brush to distinguish it from the flame sheet of laminar flames), as shown in Fig. 13-6. The presence of this thick reaction zone suggests that the fuel and the reactant tend to coexist due to strong turbulent mixing. In other words, the mass fraction profiles of the fuel and the oxygen overlap. As a result of this overlapping, physical variables such as $Y_f(x, y)$, $Y_{ox}(x, y)$, and $h(x, y)$ cannot be as easily recovered from the Shvab-Zeldovich variables as in the laminar flame case. The difference between the structures of a turbulent boundary-layer flame and the laminar counterpart is shown in Fig. 13-6.

One approach to this recovery is the so-called *K-ϵ-g* model [30-37]. In this model, the time mean squared fluctuation of the Shvab-Zeldovich variable *g*, defined as

$$g = \overline{J'^2} , \tag{13-37}$$

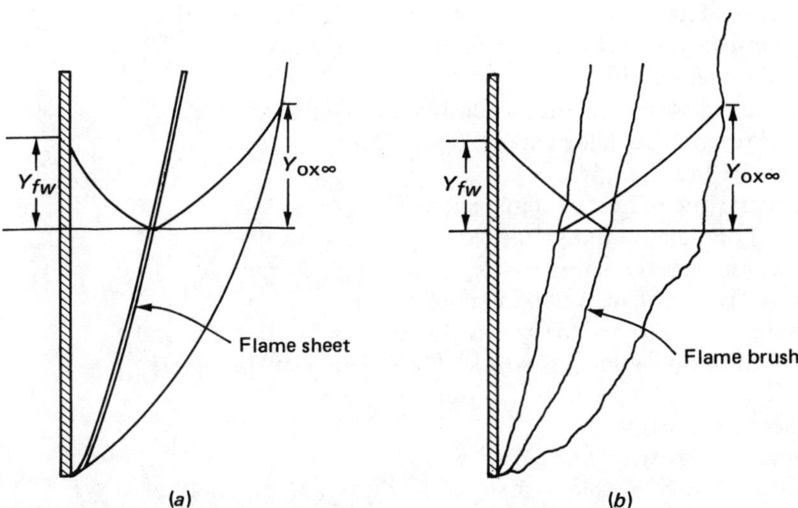

Figure 13-6 Comparison of the flame structures between the turbulent boundary-layer flame (*b*) and the laminar counterpart (*a*).

is sought along with K and ϵ (see Section 12-4). In Eq. (13-37), J' is the fluctuation of J defined as, for example,

$$J = \frac{\gamma - \gamma_\infty}{\gamma_0 - \gamma_\infty}.$$

In turbulent jet flames, the subscripts ∞ and 0 denote the ambient and the nozzle exit, respectively. With the nodal solutions of J and g, the nodal probability density functions can be determined. Subsequently, the overlapping physical variable distributions such as h, Y_f, and Y_{ox} can be recovered. This paragraph is merely meant to make readers aware of the existing models for analyzing turbulent boundary-layer diffusion flames. For a thorough understanding of the analysis, interested readers should consult the references cited.

SYMBOLS

a	absorption coefficient, m^{-1}
A	frequency (or preexponential) factor, m^3/molecule·s
Da	Damkohler number ($= t_{residence}/t_{reaction}$)
D_3	dimensionless parameter ($= q_{ox} Y_{ox\,\infty}/h_w$)
e	internal energy, J/kg
e_b	blackbody emissive power, W/m^2
E	activation energy, J/gmol
$E_n(t)$	exponential function [$= \int_0^1 \mu^{n-2} \exp(-t/\mu)\, d\mu$]
g	time average of squared fluctuation of Shvab-Zeldovich variables ($= \overline{J'^2}$)
J'	fluctuation of J
J_1	normalized Shvab-Zeldovich variable [$= (\gamma - \gamma_\infty)/(\gamma_w - \gamma_\infty)$]
J_2	normalized Shvab-Zeldovich variable [$= (\beta - \beta_\infty)/(\beta_w - \beta_\infty)$]
k_2	rate constant for second-order reaction, m^3/molecule·s
l	length of fuel slab, m
$\dot{m}'''$	volumetric mass generation (or consumption) rate, kg/m^3·s
q_{ox}	heat of reaction per kilogram of oxygen, J/kg
$\dot{q}''$	conductive heat flux, W/m^2
$\dot{q}_c'''$	volumetric rate of heat of combustion, W/m^3
$\dot{q}_r'''$	gradient of radiative flux, W/m^3
r	mass consumption number ($= Y_{fw}\nu_{ox}M_{ox}/Y_{ox\infty}\nu_f M_f$)
$\tilde{R}$	universal gas constant, 1.986 cal/gmol·K
$R(x, t)$	prescribed boundary of the system domain as a function of x and t
w	transformed flow velocity {$= [u \int_0^y (\partial\rho/\partial x)\, dy + \rho v]/\rho_\infty$}, m/s
z	transformed y coordinate [$= \int_0^y (\rho/\rho_\infty)\, dy$], m
Y	species mass fraction
α	thermal diffusivity ($= k/\rho c_p$), m^2/s
β	Shvab-Zeldovich variable ($= Y_f/\nu_f M_f - Y_{ox}/\nu_{ox}M_{ox}$), gmol/kg
γ	Shvab-Zeldovich variable ($= -h/q_{ox}\nu_{ox}M_{ox} - Y_{ox}/\nu_{ox}M_{ox}$), gmol/kg
ϵ_w	emissivity at the wall

η	dimensionless variable $[= r/R(x, t)]$
θ_r	temperature ratio $(= T_\infty/T_w)$
κ	optical path $(= ay)$
μ	viscosity, kg/m·s, or dummy variable in the exponential function
ν	stoichiometric coefficient

Subscripts

f	fuel
fl	flame stand-off distance
N	nitrogen
ox	oxygen
p	product
w	wall

REFERENCES

1. B. Lewis and G. von Elbe, *Combustion, Flames and Explosions of Gases*, Academic, New York, 1961.
2. F. A. Williams, *Combustion Theory*, Addison-Wesley, Reading, Mass., 1965.
3. J. N. Bradley, *Flame and Combustion Phenomena*, Chapman & Hall, London, 1972.
4. A. M. Kanury, *Introduction to Combustion Phenomena*, Gordon & Breach, New York, 1975.
5. D. B. Spalding, *Combustion and Mass Transfer*, Pergamon, London, 1979.
6. K. Schofield, An Evaluation of Kinetic Rate Data for Reaction of Neutrals of Atmospheric Interest, *J. Planet, Space Sci.*, vol. 15, pp. 643–670, 1967.
7. C. K. Law, On the Stagnation-Point Ignition of a Premixed Combustible, *Int. J. Heat Mass Transfer*, vol. 21, pp. 1363-1368, 1978.
8. G. R. Otey and H. A. Dwyer, Numerical Study of the Interaction of Fast Chemistry and Diffusion, *AIAA J.*, vol. 17, pp. 606–613, 1979.
9. E. Saatdjian and J. P. Caltagirone, Natural Convection in a Porous Layer under the Influence of an Exothermic Decomposition Reaction, *ASME J. Heat Transfer*, vol. 102, pp. 654-658, 1980.
10. K. H. Winters, J. Rae, C. P. Jackson, and K. A. Cliffe, The Finite Element Method for Laminar Flow with Chemical Reaction, *Int. J. Numer. Methods Eng.*, vol. 17, pp. 239-253, 1981.
11. S. P. Burke and T. E. W. Schumann, Diffusion Flames, *Ind. Eng. Chem.*, vol. 20, pp. 998-1004, 1928.
12. D. E. Negrelli, J. R. Lloyd, and J. L. Novotny, A Theoretical and Experimental Study of Radiation-Convection Interaction in a Diffusion Flame, *ASME J. Heat Transfer*, vol. 99, pp. 212-220, 1977.
13. T. Ahmad and G. M. Faeth, An Investigation of the Laminar Overfire Region along Upright Surfaces, *ASME J. Heat Transfer*, vol. 100, pp. 112-119, 1978.
14. T. M. Shih and P. J. Pagni, Laminar Mixed-Mode, Forced and Free, Diffusion Flames, *ASME J. Heat Transfer*, vol. 100, pp. 253-259, 1978.
15. C. M. Kinoshita and P. J. Pagni, Laminar Wake Flame Heights, *ASME J. Heat Transfer*, vol. 102, pp. 104-109, 1980.
16. C. N. Liu and T. M. Shih, Laminar, Mixed-Convection, Boundary-Layer, Nongray-radiative, Diffusion Flames, *ASME J. Heat Transfer*, vol. 102, pp. 724-730, 1980.
17. K. V. Liu, J. R. Lloyd, and K. T. Yang, An Investigation of a Laminar Diffusion Flame Adjacent to a Vertical Flat Plate Burner, *Int. J. Heat Mass Transfer*, vol. 24, pp. 1959-1970, 1981.

18. W. B. Bush and S. F. Fink, On Diffusion Flames in Turbulent Shear Flows: Modeling Reactant Consumption in a Circular Fuel Jet, *AIAA J.*, vol. 19, pp. 372–379, 1981.
19. L. Howarth, Concerning the Effect of Compressibility on Laminar Boundary Layers and Their Separation, *Proc. R. Soc. London Ser. A*, vol. 194, pp. 16–42, 1948.
20. A. Dorodnitzyn, Laminar Boundary Layer in Compressible Fluid, *C.R. (Doklady) Akad. Nauk, USSR*, vol. 34, pp. 213–219, 1942.
21. H. W. Emmons, The Film Combustion of Liquid Fuel, *Z. Angew. Math. Mech.*, vol. 36, pp. 60–71, 1956.
22. F. J. Kosdon, F. A. Williams, and C. Burman, Combustion of Vertical Cellulosic Cylinders in Air, in *Twelfth Symposium (International) on Combustion*, pp. 253–264, Combustion Institute, Pittsburgh, 1969.
23. J. S. Kim, J. deRis, and F. W. Krosser, Laminar Free-Convection Burning of Fuel Surface, in *Thirteenth Symposium (International) on Combustion*, pp. 949–961, Combustion Institute, Pittsburgh, 1971.
24. J. S. Kim, J. De Ris, and F. W. Kroesser, Laminar Burning between Parallel Fuel Surfaces, *Int. J. Heat Mass Transfer*, vol. 17, pp. 439–451, 1974.
25. R. Siegel and J. R. Howell, *Thermal Radiation Heat Transfer*, p. 680, McGraw-Hill, New York, 1972.
26. E. M. Sparrow and R. D. Gess, *Radiation Heat Transfer*, pp. 212–222, Hemisphere, Washington, D.C., and McGraw-Hill, New York, 1978.
27. A. Vranos, J. E. Faucher, and W. E. Curtis, Turbulent Mass Transport and Rates of Reaction in a Confined Hydrogen-Air Diffusion Flame, in *Twelfth Symposium (International) on Combustion*, pp. 1051–1057, Combustion Institute, Pittsburgh, 1969.
28. J. H. Kent and R. W. Bilger, Turbulent Diffusion Flames, in *Fourteenth Symposium (International) on Combustion*, pp. 615–624, Combustion Institute, 1973.
29. R. W. Bilger and J. H. Kent, Concentration Fluctuations in Turbulent Jet Diffusion Flames, *Combust. Sci. Technol.*, vol. 9, pp. 25–29, 1974.
30. D. B. Spalding, Concentration Fluctuations in a Round Turbulent Free Jet, *Chem. Eng. Sci.*, vol. 26, pp. 95–107, 1971.
31. S. E. Elghobashi and W. M. Pun, A Theoretical and Experimental Study of Turbulent Diffusion Flames in Cylindrical Furnaces, in *Fifteenth Symposium (International) on Combustion*, pp. 1353–1364, Combustion Institute, Pittsburgh, 1974.
32. E. E. Khalil, D. B. Spalding, and J. H. Whitelaw, The Calculation of Local Flow Properties in Two-dimensional Furnaces, *Int. J. Heat Mass Transfer*, vol. 18, pp. 775–791, 1975.
33. F. C. Lockwood and A. S. Naguib, The Prediction of the Fluctuations in the Properties of Free, Round-Jet, Turbulent, Diffusion Flames, *Combust. Flame*, vol. 24, p. 109, 1975.
34. T. M. Shih and P. J. Pagni, Analytic Incorporation of Probability Density Functions in Turbulent Flames, *Int. J. Heat Mass Transfer*, vol. 21, pp. 818–821, 1978.
35. C. J. Chen and C. H. Chen, On Prediction and Unified Correlation for Decay of Vertical Buoyancy Jets, *ASME J. Heat Transfer*, vol. 101, pp. 532–537, 1979.
36. C. J. Chen and C. P. Nikitopoulos, On the Near Field Characteristics of Axisymmetric Turbulent Buoyant Jets in a Uniform Environment, *Int. J. Heat Mass Transfer*, vol. 22, pp. 245–255, 1979.
37. S. E. Elghobashi and A. T. Wassel, The Effect of Turbulent Heat Transfer on the Propagation of an Optical Beam across Supersonic Boundary/Shear Layers, *Int. J. Heat Mass Transfer*, vol. 23, pp. 1229–1241, 1981.

PROBLEMS

13-1 Repeat Example 13-3 if the species transport is governed by

$$\rho v \frac{dY_i}{dy} = \rho D \frac{d^2 Y_i}{dy^2} + \dot{m}_i''' ,$$

where v is a constant. When v is reduced to zero, are the results simplified to those given in Example 13-3?

13-2 Consider again Example 13-3. If the boundary conditions for the temperature are prescribed as

$$T(0) = T_w \quad \text{and} \quad T(L) = T_\infty ,$$

find the temperature profile. Estimate the energy release rate due to combustion.

13-3 (*a*) Derive Eq. (13-26).
(*b*) If $r = 1$ and $J_2 = 0.5$, what are the values of Y_f and Y_{ox}?

13-4 If $J_1 = J_2 = J$ for the nonradiative flame, transform Eqs. (13-19), (13-20), and (13-26) into a set of two similarity equations of which $f(\eta)$ and $J(\eta)$ are the two similarity functions.

13-5 Derive Eq. (13-30).

PART FOUR

NUMERICAL ANALYSES

CHAPTER

FOURTEEN

SPACES AND ERROR BOUNDS

In Chapter 4, we discussed the accuracy of various schemes and mentioned that the accuracy of finite-element schemes may be examined by using error bound analysis. Since this requires some knowledge of functional analysis, it was omitted there. In this chapter we will return to this subject, which we hope will prove beneficial to readers who have not had the opportunity to study functional analysis. Numerous examples will be presented to enhance the readers' understanding of the definitions.

In Section 14-1, definitions of the terminology are introduced. Each definition is accompanied by a few examples.

In Section 14-2, the definitions are used to expedite the derivation of an upper error bound for the approximate finite-element solution of a linear positive-definite differential equation. This derivation is the first step of the procedure leading to the error bound expression in terms of the mesh size h.

The second step of the procedure leading to the error estimate is described in Section 14-3. Error bounds of the one-dimensional quadratic interpolation function and the two-dimensional linear interpolation function, considered as two special cases of the finite-element approximate functions, are derived. The result is expressed in terms of h. The combination of Sections 14-1 and 14-2 constitutes an elementary, but complete, derivation of the error bound (or the accuracy or convergence rate) of a certain finite-element scheme.

14-1 DEFINITIONS AND EXAMPLES

For convenience, we will consider all the definitions and examples to be one-dimensional with x as the independent variable, $x \in [a, b]$. This restriction can generally be removed to include multidimensional cases with slight modifications.

14-1*a* Space

A space is a set of elements that are generally functions or vectors. We will restrict our interest to *bounded* elements.

Example 14-1 The set

$$\{x^2, 2-x, \sin x, e^x, \ldots; \quad x \in [1, 10]\}$$

is a space. Functions such as $\log(x-2)$ and $1/(x-3)$ do not belong to this space.

14-1*b* Linear Space

Let $\phi_1, \phi_2, \ldots$ be bounded functions of x and $a_1, a_2, \ldots$ be pure numbers. A space $X = \{\phi_1, \phi_2, \ldots\}$ is said to be linear if its elements satisfy:

$$\text{(i)}\ \ \phi_1 + \phi_2 \in X; \qquad \text{(ii)}\ \ \phi_1 + \phi_2 = \phi_2 + \phi_1 ;$$

$$\text{(iii)}\ \ (\phi_1 + \phi_2) + \phi_3 = \phi_1 + (\phi_2 + \phi_3); \qquad \text{(iv)}\ \ a_1\phi_1 \in X;$$

$$\text{(v)}\ \ a_1(a_2\phi_1) = (a_1a_2)\phi_1 ; \qquad \text{(vi)}\ \ a_1(\phi_1 + \phi_2) = a_1\phi_1 + a_1\phi_2 ;$$

$$\text{(vii)}\ \ (a_1 + a_2)\phi_1 = a_1\phi_1 + a_2\phi_1 .$$

Statements (i)-(iii) are related to additive laws and statements (iv)-(vii) to multiplicative laws.

Example 14-2 The set of all real three-dimensional vectors (called Euclidean space)

$$\bar{x}_1 = 3i + \tfrac{1}{4}j + 5k , \qquad \bar{x}_2 = 2i + 0.2j + \sqrt{2}k , \ldots$$

where i, j, and k are unit vectors, and the space in Example 14-1 are both linear spaces.

14-1*c* $C^0, C^1, \ldots, C^n$ Spaces

The space that contains kth-order continuous derivatives on $[a, b]$, $k \leqslant n$, is denoted by $C^n[a, b]$.

Example 14-3 The functions of Fig. 14-1, *a* and *b*, belong to C^0 and C^1, respectively.

14-1*d* Normed Space

A norm is a measure of the size of a function $\|\phi\|$ or of the "distance" between two functions $\|\phi_1 - \phi_2\|$. It satisfies

(i) $\|\phi_1\| \geqslant 0$; the equality holds if and only if $\phi_1 = 0$.
(ii) $\|\phi_1 + \phi_2\| \leqslant \|\phi_1\| + \|\phi_2\|$; this is called the triangle inequality.
(iii) $|a_1| \, \|\phi_1\| = \|a_1\phi_1\|$.

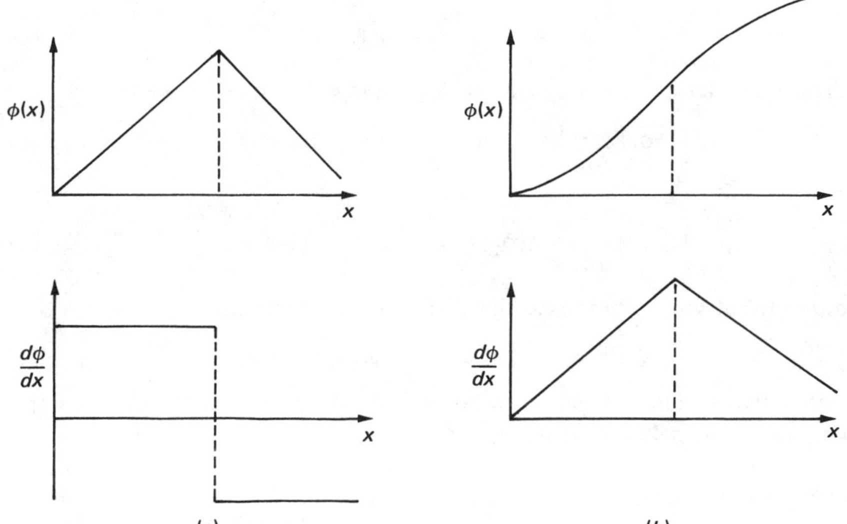

Figure 14-1 Functions belonging to C^0 (*a*) and C^1 (*b*).

A space whose elements satisfy the axioms above is called a normed space. Note that ϕ is a function and $\|\phi\|$ is a real number.

Example 14-4 Is the two-dimensional Euclidean space a normed space?

Solution: For convenience, we take two arbitrary elements (vectors)

$$\phi_1 = i + 2j \qquad \text{and} \qquad \phi_2 = 3i - j\,. \tag{a}$$

The lengths, or norms, of ϕ_1 and ϕ_2 are

$$\|\phi_1\| = (1^2 + 2^2)^{1/2} = \sqrt{5} \tag{b}$$

and

$$\|\phi_2\| = (3^2 + 1^2)^{1/2} = \sqrt{10}\,, \tag{c}$$

while the norm of the sum $\phi_1 + \phi_2$ is

$$\|\phi_1 + \phi_2\| = \|4i + j\| = \sqrt{17}\,, \tag{d}$$

which is smaller than $\sqrt{5} + \sqrt{10}$. Definition (ii) is thus satisfied. Definitions (i) and (iii) are obvious. Therefore, the Euclidean space is a normed space.

Example 14-5 Is the maximum of $|\phi(x)|$, where $\phi \in C^0\,[0, 1]$, a norm of $\phi(x)$?

Solution: Let ϕ_1 and ϕ_2 be two arbitrary elements in $C^0\,[0, 1]$. Then

$$\begin{aligned}\|\phi_1 + \phi_2\| &= \max_{0\leqslant x\leqslant 1} |\phi_1 + \phi_2| \leqslant \max_{0\leqslant x\leqslant 1} \{|\phi_1| + |\phi_2|\} \\ &\leqslant \max_{0\leqslant x\leqslant 1} |\phi_1| + \max_{0\leqslant x\leqslant 1} |\phi_2|\,.\end{aligned} \tag{a}$$

Therefore,

$$\|\phi_1 + \phi_2\| \leqslant \|\phi_1\| + \|\phi_2\| \, . \tag{b}$$

For illustration, let us arbitrarily choose $\phi_1(x)$ and $\phi_2(x)$ to be

$$\phi_1(x) = x^2 \qquad \text{and} \qquad \phi_2(x) = \sin \pi x \, . \tag{c}$$

Then,

$$\|\phi_1\| = \max_{0 \leqslant x \leqslant 1} |\phi_1| = 1 \qquad \text{and} \qquad \|\phi_2\| = 1 \, . \tag{d}$$

Condition (ii) of the definition for the norm is satisfied because

$$\|\phi_1 + \phi_2\| = \tfrac{5}{4} < 1 + 1 = \|\phi_1\| + \|\phi_2\| \, . \tag{e}$$

The other two conditions can be easily proved. Therefore, max $|\phi(x)|$ is a norm of $\phi(x)$, sometimes called Tchebycheff norm.

Example 14-6 Define L_p norm (L stands for Lebesque) as

$$\|\phi\|_p = \left(\int_0^1 |\phi(x)|^p \, dx \right)^{1/p} .^\dagger \tag{a}$$

Prove that $\|\phi\|_\infty$ is a Tchebycheff norm.

Solution: We may use the trapezoidal rule to approximate the given integral as

$$\int_0^1 |\phi|^p \, dx \approx h(|\phi_1|^p + 2|\phi_2|^p + \cdots + 2|\phi_{N-1}|^p + |\phi_N|^p) \, , \tag{b}$$

where $h = 1/N$ and ϕ_k is the nodal value at $x = kh$. If ϕ_k is the maximum of all the nodal values, $1 \leqslant k \leqslant N$, then

$$\int_0^1 |\phi|^p \, dx \approx 2h|\phi_k|^p \left(\frac{|\phi_1|^p}{2|\phi_k|^p} + \frac{|\phi_2|^p}{|\phi_k|^p} + \cdots + 1 + \cdots + \frac{|\phi_{N-1}|^p}{|\phi_k|^p} + \frac{|\phi_N|^p}{2|\phi_k|^p} \right) .$$

When p approaches infinity, all the terms except unity will vanish since an infinite power of a number less than unity is negligibly small. Thus,

$$\int_0^1 |\phi|^p \, dx \approx 2h|\phi_k|^p \tag{c}$$

†Sometimes $\|\phi\|_p$ is written as $\|\phi\|_{L_p}$. In this book we will not clarify the difference between the Lebesque integral and the Riemann integral. See [1] concerning the difference.

or $$\|\phi\|_p \approx (2h)^{1/p} |\phi_k| \approx |\phi_k| , \qquad (d)$$

which completes the proof. Therefore, the Tchebycheff norm can also be written as $\|\phi\|_\infty$.

14-1*e* Cauchy Sequence

$\{\phi_1, \phi_2, \ldots\}$ is a Cauchy sequence if $\|\phi_n - \phi_m\|$ tends to zero as m and n approach infinity independently. The type of the norm must be specified for every Cauchy sequence.

Example 14-7 Is $\{1, t/2, (t/2)^2, \ldots, (t/2)^n, \ldots\}$ with respect to $L_\infty[0, 1]$ norm a Cauchy sequence?

Solution: By means of the triangle inequality, we obtain

$$\|\phi_n - \phi_m\|_{L_\infty} \leqslant \|\phi_n\|_{L_\infty} + \|-\phi_m\|_{L_\infty} = (\tfrac{1}{2})^n + (\tfrac{1}{2})^m . \qquad (a)$$

Therefore,

$$\|\phi_n - \phi_m\|_{L_\infty} \to 0 \quad \text{as } m, n \to \infty , \qquad (b)$$

and the given sequence is a Cauchy sequence.

Example 14-8 Is the sequence $\{1, t, t^2, \ldots, t^n, \ldots\}$ with respect to $L_2[0, 1]$ norm a Cauchy sequence?

Solution:

$$\|\phi_n - \phi_m\|_{L_2} = \left[\int_0^1 (t^n - t^m)^2 \, dt\right]^{1/2}$$

$$= \left(\frac{1}{2n+1} - \frac{2}{m+n+1} + \frac{1}{2m+1}\right)^{1/2} . \qquad (a)$$

Clearly,

$$\|\phi_n - \phi_m\|_{L_2} \to 0 \quad \text{as } m, n \to \infty . \qquad (b)$$

The given sequence is therefore a Cauchy sequence. It is interesting to note that the same sequence is not a Cauchy sequence with respect to $L_\infty[0, 1]$.

14-1*f* Complete Space

A space X is said to be complete if $\phi_1, \phi_2, \ldots$ is a Cauchy sequence of elements in X and if the space contains an element $\phi(x)$, called the limit of the sequence, such that $\|\phi_n - \phi\| \to 0$ as $n \to \infty$. If the space is complete, we should be able to express any function in the space in terms of a linear combination of the elements in a Cauchy sequence.

Example 14-9 Consider the sequence $\{1, t, t(1-t), t(1-t^2), \ldots\}$ in $L_2[0, 1]$ space with L_2 norm. Prove that this sequence is a Cauchy sequence. Does $L_2[0,1]$ contain a limit of this sequence?

Solution: To determine whether this sequence is a Cauchy sequence, we derive

$$\|\phi_n - \phi_m\|_{L_2}^2 = \int_0^1 [t(1-t^n) - t(1-t^m)]^2 \, dt$$

$$= \frac{1}{2m+3} - \frac{2}{m+n+2} + \frac{1}{2n+3} \,. \tag{a}$$

As m and n approach infinity, the norm given in Eq. (*a*) approaches zero. Therefore the given sequence is a Cauchy sequence. The limit of the sequence $\phi(t)$ can be written as

$$\phi(t) = \begin{cases} t\,, & 0 \leqslant t < 1 \\ 0\,, & t = 1\,, \end{cases} \tag{b}$$

which is integrable on $[0, 1]$ and is thus an element of $L_2[0, 1]$.

Example 14-10 It is known that the space $L_2[0, 1]$ with L_2 norm is complete. Demonstrate the completeness by considering the sequence

$$\{1, t, t^2, \ldots, t^n, \ldots\}\,.$$

Solution: First we will determine whether the sequence is a Cauchy sequence. Since

$$\|t^n - t^m\|_{L_2}^2 = \int_0^1 (t^n - t^m)^2 \, dt = \frac{1}{2n+1} - \frac{2}{m+n+1} + \frac{1}{2m+1}$$

and the L_2 norm approaches zero as m and n approach infinity, it is a Cauchy sequence. Next, the limit of the sequence $\phi(t)$ can be written as

$$\phi(t) = \begin{cases} 0\,, & 0 \leqslant t < 1 \\ 1\,, & t = 1\,, \end{cases}$$

which belongs to $L_2[0, 1]$ since its Lebesque integral exists. We have therefore proved that the limit of the given Cauchy sequence is an element of $L_2[0, 1]$. It may be difficult to prove that the space $L_2[0, 1]$ with L_2 norm is complete since we need to demonstrate that *all* the Cauchy sequences in $L_2[0, 1]$ have limits that are also contained in $L_2[0, 1]$.

Example 14-11 Use the sequence formed by the familiar pyramid function

$$n(x) = \begin{cases} 0\,, & x \leqslant -\dfrac{1}{n} \\ 1 + \mathrm{nx}\,, & -\dfrac{1}{n} \leqslant x \leqslant 0 \\ 1 - nx\,, & 0 \leqslant x \leqslant \dfrac{1}{n} \\ 0\,, & x \geqslant \dfrac{1}{n} \end{cases} \tag{a}$$

to demonstrate that $L_2\,[-10, 10]$ is complete with L_2 norm.

Solution: Since

$$\begin{aligned} \|\phi_n - \phi_m\|^2_{L_2} = &\int_{-1/m}^{-1/n} [0 - (1 + mx)]^2\, dx + \int_{-1/n}^{0} [(1 + nx) - (1 + mx)]^2\, dx \\ &+ \int_{0}^{1/n} [(1 - nx) - (1 - mx)]^2\, dx \\ &+ \int_{1/n}^{1/m} [0 - (1 - mx)]^2\, dx = \frac{2(n - m)^2}{3mn^2}\,, \end{aligned} \tag{b}$$

we have

$$\|\phi_n - \phi_m\|_{L_2} \to 0 \qquad \text{as } m, n \to \infty\,. \tag{c}$$

Furthermore, by the triangle inequality given in Section 14-1d, it follows that

$$\|\phi_n - \phi\| \leqslant \|\phi_n\| + \|\phi\|\,. \tag{d}$$

But $\|\phi_n\|_{L_2}$ can be evaluated as $(2/3n)^{1/2}$, which approaches zero as $n \to \infty$, and the limit of the sequence ϕ vanishes everywhere except $\phi(0) = 1$. Therefore, $\|\phi_n - \phi\| \to 0$.

Example 14-12 Prove that the $C^0\,[-1, 1]$ with L_2 norm is not complete.

Solution: To prove that the given space is incomplete, we need only to find a Cauchy sequence whose limit does not belong to the space. As long as there exists one such sequence, the proof is sufficient. Consider

$$\phi_n(x) = \begin{cases} -1\,, & -1 \leqslant x \leqslant -\dfrac{1}{n} \\ nx\,, & -\dfrac{1}{n} \leqslant x \leqslant \dfrac{1}{n} \\ 1\,, & \dfrac{1}{n} \leqslant x \leqslant 1\,. \end{cases} \tag{a}$$

If the limit of the sequence $\{\phi_1(x), \phi_2(x), \ldots, \phi_n(x)\}$ does not belong to $C^0[-1, 1]$, then the given space is incomplete. First, let n and m be large numbers and $n > m$. Then

$$\int_{-1}^{1} \phi_n \phi_m \, dx = \int_{-1}^{-1/m} (-1)(-1)\, dx + \int_{-1/m}^{-1/n} (-1)(mx)\, dx$$

$$+ \int_{-1/n}^{1/n} (nx)(mx)\, dx + \int_{1/n}^{1/m} (1)(mx)\, dx + \int_{1/m}^{1} (1)(1)\, dx$$

$$= 2 - \frac{m}{3n^2} - \frac{1}{m} \tag{b}$$

approaches zero as m and n tend to infinity. Likewise, we can prove that

$$\int_{-1}^{1} \phi_n^2 \, dx\,, \quad \int_{-1}^{1} \phi_m^2 \, dx \to 0 \quad \text{as } m, n \to \infty\,. \tag{c}$$

Based on Eqs. (b) and (c), we demonstrated that

$$\|\phi_n - \phi_m\|_{L_2}^2 = \int_{-1}^{1} (\phi_n - \phi_m)^2 \, dx \to 0 \quad \text{as } m, n \to \infty\,,$$

which implies that the given sequence is a Cauchy sequence. However, we are not able to find a limit of this sequence ϕ that belongs to $C^0[-1, 1]$ since this limit is discontinuous at $x = 0$ as n approaches ∞. Thus, the chosen set is incomplete and $C^0[-1, 1]$ with L_2 norm that contains this set is incomplete.

14-1g Banach Space

A space is called a Banach space if it is linear, normed, and complete [2].

Example 14-13 Is the space $C^0[-1, 1]$ with L_∞ norm a Banach space?

Solution: We will skip the proof for linearity and refer to Example 14-5 for the proof that $C^0[-1, 1]$ is a normed space. To illustrate that the space is normed, let us consider two elements

$$\phi_1 = e^x \quad \text{and} \quad \phi_2 = 2 - x\,, \quad \phi_1, \phi_2 \in C^0[-1, 1]\,. \tag{a}$$

The sum of ϕ_1 and ϕ_2

$$\phi_1 + \phi_2 = e^x + 2 - x \tag{b}$$

is also continuous and therefore belongs to $C^0[-1, 1]$. We now begin by writing

$$\|\phi_1\|_{L_\infty} = e > 0 \quad \text{and} \quad \|\phi_2\|_{L_\infty} = 3 > 0 \,. \tag{c}$$

Therefore, axiom (i) of the definition for the norm is satisfied. Next, the norm of Eq. (*b*)

$$\|e^x + 2 - x\|_{L_\infty} = e + 1 \tag{d}$$

is evaluated at $x = 1$. Clearly, on the basis of Eqs. (*c*) and (*d*),

$$\|\phi_1 + \phi_2\|_{L_\infty} < \|\phi_1\|_{L_\infty} + \|\phi_2\|_{L_\infty} \,. \tag{e}$$

Definition (ii) of the norm is satisfied. Finally, for a pure number a, we have

$$\|ae^x\| = a\|e^x\| = ae \,, \qquad \|a(2-x)\| = a\|2-x\| = 3a \,, \tag{f}$$

which implies definition (iii). The next task is to prove that the given space is complete. Let $\{\phi_n\}$ be a Cauchy sequence in $C^0[-1, 1]$. Then for large m and n we have

$$\|\phi_n(x) - \phi_m(x)\|_{L_\infty} \to 0$$

or

$$\|\phi_n(x) - \phi_m(x)\|_{L_\infty} < \epsilon \qquad \text{for all } x \in [-1, 1] \,.$$

Based on a result in calculus [3], it then follows that the sequence converges uniformly to a limit that is continuous and belongs to $C^0[-1, 1]$.

Example 14-14 Is the space $L_2[-1, 1]$ with L_2 norm a Banach space?

Solution: We will refer to Eq. (*a*) in Example 14-12 to examine the completeness of the given space. We demonstrated in Example 14-12 that

$$\|\phi_n - \phi_m\|_{L_2} \to 0 \qquad \text{as } m, n \to \infty \,. \tag{a}$$

In addition, the limit of the sequence

$$\phi(x) = \begin{cases} -1 \,, & x \leqslant 0 \\ 1 \,, & x \geqslant 0 \end{cases} \tag{b}$$

is squarely integrable (but discontinuous); therefore it belongs to $L_2[-1, 1]$. For other Cauchy sequences that belong to $L_2[-1, 1]$, it is always possible to prove that they have limits in $L_2[-1, 1]$. Since the given space is linear and normed, it is a Banach space.

14-1*h* Linearly Independent Set

The set $\{\phi_1, \phi_2, \ldots, \phi_n\}$ is said to be linearly independent on $[a, b]$ if and only if the equation

$$a_1\phi_1(x) + a_2\phi_2(x) + \cdots + a_n\phi_n(x) = 0 \tag{14-1}$$

implies that

$$a_1 = a_2 = \cdots = a_n = 0$$

for all $x \in [a, b]$.

Example 14-15 The set $\{1, x, x^2, \ldots\}$ is linearly independent.

Example 14-16 Are the three elements

$$\phi_1(x) = 3x + 1\,, \quad \phi_2(x) = 2x + 3\,, \quad \text{and} \quad \phi_3(x) = x - 1$$

linearly independent?

Solution: A linear combination of the three functions can be written, according to Eq. (14-1), as

$$a_1(3x + 1) + a_2(2x + 3) + a_3(x - 1) = 0\,. \tag{a}$$

Equation (*a*) holds for all x, not only when

$$a_1 = a_2 = a_3 = 0 \tag{b}$$

but also when, for example,

$$a_1 = -\tfrac{5}{7}\,, \quad a_2 = \tfrac{4}{7}\,, \quad \text{and} \quad a_3 = 1\,. \tag{c}$$

Therefore, the three elements are linearly dependent.

14-1*i* Basis Functions

$\{\phi_1, \phi_2, \ldots\}$ is a set of basis functions in the space X if (1) all the members of the set are linearly independent and (2) every element in X can be expressed as a linear combination of ϕ_k. This set is then said to be a basis for X.

Condition 2 is equivalent to asserting that, for any $\phi \in X$ and $\epsilon > 0$, there exist an n and scalars $a_1, a_2, \ldots, a_n$ such that

$$\left\| \phi - \sum_{j=1}^{n} a_j\phi_j \right\| < \epsilon\,, \tag{14-2}$$

which a Cauchy sequence in a complete space generally satisfies.

Example 14-17 Is the set $\{1,\ t,\ t(1-t),\ t(1-t^2), \ldots,\ t(1-t^n), \ldots\}$ a basis in $L_2[0, 1]$?

Solution: The proof of the linear independence of the members can follow Example 14-16. Regarding criterion 2 for a legitimate basis, we found from Example 14-9 that the set is a Cauchy sequence. If interest is restricted to L_2 norm, the set satisfies Eq. (14-2) and is a basis in $L_2[0, 1]$.

Example 14-18 The pyramid functions given in Example 14-11 are basis functions because in the example the set was proved to be a Cauchy sequence since the relation

$$a_l \phi_l + a_m \phi_m + a_n \phi_n = 0 \tag{a}$$

is valid if and only if $a_l = a_m = a_n = 0$. This assertion can be demonstrated as follows. Let $l > m > n$. At $x = \xi_1, \xi_2$, and ξ_3 locations shown in Fig. 14-2, Eq. (*a*) becomes

$$a_n(1 + n\xi_1) = 0 \,, \tag{b}$$

$$a_m(1 + m\xi_2) + a_n(1 + n\xi_2) = 0 \,, \tag{c}$$

and

$$a_l(1 + l\xi_3) + a_m(1 + m\xi_3) + a_n(1 + n\xi_3) = 0 \,. \tag{d}$$

The fact that the value of the determinant is nonzero suggests that the only simultaneous solution of Eqs. (*b*)-(*d*) is a basis in $L_2(-\infty, \infty)$.

$$a_l = a_m = a_n = 0 \,. \tag{e}$$

Therefore, the given set is a basis in $L_2(-\infty, \infty)$.

Example 14-19 The members

$$\left\{ x\left(x - \frac{2}{3}\right), x\left(x^2 - \frac{1}{3}\right), x(x^3 - 0), x\left(x^4 + \frac{1}{3}\right), \ldots, x\left[x^n + \frac{n-3}{3}\right], \ldots \right\}$$

are not basis functions in $L_2[0, 1]$. Although they are linear independent, the limit of the sequence is not squarely integrable and therefore does not belong to $L_2[0, 1]$. It is not likely that Eq. (14-2) can be satisfied.

14-1*j* Support

The support of a function $\phi(x)$ is the interval $[a, b]$ on which $\phi(x)$ is nonzero. We use the notation sup(ϕ) to denote the interval $[a, b]$ or the elements shown in Fig. 5-3.

Example 14-20 The pyramid basis function

$$\phi_j(x) = \begin{cases} \dfrac{x - x_{j-1}}{h}, & x \in [x_{j-1}, x_j] \\[2ex] \dfrac{x_{j+1} - x}{h}, & x \in [x_j, x_{j+1}] \\[2ex] 0, & \text{elsewhere} \end{cases}$$

has the support $[x_{j-1}, x_{j+1}]$. When the basis functions have narrow local supports, the resulting matrix system is likely to be diagonally banded.

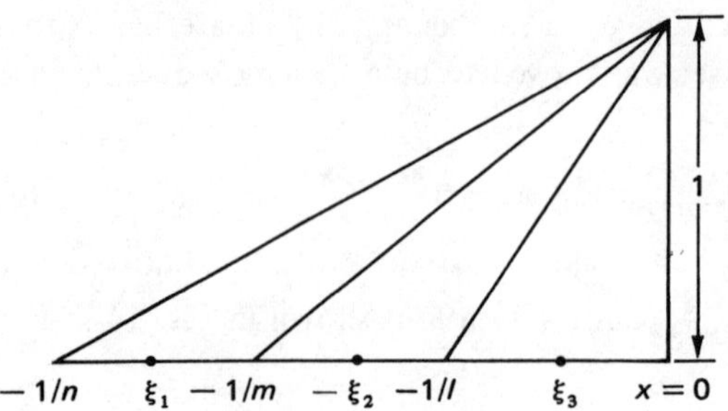

Figure 14-2 Three pyramid functions with three x locations.

14-1*k* *n*-Dimensional Subspace

An n-dimensional subspace is a space spanned by a finite number n of basis functions.

Example 14-21 Suppose the function $\phi(x)$ can be approximated by

$$\tilde{\phi}(x) = e_1 \phi_1(x) + e_2 \phi_2(x) + \cdots + e_n \phi_n(x) . \tag{a}$$

If $\phi_1(x)$, $\phi_2(x), \ldots, \phi_n(x)$ all belong to $C^0[0, 1]$, then the space containing all $\tilde{\phi}(x)$ is an n-dimensional subspace of $C^0[0, 1]$.

14-1*l* Inner Product

Let X be a real linear space. With each $\phi_1, \phi_2 \in X$ we associate a real number, denoted by (ϕ_1, ϕ_2) and called the inner product of ϕ_1 and ϕ_2, such that

(i) $(\phi_1, \phi_1) \geqslant 0$; the equality holds if and only if $\phi_1 = 0$;
(ii) $(\phi_1, \phi_2) = (\phi_2, \phi_1)$;
(iii) $(a\phi_1, \phi_2) = a(\phi_1, \phi_2)$;
(iv) $(\phi_1 + \phi_2, \phi_3) = (\phi_1, \phi_3) + (\phi_2, \phi_3)$.

If $(\phi_1, \phi_2) = 0$, then ϕ_1 and ϕ_2 are said to be orthogonal to each other.

Example 14-22 The expression

$$(\phi_1, \phi_2) = \int_a^b \phi_1 \phi_2 \, dx \tag{a}$$

can be proved to satisfy criteria (i)–(iv); therefore it is an inner product of ϕ_1 and ϕ_2. This inner product has been used throughout this book.

Example 14-23 The expression

$$(\phi_1, \phi_2) = \int_a^b \left(\phi_1 \phi_2 + \frac{d\phi_1}{dx} \frac{d\phi_2}{dx} \right) dx \tag{a}$$

is an inner product of ϕ_1 and ϕ_2.

Example 14-24 Is the expression

$$(\phi_1, \phi_2) = \frac{d}{dx}(\phi_1 \phi_2)$$

an inner product of ϕ_1 and ϕ_2?

Solution: The expression satisfies criteria (ii), (iii), and (iv), but only partially satisfies criterion (i) since ϕ_1 can be a nonzero constant and the condition $(\phi_1, \phi_1) = 0$ remains valid. Therefore, the expression is not an inner product.

14-1*m* Weak (Generalized) Form and Bilinear Form

An inner product defined as

$$(L\widetilde{\phi}, v) = (f, v) \qquad \frac{d\widetilde{\phi}^2}{dx^2}, \frac{d\widetilde{\phi}}{dx}, \widetilde{\phi}, v \in L_2[a, b] \tag{14-3a}$$

is said to be a weak form of the differential equation of second order

$$L\phi = f(x), \qquad \phi \in C^2[a, b].$$

The adjective "weak" is to convey the idea that the differentiability required for the approximate solution $\widetilde{\phi}(x)$ is weaker than that of the exact solution $\phi(x)$; the adjective "generalized" means that the space L_2 to which $\widetilde{\phi}(x)$ and its first and second derivatives belong is larger or more generalized than the space C^2 to which $\phi(x)$ belongs.

If Eq. (14-3*a*) can be integrated by parts to take the form

$$a(\widetilde{\phi}, v) = (f, v), \qquad \widetilde{\phi}, \frac{d\widetilde{\phi}}{dx}, v, \frac{dv}{dx} \in L_2[a, b] \tag{14-3b}$$

where $a(\widetilde{\phi}, v)$ is a bilinear form defined as

$$a(\widetilde{\phi}_1 + \widetilde{\phi}_2, v) = a(\widetilde{\phi}_1, v) + a(\widetilde{\phi}_2, v) \tag{14-4a}$$

and

$$a(\widetilde{\phi}, v_1 + v_2) = a(\widetilde{\phi}, v_1) + a(\widetilde{\phi}, v_2), \tag{14-4b}$$

then Eq. (14-3*b*) is also a weak form of the original differential equation and is even weaker than Eq. (14-3*a*).

Example 14-25 The exact solution to Eq. (1-1) should be at least twice differentiable, but the approximate solution to Eq. (1-25) needs to be only a continuous function (its first derivative does not exist at the nodal points). Incidentally, Eq. (1-25) is a bilinear form. So is Eq. (*a*) of Example 14-23.

14-1*n* Hilbert Space

A complete inner product space is referred to as a Hilbert space [4]. In an inner product space, we define the norm by

$$\|\phi_1\| = (\phi_1, \phi_1)^{1/2}, \tag{14-4}$$

which is recognized as the L_2 norm. Therefore, every inner product space is also a normed space. The converse, however, is not true. For example, a normed space may contain the maximum (Tchebycheff) norm, and it is not always possible to define an inner product such that

$$(\phi_1, \phi_1)^{1/2} = \|\phi_1\|_\infty . \tag{14-5}$$

Note also that a Hilbert space is a Banach space since it is linear, normed, and complete.

Example 12-26 $L_2\,[0, 1]$ space is a Hilbert space.

14-1*o* Schwarz (or Cauchy-Schwarz) Inequality

For all $\phi_1(x)$ and $\phi_2(x)$ in an inner product space $[a, b]$, the inequality

$$|(\phi_1, \phi_2)| \leqslant (\phi_1, \phi_1)^{1/2} (\phi_2, \phi_2)^{1/2} \tag{14-6}$$

is always true. The proof of Eq. (14-6) can be found in [5]. Based on Eq. (*a*) of Example 14-22 and Eq. (*a*) of Example 14-23, respectively, we can rewrite Eq. (14-6) as

$$\left| \int_a^b \phi_1 \phi_2 \, dx \right| \leqslant \left(\int_a^b \phi_1^2 \, dx \right)^{1/2} \left(\int_a^b \phi_2^2 \, dx \right)^{1/2} \tag{14-7a}$$

and

$$\left| \int_a^b \left(\phi_1 \phi_2 + \frac{d\phi_1}{dx} \frac{d\phi_2}{dx} \right) dx \right| \leqslant \left\{ \int_a^b \left[\phi_1^2 + \left(\frac{d\phi_1}{dx} \right)^2 \right] dx \right\}^{1/2} \times \left\{ \int_a^b \left[\phi_2^2 + \left(\frac{d\phi_2}{dx} \right)^2 \right] dx \right\}^{1/2} . \tag{14-7b}$$

Equations (14-6) and (14-7) are frequently used in error bound analysis.

Example 12-27 Choose $\phi_1(x) = x$ and $\phi_2(x) = \sin x$, $x \in [0, \pi]$, to show Eq. (14-7*a*).

Solution: We evaluate

$$\left| \int_0^\pi x \sin x \, dx \right| = \pi , \tag{a}$$

$$\left(\int_0^\pi x^2 \, dx \right)^{1/2} = \left(\frac{\pi^3}{3} \right)^{1/2} , \tag{b}$$

and

$$\left(\int_0^{\pi} \sin^2 x\, dx\right)^{1/2} = \left(\frac{\pi}{2}\right)^{1/2} . \tag{c}$$

Therefore,

$$\left|\int_0^{\pi} x \sin x\, dx\right| = \pi < 1.283\pi = \left(\int_0^{\pi} x^2\, dx\right)^{1/2} \left(\int_0^{\pi} \sin^2 x\, dx\right)^{1/2} . \tag{d}$$

14-1*p* Sobolev Space

A Sobolev space $W_p^m [a, b]$ is a space in which all the elements $\phi(x)$ are such that

$$\frac{d^{\alpha}\phi}{dx^{\alpha}} \in L_p [a, b] \qquad \text{for all } \alpha \leqslant m . \tag{14-8}$$

In this book, we will focus on the special case $p = 2$ and $m = 1$. The norm of such a Sobolev space $W_2^1 [a, b]$ is defined as

$$\|\phi\|_{W_2^1 [a, b]} = \left\{\int_a^b \left[\phi^2 + \left(\frac{d\phi}{dx}\right)^2\right] dx\right\}^{1/2} , \tag{14-9}$$

where the subscript 2 denotes that the function and the derivatives should be squarely integrated and that the square root of the resulting integral should then be taken. The superscript 1 denotes the order of the highest derivative in the integrand.

Example 14-28 Evaluate the Sobolev norms of

$$\phi_1 (x) = x^2 \qquad \text{and} \qquad \phi_2 (x) = \sin x \qquad \text{in } W_2^1 [0, 1] .$$

Solution: According to Eq. (14-9),

$$\|\phi_1\|_{W_2^1 [0,1]} = \left[\int_0^1 (x^4 + 4x^2)\, dx\right]^{1/2} = 1.238 \tag{a}$$

and

$$\|\phi_2\|_{W_2^1 [0,1]} = \left[\int_0^1 (\sin^2 x + \cos^2 x)\, dx\right]^{1/2} = 1 . \tag{b}$$

14-1*q* Positive-Definite Operator

A linear operator L is said to be positive-definite if and only if

$$(L\phi, \phi) \geqslant \gamma \|\phi\|^2_{W_2^1 [0,1]} , \qquad \gamma > 0 . \tag{14-10}$$

Example 12-29 Is the linear operator

$$L = -\frac{d^2}{dx^2} - \frac{d}{dx} + 2 \tag{a}$$

a positive-definite operator?

Solution: Using integration by parts and assuming that the function $\phi(x)$ vanishes on the boundaries $x = a, b$, we obtain

$$(L\phi, \phi) = \int_a^b \left[\left(\frac{d\phi}{dx}\right)^2 - \phi\left(\frac{d\phi}{dx}\right) + 2\phi^2\right] dx\,. \tag{b}$$

But in light of the Schwarz inequality we have

$$\int_a^b \phi \frac{d\phi}{dx}\, dx \leqslant \left|\int_a^b \phi \frac{d\phi}{dx}\, dx\right| \leqslant \left(\int_a^b \phi^2\, dx\right)^{1/2} \left[\int_a^b \left(\frac{d\phi}{dx}\right)^2 dx\right]^{1/2}. \tag{c}$$

Since, for $a_1, a_2 \geqslant 0$, the inequality

$$a_1^{1/2} \cdot a_2^{1/2} \leqslant \tfrac{1}{2}(a_1 + a_2) \tag{d}$$

always holds, Eq. (c) becomes

$$\int_a^b \phi \frac{d\phi}{dx}\, dx \leqslant \frac{1}{2}\left[\int_a^b \phi^2\, dx + \int_a^b \left(\frac{d\phi}{dx}\right)^2 dx\right]. \tag{e}$$

Combining Eq. (e) and Eq. (b) yields

$$(L\phi, \phi) \geqslant \int_a^b \left[\frac{1}{2}\left(\frac{d\phi}{dx}\right)^2 + \frac{3}{2}\phi^2\right] dx \geqslant \frac{1}{2}\, \|\phi\|^2_{W_2^1[0,1]}\,. \tag{f}$$

It thus can be concluded that the given operator is positive-definite.

Having briefly defined some important terminology, we are in a better position to study error bound analysis, which is introduced next.

14-2 ERROR BOUNDS OF THE FINITE-ELEMENT SOLUTION

For ease of presentation we will first consider a simple one-dimensional problem described by

$$L\phi = -\frac{d^2\phi}{dx^2} + a(x)\phi = f(x), \quad \phi(0) = \phi(1) = 0\,, \tag{14-11}$$

where $a(x)$ is a known positive-valued function such that

$$0<\beta_1 \leqslant a(x) \leqslant \beta_2 < \infty\,, \qquad x \in [0,1]\,. \tag{14-12}$$

An inner product can be written as

$$(L\phi, v) = (f, v)\,, \qquad v \in \overset{\circ}{W}{}_2^1\,, \tag{14-13}$$

where the symbol "○" above W means that all the test functions v vanish on the boundaries. Now assume that

$$\tilde{\phi}(x) = \sum_{j=1}^{J-1} N_j(x)\phi_j \tag{14-14}$$

is a finite-element approximate solution. Corresponding to Eq. (14-13) we have

$$(L\tilde{\phi}, v) = (f, v)\,, \qquad v \in S^h \subset \overset{\circ}{W}{}_2^1\,, \tag{14-15}$$

where S^h is a J-dimensional subspace and $J = 1/h$. We then define the error $\tilde{e}(x)$ of the approximate solution $\tilde{\phi}(x)$ given in Eq. (14-14) as

$$\tilde{e}(x) = \phi(x) - \tilde{\phi}(x)\,. \tag{14-16}$$

where $\phi(x)$ is the exact solution of Eq. (14-11). Subtracting Eq. (14-15) from Eq. (14-13) yields

$$(L\tilde{e}, v) = 0\,. \tag{14-17}$$

Next, based on definition (iv) of the inner product, we write

$$(L\tilde{e}, \tilde{e}) = (L\tilde{e}, \tilde{e} + \bar{\phi} - \bar{\phi}) = (L\tilde{e}, \phi - \bar{\phi}) + (L\tilde{e}, \bar{\phi} - \tilde{\phi})\,, \tag{14-18}$$

where $\bar{\phi}(x)$ is the interpolation function that coincides with the exact solution $\phi(x)$ pointwise. Since both $\bar{\phi}$ and $\tilde{\phi}$ belong to S^h and

$$\bar{\phi} - \tilde{\phi} \in S^h\,,$$

the last inner product given in Eq. (14-18) vanishes according to Eq. (14-17). Therefore, Eq. (14-18) is reduced to

$$(L\tilde{e}, \tilde{e}) = (L\tilde{e}, \phi - \bar{\phi}) \tag{14-19}$$

or, after integration by parts,

$$\int_0^1 \left[\left(\frac{d\tilde{e}}{dx}\right)^2 + a\tilde{e}^2\right] dx = \int_0^1 \left[\frac{d\tilde{e}}{dx}\frac{d}{dx}(\phi - \bar{\phi}) + a\tilde{e}(\phi - \bar{\phi})\right] dx\,. \tag{14-20}$$

For clarity of presentation, we will temporarily designate the right-hand side of Eq. (14-20) as I_1. Following Eq. (14-7a), the Schwarz inequality, we obtain

$$I_1 \leqslant \left[\int_0^1 \left(\frac{d\tilde{e}}{dx}\right)^2 dx\right]^{1/2} \left\{\int_0^1 \left[\frac{d}{dx}(\phi - \bar{\phi})\right]^2 dx\right\}^{1/2} + a_{\max}\left(\int_0^1 \tilde{e}^2\, dx\right)^{1/2} \left[\int_0^1 (\phi - \bar{\phi})^2\, dx\right]^{1/2}\,. \tag{14-21}$$

Since, by the definition of the Sobolev norm $\|\tilde{e}\|$, it is evident that

$$\frac{\left[\int_0^1 (d\tilde{e}/dx)^2\,dx\right]^{1/2}}{\|\tilde{e}\|} \leqslant 1 \tag{14-22a}$$

and

$$\frac{\left(\int_0^1 \tilde{e}^2\,dx\right)^{1/2}}{\|\tilde{e}\|} \leqslant 1\,, \tag{14-22b}$$

Eq. (14-21) can be rewritten as

$$I_1 \leqslant \|\tilde{e}\| \left\{\int_0^1 \left[\frac{d}{dx}(\phi-\bar{\phi})\right]^2 dx\right\}^{1/2} + a_{\max}\|\tilde{e}\| \left[\int_0^1 (\phi-\bar{\phi})^2\,dx\right]^{1/2}. \tag{14-23}$$

Carrying out one more step, we derive

$$I_1 \leqslant \max\{1,\beta_2\}\|\tilde{e}\| \left(\left\{\int_0^1 \left[\frac{d}{dx}(\phi-\bar{\phi})\right]^2 dx\right\}^{1/2} + \left[\int_0^1 (\phi-\bar{\phi})^2\,dx\right]^{1/2}\right). \tag{14-24}$$

Since, for $a_1, a_2 \geqslant 0$, the inequality

$$a_1^{1/2} + a_2^{1/2} \leqslant \sqrt{2}(a_1 + a_2)^{1/2} \tag{14-25}$$

is always true, Eq. (14-20) finally becomes

$$\int_0^1 \left[\left(\frac{d\tilde{e}}{dx}\right)^2 + a\tilde{e}^2\right] dx \leqslant \max\{1,\beta_2\}\sqrt{2}\|\tilde{e}\|\cdot\|\phi-\bar{\phi}\|\,. \tag{14-26}$$

On the other hand, seeking the lower bound of $\|L\tilde{e},\tilde{e}\|$ leads to

$$\int_0^1 \left[\left(\frac{d\tilde{e}}{dx}\right)^2 + a\tilde{e}^2\right] dx \geqslant \min\{1,\beta_1\} \int_0^1 \left[\left(\frac{d\tilde{e}}{dx}\right)^2 + \tilde{e}^2\right] dx$$

$$= \min\{1,\beta_1\}\|\tilde{e}\|^2\,, \tag{14-27}$$

which incidentally implies that the operator given in Eq. (14-11) is positive-definite according to Eq. (14-10). Joint use of Eqs. (14-26) and (14-27) yields

$$\|\phi-\tilde{\phi}\| \leqslant C\|\phi-\bar{\phi}\|\,, \tag{14-28}$$

where

$$C = \sqrt{2}\,\frac{\max\{1,\beta_2\}}{\min\{1,\beta_1\}}\,. \tag{14-29}$$

Equation (14-28) indicates that the error norm of the finite-element solution is always smaller than that of an interpolation function multiplied by a constant. This equation is useful because it is relatively easy to find an upper bound in terms of the mesh size h for the error norm $\|\phi - \bar{\phi}\|$. We showed this procedure in Section 4-3c for the linear interpolant and will perform a similar derivation in the following section for the quadratic interpolant. Now let us pause here and present an example to assess the validity of Eq. (14-28).

Example 14-30 We are given a simple one-dimensional problem governed by

$$-\frac{d^2\phi}{dx^2} + \phi = -x \tag{a}$$

subject to

$$\phi(0) = \phi(1) = 0 \,. \tag{b}$$

Use the Galerkin finite-element solution with $h = \frac{1}{2}$ to prove Eq. (14-28).

Solution: The exact solution to Eqs. (*a*) and (*b*) is found to be

$$\phi(x) = \frac{e^x - e^{-x}}{e - e^{-1}} - x \,. \tag{c}$$

Hence

$$\phi(\tfrac{1}{2}) = -0.05659 \,. \tag{d}$$

With the aid of Table 1-4, the Galerkin nodal solution at $x = \frac{1}{2}$ is found to be

$$\phi_1 = -0.05769 \,. \tag{e}$$

Consequently,

$$\tilde{\phi}(x) = \begin{cases} 2\phi_1 x \,, & x \in [0, \frac{1}{2}] \\ 2\phi_1(1 - x) \,, & x \in [\frac{1}{2}, 1] \,. \end{cases} \tag{f}$$

By the definition of the Sobolev norm, we obtain

$$\begin{aligned} \|\phi - \tilde{\phi}\|^2 &= \int_0^1 \left\{ \left[\frac{d}{dx}(\phi - \tilde{\phi}) \right]^2 + (\phi - \tilde{\phi})^2 \right\} dx \\ &= \int_0^{1/2} [(ae^x + ae^{-x} - 1 - 2\phi_1)^2 + (ae^x - ae^{-x} - x - 2\phi_1 x)^2]\, dx \\ &\quad + \int_{1/2}^1 [(ae^x + ae^{-x} - 1 + 2\phi_1)^2 \\ &\quad + (ae^x - ae^{-x} - x + 2\phi_1 x - 2\phi_1)^2]\, dx = 0.029293 \,, \end{aligned} \tag{g}$$

where the constant a denotes $(e - e^{-1})^{-1}$. Using the exact solution at $x = \frac{1}{2}$ given in Eq. (*d*), we obtain

$$\bar{\phi}(x) = \begin{cases} 2x\phi(\frac{1}{2}), & x \in [0, \frac{1}{2}] \\ 2(1-x)\phi(\frac{1}{2}), & x \in [\frac{1}{2}, 1]. \end{cases} \tag{h}$$

Following Eq. (*g*), we find

$$\|\phi - \bar{\phi}\|^2 = 0.028688. \tag{i}$$

Finally, the constant C in Eqs. (14-28) and (14-29) is simply $\sqrt{2}$. Therefore, it is demonstrated that

$$\|\phi - \tilde{\phi}\| < C\|\phi - \bar{\phi}\| \tag{j}$$

for this particular problem.

Readers who are interested in pursuing this topic further may wish to consult [7, 8] for a posteriori error estimates of one-dimensional finite-element solutions and [9, 10] for error bounds of variable meshes.

The aforementioned analysis can be readily extended to a two-dimensional problem. Consider

$$L\phi = -\nabla^2\phi + b(x,y)\phi = g(x,y) \quad \text{in } \Omega, \tag{14-30}$$
$$\phi = 0 \quad \text{on } \partial\Omega,$$

where $b(x, y)$ is a positive-valued function bounded by

$$0 < \gamma_1 \leqslant b(x,y) \leqslant \gamma_2 < \infty. \tag{14-31}$$

The derivation from Eq. (14-13) to Eq. (14-19) can be followed exactly. Hence,

$$\iint_\Omega \left[\left(\frac{\partial \tilde{e}}{\partial x}\right)^2 + \left(\frac{\partial \tilde{e}}{\partial y}\right)^2 + b\tilde{e}^2\right] dx\,dy = \iint_\Omega \left[\frac{\partial \tilde{e}}{\partial x}\frac{\partial}{\partial x}(\phi - \bar{\phi}) + \frac{\partial \tilde{e}}{\partial y}\frac{\partial}{\partial y}(\phi - \bar{\phi}) + b\tilde{e}(\phi - \bar{\phi})\right] dx\,dy. \tag{14-32}$$

Again, we will designate the right-hand side of Eq. (14-32) as I_2 for convenience. Based on the Schwarz inequality,

$$I_2 \leqslant \left[\iint_\Omega \left(\frac{\partial \tilde{e}}{\partial x}\right)^2 dx\,dy\right]^{1/2} \left\{\iint_\Omega \left[\frac{\partial}{\partial x}(\phi - \bar{\phi})\right]^2 dx\,dy\right\}^{1/2} + \left[\iint_\Omega \left(\frac{\partial \tilde{e}}{\partial y}\right)^2 dx\,dy\right]^{1/2} \left\{\iint_\Omega \left[\frac{\partial}{\partial y}(\phi - \bar{\phi})\right]^2 dx\,dy\right\}^{1/2} + \gamma_2 \left(\iint_\Omega \tilde{e}^2\,dx\,dy\right)^{1/2} \left[\iint_\Omega (\phi - \bar{\phi})^2\,dx\,dy\right]^{1/2}. \tag{14-33}$$

Noting that the Sobolev norm $\|\tilde{e}\|$ is defined as

$$\|\tilde{e}\| = \left\{ \iint_\Omega \left[\left(\frac{\partial \tilde{e}}{\partial x}\right)^2 + \left(\frac{\partial \tilde{e}}{\partial y}\right)^2 + \tilde{e}^2 \right] dx\, dy \right\}^{1/2} , \tag{14-34}$$

we realize that

$$\frac{\left[\iint_\Omega (\partial \tilde{e}/\partial x)^2\, dx\, dy\right]^{1/2}}{\|\tilde{e}\|} \leqslant 1 , \tag{14-35a}$$

$$\frac{\left[\iint_\Omega (\partial \tilde{e}/\partial y)^2\, dx\, dy\right]^{1/2}}{\|\tilde{e}\|} \leqslant 1 , \tag{14-35b}$$

and

$$\frac{\left(\iint_\Omega \tilde{e}^2\, dx\, dy\right)^{1/2}}{\|\tilde{e}\|} \leqslant 1 . \tag{14-35c}$$

Therefore, Eq. (14-33) becomes

$$I_2 \leqslant \max\,\{1, \gamma_2\} \|\tilde{e}\| \left(\left\{ \iint_\Omega \left[\frac{\partial}{\partial x} (\phi - \bar{\phi}) \right]^2 dx\, dy \right\}^{1/2} + \left\{ \iint_\Omega \left[\frac{\partial}{\partial y} (\phi - \bar{\phi}) \right]^2 dx\, dy \right\}^{1/2} + \left[\iint_\Omega (\phi - \bar{\phi})^2\, dx\, dy \right]^{1/2} \right) . \tag{14-36}$$

In order to relate the quantity in the parentheses to the Sobolev norm, we use the inequality

$$a^{1/2} + b^{1/2} + c^{1/2} \leqslant \sqrt{3}(a + b + c)^{1/2} \tag{14-37}$$

for all $a, b, c, > 0$. Equation (14-32) finally becomes

$$\iint_\Omega \left[\left(\frac{\partial \tilde{e}}{\partial x}\right)^2 + \left(\frac{\partial \tilde{e}}{\partial y}\right)^2 + b\tilde{e}^2 \right] dx\, dy \leqslant \max\,\{1, \gamma_2\}\, \sqrt{3} \|\tilde{e}\|\, \|\phi - \bar{\phi}\| . \tag{14-38}$$

To seek the lower bound for the integral in Eq. (14-38), we assert that the operator in Eq. (14-30) is positive-definite since

$$(L\tilde{e}, \tilde{e}) \geqslant \min\,\{1, \gamma_1\} \|\tilde{e}\|^2 . \tag{14-39}$$

Combining Eqs. (14-38) and (14-39) leads to

$$\|\tilde{e}\| \leqslant D \|\phi - \bar{\phi}\| , \tag{14-40}$$

where

$$D = \sqrt{3}\,\frac{\max\{1, \gamma_2\}}{\min\{1, \gamma_1\}}. \tag{14-41}$$

Before ending this section, we note that Eqs. (14-28) and (14-40) are both subject to the condition that the operator of the governing equations must be linear and positive-definite. For nonlinear or indefinite operators, error bound theories have not yet been well developed and the result remains inconclusive.

14-3 ERROR BOUNDS OF QUADRATIC INTERPOLATION FUNCTIONS (ONE-DIMENSIONAL)

The error bound Eqs. (14-28) and (14-40) are merely intermediate results since they express neither the accuracy nor the convergence rate of a finite-element scheme. To make them useful, we need to seek, by means of calculus, the upper bound of $\|\phi - \bar{\phi}\|$ that we hope can be expressed in terms of the mesh size h. In Section 4-3 we derived such a bound for one-dimensional *linear* interpolation functions. In the present section we will consider *quadratic* interpolation functions. Figure 14-3 can be consulted for the following analysis.

Define the error $\bar{e}$ as

$$\bar{e} = \phi - \bar{\phi}. \tag{14-42}$$

Since $\bar{\phi}(x)$ coincides with $\phi(x)$ pointwise at $x = x_{j-1}, x_j, x_{j+1}$, it follows that

$$\bar{e}(x_{j-1}) = \bar{e}(x_j) = \bar{e}(x_{j+1}) = 0. \tag{14-43}$$

In light of Rolle's theorem in calculus, there exist $\xi_1 \in (x_{j-1}, x_j)$ and $\xi_2 \in (x_j, x_{j+1})$ such that

$$\left(\frac{d\bar{e}}{dx}\right)_{x=\xi_1} = \left(\frac{d\bar{e}}{dx}\right)_{x=\xi_2} = 0. \tag{14-44a}$$

Furthermore, since $\xi_1 \neq \xi_2$, there also exists $\xi_3 \in (\xi_1, \xi_2)$ such that

$$\left(\frac{d^2\bar{e}}{dx^2}\right)_{x=\xi_3} = 0. \tag{14-44b}$$

Consequently, for $\eta_1 \in (x, x_j)$, $\eta_2 \in (\eta_1, \xi_1)$, and $\eta_3 \in (\eta_2, \xi_3)$, as shown in Fig. 14-3, we derive step by step that

$$\bar{e}(x) = \bar{e}(x) - \bar{e}(x_j) \tag{14-45a}$$

$$= (x - x_j)\left(\frac{d\bar{e}}{dx}\right)_{x=\eta_1} \tag{14-45b}$$

$$= (x - x_j)\left[\left(\frac{d\bar{e}}{dx}\right)_{x=\eta_1} - \left(\frac{d\bar{e}}{dx}\right)_{x=\xi_1}\right] \tag{14-45c}$$

$$= (x - x_j)(\eta_1 - \xi_1)\left(\frac{d^2\bar{e}}{dx^2}\right)_{x=\eta_2} \tag{14-45d}$$

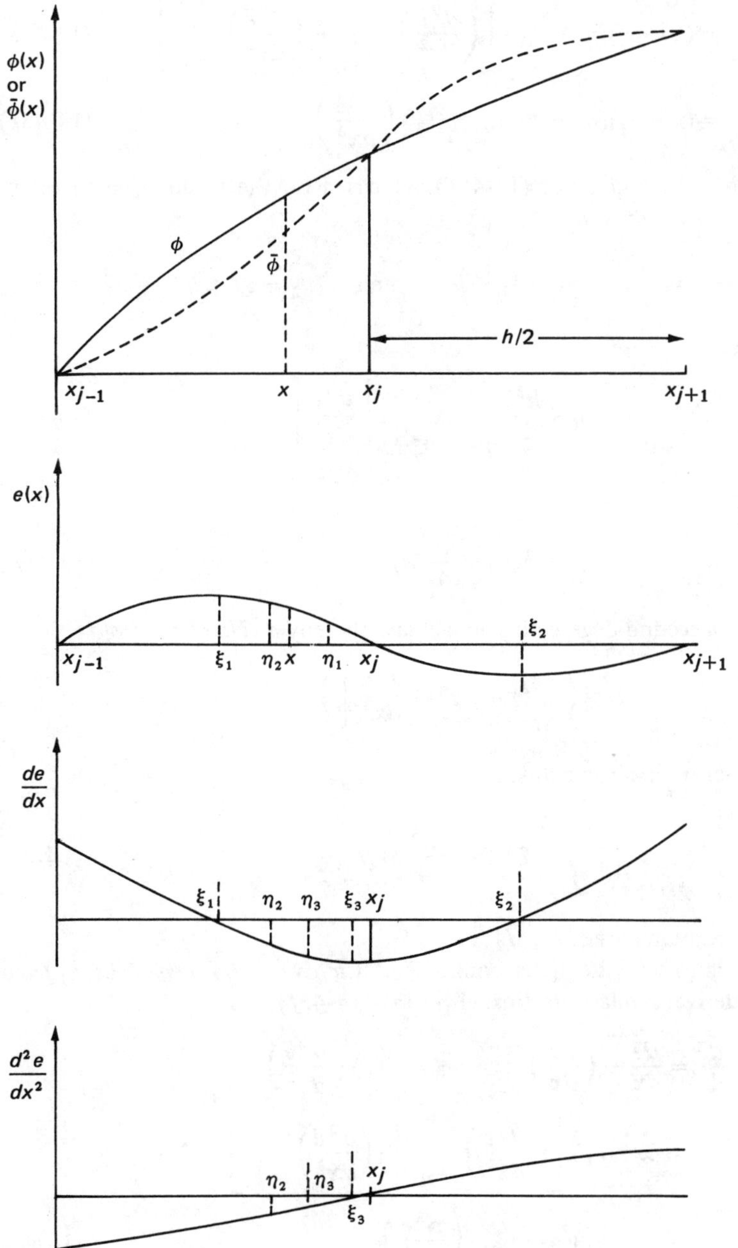

Figure 14-3 Error bound e of a quadratic interpolation function and its derivatives.

$$\bar{e}(x) = (x - x_j)(\eta_1 - \xi_1)\left[\left(\frac{d^2\bar{e}}{dx^2}\right)_{x=\eta_2} - \left(\frac{d^2\bar{e}}{dx^2}\right)_{x=\xi_3}\right] \tag{14-45e}$$

$$= (x - x_j)(\eta_1 - \xi_1)(\eta_2 - \xi_3)\left(\frac{d^3\bar{e}}{dx^3}\right)_{x=\eta_3}. \tag{14-45f}$$

Equations (14-45*b*), (14-45*d*), and (14-45*f*) are based on the mean-value theorem. Since

$$x - x_j \leqslant \frac{h}{2}, \qquad \eta_1 - \xi_1 < h, \qquad \text{and} \qquad \eta_2 - \xi_3 < h,$$

Eq. (14-45*f*) leads to

$$|\bar{e}(x)| \leqslant \frac{h^3}{2} \max_{x_{j-1} \leqslant x \leqslant x_j} \left|\frac{d^3\bar{e}}{dx^3}\right| \tag{14-46a}$$

or

$$\bar{e}^2(x) \leqslant \frac{h^6}{4} M_j, \tag{14-46b}$$

where, since $\bar{\phi}(x)$ is a second-degree polynomial and thereby $d^3\bar{e}/dx^3 = d^3\phi/dx^3$,

$$M_j = \left(\max_{x_{j-1} \leqslant x \leqslant x_j} \left|\frac{d^3\phi}{dx^3}\right|\right)^2.$$

Accounting for other intervals, we obtain

$$\int_0^L \bar{e}^2\, dx \leqslant \frac{h^6}{4} M_k L, \tag{14-47}$$

where M_k is the maximum along $M_1, M_2, \ldots, M_{J-1}$.

The next task is to seek the upper bound of $\int_0^L (d\bar{e}/dx)^2\, dx$. For $\eta_4 \in (x, \xi_1)$ and $\eta_5 \in (\eta_4, \xi_3)$, we derive, similarly to Eqs. (14-45*b*)-(14-45*f*), that

$$\frac{d\bar{e}}{dx} = \frac{d\bar{e}}{dx} - \left(\frac{d\bar{e}}{dx}\right)_{x=\xi_1} = (x - \xi_1)\left(\frac{d^2\bar{e}}{dx^2}\right)_{x=\eta_4}$$

$$= (x - \xi_1)\left[\left(\frac{d^2\bar{e}}{dx^2}\right)_{x=\eta_4} - \left(\frac{d^2\bar{e}}{dx^2}\right)_{x=\xi_3}\right]$$

$$= (x - \xi_1)(\eta_4 - \xi_3)\left(\frac{d^3\bar{e}}{dx^3}\right)_{x=\eta_5} \tag{14-48}$$

In terms of the mesh size h, Eq. (14-48) can be written as

$$\left|\frac{d\bar{e}}{dx}\right| \leqslant \frac{h^2}{2} \max_{x_{j-1} \leqslant x \leqslant x_j} \left|\frac{d^3\phi}{dx^3}\right| \tag{14-49}$$

or

$$\int_0^L \left(\frac{d\bar{e}}{dx}\right)^2 dx \leqslant \frac{h^4}{4} M_k L \,. \tag{14-50}$$

Combining Eqs. (14-47) and (14-50) yields

$$\|\bar{e}\| = \left\{ \int_0^L \left[\bar{e}^2 + \left(\frac{d\bar{e}}{dx}\right)^2 \right] dx \right\}^{1/2} \leqslant \frac{h^2}{2} [M_k L(1 + h^2)]^{1/2} \,. \tag{14-51}$$

With such considerable effort, we finally are able to link Eq. (14-28) with Eq. (14-51) to arrive at the error bound equation

$$\|\phi - \tilde{\phi}\| \leqslant Dh^2 \,, \tag{14-52}$$

where

$$D = [M_k L(1 + h^2)]^2 \, \frac{\max\{1, \beta_2\}}{\min\{1, \beta_1\}\sqrt{2}} \,.$$

14-4 ERROR BOUNDS OF BILINEAR INTERPOLATION FUNCTIONS (TWO-DIMENSIONAL)

For two-dimensional problems, the mathematics involved in deriving an inequality corresponding to Eq. (14-51) is similar to, but slightly more involved than, that for the one-dimensional problem. We will show the details in Appendix C and simply write the final result:

$$\|\phi - \bar{\phi}\| \leqslant Eh \,, \tag{14-53}$$

where $$E = 2A^{1/2}(\tfrac{9}{4}h^2 M_1^2 + M_2^2 + M_3^2)^{1/2} \,.$$

Here A denotes the area of the triangle; M_1, M_2, and M_3 denote, respectively, the maximum values given in Eqs. (C-8), (C-11*a*), and (C-11*b*). For computable error bounds of solutions to Poisson's equation, readers may consult [11].

SYMBOLS

C^n	space that contains kth-order continuous derivatives, $k \leqslant n$
$\tilde{e}$	error of the approximate solution $\phi - \tilde{\phi}$
$\bar{e}$	error of the interpolation function $\phi - \bar{\phi}$
L	linear differential operator
$L_p[a, b]$	Lebesque space that contains all the functions satisfying $\{\int_a^b [\phi(x)]^p \, dx\}^{1/p} < \infty$
M_j	$[\max_{x_{j-1} \leqslant x \leqslant x_j} \lvert d^3\phi/dx^3 \rvert]^2$
S^h	J-dimensional subspace, where $h = 1/J$

$v(x)$ test function (or weighting function)
W_p^m Sobolev space [Eq. (14-8)]
β_1, β_2 lower and upper bounds of $a(x)$ [Eq. (14-12)]

Superscripts

$\sim$ approximation
$-$ interpolation

REFERENCES

1. K. Yoshida, *Functional Analysis*, p. 40, Springer-Verlag, Berlin, 1978.
2. S. Banach, *Theorie des Operations Linearires*, Hafner, New York, 1932.
3. W. Rudin, *Principles of Mathematical Analysis*, McGraw-Hill, New York, 1964.
4. D. Hilberg, Wesen und Ziele einer Analysis der unendlich vielen unabhangigen Variablen, *Rend. Circ. Mat. Palermo*, vol. 27, pp. 59–74, 1909.
5. P. Linz, *Theoretical Numerical Analysis*, p. 12, Wiley-Interscience, New York, 1979.
6. S. L. Sobolev, Sur un theoreme d'analyse fonctionnelle, *Math. Sb.*, vol. 45, pp. 471–496, 1938.
7. I. Babuska and W. C. Rheinboldt, A Posteriori Error Estimates for the Finite Element Method, *Int. J. Numer. Methods Eng.*, vol. 12, pp. 1597–1615, 1978.
8. I. Babuska and W. C. Rheinboldt, A Posteriori Error Analysis of Finite Element Solutions for One-dimensional Problems, *SIAM (Soc. Ind. Appl. Math.) J. Numer. Anal.*, vol. 18, pp. 565–589, 1981.
9. G. F. Carey and D. L. Humphrey, Mesh Refinement and Iterative Solution Methods for Finite Element Computations, *Int. J. Numer. Methods Eng.*, vol. 17, pp. 1717–1734, 1981.
10. P. Jamet, Stability and Convergence of a Generalized Crank-Nicolson Scheme on a Variable Mesh for the Heat Equation, *SIAM (Soc. Ind. Appl. Math.) J. Numer. Anal.*, vol. 17, pp. 530–539, 1980.
11. R. R. Barnhill, J. H. Brown, N. McQueen, and A. R. Mitchell, Computable Finite Element Error Bounds for Poissons' Equation, *Int. J. Numer. Methods Eng.*, vol. 11, pp. 593–603, 1977.

PROBLEMS

14-1 Let X be the Euclidean space spanned by the unit vectors i, j, and k. Each $\phi \in X$ can be written as

$$\phi = x_1 i + x_2 j + x_3 k\,.$$

Prove that

$$\|\phi\|_2 = (x_1^2 + x_2^2 + x_3^2)^{1/2} \quad \text{and} \quad \|\phi\|_1 = |x_1| + |x_2| + |x_3|$$

are norms.

14-2 Evaluate

$$\|e^x\|_{10} = \left[\int_0^1 (e^x)^{10}\,dx\right]^{1/10}.$$

Is the value close to the maximum norm?

14-3 Which of the spaces

(*a*) $C^1\,[0,1]$ with norm $\|\phi\| = \|\phi\|_\infty$
(*b*) $C^1\,[0,1]$ with norm $\|\phi\| = \|\phi\|_\infty + \|d\phi/dx\|_\infty$

is a Banach space?

14-4 For $\phi_1, \phi_2 \in$ Hilbert space, prove that

$$\|\phi_1 + \phi_2\|^2 + \|\phi_1 - \phi_2\|^2 = 2\|\phi_1\|^2 + 2\|\phi_2\|^2 .$$

Does this relation hold for $C^0\,[a,b]$ with maximum norm?

14-5 Show that the sequence $\{\sin \pi x, \sin 2\pi x, \ldots, \sin n\pi x, \ldots\}$ is a Cauchy sequence with respect to L_1 norm but is not with respect to L_2 norm.

14-6 In reference to Example 14-11, show that $C^0\,(-\infty, \infty)$ is incomplete with L_2 norm.

14-7 A linear space X is said to be a metric space if every element $\phi_1, \phi_2 \in X$ is associated with a positive number $d(\phi_1, \phi_2)$ such that

(*a*) $d(\phi_1, \phi_2) \geqslant 0$; equality holds if and only if $\phi_1 = \phi_2$;
(*b*) $d(\phi_1, \phi_2) = d(\phi_2, \phi_1)$;
(*c*) $d(\phi_1, \phi_2) \leqslant d(\phi_1, \phi_3) + d(\phi_3, \phi_2)$;

Show that every normed space is also a metric space if $d(\phi_1, \phi_2)$ represents $\|\phi_1 - \phi_2\|$.

14-8 Based on Eq. (14-7*a*), prove Eq. (14-7*b*).

14-9 In space $C^1\,[0,1]$, is

$$(\phi_1, \phi_2) = \int_0^1 \frac{d\phi_1}{dx}\frac{d\phi_2}{dx}\,dx + \phi_1(0)\phi_2(0)$$

an inner product?

14-10 Prove Eq. (14-7*b*), using $\phi_1(x) = x^2$ and $\phi_2(x) = \sin x$.

14-11 Prove that the operator $L = -d^2/dx^2 + a\,d/dx + 2$ is a positive-definite operator regardless of the sign of a.

CHAPTER

FIFTEEN

COMPARISON OF FINITE DIFFERENCE METHOD AND FINITE ELEMENT METHOD

Before solving a physical problem, one of the key questions to be answered is, Which numerical method should we choose? It is difficult to answer this question since the requirements of different tasks vary; while high accuracy may be emphasized in one task, inexpensiveness may be the first priority in another. Nevertheless, in this chapter we will attempt to establish some general guidelines by comparing the finite-difference method and the finite-element method in several respects. Such guidelines inevitably involve the author's personal opinion. Further, numerous novel ideas and techniques are now being developed. Hence these guidelines are meant to be only general suggestions and not strict rules.

In Section 15-1, the methods are compared with respect to the smoothness of the approximate solutions. Finite-element solutions require less smoothness and therefore the spaces that contain these solutions are generally larger than their finite-difference counterparts.

In Section 15-2 we consider the numerical instability caused by either a relatively large convective term in the steady-state equation or the adoption of an improper mesh size ratio in the parabolic transient equation. The finite-element method provides the option of choosing an optimal asymmetric weighting function such that both stability and accuracy can be attained. On the other hand, the restriction of the mesh size ratio to guarantee numerical stability in the explicit schemes is more severe for the finite-element method.

The accuracy of the numerical schemes is considered in Section 15-3. For the finite-difference method, the accuracy can be readily examined by the order of the truncation errors. For the finite-element method, since no Taylor's series expansion is used throughout the formulation, accuracy is inspected by an error bound equation.

In Section 15-4, discretization schemes of higher-order accuracy are discussed. We have established that it is generally possible to derive n-point relations by linearly combining the Taylor's series of adjacent unknowns expanded about the central grid point j. The corresponding n-point relations obtained by the finite-element formulation yield larger errors in the absolute nodal solution (but not in the Sobolev norm) if the weighted coefficients are found to have numerical values different from those obtained by the finite-difference formulation.

In Section 15-5, the capability of dealing with irregular geometry of the system domain is examined. The finite-element method appears to be flexible, since the region near the boundary can be triangulated with high conformity.

Section 15-6 concerns the capabilities of the two methods in tackling nonlinear problems. Incorporation with mixed boundary conditions is considered in Section 15-7. Finally, it is shown in Section 15-8 that the numerical results obtained with a nonuniform grid can sometimes be more accurate than those obtained with a uniform grid. Brief concluding remarks are presented in Section 15-9.

15-1 SMOOTHNESS

One of the major differences between the finite-difference method and the finite-element method is in the smoothness of the approximate solution to the differential equation. The former method generally requires that the approximation solution belong to C^{2k}, where $2k$ is the order of the differential equation, whereas the latter requires only that it belong to W_2^k (see Section 14-1p). We will illustrate this difference by considering the following two linear ordinary differential equations:

$$L_1\phi = -\frac{d^2\phi}{dx^2} + f_1\left(\frac{d\phi}{dx}, \phi, x\right) = 0\,, \qquad x \in [a, b] \tag{15-1a}$$

and

$$L_2\phi = \frac{d^4\phi}{dx^4} + f_2\left(\frac{d^3\phi}{dx^3}, \frac{d^2\phi}{dx^2}, \frac{d\phi}{dx}, \phi, x\right) = 0 \tag{15-1b}$$

subject to certain prescribed boundary conditions.

15-1a Finite-Difference Method

As mentioned in the preceding chapters, finite-difference methods employ the Taylor's series expansion. To approximate the second derivative $d^2\phi/dx^2$, we first expand $\phi(x_{j+1})$ about $\phi(x_j)$ by

$$\phi(x_{j+1}) = \phi(x_j) + \left(\frac{d\phi}{dx}\right)_{x_j} h + \left(\frac{d^2\phi}{dx^2}\right)_{x_j} \frac{h^2}{2} + \text{higher-order terms}\,.$$

Truncating the higher-order terms leads to the approximation solution

$$\tilde{\phi}(x_{j+1}) = \tilde{\phi}(x_j) + \left(\frac{d\tilde{\phi}}{dx}\right)_{x_j} h + \left(\frac{d^2\tilde{\phi}}{dx^2}\right)_{x_j} \frac{h^2}{2}, \tag{15-2a}$$

which suggests that $\tilde{\phi}(x) \in C^2[a, b]$ must have at least a second derivative. Similarly, we write

$$\tilde{\phi}(x_{j-1}) = \tilde{\phi}(x_j) - \left(\frac{d\tilde{\phi}}{dx}\right)_{x_j} h + \left(\frac{d^2\tilde{\phi}}{dx^2}\right)_{x_j} \frac{h^2}{2}. \tag{15-2b}$$

Then rearranging Eqs. (15-2*a*) and (15-2*b*) leads to the familiar three-point relation

$$\frac{d^2\tilde{\phi}}{dx^2} = \frac{1}{h^2}(\phi_{j-1} - 2\phi_j + \phi_{j+1}).$$

To approximate the fourth derivative $d^4\phi/dx^4$, we write

$$\tilde{\phi}(x_{j+k}) = \tilde{\phi}(x_j) + \left(\frac{d\tilde{\phi}}{dx}\right)_{x_j}(kh) + \left(\frac{d^2\tilde{\phi}}{dx^2}\right)_{x_j}\frac{(kh)^2}{2} + \left(\frac{d^3\tilde{\phi}}{dx^3}\right)_{x_j}\frac{(kh)^3}{6}$$
$$+ \left(\frac{d^4\tilde{\phi}}{dx^4}\right)_{x_j}\frac{(kh)^4}{24}, \quad k = 1, 2 \tag{15-3a, b}$$

and

$$\tilde{\phi}(x_{j-k}) = \tilde{\phi}(x_j) - \left(\frac{d\tilde{\phi}}{dx}\right)_{x_j}(kh) + \left(\frac{d^2\tilde{\phi}}{dx^2}\right)_{x_j}\frac{(kh)^2}{2} - \left(\frac{d^3\tilde{\phi}}{dx^3}\right)_{x_j}\frac{(kh)^3}{6}$$
$$+ \left(\frac{d^4\tilde{\phi}}{dx^4}\right)_{x_j}\frac{(kh)^4}{24}. \tag{15-3c, d}$$

It can be seen from Eqs. (15-3*a*)–(15-3*d*) that $C^4[a, b]$ continuity of $\tilde{\phi}(x)$ is required. Eliminating $(d^\nu\phi/dx^\nu)_{x_j}$, $\nu = 1, 2, 3$, from Eqs. (15-3*a*)–(15-3*d*) eventually yields the five-point relation

$$\left(\frac{d^4\tilde{\phi}}{dx^4}\right)_{x_j} = \frac{1}{h^4}(\phi_{j-2} - 4\phi_{j-1} + 6\phi_j - 4\phi_{j+1} + \phi_{j+2}). \tag{15-4}$$

It is noteworthy that, after the nodal unknowns are computed, the approximate solution in the entire domain, other than at those discrete points, remains unavailable unless additional interpolation is made.

Example 15-1 Suppose that a finite-difference scheme yields the nodal solution as

$$\tilde{\phi}(3) = 1, \quad \tilde{\phi}(4) = 2, \quad \text{and} \quad \tilde{\phi}(5) = 5. \tag{a}$$

Find $\tilde{\phi}(4.5)$ by use of Eq. (15-2*a*). Let $x_j = 4$ and $h = 1$.

Solution The first and second derivatives can be computed, respectively, by

$$\left(\frac{d\tilde{\phi}}{dx}\right)_{x=4} = \frac{\tilde{\phi}(5) - \tilde{\phi}(3)}{5 - 3} = 2 \tag{b}$$

and

$$\left(\frac{d^2\tilde{\phi}}{dx^2}\right)_{x=4} = \tilde{\phi}(3) - 2\tilde{\phi}(4) + \tilde{\phi}(5) = 2. \tag{c}$$

Substituting Eqs. (b) and (c) into Eq. (15-2a) yields

$$\tilde{\phi}(4.5) = \tilde{\phi}(4) + \left(\frac{d\tilde{\phi}}{dx}\right)_{x=4}(0.5) + \left(\frac{d^2\tilde{\phi}}{dx^2}\right)_{x=4}\frac{(0.5)^2}{2} = 3.25\,, \qquad (d)$$

which is slightly different from the average of $\tilde{\phi}(4)$ and $\tilde{\phi}(5)$.

15-1b Finite-Element Method

We will exclusively speak of the Galerkin formulation here. Let the approximate function $\tilde{\phi}(x)$ be expressed by

$$\tilde{\phi}(x) = \sum_{j=0}^{J} v_j(x)\phi_j\,, \qquad v_j(a) = v_j(b) = 0 \qquad \text{if } v_j \in W_2^1$$

$$v_j(a) = \frac{dv_j(a)}{dx} = v_j(b) = \frac{dv_j(b)}{dx} = 0 \qquad \text{if } v_j \in W_2^2\,,$$

where W denotes the Sobolev space (see Section 14-1p). Integrating the inner product $(L_1\tilde{\phi}, v_j)$ by parts yields

$$(L_1\tilde{\phi}, v_j) = \int_a^b \frac{d\tilde{\phi}}{dx}\frac{dv_j}{dx}\,dx + \int_a^b f_1 v_j\,dx = 0\,. \qquad (15\text{-}5)$$

It is seen from Eq. (15-5) that the requirement of smoothness of $\tilde{\phi}(x)$ is much relaxed; $\tilde{\phi}(x)$ only needs to be such that $d\tilde{\phi}/dx$ is integrable, i.e., $\tilde{\phi}(x) \in W_2^1\,[a, b]$.

Next, integrating the inner product $(L_2\tilde{\phi}, v_j)$ by parts twice yields

$$(L_2\tilde{\phi}, v_j) = \int_a^b \frac{d^4\tilde{\phi}}{dx^4}v_j\,dx + \int_a^b f_2 v_j\,dx = \int_a^b \frac{d^2\tilde{\phi}}{dx^2}\frac{d^2v_j}{dx^2}\,dx + \int_a^b f_2 v_j\,dx^{\dagger}$$
$$= 0\,, \qquad (15\text{-}6)$$

which dictates that $d^2\tilde{\phi}/dx^2$ is integrable, i.e., $\tilde{\phi} \in W_2^2\,[a, b]$.

Example 15-2 Is the quadratic spline

$$S(x) = \begin{cases} x^2\,, & x \in [0, 1] \\ 1 + 2(x-1) - (x-1)^2\,, & x \in [1, 2] \\ 2 - (x-2)^2\,, & x \in [2, 3] \\ (4-x)^2\,, & x \in [3, 4] \end{cases} \qquad (a)$$

an element in $W_2^2\,[0, 4]$?

†If $d^3\tilde{\phi}/dx^3$ is contained in the function f_2, it must also be integrated by parts at least once.

Solution If $S(x)$ belongs to $W_2^2\,[0,4]$, we must have

$$\int_0^4 \left[S^2 + \left(\frac{dS}{dx}\right)^2 + \left(\frac{d^2S}{dx^2}\right)^2 \right] dx < \infty . \tag{b}$$

Answering the question is equivalent to examining whether Eq. (*b*) is valid. Differentiating Eq. (*a*) twice yields

$$\frac{d^2S}{dx^2} = \begin{cases} 2\,, & x \in [0,1] \\ -2\,, & x \in [1,2] \\ -2\,, & x \in [2,3] \\ 2\,, & x \in [3,4] \end{cases} . \tag{c}$$

Although d^2S/dx^2 does not exist at $x = 1$ and $x = 3$, its integral $\int_0^4 (d^2S/dx^2)\,dx$ or the integral of the squared expression $\int_0^4 (d^2S/dx^2)^2\,dx$ does, i.e.,

$$\int_0^4 \left(\frac{d^2S}{dx^2}\right)^2 dx = 16 < \infty .$$

Further, since the quadratic spline has a continuous first derivative, it follows that Eq. (*b*) is valid. Therefore, $S(x)$ in Eq. (*a*) belongs to $W_2^2\,[0,4]$.

Remarks:

1. The finite-element method that incorporates evaluation of the inner-product integral enables us to seek an approximate solution in large spaces. We are provided with more functions in large spaces and therefore are in a better position to seek admissible functions.
2. In the finite-element method, an approximate solution throughout the entire domain automatically becomes available as soon as the nodal variables are obtained. In the finite-difference method, the solution is strictly discrete; we are seeking the unknowns only at the grid points, not throughout the entire domain. To obtain the solution within a computational molecule, we need to use, for example, interpolation and perform additional calculations.

15-2 NUMERICAL INSTABILITY

At least two types of numerical instability may arise due to the discretization. One is associated with a large convective term in the streamwise diffusion equation. The other concerns the adoption of an improper mesh size ratio in the transient equation. We will compare the instabilities of both the finite-difference equation and the finite-element equation in terms of these two types.

15-2*a* Convective Numerical Instability

We recall (Section 8-3) that when the one-dimensional streamwise diffusion equation

$$\frac{d^2\phi}{dx^2} - \frac{u}{\alpha}\frac{d\phi}{dx} = 0 \tag{15-7}$$

was discretized by the central finite-difference scheme

$$\frac{d\phi}{dx} = \frac{1}{2h}(\phi_{j+1} - \phi_{j-1}) + O(h^2), \tag{15-8a}$$

numerical instability arose if $\mathrm{Pe} > 2$, where $\mathrm{Pe} = uh/\alpha$. This instability can be removed by using the upwind finite-difference scheme

$$\frac{d\phi}{dx} = \frac{1}{h}(\phi_j - \phi_{j-1}) + O(h). \tag{15-8b}$$

But unfortunately, the accuracy of this scheme, as shown in Eq. (15-8*b*), is reduced to $O(h)$. Other finite-difference treatments, such as the Arakawa scheme [1-3], have been developed to achieve both accuracy and stability.

On the other hand, through the finite-element formulation with the use of asymmetric weighting functions (Section 8-3*b*), defined as

$$W_j(x) = \begin{cases} N_j(x) + \mu n_j(x), & x \in [x_{j-1}, x_j], \\ N_j(x) - \mu n_j(x), & x \in [x_j, x_{j+1}], \\ 0 & \text{elsewhere}, \end{cases} \tag{15-9}$$

the first derivative can be discretized to

$$\frac{\int_{x_{j-1}}^{x_{j+1}} (d\tilde{\phi}/dx) W_j \, dx}{\int_{x_{j-1}}^{x_{j+1}} W_j \, dx} = -\frac{1}{2h}(1+\mu)\phi_{j-1} + \frac{\mu}{h}\phi_j + \frac{1}{2h}(1-\mu)\phi_{j+1}, \tag{15-10}$$

where $n_j(x) = 3(x - x_{j-1})(x_{j+1} - x)/h^2$ and μ is a constant to be optimally chosen to achieve both stability and accuracy. If μ is chosen to be 0 and 1, Eq. (15-10) degenerates to Eqs. (15-8*a*) and (15-8*b*), respectively. Therefore, Eq. (15-10) appears to be more general than the other two. In fact, it was shown in Section 8-3*b* that, as long as

$$\mu > \frac{\mathrm{Pe} - 2}{\mathrm{Pe}},$$

the discretized form of Eq. (15-7) is stable. Thus, there may exist an optimal value of μ other than 0 and 1 such that both accuracy and stability of the discretization scheme are achieved. This optimal value can be determined by Eq. (15-7) and generally lies between 0 and 1.

Example 15-3 Consider the function $\phi(x) = e^{\text{Pe}\,x}$, which is an exact solution of Eq. (15-7). Compare the accuracies of the numerical values $(d\phi/dx)_{x=1}$ obtained by using Eqs. (15-8*b*) and (15-10). Let Pe = 3 and $h = 0.1$.

Solution First, the exact value is evaluated as

$$\left(\frac{d\phi}{dx}\right)_{x=1} = 3e^3 = 60.255\,. \tag{a}$$

Equation (15-8*b*) leads to

$$\frac{1}{h}(\phi_j - \phi_{j-1}) = \frac{e^3 - e^{2.7}}{0.1} = 52.058\,, \tag{b}$$

which deviates greatly from the exact value given in Eq. (*a*). Next, Eq. (15-10) is evaluated as

$$\left(\frac{d\phi}{dx}\right)_{x=1} \approx -5(1+\mu)e^{2.7} + 10\mu e^3 + 5(1-\mu)e^{3.3} = 61.1643 - 9.1117\mu\,. \tag{c}$$

If μ is chosen to be 0.1, then Eq. (*c*) approaches the exact value given in Eq. (*a*). Unfortunately, this choice is not allowed because the stability criterion demands that, at least,

$$\mu > \frac{\text{Pe} - 2}{\text{Pe}} = \frac{1}{3}\,. \tag{d}$$

Therefore, we arbitrarily choose a value of $\mu \in [\frac{1}{3}, 1]$, say 0.5, and Eq. (*c*) yields

$$\left(\frac{d\phi}{dx}\right)_{x=1} \approx 56.6085\,. \tag{e}$$

Equation (*e*) is closer to the exact value than Eq. (*b*).

An alternative to selection of the asymmetric weighting function in Eq. (15-9) is to find the solutions that satisfy the adjoint equation of Eq. (15-7)

$$\frac{d^2 W_j}{dx^2} + \frac{u}{\alpha}\frac{dW_j}{dx} = 0 \tag{15-11}$$

subject to

$$W_j(x_{j-1}) = 0 \quad \text{and} \quad W_j(x_j) = 1 \quad \text{if } x \in [x_{j-1}, x_j]$$

and to

$$W_j(x_j) = 1 \quad \text{and} \quad W_j(x_{j+1}) = 0 \quad \text{if } x \in [x_j, x_{j+1}]\,.$$

The rationale for choosing such an asymmetric weighting function was presented in Section 8-3*d*. With this choice, the finite-element discretization equation was shown to be stable as well as accurate.

15-2*b* Transient Instability

When the one-dimensional transient heat conduction equation

$$\frac{\partial \phi}{\partial t} = \alpha \frac{\partial^2 \phi}{\partial x^2}$$

is discretized, the stability criterion is

$$r = \frac{\alpha\, \Delta t}{h^2} \leqslant \frac{1}{2}$$

for the explicit finite-difference scheme, as indicated by Eq. (4-28), and

$$r \leqslant \tfrac{1}{6}$$

for the "explicit" finite-element scheme, as indicated by Eq. (5-30). Since the latter criterion is subject to a rather severe restriction on r and the corresponding scheme is not really explicit, we generally adopt the implicit finite-element method to discretize a transient equation.

For two-dimensional transient problems described by

$$\frac{\partial \phi}{\partial t} = \alpha \left(\frac{\partial^2 \phi}{\partial x^2} + \frac{\partial^2 \phi}{\partial y^2} \right),$$

the explicit finite-difference scheme with uniform mesh size is stable if

$$r \leqslant \tfrac{1}{4}$$

and the explicit finite-element method with square elements is stable if

$$r \leqslant \tfrac{1}{12} .$$

Remarks:

1. Finite-element explicit schemes are generally subject to stricter stability criteria than their finite-difference counterparts.
2. After Galerkin formulation with the pyramid weighting function, the one-dimensional transient equation can be discretized to

$$\frac{1}{6} \frac{d\phi_{j-1}}{dt} + \frac{2}{3} \frac{d\phi_j}{dt} + \frac{1}{6} \frac{d\phi_{j+1}}{dt} = r\phi_{j-1} - 2r\phi_j + r\phi_{j+1} . \qquad (15\text{-}12)$$

Using the implicit scheme, we can further discretize Eq. (15-12) to

$$(1 - 6r)\phi_{j-1}^{(n)} + (4 + 12r)\phi_j^{(n)} + (1 - 6r)\phi_{j+1}^{(n)} = \phi_{j-1}^{(n-1)} + 4\phi_j^{(n-1)} + \phi_{j+1}^{(n-1)}, \qquad (15\text{-}13)$$

whose matrix form is by no means more complicated than that of the explicit scheme. Furthermore, letting

$$\phi_j^{(n)} = A_m \xi^n \exp{(imj\pi h)}$$

in the von Neumann stability analysis, we can derive

$$-1 \leqslant \xi = \frac{2 + \cos\theta}{(1 - 6r)\cos\theta + 2 + 6r} \leqslant 1\,,$$

where $\theta = m\pi h$. Since this equation is valid for any r, Eq. (15-13) is unconditionally stable. For these two reasons, it is preferable that the semidiscrete Galerkin Eq. (15-12) be discretized by using the implicit scheme.

3. When both the transient term and the convective term are present in the same governing equation, e.g., the linearized Burgers equation

$$\frac{\partial\phi}{\partial t} = \alpha\frac{\partial^2\phi}{\partial x^2} - u\frac{\partial\phi}{\partial x}\,, \tag{15-14}$$

the mathematics of the stability analysis becomes slightly more complicated. Readers may consult [4-10] for details.

15-3 ACCURACY AND ERROR BOUNDS

In the finite-difference method, accuracy is judged by the truncation error of the Taylor's series expansion

$$L\phi = \tilde{L}\phi_{j,k}^{(n)} + O(h^{\nu}, (\Delta t)^{\mu})\,. \tag{15-15}$$

The scheme is said to be accurate if the magnitude of the second term $O(\cdot)$ is small. The accuracy can be increased in the following ways:

1. Reducing the mesh sizes h and Δt (if the difference scheme is consistent). But this reduction is penalized by the increase in computer memory storage and the round-off error.
2. Choosing proper interval size ratios such that the higher-order derivatives are nearly or completely canceled by each other. See the treatments for the transient one-dimensional heat conduction equation in Section 4-4 and for the boundary-layer equation in Section 6-4*c*.
3. Adopting higher-order analyses by considering many-point relations such as the five-point relation for second ordinary derivatives [Eqs. (1-8*a*) and (1-8*b*)] and the nine-point and 21-point relations for the Laplace equation [Eq. (15-18)].
4. Adopting higher-order analyses by requiring smoother continuities and considering more nodal unknowns. The number of nodal points, however, remains the same as in second-order analyses. See Example 1-1 and Section 5-2*b*. Generally, the discretized equation is arranged in a block matrix system for efficient solution by a block matrix solver. Similar approaches are operator compact implicit (OCI) schemes [11-13] which are sometimes called Mehrstellen schemes.

For finite-element methods, we can no longer speak of the truncation errors (unless it is defined in the broad sense [14]). Instead, we judge the accuracy of the finite-element solution $\tilde{\phi}$ by the error bounds measured by norms of a certain type (Section 14-1):

$$\|e\| = \|\phi - \tilde{\phi}\| \leqslant Mh^{\nu} . \tag{15-16}$$

Loosely speaking, if the value of Mh^{ν} is small, the scheme is accurate. This accuracy can be increased by:

1. Subdividing the domain into finer mesh systems.
2. Raising the power ν by means of higher-order C^0 and C^1 elements (see Sections 3-1 and 14-4). This treatment also increases the rate of convergence, as discussed in Section 4-3*c*.

Remarks:

1. It is relatively easy to compare the accuracy of a finite-difference scheme with that of another finite-difference scheme based simply on the magnitude of the truncation error. It is also straightforward to compare the error bound of a finite-element scheme with that of another finite-element method by using Eq. (15-16). For example, the higher-order formula

$$\frac{d^2\phi}{dx^2} = \frac{1}{12h^2}(-\phi_{j-2} + 16\phi_{j-1} - 30\phi_j + 16\phi_{j+1} - \phi_{j+2}) + O(h^4)$$

is more accurate than the second-order formula

$$\frac{d^2\phi}{dx^2} = \frac{1}{h^2}(\phi_{j-1} - 2\phi_j + \phi_{j+1}) + O(h^2) .$$

Also, the error bound

$$\|\phi - \tilde{\phi}\| \leqslant M_1 h^2 ,$$

obtained by using the one-dimensional quadratic element, has a higher power of h than the error bound

$$\|\phi - \tilde{\phi}\| \leqslant M_2 h ,$$

obtained by using the one-dimensional linear element [15].

2. It is, however, difficult to compare the accuracy of a finite-*difference* scheme with that of a finite-*element* scheme unless the exact solution is available.

Example 15-4 We assume that the finite-difference solution Eq. (1-6) can be represented by

$$\begin{aligned}\tilde{\phi}_{FD} &= \phi_1 N_1(x) + \phi_2 N_2(x) + \phi_3 N_3(x) + N_4(x)\\ &= 0.0582N_1(x) + 0.1552N_2(x) + 0.3944N_3(x) + N_4(x)\end{aligned} \tag{a}$$

and rewrite the finite-element solution Eq. (1-30*b*) as

$$\tilde{\phi}_{FE} = 0.0444N_1(x) + 0.1269N_2(x) + 0.3563N_3(x) + N_4(x) . \tag{b}$$

Compare $\|\phi - \tilde{\phi}_{FD}\|_{L_2}$ with $\|\phi - \tilde{\phi}_{FE}\|_{L_2}$, where ϕ is the exact solution of Eq. (1-1).

Solution The exact solution of Eq. (1-1) is

$$\phi(x) = a(e^x - e^{-2x}), \tag{c}$$

where

$$a = \frac{1}{e^4 - e^{-8}}.$$

Therefore, with the coordinate transformation we derive

$$\|\phi - \tilde{\phi}\|_{L_2}^2 = \int_0^4 [\phi(x) - \tilde{\phi}(x)]^2\, dx$$

$$= \int_0^1 [a(e^\xi - e^{-2\xi}) - \phi_1 \xi]^2\, d\xi + \int_0^1 [a(e^{1+\xi} - e^{-2-2\xi})$$

$$- \phi_1 - \xi(\phi_2 - \phi_1)]^2\, d\xi + \int_0^1 [a(e^{2+\xi} - e^{-4-4\xi})$$

$$- \phi_2 - \xi(\phi_3 - \phi_2)]^2\, d\xi + \int_0^1 [a(e^{3+\xi} - e^{-6-6\xi}) - \phi_3$$

$$- \xi(1 - \phi_3)]^2\, d\xi. \tag{d}$$

Substituting the values of ϕ_1, ϕ_2, and ϕ_3, given in Eqs. (*a*) and (*b*), into Eq. (*d*) yields

$$\|\phi - \tilde{\phi}_{FD}\|_{L_2}^2 = 0.00723 \tag{e}$$

and

$$\|\phi - \tilde{\phi}_{FE}\|_{L_2}^2 = 0.00286. \tag{f}$$

On the basis of L_2 norm, we conclude that the finite-element scheme Eq. (1-30*b*) is more accurate than the finite-difference scheme Eq. (1-6). In fact, in terms of Sobolev norm, the same conclusion also holds.

15-4 DISCRETIZATION SCHEMES OF HIGHER-ORDER ACCURACY

Higher-order equations such as the 21-point relation can be derived from either the finite-difference formulation or the finite-element formulation. In this section we will present the derivation of both equations. Although use of the 21-point relation may be impractical, such a derivation can be readily extended to relations of fewer points, as illustrated in Examples 15-5 and 15-6. The finite-difference formulation is to be presented first.

15-4*a* Finite-Difference Formulation

Figure 15-1 shows the computational molecule containing the central grid point j and the 20 surrounding grid points. A typical nodal unknown ϕ_{NW} can be expanded about ϕ_j in Taylor's series as

$$\phi_{NW} = \phi_j + \left[h\left(-\frac{\partial}{\partial x} + \frac{\partial}{\partial y}\right) + \frac{h^2}{2}\left(-\frac{\partial}{\partial x} + \frac{\partial}{\partial y}\right)^2 + \frac{h^3}{6}\left(-\frac{\partial}{\partial x} + \frac{\partial}{\partial y}\right)^3 + \frac{h^4}{24}\left(-\frac{\partial}{\partial x} + \frac{\partial}{\partial y}\right)^4 + \frac{h^5}{120}\left(-\frac{\partial}{\partial x} + \frac{\partial}{\partial y}\right)^5 \right] \phi(x,y) + O(h^6). \tag{15-15}$$

Multiplying Eq. (15-15) by a constant a_1 and rearranging the result leads to

$$a_1(\phi_{NW} - \phi_j) = -a_1 h \frac{\partial \phi}{\partial x} + a_1 h \frac{\partial \phi}{\partial y} + \frac{a_1 h^2}{2} \frac{\partial^2 \phi}{\partial x^2} + \frac{a_1 h^2}{2} \frac{\partial^2 \phi}{\partial y^2} - \frac{a_1 h^2}{2} \frac{\partial^2 \phi}{\partial x \, \partial y} + \cdots - \frac{a_1 h^5}{120} \frac{\partial^5 \phi}{\partial x^5} + \frac{a_1 h^5}{120} \frac{\partial^5 \phi}{\partial y^5}. \tag{15-16a}$$

Similarly,

$$a_2(\phi_{NWW} - \phi_j) = -a_2 h \frac{\partial \phi}{\partial x} + \frac{a_2 h}{2} \frac{\partial \phi}{\partial y} + \frac{a_2 h^2}{2} \frac{\partial^2 \phi}{\partial x^2} + \frac{a_2 h^2}{8} \frac{\partial^2 \phi}{\partial y^2} - \frac{a_2 h^2}{4} \frac{\partial^2 \phi}{\partial x \, \partial y} + \cdots - \frac{a_2 h^5}{120} \frac{\partial^5 \phi}{\partial x^5} + \frac{a_2 h^5}{120 \times 32} \frac{\partial^5 \phi}{\partial y^5}, \tag{15-16b}$$

$$a_3(\phi_W - \phi_j) = -a_3 h \frac{\partial \phi}{\partial x} + 0 + \frac{a_3 h^2}{2} \frac{\partial^2 \phi}{\partial x^2} + 0 + 0 + \cdots - \frac{a_3 h^5}{120} \frac{\partial^5 \phi}{\partial x^5} + 0, \tag{15-16c}$$

. . . . . .

$$a_{20}(\phi_e - \phi_j) = a_{20} \frac{h}{2} \frac{\partial \phi}{\partial x} + 0 + \frac{a_{20} h^2}{8} \frac{\partial^2 \phi}{\partial x^2} + 0 + 0 + \cdots + \frac{a_{20} h^5}{120 \times 32} \frac{\partial^5 \phi}{\partial x^5} + 0. \tag{15-16t}$$

The sum of the coefficients of $\partial\phi/\partial x$, for example, is

$$h(-a_1 - a_2 - a_3 + \cdots + \tfrac{1}{2} a_{20}).$$

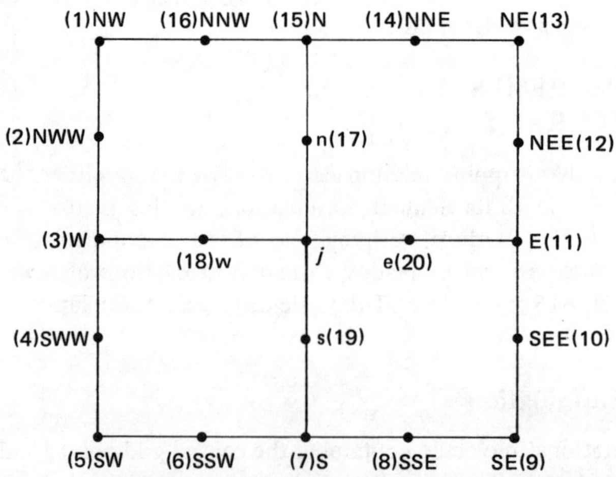

Figure 15-1 Quadratic square element involving 21 grid points.

The numerical values of these undetermined coefficients $a_1, a_2, a_3, \ldots, a_{20}$ can be evaluated by 20 equations:

$$-a_1 - a_2 - a_3 + \cdots + \frac{a_{20}}{2} = 0\,,$$

$$a_1 + \frac{a_2}{2} + 0 + \cdots + 0 = 0\,,$$

$$\frac{a_1}{2} + \frac{a_2}{2} + \frac{a_3}{2} + \cdots + \frac{a_{20}}{8} = 1\,, \qquad (15\text{-}17a\text{–}t)$$

$$\frac{a_1}{2} + \frac{a_2}{8} + 0 + \cdots + 0 = 1\,,$$

$$-\frac{a_1}{2} - \frac{a_2}{4} + 0 + \cdots + 0 = 0\,,$$

... ...

$$-\frac{a_1}{120} - \frac{a_2}{120} - \frac{a_3}{120} + \cdots + \frac{a_{20}}{120 \times 32} = 0\,,$$

and

$$\frac{a_1}{120} + \frac{a_2}{120 \times 32} + 0 + \cdots + 0 = 0\,.$$

Note that only the coefficients of $\partial^2\phi/\partial x^2$ and $\partial^2\phi/\partial y^2$ are set to unity, and others are forced to vanish. The calculation can be facilitated by letting

$$a_1 = a_5 = a_9 = a_{13}\,,$$

$$a_3 = a_7 = a_{11} = a_{15}\,,$$

$$a_2 = a_4 = a_6 = a_8 = a_{10} = a_{12} = a_{14} = a_{16}\,,$$

and

$$a_{17} = a_{18} = a_{19} = a_{20}\,.$$

Use of this set of equations leads to

$$\frac{a_1 \sum_{p=NW}^{NE} \phi_p + a_3 \sum_{q=W}^{N} \phi_q + a_2 \sum_{r=NWW}^{NNW} \phi_r + a_{17} \sum_{t=w}^{n} \phi_t - 4(a_1 + a_3 + 2a_2 + a_{17})\phi_j}{h^2}$$

$$= \frac{\partial^2\phi}{\partial x^2} + \frac{\partial^2\phi}{\partial y^2} + O(h^4)\,, \qquad (15\text{-}18)$$

where $a_1 = \frac{2}{3}$, $a_2 = -\frac{4}{3}$, $a_3 = 1$, and $a_{17} = 8$.

By use of the method of undetermined coefficients, any discretization scheme representing an n-point relation can, in principle, be obtained. We also note that the sum of the coefficients on the left-hand side of Eq. (15-18) should be equal to zero because the equation is also valid for the case of uniform ϕ.

Example 15-5 Derive the five-point relation for the Laplace equation by means of the method of undetermined coefficients.

Solution The positions W, S, E, and N are designated as 3, 7, 11, and 15, respectively. We need to take only the leading four terms in Eq. (15-15), namely $\partial/\partial x$, $\partial/\partial y$, $\partial^2/\partial x^2$, and $\partial^2/\partial y^2$, and write

$$a_3(\phi_W - \phi_j) = -a_3 h \frac{\partial \phi}{\partial x} + 0 + \frac{a_3 h^2}{2} \frac{\partial^2 \phi}{\partial x^2} + 0\,, \tag{a}$$

$$a_7(\phi_S - \phi_j) = 0 - a_7 h \frac{\partial \phi}{\partial y} + 0 + \frac{a_7 h^2}{2} \frac{\partial^2 \phi}{\partial y^2}\,, \tag{b}$$

$$a_{11}(\phi_E - \phi_j) = a_{11} h \frac{\partial \phi}{\partial x} + 0 + \frac{a_{11} h^2}{2} \frac{\partial^2 \phi}{\partial x^2} + 0\,, \tag{c}$$

and

$$a_{15}(\phi_N - \phi_j) = 0 + a_{15} h \frac{\partial \phi}{\partial y} + 0 + \frac{a_{15} h^2}{2} \frac{\partial^2 \phi}{\partial y^2}\,. \tag{d}$$

In order to obtain the desired discretization equation

$$a_3\phi_W + a_7\phi_S + a_{11}\phi_E + a_{15}\phi_N - (a_3 + a_7 + a_{11} + a_{15})\phi_j = h^2\left(\frac{\partial^2 \phi}{\partial x^2} + \frac{\partial^2 \phi}{\partial y^2}\right), \tag{e}$$

we must have

$$-a_3 + a_{11} = 0\,, \tag{f}$$

$$-a_7 + a_{15} = 0\,, \tag{g}$$

$$\frac{a_3}{2} + \frac{a_{11}}{2} = 1\,, \tag{h}$$

and

$$\frac{a_7}{2} + \frac{a_{15}}{2} = 1\,. \tag{i}$$

Solving Eqs. (f)–(i) yields

$$a_3 = a_7 = a_{11} = a_{15} = 1\,. \tag{j}$$

Consequently, Eq. (e) becomes

$$\frac{\partial^2 \phi}{\partial x^2} + \frac{\partial^2 \phi}{\partial y^2} = \frac{1}{h^2}\left(\sum_{p=W}^{N} \phi_p - 4\phi_j\right), \tag{k}$$

which is the familiar five-point relation for the Laplace equation.

Example 15-6 Derive the five-point relation for $d^2\phi/dx^2$ by means of the method of undetermined coefficients.

Solution In Eq. (15-15), we retain only the terms from first to fourth derivatives with respect to x. In reference to Fig. 15-1, the positions W, E, w, and e are designated as 3, 11, 18, and 20, respectively. Letting the coefficients of $d\phi/dx, d^3\phi/dx^3$, and $d^4\phi/dx^4$ vanish and that of $d^2\phi/dx^2$ be equal to unity, we obtain

$$-a_3 + a_{11} - \frac{a_{18}}{2} + \frac{a_{20}}{2} = 0\,, \tag{a}$$

$$\frac{a_3}{2} + \frac{a_{11}}{2} + \frac{a_{18}}{8} + \frac{a_{20}}{8} = 1\,, \tag{b}$$

$$-\frac{a_3}{6} + \frac{a_{11}}{6} - \frac{a_{18}}{48} + \frac{a_{20}}{48} = 0\,, \tag{c}$$

and

$$\frac{a_3}{24} + \frac{a_{11}}{24} + \frac{a_{18}}{384} + \frac{a_{20}}{384} = 0\,. \tag{d}$$

The simultaneous solution of Eqs. (*a*)-(*d*) is

$$a_3 = a_{11} = -\tfrac{1}{3}\,, \quad \text{and} \quad a_{18} = a_{20} = \tfrac{16}{3}\,. \tag{e}$$

Consequently, the five-point relation becomes

$$\frac{d^2\phi}{dx^2} = \frac{1}{3h^2}(-\phi_W + 16\phi_w - 30\phi_j + 16\phi_e - \phi_E)\,. \tag{f}$$

Since the interval size given here is half of that given in Eq. (1-8*a*) previously, Eq. (*f*) is the same as Eq. (1-8*a*). Furthermore, if the numerical values of a_3, a_{11}, a_{18}, and a_{20} are used to calculate the coefficients of $\partial^5\phi/\partial x^5$ and $\partial^6\phi/\partial x^6$, respectively, we find that

$$\frac{h^5}{120}\frac{\partial^5\phi}{\partial x^5}\left(a_3 + a_{11} - \frac{a_{18}}{32} + \frac{a_{20}}{32}\right) = 0 \tag{g}$$

and

$$\frac{h^6}{720}\frac{\partial^6\phi}{\partial x^6}\left(a_3 + a_{11} + \frac{a_{18}}{64} + \frac{a_{20}}{64}\right) = \frac{-h^6}{1440}\frac{\partial^6\phi}{\partial x^6}\,. \tag{h}$$

This indicates that the five-point relation Eq. (*f*) has accuracy $O(h^4)$.

15-4*b* Finite-Element Formulation

In reference to the preceding example, it is interesting to examine the accuracy of the five-point relation, Eq. (*f*) in Example 3-2, derived by the finite-element method. We rewrite

$$\frac{\int_{x_W}^{x_E}(d^2\phi/dx^2)N_j\,dx}{\int_{x_W}^{x_E}N_j\,dx} = \frac{1}{h^2}(-\phi_W + 8\phi_w - 14\phi_j + 8\phi_e - \phi_E)\,. \tag{15-19}$$

Matching with the designated numbers shown in Fig. 15-1 yields

$$a_3 = a_{11} = -1\ , \quad a_{18} = a_{20} = 8\ , \tag{15-20}$$

which satisfy Eqs. (*a*)-(*c*), but not Eq. (*d*), in Example 15-6.

Next, we will turn our attention to the finite-element formulation that leads to the 21-point relation.

If we approximate the interpolant inside the isoparametric element shown in Fig. 15-2 in terms of the eight nodal values as

$$\tilde{\phi}(x,y) = \sum_{m=a}^{h} N_m(x,y)\phi_m\ , \tag{15-21}$$

it can be derived after considerable algebra that

$$N_p = \frac{(1+\xi\xi_p)(1+\eta\eta_p)(\xi\xi_p+\eta\eta_p-1)}{4}\ , \qquad p = a,c,e,g \tag{15-22a–d}$$

$$N_q = \frac{(1-\xi^2)(1+\eta\eta_q)}{2}\ , \qquad q = b,f \tag{15-22e, f}$$

and

$$N_r = \frac{(1+\xi\xi_r)(1-\eta^2)}{2}\ , \qquad r = d,h \tag{15-22g, h}$$

where $\xi = 2(x-x^*)/h$ and $\eta = 2(y-y^*)/h$ are isoparametric coordinates ranging from -1 to 1; x^* and y^* are the coordinates of the element center. Following the derivation of Eq. (5-51), we obtain

$$(-\nabla^2\tilde{\phi}, N_j) = K_{NW} + K_{SW} + K_{SE} + K_{NE} = 0\ , \tag{15-23}$$

where

$$K_p = \iint_{e_p} \left(\frac{\partial\tilde{\phi}}{\partial x}\frac{\partial N_j}{\partial x} + \frac{\partial\tilde{\phi}}{\partial y}\frac{\partial N_j}{\partial y} \right) dx\, dy\ , \qquad p = NW, SW, SE, NE\ . \tag{15-24}$$

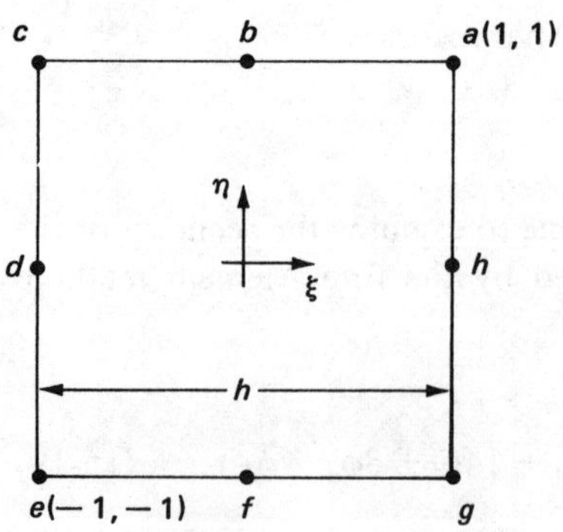

Figure 15-2 Isoparametric quadratic square element.

Table 15-1 Convenient result that may be used in deriving the 20-point discretized equation

$$E = \frac{\partial \phi}{\partial x}\frac{\partial N_j}{\partial x} + \frac{\partial \phi}{\partial y}\frac{\partial N_j}{\partial y} \quad \text{and} \quad A_{pq} = \int_{-1}^{1}\int_{-1}^{1}\left(\frac{\partial N_p}{\partial \xi}\frac{\partial N_q}{\partial \xi} + \frac{\partial N_p}{\partial \eta}\frac{\partial N_q}{\partial \eta}\right) d\xi\, d\eta$$

$\iint_{e_{NW}} E\,dx\,dy$	$A_{ag}\phi_N + A_{bg}\phi_{NNW} + A_{cg}\phi_{NW} + A_{dg}\phi_{NWW} + A_{eg}\phi_W + A_{fg}\phi_w + A_{gg}\phi_j + A_{hg}\phi_n$
$\iint_{e_{SW}} E\,dx\,dy$	$A_{aa}\phi_j + A_{ba}\phi_w + A_{ca}\phi_W + A_{da}\phi_{SWW} + A_{ea}\phi_{SW} + A_{fa}\phi_{SSW} + A_{ga}\phi_S + A_{ha}\phi_s$
$\iint_{e_{SE}} E\,dx\,dy$	$A_{ac}\phi_E + A_{bc}\phi_e + A_{cc}\phi_j + A_{dc}\phi_s + A_{ec}\phi_S + A_{fc}\phi_{SSE} + A_{gc}\phi_{SE} + A_{hc}\phi_{SEE}$
$\iint_{e_{NE}} E\,dx\,dy$	$A_{ae}\phi_{NE} + A_{be}\phi_{NNE} + A_{ce}\phi_N + A_{de}\phi_n + A_{ee}\phi_j + A_{fe}\phi_e + A_{ge}\phi_E + A_{he}\phi_{NEE}$
$A_{aa}, A_{cc}, A_{ee}, A_{gg}$	$\frac{52}{45}$
A_{ag}	$\frac{1}{2}$
A_{hg}	$-\frac{37}{45}$
A_{ae}	$\frac{23}{45}$
A_{bg}	$-\frac{23}{45}$

For bookkeeping, these integrals are derived and expressed in terms of the nodal unknowns and the result is given in Table 15-1. Using this table, we finally obtain the algebraic equation

$$(-\nabla^2\tilde{\phi}, N_j) = 4A_{aa}\phi_j + 2A_{ag}(\phi_W + \phi_S + \phi_E + \phi_N) + 2A_{hg}(\phi_w + \phi_s + \phi_e + \phi_n)$$
$$+ A_{ae}(\phi_{NW} + \phi_{SW} + \phi_{SE} + \phi_{NE}) + A_{bg}(\phi_{NWW} + \cdots + \phi_{NNW}) = 0\,. \tag{15-25}$$

In order to compare Eq. (15-25) with Eq. (15-18), we evaluate

$$(1, N_j) = \iint_{e_{\text{sup}}} N_j\,dx\,dy = -\frac{h^2}{3}\,. \tag{15-26}$$

Then dividing Eq. (15-25) by Eq. (15-26) for normalization yields

$$\frac{(\nabla^2\tilde{\phi},N_j)}{(1,N_j)}=\frac{\frac{23}{15}\sum_{p=NW}^{NE}\phi_p+3\sum_{q=W}^{N}\phi_q-\frac{23}{15}\sum_{r=NWW}^{NNW}\phi_r-\frac{74}{15}\sum_{t=n}^{w}\phi_t+\frac{208}{15}\phi_j}{h^2}=0\,, \tag{15-27}$$

which is different from Eq. (15-18). If we further simplify Eq. (15-27) to a 5-point relation for the one-dimensional system, the result is also different from Eq. (15-19).

Example 15-7 Check the validity of Eq. (15-27) by setting all the nodal unknowns equal.

Solution Let

$$\phi_{NW}=\phi_{NWW}=\cdots=\phi_e=\phi_j\,. \tag{a}$$

Then the right-hand side of Eq. (15-27) becomes

$$\phi_j(\tfrac{23}{15}\times 4+3\times 4-\tfrac{23}{15}\times 8-\tfrac{74}{15}\times 4+\tfrac{208}{15})=0\,. \tag{b}$$

This result is a special case when the dependent variable is uniform throughout the field. Clearly, application of the Laplace operator to a constant should result in zero, as expected.

Example 15-8 Use the Taylor's series to expand the 20 nodal values about ϕ_j and linearly combine these expressions in the manner indicated by Eq. (15-27). What are the coefficients of $\partial^2\phi/\partial x^2$ and $\partial^2\phi/\partial y^2$?

Solution According to Fig. 15-1, and with emphasis on the second derivatives, we write the 20 expressions

$$\phi_{NW}=\phi_j+\cdots+\frac{h^2}{2}\frac{\partial^2\phi}{\partial x^2}+\frac{h^2}{2}\frac{\partial^2\phi}{\partial y^2}+\cdots, \tag{a}$$

. . . . . .

$$\phi_e=\phi_j+\cdots+\frac{h^2}{8}\frac{\partial^2\phi}{\partial x^2}+0\cdot\frac{\partial^2\phi}{\partial y^2}+\cdots. \tag{t}$$

These expressions are then linearly combined according to Eq. (15-27); the coefficients of $\partial^2\phi/\partial x^2$ and $\partial^2\phi/\partial y^2$ are written below, respectively, as

$$\tfrac{23}{15}(\tfrac{1}{2}+\tfrac{1}{2}+\tfrac{1}{2}+\tfrac{1}{2})+3(\tfrac{1}{2}+0+\tfrac{1}{2}+0)-\tfrac{23}{15}(\tfrac{1}{2}+\tfrac{1}{2}+\tfrac{1}{8}+\tfrac{1}{8}+\tfrac{1}{2}+\tfrac{1}{2}+\tfrac{1}{8}+\tfrac{1}{8})$$
$$-\tfrac{74}{15}(0+\tfrac{1}{8}+0+\tfrac{1}{8})+\tfrac{208}{15}\times 0=1\,, \tag{u}$$

and

$$\tfrac{23}{15}(\tfrac{1}{2}+\tfrac{1}{2}+\tfrac{1}{2}+\tfrac{1}{2})+3(0+\tfrac{1}{2}+0+\tfrac{1}{2})-\tfrac{23}{15}(\tfrac{1}{8}+\tfrac{1}{8}+\tfrac{1}{2}+\tfrac{1}{2}+\tfrac{1}{8}+\tfrac{1}{8}+\tfrac{1}{2}+\tfrac{1}{2})$$
$$-\tfrac{74}{15}(\tfrac{1}{8}+0+\tfrac{1}{8}+0)+\tfrac{208}{15}\times 0=1\,, \tag{v}$$

where the sequence of the summation sign is counted counterclockwise.

Example 15-9 Follow the procedure given in Example 15-8 to find the coefficient of $\partial^4 \phi/\partial x^4$.

Solution The Taylor's series of the 20 nodal values expanded about ϕ_j are

$$\phi_{NW} = \phi_j + \cdots + \frac{h^4}{24} \frac{\partial^4 \phi}{\partial x^4} + \cdots, \tag{a}$$

$$\cdots \qquad \cdots$$

$$\phi_e = \phi_j + \cdots + \frac{h^4}{24 \times 16} \frac{\partial^4 \phi}{\partial x^4} + \cdots. \tag{t}$$

Combining Eqs. (*a*)–(*t*) linearly with the guideline of Eq. (15-27) generates

$$\begin{aligned} h^2 [&\tfrac{23}{15}(\tfrac{1}{24} + \tfrac{1}{24} + \tfrac{1}{24} + \tfrac{1}{24}) + 3(\tfrac{1}{24} + 0 + \tfrac{1}{24} + 0) \\ &- \tfrac{23}{15}(\tfrac{1}{24} + \tfrac{1}{24} + \tfrac{1}{384} + \tfrac{1}{384} + \tfrac{1}{24} + \tfrac{1}{24} + \tfrac{1}{384} + \tfrac{1}{384}) \\ &- \tfrac{74}{15}(0 + \tfrac{1}{384} + 0 + \tfrac{1}{384}) + \tfrac{204}{15} \times 0] = \tfrac{5}{24} h^2, \end{aligned} \tag{u}$$

which suggests that Eq. (15-27) is of $O(h^2)$ accuracy.

Remarks:

1. Any finite-difference scheme representing an n-point relation can, in principle, be obtained by the method of undetermined coefficients.
2. The Galerkin finite-element method can also generate the n-point relation. During the actual computation of the matrix system, a 13-point relation for the half-way point, in addition to Eq. (15-25), is needed. If an iteration scheme is adopted, it is necessary to alternate the use of these two equations.

15-5 IRREGULAR GEOMETRY OF THE SCHEME

When the geometry of the system is irregular, such as the one shown in Fig. 15-3, special treatments are needed to account for the irregularity of the boundary. In this

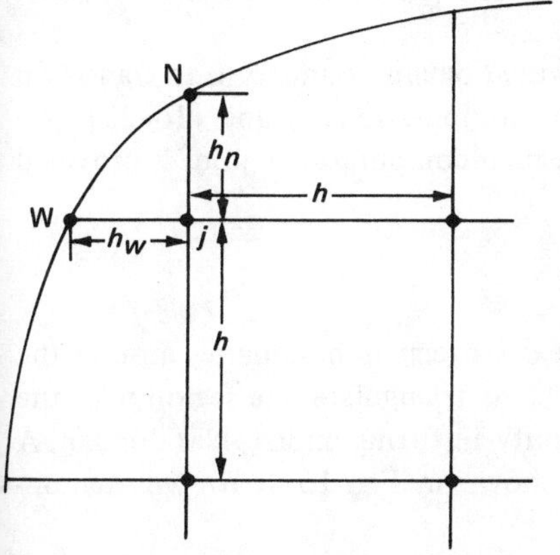

Figure 15-3 Mesh system near irregular boundary.

section, we will assume that $\phi(x, y)$ is prescribed on the boundary and the partial differential equation is linear.

15-5*a* Finite-Difference Method

One way to account for the irregularity of the boundary is to expand $\phi(x, y)$ in a Taylor's series about the point j, temporarily taken to be the origin, as

$$\phi(x,y) = \phi_j + x\left(\frac{\partial\phi}{\partial x}\right)_j + y\left(\frac{\partial\phi}{\partial y}\right)_j + \frac{x^2}{2}\left(\frac{\partial^2\phi}{\partial x^2}\right)_j + \frac{y^2}{2}\left(\frac{\partial^2\phi}{\partial y^2}\right)_j + \cdots. \tag{15-28}$$

Substituting the coordinates of the points $W, S, E,$ and N into Eq. (15-28) yields

$$\begin{bmatrix} -h_W & 0 & \frac{1}{2}h_W^2 & 0 \\ 0 & -h & 0 & \frac{1}{2}h^2 \\ h & 0 & \frac{1}{2}h^2 & 0 \\ 0 & h_N & 0 & \frac{1}{2}h_N^2 \end{bmatrix} \begin{Bmatrix} \left(\frac{\partial\phi}{\partial x}\right)_j \\ \left(\frac{\partial\phi}{\partial y}\right)_j \\ \left(\frac{\partial^2\phi}{\partial x^2}\right)_j \\ \left(\frac{\partial^2\phi}{\partial y^2}\right)_j \end{Bmatrix} = \begin{Bmatrix} \phi_W - \phi_j \\ \phi_S - \phi_j \\ \phi_E - \phi_j \\ \phi_N - \phi_j \end{Bmatrix}. \tag{15-29}$$

The matrix system can be inverted to yield the expressions for $(\partial\phi/\partial x)_j$, $(\partial\phi/\partial y)_j$, $(\partial^2\phi/\partial x^2)_j$, and $(\partial^2\phi/\partial y^2)_j$ in terms of five nodal functions. For example,

$$\left(\frac{\partial\phi}{\partial x}\right)_j = \frac{h_W}{h^2 + hh_W}\phi_E + \frac{h - h_W}{hh_W}\phi_j - \frac{h}{h_W^2 + hh_W}\phi_W \tag{15-30a}$$

and

$$\left(\frac{\partial^2\phi}{\partial x^2}\right)_j = \frac{2}{h^2 + hh_W}\phi_E - \frac{2}{hh_W}\phi_j + \frac{2}{h_W^2 + hh_W}\phi_W\,. \tag{15-30b}$$

Observe that Eqs. (15-30*a*) and (15-30*b*), respectively, reduce to the familiar central-difference scheme and three-point relation if $h = h_W$. Therefore, for finite-difference discretization of the differential equations at the nodal points adjacent to the boundary, Eqs. (15-30*a*) and (15-30*b*) can be used.

Another way to deal with the irregularity of the domain boundary is to transform the irregular physical plane into the rectangular computational plane [16–20]. An excellent review of techniques for generating irregular computational grids is provided in [21].

15-5*b* Finite-Element Method

In the finite-element formulation, when there exist irregular boundaries around the physical domain, triangular elements can be used to triangulate the region near the boundary since these elements have high conformity in fitting an irregular domain. A typical partially triangulated irregular region is shown in Fig. 15-4. We will demon-

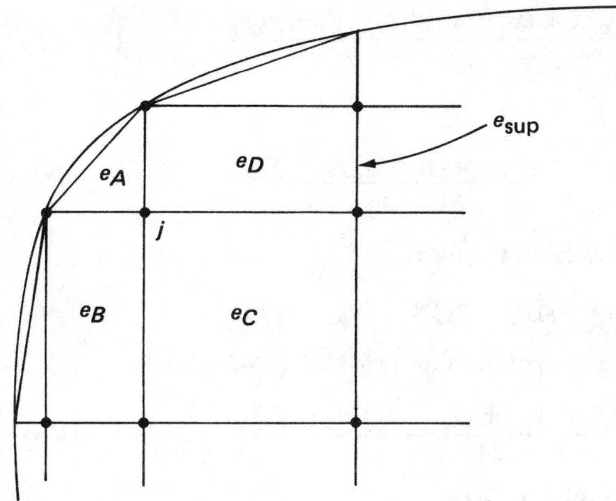

Figure 15-4 Partially triangulated irregular region; e_{sup} includes e_A, e_B, e_C, and e_D.

strate the Galerkin discretization of the Laplace equation for grid point j surrounded by three regular rectangular elements and one triangular element.

A weak Galerkin form of the Laplace equation can be written as

$$(-\nabla^2 \tilde{\phi}, N_j) = \iint_{e_{sup}} \left(\frac{\partial \tilde{\phi}}{\partial x} \frac{\partial N_j}{\partial x} + \frac{\partial \tilde{\phi}}{\partial y} \frac{\partial N_j}{\partial y} \right) dx\, dy = 0 , \qquad (15\text{-}31)$$

where $N_j(x, y)$ is a piecewise basis function having local support in elements e_A, e_B, e_C, and e_D. The double integral in Eq. (15-31) can be evaluated separately in the four elements. For example, in the triangular element e_A, we derive

$$\iint_{e_A} \left(\frac{\partial \tilde{\phi}}{\partial x} \frac{\partial N_j}{\partial x} + \frac{\partial \tilde{\phi}}{\partial y} \frac{\partial N_j}{\partial y} \right) dx\, dy = B_{bb}\phi_j + B_{cb}\phi_N + B_{ab}\phi_W , \qquad (15\text{-}32)$$

where

$$B_{pb} = \iint_{e_A} \left(\frac{\partial N_p}{\partial x} \frac{\partial N_b}{\partial x} + \frac{\partial N_p}{\partial y} \frac{\partial N_b}{\partial y} \right) dx\, dy , \quad p = a, b, c . \quad (15\text{-}33a, b, c)$$

It is desirable from the viewpoint of computer programming that B_{pb} be an explicit function of the coordinates of three vertices. With straightforward algebra, an approximate solution $\tilde{\phi}(x, y)$ within a triangle e_{abc} can be expressed in terms of the three nodal variables ϕ_a, ϕ_b, and ϕ_c as

$$\tilde{\phi}(x, y) = N_a(x, y)\phi_a + N_b(x, y)\phi_b + N_c(x, y)\phi_c , \qquad (15\text{-}34)$$

where

$$N_a(x, y) = \frac{(y_b - y_c)x + (x_c - x_b)y + x_b y_c - x_c y_b}{A} , \qquad (15\text{-}35a)$$

$$N_b(x,y) = \frac{(y_c - y_a)x + (x_a - x_c)y + x_c y_a - x_a y_c}{A}, \tag{15-35b}$$

and

$$N_c(x,y) = \frac{(y_a - y_b)x + (x_b - x_a)y + x_a y_b - x_b y_a}{A}. \tag{15-35c}$$

Here A is twice the triangular area and takes the form

$$A = x_a(y_b - y_c) + x_b(y_c - y_a) + x_c(y_a - y_b). \tag{15-36}$$

Based on Eqs. (15-35a)-(15-35c), we can rewrite Eqs. (15-33a)-(15-33c) as

$$B_{ab} = \frac{(y_b - y_c)(y_c - y_a) + (x_c - x_b)(x_a - x_c)}{2A}, \tag{15-37a}$$

$$B_{bb} = \frac{(y_c - y_a)^2 + (x_a - x_c)^2}{2A}, \tag{15-37b}$$

and

$$B_{cb} = \frac{(y_a - y_b)(y_c - y_a) + (x_b - x_a)(x_a - x_c)}{2A}. \tag{15-37c}$$

The fact that the coefficients B_{ab}, B_{bb}, and B_{cb} are expressions in the coordinates x_p and y_p, $p = a, b, c$, renders computer programming of the finite-element formulation very efficient.

Example 15-10 A local region is triangulated by seven triangular finite elements, as shown in Fig. 15-5. Use the Galerkin method to discretize the Laplace equation into an algebraic equation containing $\phi_j, \phi_1, \ldots, \phi_7$.

Solution The inner product $(-\nabla^2 \tilde{\phi}, N_j)$ can be integrated by parts to yield

$$(-\nabla^2 \tilde{\phi}, N_j) = \iint_{e_{\text{sup}}} \left(\frac{\partial \tilde{\phi}}{\partial x} \frac{\partial N_j}{\partial x} + \frac{\partial \tilde{\phi}}{\partial y} \frac{\partial N_j}{\partial y} \right) dx\, dy = K_{\text{I}} + K_{\text{II}} + \cdots + K_{\text{VII}}, \tag{a}$$

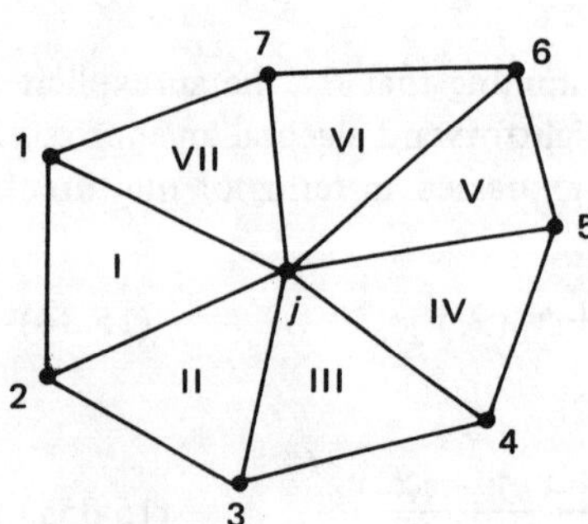

Figure 15-5 Local region that is triangulated into seven finite elements.

where

$$K_{\mathrm{I}} = \iint_e \left(\frac{\partial \tilde{\phi}}{\partial x} \frac{\partial N_j}{\partial x} + \frac{\partial \tilde{\phi}}{\partial y} \frac{\partial N_j}{\partial y} \right) dx\, dy \tag{b}$$

and so forth. Based on Eq. (15-32), we simplify Eq. (*b*) to

$$K_{\mathrm{I}} = B_{jj}\phi_j + B_{1j}\phi_1 + B_{2j}\phi_2 , \tag{c}$$

where, according to the correspondence $2 \to a, j \to b$, and $1 \to c$,

$$B_{jj} = \frac{(y_1 - y_2)^2 + (x_2 - x_1)^2}{2A_{\mathrm{I}}}, \tag{d}$$

$$B_{1j} = \frac{(y_2 - y_j)(y_1 - y_2) + (x_j - x_2)(x_2 - x_1)}{2A_{\mathrm{I}}}, \tag{e}$$

and

$$B_{2j} = \frac{(y_j - y_1)(y_1 - y_2) + (x_1 - x_j)(x_2 - x_1)}{2A_{\mathrm{I}}}. \tag{f}$$

Here A_{I} is twice the triangular area of element e_{j12} and can be expressed by

$$A_{\mathrm{I}} = x_2(y_j - y_1) + x_j(y_1 - y_2) + x_1(y_2 - y_1). \tag{g}$$

We can also write the expressions for $K_{\mathrm{II}}, \ldots, K_{\mathrm{VII}}$ as in Eq. (*c*). Finally, we derive the algebraic equation

$$(-\nabla^2 \tilde{\phi}, N_j) = (B_{jj,\mathrm{I}} + B_{jj,\mathrm{II}} + \cdots + B_{jj,\mathrm{VII}})\phi_j + \sum_{k=1}^{7} B_{kj}\phi_k . \tag{h}$$

15-6 DISCRETIZATION OF NONLINEAR TERMS

It is difficult to express nonlinear derivatives in a general form since nonlinearity can arise due to infinitely many possible arrangements. For convenience of presentation, we will consider two typical nonlinear terms

$$\frac{d}{dx}\left(\phi \frac{d\phi}{dx}\right) \quad \text{and} \quad \left(\frac{d\phi}{dx}\right)\left(\frac{d^2\phi}{dx^2}\right).$$

15-6*a* Finite-Difference Method

In the finite-difference formulation, the discretizations of these two terms are similar and straightforward. Immediately, we may write

$$\frac{d}{dx}\left(\tilde{\phi} \frac{d\tilde{\phi}}{dx}\right)_{x_j} = \phi_j \left(\frac{d^2\tilde{\phi}}{dx^2}\right)_{x_j} + \left(\frac{d\tilde{\phi}}{dx}\right)^2_{x_j} = \frac{\phi_j(\phi_{j-1} - 2\phi_j + \phi_{j+1})}{h^2} + \frac{(\phi_{j+1} - \phi_{j-1})^2}{4h^2} \tag{15-38}$$

and

$$\left(\frac{d\tilde{\phi}}{dx}\right)\left(\frac{d^2\tilde{\phi}}{dx^2}\right) = \left(\frac{\phi_{j+1} - \phi_{j-1}}{2h}\right)\left(\frac{\phi_{j+1} - 2\phi_j + \phi_{j-1}}{h^2}\right), \tag{15-39}$$

and the discretization is accomplished.

15-6*b* Finite-Element Method

In the Galerkin finite-element formulation, there is a significant distinction between the discretizations of the two nonlinear terms. To discretize $(d/dx)(\phi\, d\phi/dx)$, we approximate $\phi(x)$ by $\Sigma_{j=0}^{J} v_j(x)\phi_j$ and integrate the inner product

$$\left(\frac{d}{dx}\left(\tilde{\phi}\frac{d\tilde{\phi}}{dx}\right), v_j\right) = \int \frac{d}{dx}\left(\tilde{\phi}\frac{d\tilde{\phi}}{dx}\right) v_j\, dx \tag{15-40}$$

by parts to yield

$$\left(\frac{d}{dx}\left(\tilde{\phi}\frac{d\tilde{\phi}}{dx}\right), v_j\right) = -\int \tilde{\phi}\frac{d\tilde{\phi}}{dx}\frac{dv_j}{dx}\, dx + \text{boundary terms}\,. \tag{15-41}$$

It is seen from Eq. (15-41) that the differentiability of $\tilde{\phi}(x)$ is weakened; i.e., $\phi(x)$ belongs to C^2, whereas $\tilde{\phi}(x)$ belongs to W_2^1.

Regarding the discretization of $(d\phi/dx)(d^2\phi/dx^2)$, the situation becomes different. We find that the inner product

$$\left(\frac{d\tilde{\phi}}{dx}\frac{d^2\tilde{\phi}}{dx^2}, v_j\right) = \int \frac{d\tilde{\phi}}{dx}\frac{d^2\tilde{\phi}}{dx^2} v_j\, dx \tag{15-42}$$

cannot weaken the differentiability of $\tilde{\phi}$ by integration by parts. Under such circumstances, the Galerkin method may not have advantages over the finite-difference method.

Remarks:

1. The discretization procedure of the finite-element method generally requires lengthier algebra than that of the finite-difference method. If the differentiability of the approximate solution can be weakened, as shown in Eq. (15-41), it may be worthwhile to adopt the Galerkin method for discretization. Otherwise, it may be wise to use the finite-difference method.
2. Sometimes the discretized nonlinear result obtained by the Galerkin method is identical to that obtained by the finite-difference method (see Section 5-1*d*).

15-7 INCORPORATION WITH MIXED BOUNDARY CONDITIONS

Consider again the classical Eq. (15-7), but now subject to

$$\phi(0) = 1 \qquad \text{and} \qquad 2\phi'(4) = \phi(4)\,. \tag{15-43}$$

As mentioned before, both the central-difference scheme and the Galerkin scheme will yield the discretized equation

$$\left(1+\frac{\mathrm{Pe}}{2}\right)\phi_{j-1}-2\phi_j+\left(1-\frac{\mathrm{Pe}}{2}\right)\phi_{j+1}=0\,,\quad j=1,2,\ldots,J-1\,. \tag{15-44}$$

However, at the boundary point ($j=J$) major differences exist between the finite-difference equation and the finite-element equation. In the central-difference scheme, a fictitious nodal unknown ϕ_{J+1} must be introduced for the discretization of Eq. (15-43). Then eliminating ϕ_{J+1} from the discretized result and Eq. (15-44) yields

$$\phi_J=\frac{4}{4-2h+h\,\mathrm{Pe}}\,\phi_{J-1}\,. \tag{15-45}$$

On the other hand, at $j=J$, the Galerkin scheme gives

$$(L\tilde{\phi},N_J)=\int_0^4\left(\frac{dN_J}{dx}+\frac{u}{\alpha}N_J\right)\left(\frac{d\tilde{\phi}}{dx}\right)dx-N_J(4)\,\frac{d\tilde{\phi}(4)}{dx}\,. \tag{15-46}$$

After simplification, Eq. (15-46) becomes

$$\phi_J=\left(\frac{2+\mathrm{Pe}}{2+\mathrm{Pe}-h}\right)\phi_{J-1}\,, \tag{15-47}$$

which *naturally* takes into account the influence of Pe through integration by parts.

Remark:

When the boundary conditions are *not* of the Dirichlet type, finite-element schemes that generate weak equations incorporate the boundary conditions into discretization spontaneously. Consequently, the characteristics of the original differential equation are better preserved and the accuracy of the numerical solution is higher (Problem 15-8).

15-8 NONUNIFORM GRID

When the field variable changes rapidly in thin boundary layers, adoption of a uniform grid may not be suitable because, for adequate resolution, several mesh points are required and the entire domain of the system may be of orders of magnitude larger than the boundary-layer thickness. In this case, a nonuniform grid that is fine within the boundary layer and coarse elsewhere may provide accurate results while keeping the computation inexpensive [22–25].

In this section we will use both a uniform grid and a nonuniform grid to solve a simple one-dimensional convection-diffusion equation by the central-difference scheme and the Galerkin scheme. Consider modified Eq. (1-1)

$$-\tfrac{1}{4}\phi''-\phi'+8\phi=0\,,\quad \phi(0)=0\quad\text{and}\quad \phi(1)=1\,, \tag{15-48}$$

whose exact solution is

$$\phi(t) = \frac{e^{4t} - e^{-8t}}{e^{4} - e^{-8}} . \tag{15-49}$$

15-8*a* Central Difference

Using a Taylor's series expansion and some algebra, we can obtain

$$\phi_j' = \frac{-\sigma^2 \phi_{j-1} + \phi_j(\sigma^2 - 1) + \phi_{j+1}}{h\sigma(1+\sigma)} + \text{higher-order terms} \tag{15-50a}$$

and

$$\phi_j'' = \frac{2[\sigma\phi_{j-1} - \phi_j(\sigma+1) + \phi_{j+1}]}{h^2\sigma(1+\sigma)} + \frac{h(1-\sigma)\phi_j'''}{3} - \frac{h^2(1-\sigma+\sigma^2)\phi_j''''}{12}$$
$$+ \text{higher-order terms} , \tag{15-50b}$$

where $\sigma = k/h$, $h = x_j - x_{j-1}$, and $k = x_{j+1} - x_j$. For a uniform grid ($\sigma = 1$), Eq. (15-50*b*) reduces to

$$\phi_j'' = \frac{\sigma_{j-1} - 2\phi_j + \phi_{j+1}}{h^2} - \frac{h^2\phi_j''''}{12} . \tag{15-51}$$

The magnitudes of the truncation errors in Eqs. (15-50*b*) and (15-51) are problem-dependent. It is likely that, with the proper choice of a nonuniform grid applied to certain problems, the truncation error in Eq. (15-50*b*) will be smaller than that in Eq. (15-51). For example, consider the exact solution to Eq. (15-48). At $t = 0.75$, we may approximate

$$\phi(0.75) \approx 0.0183e^3 \approx \phi'(0.75) \approx \phi''(0.75) \approx \phi'''(0.75) \approx \phi''''(0.75) .$$

With these approximations, the truncation error in the nonuniform grid case with $h = 1.25$ and $\sigma = 0.6$ is 0.02488, whereas the error in the uniform grid case with $h = 1$ is 0.03062. The numerical results are listed in Table 15-2. It is seen that, at $t = 0.75$. the solution for the nonuniform grid is more accurate than that for the uniform grid.

Table 15-2 Numerical solutions to Eq. (15-48) for uniform and nonuniform grids; at $t = 0.75$, solutions for both grids are available for comparison

Scheme	$t = 0.25$	$t = 0.4375$	$t = 0.5$	$t = 0.75$	$t = 0.9375$
Exact	0.0473	0.1049	0.1350	0.3678	0.7788
Finite difference (uniform grid)	0.0582		0.1552	0.3944	
Finite difference (nonuniform grid)		0.1184		0.3720	0.7740
Finite element (uniform grid)	0.044		0.1269	0.3563	
Finite element (nonuniform grid)		0.0913		0.3486	0.770

15-8b Galerkin Finite-Element

In the finite-element formulation, we should not speak of truncation errors. Instead, we will present a very elementary example to illustrate the advantage of adopting a nonuniform grid. Suppose it is desired to approximate the function $\phi(x) = x^2(1-x)$, $x \in [0, 1]$, by a pyramid function

$$\tilde{\phi}(x) = \begin{cases} \dfrac{x\phi^*}{h} = N^-(x)\phi^*, & x \in [0, h] \\[2ex] \dfrac{(1-x)\phi^*}{1-h} = N^+(x)\phi^*, & x \in [h, 1] . \end{cases}$$

Forcing the residual orthogonal to the basis functions, we obtain

$$\int_0^h [\phi(x) - \tilde{\phi}(x)]N^-(x)\,dx + \int_h^1 [\phi(x) - \tilde{\phi}(x)]N^+(x)\,dx = 0 . \tag{15-52}$$

Substitution of $\phi^* = h^2 - h^3$ into Eq. (15-52) yields

$$\frac{h^2}{3} - \frac{7}{12}h^3 + \frac{1}{5}h^4 - \frac{1}{1-h}\left(\frac{1}{30} - \frac{h^3}{3} + \frac{h^4}{2} - \frac{h^5}{5}\right) = 0 ,$$

of which one of the roots is $h \approx \frac{2}{3}$ (and not $\frac{1}{2}$). The discretized form of Eq. (15-48) is

$$-\left(\frac{1}{h} - 2 - \frac{16h}{3}\right)\phi_{j-1} + \left[\frac{1}{h} + \frac{1}{k} + \frac{32}{3}(h+k)\right]\phi_j - \left(\frac{1}{k} + 2 - \frac{16k}{3}\right)\phi_{j+1} = 0$$

and its solutions for both uniform and nonuniform grids are also listed in Table 15-2. Again, we find that the nonuniform grid solution is more accurate in the boundary layer (vicinity of $t = 1$).

15-9 CONCLUSION

In this chapter we have attempted, in a rather elementary way, to identify the similarities and differences between finite-difference schemes and finite-element schemes in several respects. Much work remains in comparing these two powerful methods in a rigorous and conclusive manner.

SYMBOLS

a	convenient symbol $[= 1/(e^4 - e^{-8})]$
$a_1, a_2, \ldots, a_{20}$	twenty undetermined constants
h_W, h_N	grid sizes between grid points j and W, j and N
K_p	convenient integrals [Eq. (15-24)]
L_1, L_2	linear operators [Eqs. (15-1*a*) and (15-1*b*)]
Pe	Peclet number $(= uh/\alpha)$

r	mesh size ratio ($= \alpha \Delta t/h^2$)
$S(x)$	quadratic splines
$v(x)$	test function or weighting function
$W_j(x)$	asymmetric weighting function
η	isoparametric coordinate $[= 2(y - y^*)/h]$
μ	constant to be optimized to achieve both stability and accuracy
ξ	amplification factor used in von Neumann analysis, or isoparametric coordinate $[= 2(x - x^*)/h]$
σ	grid size ratio ($= k/h$)

Superscript

(n)	at time $n\,\Delta t$

REFERENCES

1. A. Arakawa, Computational Design for Long-Term Numerical Integration of the Equations of Fluid Motion: Two-dimensional Incompressible Flow, Part I, *J. Comput. Phys.*, vol. 1, pp. 119–143, 1966.
2. P. J. Roache, *Computational Fluid Dynamics*, pp. 105–106, Hermosa, Albuquerque, N.M., 1972.
3. Y. K. Sasaki and J. N. Reddy, A Comparison of Stability and Accuracy of Some Numerical Models of Two-dimensional Circulation, *Int. J. Numer. Methods Eng.*, vol. 16, pp. 149–170, 1980.
4. R. D. Richtmyer and K. W. Morton, *Difference Methods for Initial-Value Problems*, pp. 195–198, Wiley-Interscience, New York, 1967.
5. A. K. Runchal, Comparative Criteria for Finite Difference Formulations for Problems of Fluid Flow, *Int. J. Numer. Methods Eng.*, vol. 11, pp. 1667–1679, 1977.
6. J. L. Siemieniuch and I. Gladwell, Analysis of Explicit Difference Methods for a Diffusion-Convection Equation, *Int. J. Numer. Methods Eng.*, vol. 12, pp. 899–916, 1978.
7. R. K. C. Chan, A Balanced Expansion Technique for Constructing Accurate Finite Difference Advection Scheme, *Int. J. Numer. Methods Eng.*, vol. 12, pp. 1131–1150, 1978.
8. K. W. Morton, Stability of Finite Difference Approximations to a Diffusion-Convection Equation, *Int. J. Numer. Methods Eng.*, vol. 15, pp. 677–683, 1980.
9. D. F. Griffiths, I. Christie, and A. R. Mitchell, Analysis of Error Growth for Explicit Difference Schemes in Conduction-Convection Problems, *Int. J. Numer. Methods Eng.*, vol. 15, pp. 1075–1081, 1980.
10. J. M. Varah, Stability Restrictions on Second Order, Three Level Finite Difference Schemes for Parabolic Equations, *SIAM J. Numer. Anal.*, vol. 17, pp. 300–309, 1980.
11. R. Hirsh, Higher Order Accurate Difference Solution of Fluid Mechanics Problems by a Compact Differencing Technique, *J. Comput. Phys.*, vol. 19, pp. 90–109, 1975.
12. E. Krause, E. H. Hirschel, and W. Kordulla, Fourth Order Mehrstellen-Integration for Three Dimensional Turbulent Boundary Layers, *Comput. Fluids*, vol. 4, pp. 77–92, 1976.
13. M. Ciment, S. Leventhal, and B. Weinberg, The Operator Compact Implicit Method for Parabolic Equations, *J. Comput. Phys.*, vol. 28, pp. 135–166, 1978.
14. M. J. P. Cullen and K. W. Morton, Analysis of Evolutionary Error in Finite Element and Other Methods, *J. Comput. Phys.*, vol. 34, pp. 245–267, 1980.
15. R. M. Smith, Finite Element Solutions of the Energy Equation at a High Peclet Number, *Comput. Fluids*, vol. 8, pp. 335–350, 1980.

16. W. H. Frey, Flexible Finite Difference Stencils from Isoparametric Finite Elements, *Int. J. Numer. Methods Eng.*, vol. 11, pp. 1653–1665, 1977.
17. B. Hunt, Finite Difference Approximations of Boundary Conditions along Irregular Boundaries, *Int. J. Numer. Methods Eng.*, vol. 12, pp. 229–235, 1978.
18. J. L. Steger, Implicit Finite Difference Simulation of Flow about Arbitrary Two-dimensional Geometries, *AIAA J.*, vol. 16, pp. 679–686, 1978.
19. S. R. Robertson, A Finite Difference Formulation of the Equation of Heat Conduction in Generalized Coordinates, *Numer. Heat Transfer*, vol. 2, pp. 61–80, 1979.
20. C. D. Mobley and R. J. Stewart, On the Numerical Generation of Boundary-Fitted Orthogonal Curvilinear Coordinate Systems, *J. Comput. Phys.*, vol. 34, pp. 124–135, 1980.
21. W. C. Thacker, A Brief Review of Techniques of Generating Irregular Computational Grids, *Int. J. Numer. Methods Eng.*, vol. 15, pp. 1335–1341, 1981.
22. A. K. Gupta, A Finite Element for Transition from a Fine to a Coarse Grid, *Int. J. Numer. Methods Eng.*, vol. 12, pp. 35–45, 1978.
23. L. Farnell, Solution of Poisson Equations on a Nonuniform Grid, *J. Comput. Phys.*, vol. 35, pp. 408–425, 1980.
24. I. diFisica, On Numerical Differentiation on a Nonuniform Grid, *J. Comput. Phys.*, vol. 39, pp. 481–483, 1981.
25. V. M. Kovenya and N. N. Yanenko, Numerical Method for Solving the Viscous Gas Equations on Moving Grids, *Comput. Fluids*, vol. 8, pp. 59–70, 1980.

PROBLEMS

15-1 (*a*) Derive Eq. (15-4) from Eqs. (15-3*a*)–(15-3*d*).
(*b*) Check the validity of Eq. (15-4) by setting $\phi(x) = x^4$ and $x_j = 2$.

15-2 Suppose a finite-difference scheme yields the nodal solution as

$$\tilde{\phi}(3) = 1\,, \qquad \tilde{\phi}(4) = 2\,, \qquad \text{and} \qquad \tilde{\phi}(5) = 8\,.$$

Find $\phi(4.5)$ according to Eq. (15-2*a*). Let $x_j = 4$ and $h = 1$.

15-3 Consider the Blasius equation

$$f''' + \tfrac{1}{2}ff'' = 0\,, \qquad f(0) = f'(0) = 0\,, \qquad \text{and} \qquad f'(\eta_\infty) = 1\,.$$

If the Galerkin method is to be used with the approximation $\tilde{f} = v_1(\eta)f_1 + v_2(\eta)f_2 + \cdots + v_J(\eta)f_J$, what kind of weighting function should we choose? After a proper choice of $v_j(\eta)$, discretize the Blasius equation.

15-4 Consider the one-dimensional streamwise diffusion equation

$$-\frac{d^2\phi}{dx^2} + \frac{u}{\alpha}\frac{d\phi}{dx} + c\phi = 0\,.$$

What is the stability criterion for the central-difference scheme? Is it less restricted than Pe $\leqslant$ 2?

15-5 Using Eq. (15-13), complete the proof that it is unconditionally stable.

15-6 Derive the seven-point relation for $d^2\phi/dx^2$ by means of the method of undetermined coefficients.

15-7 Utilize Table 15-1 to derive Eq. (15-25).

15-8 Using the Thomas algorithm (Section 5-1*b*) with $J = 20$ and Pe $= 0.5$, compute Eq. (15-44) along with, respectively, Eqs. (15-45) and (15-47). In comparison with the exact solution, which scheme yields more accurate results?

APPENDIX

A

TABLE OF BLASIUS SIMILARITY SOLUTION

η	$f(\eta)$	$f'(\eta)$	$f''(\eta)$
.00	.0000000	.0000000	.3320584
.02	.0000664	.0066412	.3320583
.04	.0002656	.0132823	.3320578
.06	.0005977	.0199235	.3320564
.08	.0010626	.0265646	.3320537
.10	.0016603	.0332056	.3320492
.20	.0066410	.0664080	.3319849
.30	.0149415	.0995989	.3318105
.40	.0265600	.1327646	.3314710
.50	.0414929	.1658858	.3309121
.60	.0597348	.1989379	.3300803
.70	.0812772	.2318909	.3289232
.80	.1061085	.2647099	.3273904
.90	.1342134	.2973549	.3254338
1.00	.1655722	.3297810	.3230083
1.10	.2001606	.3619394	.3200727
1.20	.2379493	.3937772	.3165904
1.30	.2789034	.4252381	.3125300
1.40	.3229823	.4562630	.3078666
1.50	.3701393	.4867906	.3025817
1.60	.4203216	.5167582	.2966646
1.70	.4734700	.5461022	.2901128
1.80	.5295190	.5747596	.2829321

η	$f(\eta)$	$f'(\eta)$	$f''(\eta)$
1.90	.5883968	.6026681	.2751375
2.00	.6500254	.6297673	.2667526
2.20	.7811945	.6813120	.2483519
2.40	.9222914	.7289836	.2280927
2.60	1.0725074	.7724568	.2064555
2.80	1.2309787	.8115114	.1840074
3.00	1.3968097	.8460463	.1613610
3.20	1.5690965	.8760834	.1391287
3.40	1.7469517	.9017632	.1178768
3.60	1.9295268	.9233316	.0980867
3.80	2.1160314	.9411199	.0801263
4.00	2.3057479	.9555202	.0642344
4.20	2.4980411	.9669590	.0505200
4.40	2.6923622	.9758727	.0389728
4.60	2.8882491	.9826853	.0294839
4.80	3.0853217	.9877912	.0218713
5.00	3.2832746	.9915436	.0159069
6.00	4.2796203	.9989742	.0024021
7.00	5.2792355	.9999226	.0002202
8.00	6.2792070	.9999969	.0000122
9.00	7.2792001	1.0000000	.0000004
10.00	8.2791915	1.0000000	.0000000
11.00	9.2791786	1.0000000	.0000000
12.00	10.2791657	1.0000000	.0000000

APPENDIX

B

COMPUTER PROGRAMS AND RESULTS

B-1 STAGGERED–GRID SCHEME AND PRIMITIVE VARIABLES USED TO EXAMINE SHEAR–DRIVEN FLOW

B-1*a* FORTRAN Listing

```
C *****************************************************************
C     THIS IS THE PROGRAM TO USE THE STAGERRED-GRID SCHEME AND THE
C     PRIMITIVE VARIABLES TO EXAMINE THE SHEAR-DRIVEN FLOW. *****
      DIMENSION P(11,11),U(11,11),V(11,11),US(11,11),VS(11,11),
     $PP(11,11),PS(11,11),AJ(11,11),BJ(11,11),PTEMP(11,11)
     $,UTEMP(11,11),VTEMP(11,11),B1(11,11)
C **RE=REYNOLDS NUMBER, N=NUMBER OF GRID NODES IN A LINE, H=GRID SIZE
C ***NIMAX=MAXIMUM NUMBER OF ITERATION CYCLES, EPS=CONVERGENCE CRITERION
C **Z1,ZP=UNDER-RELAXATION PARAMETERS
      RE=10.
      N=11
      NM1=N-1
      NM2=N-2
      H=1./NM1
      HRER=1./(H*RE)
      NIMAX=600
      S1=1.E-8
      EPS=.005
      Z1=.5
      ZP=.8
      DO 10 J=1,N
      VS(N,J)=1.
      DO 10 I=1,N
10    PS(I,J)=0
      NI=0
      LL=3
      LP=5
1     CONTINUE
```

```
      KU=0
      KV=0
      KP=0
C     COMPUTE U.  **************
      DO 20 IU=1,LL
      DO 20 J=2,NM1
      UE=.5*(US(1,J)+US(2,J))
      VN=.5*(VS(1,J)+VS(2,J))
      VSS=.5*(VS(1,J-1)+VS(2,J-1))
      AJ(1,J)=5.*HRER+2.*UE+.5*(VN-VSS)
      AN=-.5*VN+HRER
      AS=.5*VSS+HRER
      AE=HRER
      US(1,J)=Z1*((AN*US(1,J+1)+AS*US(1,J-1)+AE*US(2,J)+PS(1,J)-PS(2,J
     $))/(AJ(1,J)+S1))+(1-Z1)*US(1,J)
      DO 25 I=2,NM2
      UE=.5*(US(I,J)+US(I+1,J))
      UW=.5*(US(I,J)+US(I-1,J))
      VN=.5*(VS(I,J)+VS(I+1,J))
      VSS=.5*(VS(I,J-1)+VS(I+1,J-1))
      AW=HRER
      AS=.5*VSS+HRER
      AE=HRER
      AN=HRER-.5*VN
      AJ(I,J)=4.*HRER+2.*(UE-UW)+.5*(VN-VSS)
25    US(I,J)=Z1*((AW*US(I-1,J)+AS*US(I,J-1)+AE*US(I+1,J)+AN*US(I,J+1)
     $+PS(I,J)-PS(I+1,J))/(AJ(I,J)+S1))+(1-Z1)*US(I,J)
      UW=.5*(US(NM1,J)+US(NM2,J))
      VN=.5*(VS(NM1,J)+VS(N,J-1))
      VSS=.5*(VS(NM1,J-1)+VS(N,J-1))
      AW=HRER
      AS=.5*VSS+HRER
      AN=HRER-.5*VN
      AJ(NM1,J)=5.*HRER-2.*UW+.5*(VN-VSS)
      US(NM1,J)=Z1*((AW*US(NM2,J)+AS*US(NM1,J-1)+AN*US(NM1,J+1)
     $+PS(NM1,J)-PS(N,J))/(AJ(NM1,J)+S1))+(1-Z1)*US(NM1,J)
20    CONTINUE
C     COMPUTE V.  **************
      DO 30 IV=1,LL
      DO 30 I=2,NM1
      UE=.5*(US(I,1)+US(I,2))
      UW=.5*(US(I-1,1)+US(I-1,2))
      VN=.5*(VS(I,1)+VS(I,2))
      BJ(I,1)=5.*HRER+2.*VN+.5*(UE-UW)
      BW=.5*UW+HRER
      BE=-.5*UE+HRER
      BN=HRER
      VS(I,1)=Z1*((BW*VS(I-1,1)+BE*VS(I+1,1)+BN*VS(I,2)+PS(I,1)-PS(I,
     $2))/(BJ(I,1)+S1))+(1-Z1)*VS(I,1)
      DO 35 J=2,NM2
      UE=.5*(US(I,J)+US(I,J+1))
      UW=.5*(US(I-1,J)+US(I-1,J+1))
      VN=.5*(VS(I,J)+VS(I,J+1))
      VSS=.5*(VS(I,J)+VS(I,J-1))
      BW=.5*UW+HRER
      BS=HRER
      BE=-.5*UE+HRER
      BN=HRER
      BJ(I,J)=4.*HRER+2.*VN-2.*VSS+.5*(UE-UW)
35    VS(I,J)=Z1*((BW*VS(I-1,J)+BS*VS(I,J-1)+BE*VS(I+1,J)+BN*VS(I,J+1)
     $+PS(I,J)-PS(I,J+1))/BJ(I,J))+(1-Z1)*VS(I,J)
      UE=.5*(US(I,NM1)+US(I,N))
      UW=.5*(US(I-1,NM1)+US(I-1,N))
      VSS=.5*(VS(I,NM1)+VS(I,NM2))
      BJ(I,NM1)=5.*HRER-2.*VSS+.5*(UE-UW)
      BW=.5*UW+HRER
      BS=HRER
      BE=-.5*UE+HRER
30    VS(I,NM1)=Z1*((BW*VS(I-1,NM1)+BS*VS(I,NM2)+BE*VS(I+1,NM1)+PS(I,
     $NM1)-PS(I,N))/(BJ(I,NM1)+S1))+(1-Z1)*VS(I,NM1)
```

```
C     COMPUTE P. **************
      DO 300 I=1,N
      DO 300 J=1,N
300   PP(I,J)=0
C     SET ALL THE PRESSURE VARIATIONS TO ZERO.  OTHERWISE, THE PRESSURE
C     WILL ACCUMULATE TO A LARGE ARBITRARY NUMBER. *******
      DO 40 IP=1,LP
      DO 40 I=2,NM1
      DO 40 J=2,NM1
      CW=1/AJ(I-1,J)
      CS=1/BJ(I,J-1)
      CE=1/AJ(I,J)
      CN=1/BJ(I,J)
      B1(I,J)=US(I-1,J)-US(I,J)+VS(I,J-1)-VS(I,J)
      CJ=CW+CS+CE+CN
40    PP(I,J)=(CW*PP(I-1,J)+CS*PP(I,J-1)+CE*PP(I+1,J)+CN*PP(I,J+1)+
     $B1(I,J))/(CJ+S1)
      DO 50 I=1,N
      DO 50 J=1,N
      IF(ABS(P(I,J))  .GT. 10.)GO TO 5
      P(I,J)=PS(I,J)+PP(I,J)
      PS(I,J)=P(I,J)
      U(I,J)=US(I,J)+(PP(I,J)-PP(I+1,J))/(AJ(I,J)+S1)
      V(I,J)=VS(I,J)+(PP(I,J)-PP(I,J+1))/(BJ(I,J)+S1)
50    CONTINUE
C     USE A CRUDE APPROXIMATION FOR THE P BOUNDARY CONDITION(ZERO
C     NORMAL PRESSURE GRADIENTS). ********
      DO 90 K=2,NM1
      PS(1,K)=PS(2,K)
      PS(N,K)=PS(NM1,K)
      PS(K,1)=PS(K,2)
90    PS(K,N)=PS(K,NM1)
C     CHECK THE CONVERGENCE. **********
      DO 60 I=1,NM1
      DO 60 J=2,NM1
      DIFU=(U(I,J)-UTEMP(I,J))/(U(I,J)+S1)
      IF(ABS(DIFU)-EPS)62,61,61
61    KU=KU+1
62    UTEMP(I,J)=U(I,J)
      US(I,J)=U(I,J)
60    CONTINUE
      DO 80 I=2,NM1
      DO 80 J=1,NM1
      DIFV=(V(I,J)-VTEMP(I,J))/(V(I,J)+S1)
      IF(ABS(DIFV)-EPS)82,81,81
81    KV=KV+1
82    VTEMP(I,J)=V(I,J)
      VS(I,J)=V(I,J)
80    CONTINUE
      DO 95 I=1,N
      DO 95 J=1,N
      IF(I .EQ. 1 .AND. J .EQ.1)GO TO 95
      IF(I .EQ. 1 .AND. J .EQ.N)GO TO 95
      IF(I .EQ. N .AND. J .EQ.1)GO TO 95
      DIFP=(PTEMP(I,J)-PS(I,J))/(PS(I,J)+S1)
      IF(ABS(DIFP)-EPS)92,91,91
91    KP=KP+1
92    PTEMP(I,J)=PS(I,J)
95    CONTINUE
      IF(NI .EQ. 40)PRINT 101,NI,KU,KV,KP
C     PRINT OUT THE RESULT. **********
      NI=NI+1
      KSUM=KU+KV+KP
      IF(KSUM .EQ. 0)GO TO 5
      IF(NI .EQ. NIMAX)GO TO 5
      GO TO 1
5     CONTINUE
      PRINT 101,NI,KU,KV,KP
      PRINT 111
      DO 70 J=N,1,-1
```

```
70      PRINT 102, (US(I,J),I=1,NM1)
        PRINT 112
        DO 71 J=NM1,1,-1
71      PRINT 102, (VS(I,J),I=1,N)
        PRINT 113
        DO 72 J=N,1,-1
72      PRINT 102, (PS(I,J),I=1,N)
C
101     FORMAT(1X,'NI=',I3,1X,'KU=',I3,1X,'KV=',I3,1X,'KP=',I3)
102     FORMAT(11F7.3)
111     FORMAT(1X,'U VALUES')
112     FORMAT(1X,'V VALUES')
113     FORMAT(1X,'P VALUES')
        END
```

B-1*b* Results

```
NI= 40 KU=   4 KV=   4 KP=117
NI=255 KU=   0 KV=   0 KP=   0
U VALUES
   .000   .000   .000   .000   .000   .000   .000   .000   .000   .000
   .001  -.007  -.024  -.047  -.075  -.106  -.133  -.140  -.105  -.031
   .000  -.012  -.033  -.061  -.093  -.124  -.143  -.136  -.092  -.029
   .001  -.009  -.026  -.048  -.071  -.090  -.099  -.088  -.055  -.018
   .004  -.001  -.008  -.018  -.027  -.034  -.036  -.030  -.018  -.007
   .007   .010   .014   .019   .024   .028   .030   .027   .017   .005
   .010   .020   .035   .052   .071   .086   .092   .082   .052   .017
   .012   .026   .048   .076   .105   .130   .144   .133   .088   .027
   .011   .026   .050   .080   .114   .147   .170   .168   .121   .034
   .008   .017   .034   .056   .083   .113   .140   .153   .127   .028
   .000   .000   .000   .000   .000   .000   .000   .000   .000   .000
V VALUES
   .000  -.009  -.013  -.016  -.018  -.017  -.008   .023   .113   .368  1.000
   .000  -.017  -.030  -.039  -.047  -.048  -.035   .015   .147   .442  1.000
   .000  -.030  -.052  -.068  -.080  -.080  -.055   .022   .191   .505  1.000
   .000  -.041  -.070  -.091  -.104  -.100  -.064   .031   .222   .542  1.000
   .000  -.046  -.079  -.102  -.115  -.109  -.067   .036   .233   .553  1.000
   .000  -.044  -.076  -.098  -.111  -.106  -.066   .032   .223   .541  1.000
   .000  -.035  -.062  -.082  -.094  -.092  -.061   .020   .192   .505  1.000
   .000  -.022  -.040  -.056  -.066  -.067  -.049   .008   .146   .444  1.000
   .000  -.008  -.018  -.026  -.033  -.035  -.027   .005   .099   .357  1.000
   .000   .001  -.002  -.005  -.007  -.007  -.001   .017   .073   .257  1.000
P VALUES
   .000 -3.225 -3.233 -3.225 -3.199 -3.146 -3.055 -2.916 -2.734 -2.578   .000
 -3.225 -3.225 -3.233 -3.225 -3.199 -3.146 -3.055 -2.916 -2.734 -2.578 -2.578
 -3.228 -3.228 -3.232 -3.224 -3.202 -3.164 -3.105 -3.025 -2.937 -2.881 -2.881
 -3.238 -3.238 -3.242 -3.236 -3.224 -3.203 -3.174 -3.136 -3.098 -3.076 -3.076
 -3.256 -3.256 -3.259 -3.258 -3.256 -3.251 -3.245 -3.235 -3.222 -3.208 -3.208
 -3.278 -3.278 -3.280 -3.283 -3.289 -3.299 -3.311 -3.321 -3.323 -3.310 -3.310
 -3.299 -3.299 -3.300 -3.306 -3.320 -3.342 -3.370 -3.398 -3.415 -3.403 -3.403
 -3.312 -3.312 -3.313 -3.323 -3.343 -3.377 -3.422 -3.473 -3.514 -3.517 -3.517
 -3.316 -3.316 -3.318 -3.330 -3.356 -3.400 -3.465 -3.548 -3.638 -3.690 -3.690
 -3.306 -3.306 -3.310 -3.322 -3.351 -3.403 -3.487 -3.614 -3.793 -4.000 -4.000
   .000 -3.306 -3.310 -3.322 -3.351 -3.403 -3.487 -3.614 -3.793 -4.000   .000
```

B-2 BIHARMONIC–STREAMFUNCTION FORMULATION USED TO EXAMINE SHEAR-DRIVEN FLOW

B-2*a* FORTRAN Listing

```
C *****************************************************************
C         THIS IS THE PROGRAM TO USE THE BIHARMONIC-STREAMFUNCTION
C      FORMULATION TO EXAMINE THE SHEAR-DRIVEN FLOW. **********
C      THIS PROGRAM IS REFERRED IN EXAMPLE 8-2 IN CHAPTER 8. ***
       DIMENSION PSI(20,20),PSIT(20,20)
       RE=50.
       N=10
       H=1./N
       NP1=N+1
       NP2=N+2
       NP3=N+3
C      NIMAX=THE MAXIMUM NUMBER OF ITERATION CYCLES. **********
       NIMAX=1000
C      S1=A SMALL REAL NUMBER FOR AVOIDING OVERFLOW PROBLEMS. *****
C      EPS=CONVERGENCE CRITERION,  Z1=UNDER-RELAXATION PARAMETER. **
       S1=1.E-8
       EPS=.001
       Z1=.5
1      CONTINUE
       KP=0
C      START THE COMPUTATION BY MEANS OF THE GAUSS-SEIDEL METHOD.***
C      WITH RELAXATION. **********
       DO 5 I=3,NP1
       DO 5 J=3,NP1
       Q1=(PSIT(I,J+1)-PSIT(I,J-1))*RE/4
       Q2=(PSIT(I-1,J)-PSIT(I+1,J))*RE/4
       CON1=(-Q1-1)*PSI(I-2,J)+(-Q2-1)*PSI(I,J-2)+(Q1-1)*PSI(I+2,J)+
      $      (Q2-1)*PSI(I,J+2)+(4*Q1+8)*PSI(I-1,J)+(4*Q2+8)*PSI
      $      (I,J-1)+(-4*Q1+8)*PSI(I+1,J)+(-4*Q2+8)*PSI(I,J+1)
      $+(-Q1+Q2-2)*PSI(I-1,J+1)-(Q1+Q2+2)*PSI(I-1,J-1)+(Q1-Q2-2)*
      $PSI(I+1,J-1)+(Q1+Q2-2)*PSI(I+1,J+1)
       PSI(I,J)=Z1*(.05*CON1)+(1-Z1)*PSI(I,J)
5      CONTINUE
C      DETERMINE THE BOUNDARY CONDITIONS FOR THE NEXT ITERATION. ****
       DO 10 J=3,NP1
       PSI(1,J)=PSI(3,J)
10     PSI(NP3,J)=PSI(NP1,J)-2*H
       DO 15 I=3,NP1
       PSI(I,1)=PSI(I,3)
15     PSI(I,NP3)=PSI(I,NP1)
C      CHECK THE CONVERGENCE. ********
       DO 20 I=1,NP3
       DO 20 J=1,NP3
       DIFF=(PSI(I,J)-PSIT(I,J))/(PSI(I,J)+S1)
       IF(ABS(DIFF)-EPS)21,22,22
22     KP=KP+1
21     PSIT(I,J)=PSI(I,J)
20     CONTINUE
       IF(NI .EQ. 5)PRINT 101,NI,KP
       IF(NI .EQ. 50)PRINT 101,NI,KP
C      PRINT THE RESULT. ******************************
       NI=NI+1
       IF(NI .EQ. NIMAX)GO TO 6
       IF(KP .EQ. 0)GO TO 6
       GO TO 1
6      PRINT 101,NI,KP
       PRINT 102
       DO 30 J=NP1,3,-1
30     PRINT 103,(PSI(I,J),I=3,NP1)
C
101    FORMAT(1X,'NI=',I3,1X,'KP=',I3)
102    FORMAT(1X,'STREAMFUNCTION VALUES')
103    FORMAT(10F8.4)
       END
```

B-2*b* Results

```
NI=  5 KP=117
NI= 50 KP=110
NI=939 KP=  0
STREAMFUNCTION VALUES
   .0001    .0008    .0022    .0047    .0090    .0159    .0262    .0381    .0403
   .0008    .0034    .0081    .0154    .0264    .0415    .0587    .0701    .0590
   .0018    .0065    .0145    .0261    .0420    .0611    .0788    .0854    .0656
   .0025    .0088    .0188    .0327    .0503    .0695    .0853    .0886    .0665
   .0027    .0094    .0197    .0336    .0503    .0676    .0810    .0833    .0634
   .0023    .0083    .0174    .0294    .0435    .0578    .0689    .0719    .0570
   .0016    .0060    .0128    .0217    .0320    .0426    .0515    .0556    .0473
   .0008    .0032    .0071    .0123    .0185    .0250    .0312    .0358    .0340
   .0001    .0009    .0022    .0040    .0062    .0087    .0115    .0147    .0170
```

APPENDIX

C

DERIVATION OF THE ERROR BOUNDS OF THE TWO-DIMENSIONAL BILINEAR INTERPOLATION FUNCTIONS

In this appendix we present the derivation of Eq. (14-53). Let $\phi(x, y) \in C^2$ be the exact solution of a certain weak equation and let $\bar{\phi}(x, y) \in C^0$ be the linear interpolating function within a right triangle. More explicitly, $\bar{\phi}(x, y)$ can be written in terms of the three nodal solutions as

$$\bar{\phi}(x,y) = N_a(x,y)\phi_a + N_b(x,y)\phi_b + N_c(x,y)\phi_c \; ,$$

where N_a, N_b, and N_c are linear functions given in Eqs. (15-35a)-(15-35c). If we define the error $\bar{e}(x,y)$ as

$$\bar{e}(x,y) = \phi(x,y) - \bar{\phi}(x,y) \; ,$$

then, since the interpolating function coincides with the exact solution at grid points a (denoting the right-angled vertex), b, and c, it follows that

$$\bar{e}(x_1,y_1) = 0 \; , \tag{C-2a}$$

$$\bar{e}(x_2,y_1) = 0 \; , \tag{C-2b}$$

and

$$\bar{e}(x_1,y_2) = 0 \; . \tag{C-2c}$$

According to Eqs. (C-2a)-(C-2c) and Rolle's theorem in calculus, there exist $x_{12} \in (x_1, x_2)$ and $y_{12} \in (y_1, y_2)$ such that

$$\left(\frac{\partial \bar{e}}{\partial x}\right)_{x_{12},y_1} = 0 \tag{C-3a}$$

and

$$\left(\frac{\partial \bar{e}}{\partial y}\right)_{x_1,y_{12}} = 0 \; . \tag{C-3b}$$

We now may write

$$\bar{e}(x,y) = \bar{e}(x,y) - \bar{e}(x_1,y_1)$$
$$= [\bar{e}(x,y) - \bar{e}(x,y_1)] + [\bar{e}(x,y_1) - \bar{e}(x_1,y_1)] . \quad \text{(C-4)}$$

Based on the mean-value theorem, there exist $\eta_1 \in (y, y_1)$ and $\xi_1 \in (x, x_1)$ such that

$$\bar{e}(x,y) = \left(\frac{\partial \bar{e}}{\partial y}\right)_{x,\eta_1} (y - y_1) + \left(\frac{\partial \bar{e}}{\partial x}\right)_{\xi_1,y_1} (x - x_1) . \quad \text{(C-5)}$$

But, recalling Eq. (C-3b), we derive

$$\left(\frac{\partial \bar{e}}{\partial y}\right)_{x,\eta_1} = \left[\left(\frac{\partial \bar{e}}{\partial y}\right)_{x,\eta_1} - \left(\frac{\partial e}{\partial y}\right)_{x_1,\eta_1}\right] + \left[\left(\frac{\partial e}{\partial y}\right)_{x_1,\eta_1} - \left(\frac{\partial e}{\partial y}\right)_{x_1,y_{12}}\right] ,$$

which, by the mean-value theorem with $y_{112} \in (\eta_1, y_{12})$, becomes

$$\left(\frac{\partial \bar{e}}{\partial y}\right)_{x,\eta_1} = \left(\frac{\partial^2 \bar{e}}{\partial x\, \partial y}\right)_{\xi_1,\eta_1} (x - x_1) + \left(\frac{\partial^2 \bar{e}}{\partial y^2}\right)_{x_1,y_{112}} (\eta_1 - y_{12}) . \quad \text{(C-6}a\text{)}$$

Also, recalling Eq. (C-3a), we derive

$$\left(\frac{\partial \bar{e}}{\partial x}\right)_{\xi_1,y_1} = \left(\frac{\partial \bar{e}}{\partial x}\right)_{\xi_1,y_1} - \left(\frac{\partial \bar{e}}{\partial x}\right)_{x_{12},y_1} = \left(\frac{\partial^2 \bar{e}}{\partial x^2}\right)_{x_{112},y_1} (\xi_1 - x_{12}) . \quad \text{(C-6}b\text{)}$$

Substituting Eqs. (C-6a) and (C-6b) into Eq. (C-5) yields

$$\bar{e}(x,y) = \left[\left(\frac{\partial^2 \bar{e}}{\partial x\, \partial y}\right)_{\xi_1,\eta_1} (x - x_1) + \left(\frac{\partial^2 \bar{e}}{\partial y^2}\right)_{x_1,y_{112}} (\eta_1 - y_{12})\right] (y - y_1)$$
$$+ \left(\frac{\partial^2 \bar{e}}{\partial x^2}\right)_{x_{112},y_1} (\xi_1 - x_{12})(x - x_1) . \quad \text{(C-7)}$$

Let h be the length of the longest side of the triangle. Since

$$\frac{\partial^2 \bar{e}}{\partial x\, \partial y} = \frac{\partial^2 \phi}{\partial x\, \partial y} ,$$

$$\frac{\partial^2 \bar{e}}{\partial x^2} = \frac{\partial^2 \phi}{\partial x^2} ,$$

and

$$\frac{\partial^2 \bar{e}}{\partial y^2} = \frac{\partial^2 \phi}{\partial y^2} ,$$

we deduce that

$$\bar{e}(x,y) \leqslant 3h^2 \max \left\{ \left|\left(\frac{\partial^2 e}{\partial x\, \partial y}\right)_{\xi_1,\eta_1}\right| , \left|\left(\frac{\partial^2 e}{\partial y^2}\right)_{x_1,y_{112}}\right| , \left|\left(\frac{\partial^2 e}{\partial x^2}\right)_{x_{112},y_1}\right| \right\}$$
$$= 3h^2 M_1 . \quad \text{(C-8)}$$

The next task is to estimate the upper bounds of $\partial\bar{e}/\partial x$ and $\partial\bar{e}/\partial y$. Utilizing Eq. (C-3*a*), we write

$$\frac{\partial\bar{e}}{\partial x}=\frac{\partial\bar{e}}{\partial x}-\left(\frac{\partial\bar{e}}{\partial x}\right)_{x_{12},y_1}$$
$$=\left[\frac{\partial\bar{e}}{\partial x}-\left(\frac{\partial\bar{e}}{\partial x}\right)_{x,y_1}\right]+\left[\left(\frac{\partial\bar{e}}{\partial x}\right)_{x,y_1}-\left(\frac{\partial\bar{e}}{\partial x}\right)_{x_{12},y_1}\right]. \tag{C-9}$$

Following the mean-value theorem, we obtain

$$\frac{\partial\bar{e}}{\partial x}=\left(\frac{\partial^2\bar{e}}{\partial x^2}\right)_{x,\eta_1}(y-y_1)+\left(\frac{\partial^2\bar{e}}{\partial x^2}\right)_{\xi_{12},y_1}(x-x_{12}), \tag{C-10}$$

where $\xi_{12}\in(x,x_{12})$. The upper bound of $\partial\bar{e}/\partial x$ therefore can be written as

$$\frac{\partial\bar{e}}{\partial x}\leqslant 2h\max\left\{\left|\left(\frac{\partial^2\bar{e}}{\partial x^2}\right)_{x,\eta_1}\right|,\left|\left(\frac{\partial^2\bar{e}}{\partial x^2}\right)_{\xi_{12},y_1}\right|\right\}$$
$$=2hM_2\,. \tag{C-11a}$$

Similarly,

$$\frac{\partial\bar{e}}{\partial y}\leqslant 2h\max\left\{\left|\left(\frac{\partial^2\bar{e}}{\partial y^2}\right)_{\xi_1,y}\right|,\left|\left(\frac{\partial^2\bar{e}}{\partial y^2}\right)_{x_1,\eta_{12}}\right|\right\}$$
$$=2hM_3\,. \tag{C-11b}$$

Combining Eqs. (C-10) and (C-11) yields

$$\|\phi-\bar{\phi}\|=\left\{\iint_\Delta\left[\bar{e}^2+\left(\frac{\partial\bar{e}}{\partial x}\right)^2+\left(\frac{\partial\bar{e}}{\partial y}\right)^2\right]dx\,dy\right\}^{1/2}\leqslant Eh\,, \tag{C-12}$$

where

$$E=2A^{1/2}\left(\tfrac{9}{4}h^2M_1^2+M_2^2+M_3^2\right)^{1/2}.$$

Here A denotes the area of the triangle; M_1, M_2, and M_3 denote, respectively, the maximum values given in Eqs. (C-8), (C-11*a*), and (C-11*b*).

NOMENCLATURE

SYMBOLS USED THROUGHOUT THE TEXT

c_p	specific heat, J/kg·K
e	element
$f(\eta)$	similarity function
h	mesh size
h_c	heat transfer coefficient, $W/m^2 \cdot K$
I	variational principle
j	index for grid point
J	largest number of j
k	thermal conductivity, W/m·K
L	linear operator
N_j	linear piecewise basis functions [Eq. (1-16)]
p	pressure, NT/m^2
Pr	Prandtl number ($= \nu/\alpha$)
t	time, s; isoparametric coordinate ($0 \leqslant t \leqslant 1$)
T	temperature, K
u	x-direction flow velocity, m/s
v	y-direction flow velocity, m/s
α	thermal diffusivity ($= k/\rho c_p$), m^2/s
δ	boundary-layer thickness, m; variation of a function
η	similarity variable
$\Lambda_{j,k}$	integrals defined and listed in Table 1-4
μ	viscosity, kg/m·s

ν	momentum diffusivity, m^2/s
ρ	density, kg/m^3
ϕ	field variable denoting u, T, etc.
ψ	stream function, m^2/s
ω	vorticity, s^{-1}; normalized stream function

Subscripts

W, S, E, N	west, south, east, north

Superscripts

$-$	previously iterated values or upstream
$\sim$	approximation of a function
*	adjoint

INDEX